SCIENCE AND CAPITALISM

Entangled Histories

EDITED BY

Lukas Rieppel, William Deringer, and Eugenia Lean

OSIRIS | 33

A Research Journal Devoted to the
History of Science and Its Cultural Influences

Osiris

Series Editors, 2018–2022

W. PATRICK MCCRAY, *University of California Santa Barbara*

SUMAN SETH, *Cornell University*

Volumes 33 to 37 aim to connect the history of science with other areas of historical scholarship. Volumes of the journal are designed to explore how, where, and why science draws upon and contributes to society, culture, and politics. The journal's editors and board members strongly encourage proposals that engage with and examine broad themes while aiming for diversity across time and space. The journal is also very interested in receiving proposals that assess the state of the history of science as a field, broadly construed, in both established and emerging areas of scholarship.

33 LUKAS RIEPPEL, EUGENIA LEAN, & WILLIAM DERINGER, EDS., *Science and Capitalism: Entangled Histories*

Series editor, 2013–2017

ANDREA RUSNOCK, *University of Rhode Island*

Volumes 28 to 32 in this series are designed to connect the history of science to broader cultural developments, and to place scientific ideas, institutions, practices, and practitioners within international and global contexts. Some volumes address new themes in the history of science and explore new categories of analysis, while others assess the "state of the field" in various established and emerging areas of the history of science.

28 ALEXANDRA HUI, JULIA KURSELL, & MYLES W. JACKSON, EDS., *Music, Sound, and the Laboratory from 1750 to 1980*
29 MATTHEW DANIEL EDDY, SEYMOUR H. MAUSKOPF, & WILLIAM R. NEWMAN, EDS., *Chemical Knowledge in the Early Modern World*
30 ERIKA LORRAINE MILAM & ROBERT A. NYE, EDS., *Scientific Masculinities*
31 OTNIEL E. DROR, BETTINA HITZER, ANJA LAUKÖTTER, & PILAR LEÓN-SANZ, EDS., *History of Science and the Emotions*
32 ELENA ARONOVA, CHRISTINE VON OERTZEN, & DAVID SEPKOSKI, EDS., *Data Histories*

Cover Illustration:

Evaporator, School of Sugar Manufacture, Glasgow and West of Scotland Technical College, c. 1930 (OP4/145, Archives and Special Collections, University of Strathclyde Library).

Acknowledgments

This volume grew out of a two-day workshop on the entangled histories of science and capitalism that took place at Columbia University's Heyman Center for the Humanities in the summer of 2016. In addition to all of the contributors to this volume, we would like to thank Dan Bouk, Catherine Burns, and Mario Biagioli for their active participation and stimulating discussion, as well as Sau-yi Fong, who was extremely helpful as the graduate student rapporteur at the workshop. We would also like to thank the Heyman Center for the Humanities, the Society of Fellows in the Humanities, the Center for Science and Society, and the Weatherhead East Asian Institute of Columbia University for the generous financial support that made this workshop possible. In addition, we would like to acknowledge Eileen Gillooly, Jonah Cardillo, and Conley Lowrance for their assistance with that event. The *Osiris* general editors, W. Patrick McCray and Suman Seth, as well as the previous general editor, Andrea Rusnock, and the entire *Osiris* Editorial Board also deserve thanks for their help, guidance, and advice at various points along the way. Finally, this volume benefitted enormously from the feedback of two anonymous referees, whose detailed, generous, and constructive comments improved the quality of each essay considerably.

Introduction:
The Entangled Histories of Science and Capitalism

by Lukas Rieppel, Eugenia Lean,§ and William Deringer#*

ABSTRACT

This volume revisits the mutually constitutive relationship between science and capitalism from the seventeenth century to the present day. Adopting a global approach, we reject the notion that either science or capitalism can be understood as stages of modernity that emerged in the West and subsequently engendered a "Great Divergence" with the rest of the world. Instead, both science and capitalism were historical institutions that arose in an imperial context of global exchange and whose entanglement has been continuously remade. Rather than seek to explain either the development of modern science as a product of economic forces or the divergence of capitalist economies as a result of technical innovation, we want to emphasize the knowledge work that has been a central feature of both modern science and capitalism across the globe.

This volume examines the relationship between two cultural institutions—science and capitalism—that have proven enormously powerful in shaping the modern world. Due to its considerable scope and significance, scholars in the history and social studies of science have debated the exact nature of that relationship at least since the 1930s and 1940s, if not before. To be sure, then, ours is a massive and complex topic, one whose elucidation far exceeds the scope of any one volume. It also admits of a seemingly endless number of interpretations. But we nonetheless believe it is well worth revisiting. That is the goal of this volume: to explore how, and the extent to which, science and capitalism have been entangled with one another—historically, epistemically, and materially.

* Department of History, Brown University, 79 Brown Street, Providence, RI 02912; lukas_rieppel@brown.edu.

§ Department of East Asian Languages and Cultures, Columbia University, New York, NY 10027; eyl2006@columbia.edu.

Program in Science, Technology, and Society, Massachusetts Institute of Technology, 77 Massachusetts Avenue, E51-188, Cambridge, MA 02139; deringer@mit.edu.

We would like to acknowledge Marwa El-Shakry, Kavita Sivaramakrishnan, and their students in the fall 2017 "Global History of Science Seminar," who generously provided feedback on this introduction. We would also like to thank the *Osiris* general editors, W. Patrick McCray and Suman Seth, for their extensive comments and valuable feedback on this introduction.

Not only are we convinced that exploring the science-capitalism nexus remains a worthwhile endeavor, but we also believe that now is a particularly opportune time to do so. For a start, recent developments within the world of science and technology have brought economic issues to the forefront of our discipline's attention. In a world full of biotech spin-offs, technology-transfer offices, and patented gene sequences, Robert K. Merton's classic account of the scientific community as one whose normative structure effectively insulates its members from the demands of the marketplace has come to seem increasingly out of touch.[1] Indeed, some scholars have begun to suspect the very nature and authority of science itself may have undergone a foundational transformation.[2] But while economists, business leaders, and politicians often celebrate these changes in the name of a brave new "innovation economy," scholars of science and technology studies have been more interested in asking how a supposedly objective and value-neutral process of knowledge making has contributed to the creation of a deeply stratified society. Judging from the rich literature on biopiracy and biocapitalism, agnotology and the social construction of ignorance, as well as the toxic effects of the chemical industry and the use and abuse of big data, it is clear that although the story is varied, complex, and context dependent, part of the answer must involve the role capital often plays in shaping the research priorities of scientists.[3] These concerns have thus given rise to calls for a more inclusive and engaged debate about knowledge in a democratic society, one that actively questions who is and is not involved in decisions about what sorts of research should be conducted, who pays for that research, and who ultimately suffers or benefits as a result.[4]

While scholars of science and technology have increasingly turned their attention to political economy, the panic of 2008 and the ensuing global debt crisis have had a

[1] Robert K. Merton, "Science and Technology in a Democratic Order," *J. Legal Polit. Sociol.* 1 (1942): 115–26; Merton, *The Sociology of Science: Theoretical and Empirical Investigations* (Chicago, 1973).

[2] See, e.g., Daniel Lee Kleinman and Steven P. Vallas, "Science, Capitalism, and the Rise of the 'Knowledge Worker': The Changing Structure of Knowledge Production in the United States," *Theory Soc.* 30 (2001): 451–92, on 481. See also Kleinman, *Impure Cultures: University Biology and the World of Commerce* (Madison, Wis., 2003); Philip Mirowski, *Science-Mart: Privatizing American Science* (Cambridge, Mass., 2011); Elizabeth Popp Berman, *Creating the Market University: How Academic Science Became an Economic Engine* (Princeton, N.J., 2012); Steven Shapin, *The Scientific Life: A Moral History of a Late Modern Vocation* (Chicago, 2008).

[3] On biocapitalism and biopiracy, see Stefan Helmreich, "Blue-Green Capital, Biotechnological Circulation and an Oceanic Imaginary: A Critique of Biopolitical Economy," *BioSocieties* 2 (2007): 287–302; Jack Ralph Kloppenburg, *First the Seed: The Political Economy of Plant Biotechnology, 1492–2000* (Cambridge, 1988); Londa Schiebinger, *Plants and Empire: Colonial Bioprospecting in the Atlantic World* (Cambridge, Mass., 2004); Vandana Shiva, *Biopiracy: The Plunder of Nature and Knowledge* (Boston, 1997); Kaushik Sunder Rajan, *Biocapital: The Constitution of Postgenomic Life* (Durham, N.C., 2006). On agnotology, see Naomi Oreskes and Erik M. Conway, *Merchants of Doubt: How a Handful of Scientists Obscured the Truth on Issues from Tobacco Smoke to Global Warming* (New York, 2010); Robert Proctor and Londa L. Schiebinger, eds., *Agnotology: The Making and Unmaking of Ignorance* (Stanford, Calif., 2008). On toxicity, see Michelle Murphy, *Sick Building Syndrome and the Problem of Uncertainty: Environmental Politics, Technoscience, and Women Workers* (Durham, N.C., 2006); Robert Proctor, *Golden Holocaust: Origins of the Cigarette Catastrophe and the Case for Abolition* (Berkeley and Los Angeles, 2011). On big data algorithms, see the 2017 volume of Osiris. For a broader critique of this history, see David F. Noble, *America by Design: Science, Technology, and the Rise of Corporate Capitalism* (New York, 1977).

[4] For classic accounts, see Donna Haraway, "Situated Knowledges: The Science Question in Feminism and the Privilege of Partial Perspective," *Feminist Stud.* 14 (1988): 575–99; Sandra G. Harding, *Whose Science? Whose Knowledge?* (Ithaca, N.Y., 1991).

similar effect within history departments as well, contributing to a wave of enthusiasm for a new brand of economic history that is often described as the history of capitalism. To the extent that it can be meaningfully distinguished from labor, business, and economic history, the "new" history of capitalism attempts to bring the lessons of social and cultural history to bear on the development of the modern economy. For such a young subfield, it has already generated a great deal of enthusiasm, even garnering front-page coverage in the *New York Times*.[5] Our aim in this volume is to take into account the lessons that have been learned from both of these historiographic traditions—the renewed attention to political economy among historians of science and technology as well as the new history of capitalism—leveraging insights from the past several decades of scholarship to revisit a classic debate about the way science and capitalism have mutually informed one another.

Before delving into the details, we want to acknowledge that we are far from the first to examine the productive but controversial relationship between these two institutions. Often, previous scholars leveraged one side of the science-capitalism dyad as an explanatory resource to account for the other. Early on, for example, Marxist historians characterized the emergence of science in early modern Europe as a direct by-product of concurrent transformations in the means of economic production. Boris Hessen articulated a particularly outspoken version of this claim when he explained the "social and economic roots of Newton's *Principia*" in explicitly materialist terms as early as 1931, whereas less than a decade later, Edgar Zilsel contended that modern science came into being when "the advance of early capitalistic society" broke down traditional class barriers between scholars and artisans. Others drew similar conclusions without invoking an explicitly Marxian logic, including the Austrian economist Joseph Schumpeter, who proclaimed that modern science was produced by "the spirit of rationalist individualism, the spirit generated by rising capitalism." More recently, Carolyn Merchant drew upon Zilsel's research to formulate a powerful feminist interpretation of the Scientific Revolution, arguing that Bacon's faith in science to give mankind "dominion" over a passive and feminine nature derived in part from capital's emergent domination of labor.[6] Despite their various differences, what united all of these authors was a shared emphasis on the material base out of which modern science developed.

[5] Jennifer Schuessler, "In History Departments, It's Up with Capitalism," *New York Times*, 6 April 2013, A1. For an introduction to the "new" history of capitalism, see Sven Beckert and Christine Desan, *American Capitalism: New Histories* (New York, 2018); Beckert, "History of American Capitalism," in *American History Now*, ed. Eric Foner and Lisa McGirr (Philadelphia, 2011); Kenneth Lipartito, "Reassembling the Economic: New Departures in Historical Materialism," *Amer. Hist. Rev.* 121 (2016): 101–39; Seth Rockman, "What Makes the History of Capitalism Newsworthy?," *J. Early Repub.* 34 (2014): 439–66; Jeffrey Sklansky, "The Elusive Sovereign: New Intellectual and Social Histories of Capitalism," *Mod. Int. Hist.* 9 (2012): 233–48; Michael Zakim and Gary John Kornblith, eds., *Capitalism Takes Command: The Social Transformation of Nineteenth-Century America* (Chicago, 2012).

[6] Boris Hessen, "The Social and Economic Roots of Newton's *Principia*," in *The Social and Economic Roots of the Scientific Revolution*, ed. Gideon Freudenthal and Peter McLaughlin (Dordrecht, 2009), 41–101; Edgar Zilsel, *The Social Origins of Modern Science*, ed. Diederick Raven, Wolfgang Krohn, and R. S. Cohen (Dordrecht, 2000), 7; Joseph A. Schumpeter, *Capitalism, Socialism, and Democracy* (New York, 1942), 124; Carolyn Merchant, *The Death of Nature: Women, Ecology and the Scientific Revolution* (New York, 1980), chap. 7. See also J. D. Bernal, *Science and Industry in the Nineteenth Century* (London, 1953).

Another classic approach tells a different story entirely, while maintaining a similar logical structure: rather than embed science within its economic context, this literature invokes science to help account for the rise of modern capitalism. Mid-twentieth-century theorists of economic "modernization" like W. W. Rostow, for example, identified "the gradual evolution of modern science and the modern scientific attitude" as a decisive factor separating vibrant, capitalist economies from less dynamic predecessors and alternatives. The generation of knowledge through scientific inquiry enabled technological advances, increased productivity, and abetted the accumulation of capital, the argument went—all of which fueled the process and ethos of growth central to the promise of capitalism. Among the foremost voices for this view was Simon Kuznets, who wrote in 1966 that "one might define modern economic growth as the spread of a system of production . . . based on the increased application of science."[7] Moreover, central to this notion of modernization was the idea that science originated in Europe and subsequently spread to the rest of the world through a process of "diffusion."[8] Thus, rather than locate the Scientific Revolution within the context of capitalism, this literature used it to explain the so-called Great Divergence between Europe and the rest of the world, about which we will have more to say in the pages that follow.

In contrast to both of these classic approaches, recent scholarship tends to frame the historical relationship between science and capitalism as a more nuanced and complex affair. Still, perhaps because of differing social and institutional networks stemming from distinct patterns in graduate training, it remains surprisingly rare to find truly symmetric analyses of the way these two institutions have developed in tandem.[9] One of the main contributions we hope to make with this volume is thus simply to further integrate the history of science with the new history of capitalism. By emphasizing powerful points of synergy between the two fields, we would like to help generate a more robust conversation across the disciplinary divide. In addition, however, we do want to offer a few more substantive contributions as well. These contributions broadly fall into three clusters or categories. First, we want to stress that the most useful way to understand the historical relationship between science and capitalism does not privilege one or the other side of the dyad, attempting to parse out the unique causal or explanatory contributions of each. Rather, we feel that it is both more important and fruitful to examine the ways in which science and capitalism have been continually coproduced in a variety of contexts and time periods. Understanding the mutually constitutive entanglement of science and capitalism is an empirical project; the best way forward is to accumulate a diverse array of examples from which more specific themes, patterns, and trends may emerge over time.

[7] W. W. Rostow, "The Stages of Economic Growth," *Econ. Hist. Rev.*, 2nd ser., 12 (1959): 1–16, on 4; more generally, Rostow, *The Stages of Economic Growth: A Non-Communist Manifesto* (Cambridge, 1962). Simon Kuznets, *Economic Growth and Structure: Selected Essays* (London, 1966), 84; also quoted in Joel Mokyr, "Innovation in Historical Perspective: Tales of Technology and Evolution," in *Technological Innovation and Economic Performance*, ed. Benn Steil, David G. Victor, and Richard R. Nelson (Princeton, N.J., 2002), 23–46, on 25. For a discussion of economic growth as the essence of capitalism, see Timothy Shenk, "Apostles of Growth," *Nation*, 5 November 2014.

[8] See George Basalla, "The Spread of Western Science," *Science* 156 (1967): 611–22.

[9] For some recent, and notable, exceptions, see Dan Bouk, *How Our Days Became Numbered* (Chicago, 2015); Eli Cook, *The Pricing of Progress: Economic Indicators and the Capitalization of American Life* (Cambridge, Mass., 2017); Jamie L. Pietruska, *Looking Forward: Prediction and Uncertainty in Modern America* (Chicago, 2017).

Second, we worry that, perhaps because of its Marxian roots, scholarship on the way science and capitalism intersect often deploys a materialist ontology that downplays the importance of thinking. In contrast, we want to emphasize the role of cognitive practices such as theorizing, calculating, and so on, in both the history of science and capitalism. But we do not simply advocate a return to the history of ideas. Instead, we look to scholarship on the material culture of science for inspiration on how to break down the neat binary between thoughts and actions, words and things, representation and reality on which the Marxian distinction between base and superstructure ultimately relies.[10] Texts, utterances, and other representational artifacts can thus be regarded as real things in the world, whereas ways of knowing can be seen as a form of cognitive labor. Not only should thinking, calculating, planning, forecasting, organizing, and theorizing all be afforded a central place in the history of capitalism, we contend, but these seemingly abstract and disembodied activities can and ought to be studied as genuine forms of practice with the power to produce far-reaching effects in surprisingly distant parts of the world. This makes it possible to denaturalize some of our culture's most authoritative knowledge claims—rendering both modern science and capitalism as a product of particular people with specific motivations informed by their local circumstances—without denying how solid and durable those knowledge claims often turn out to be.[11]

Last but not least, this volume seeks to challenge older assumptions about the spaces and places in which both science and capitalism developed. Here our aim is to do more than simply make evenhanded comparisons between different parts of the world. Instead, we want to insist that neither science nor capitalism can properly be said to have "originated" in any particular place whatsoever, geographic or otherwise. As a great deal of recent scholarship has been at pains to demonstrate, both institutions were continually produced and re-produced through a global process of circulation. Thus, in addition to moving beyond the laboratory and factory floor, this volume seeks to treat both science and capitalism as transregional, indeed global, phenomena. Further, in line with our desire to denaturalize both capitalist markets and scientific knowledge, we resist the temptation to treat either science or capitalism as a universal category, always and everywhere the same. Instead, both are historically contingent products of local practices. Of course, we do not deny that both have acquired considerable epistemic prestige and became geographically widespread. For that reason, much of this volume is geared to addressing the way science and capitalism rose to such power over the past several centuries, but without making teleological claims of

[10] The literature on new materialism is huge and getting bigger by the minute. In the history of science, we find the turn to study material culture especially useful. See, e.g., Peter Galison, *Image and Logic: A Material Culture of Microphysics* (Chicago, 1997); Hans-Jörg Rheinberger, *Toward a History of Epistemic Things: Synthesizing Proteins in the Test Tube* (Stanford, Calif., 1997); Pamela Smith, *The Body of the Artisan: Art and Experience in the Scientific Revolution* (Chicago, 2004).

[11] For more on the way scientific theorizing in particular can be studied as a form of cognitive practice, see, e.g., David Kaiser, *Drawing Theories Apart: The Dispersion of Feynman Diagrams in Postwar Physics* (Chicago, 2005); Hélène Mialet, *Hawking Incorporated: Stephen Hawking and the Anthropology of the Knowing Subject* (Chicago, 2012); Andrew Warwick, *Masters of Theory: Cambridge and the Rise of Mathematical Physics* (Chicago, 2003). For an attempt to articulate a radically materialist metaphysics, see Gilles Deleuze and Félix Guattari, *A Thousand Plateaus: Capitalism and Schizophrenia* (Minneapolis, 1987); Graham Harman, *Tool-Being: Heidegger and the Metaphysics of Objects* (Chicago, 2002); Quentin Meillassoux, *After Finitude: An Essay on the Necessity of Contingency* (London, 2008).

inherently progressive development (or regressive fall from grace, as the case may be).[12]

The rest of this introduction spells out these scholarly interventions in further detail. But first, one major caveat needs to be made explicit: because we regard science and capitalism as historical entities that are continually enacted in practice, we deliberately resist the temptation to offer a stable definition of either. Instead, we treat both as objects of empirical study. The point here is not to drain the key words in our title of their meaning, insisting that all knowledge is scientific or every economy capitalist. Rather, it is to treat both cultural institutions as *historically* constituted entities. That said, we do recognize several strands of similarity that create a kind of family resemblance between different ways the political economy of modern capitalism has been enacted and the epistemic ideals of modern science performed. For example, capitalist societies are often described as ones in which markets play a central role as the principal mechanism to coordinate between supply and demand. According to Karl Polanyi, rather than embed the marketplace within a broader set of cultural institutions, modern capitalism reframes all manner of social interactions as market transactions.[13] For that reason, capitalist societies tend to regard individual liberty as sacrosanct. They also feature strong legal regimes to protect private property and to enforce contracts. Finally, capitalism extends the commodity-form to nearly all aspects of life, including "intellectual property." But Polanyi's emphasis on the cash nexus as a mechanism for coordinating the circulation of commodities is not the only way to understand what is specific about the political economy of modern capitalism. A different but equally longstanding tradition primarily regards capitalism as an engine for the accumulation of wealth.[14] On this view, capitalism should be understood as a means to generate sustained growth by systematically using a portion of today's profits to fund tomorrow's productive enterprises. Money only becomes capital once it has been invested to expand the means of production. Insofar as it functions as a technology for shaping the future, capital thus confers immense social and political power. Moreover, absent a means of redistribution, capitalist economies tend to concentrate wealth and often produce high levels of inequality. To borrow Thomas Piketty's evocative phrase, in capitalism, "the past devours the future."[15] But some-

[12] For an explicit defense of global history as an appropriate methodological framework for writing the history of capitalism, see, e.g., Sven Beckert, "Emancipation and Empire: Reconstructing the Worldwide Web of Cotton Production in the Age of the American Civil War," *Amer. Hist. Rev.* 109 (2004): 1405–38; Beckert, "From Tuskegee to Togo: The Problem of Freedom in the Empire of Cotton," *J. Amer. Hist.* 92 (2005): 498–526. For more on the importance of circulation to the history of science, see, e.g., Aileen Fyfe and Bernard Lightman, *Science in the Marketplace: Nineteenth-Century Sites and Experiences* (Chicago, 2007); David N. Livingstone, *Putting Science in Its Place: Geographies of Scientific Knowledge* (Chicago, 2003); Kapil Raj, *Relocating Modern Science: Circulation and the Construction of Knowledge in South Asia and Europe, 1650–1900* (Houndmills, 2007); James Secord, "Knowledge in Transit," *Isis* 95 (2004): 654–72.

[13] See Karl Polanyi, *The Great Transformation* (New York, 1944). On the question of valuation in particular, see also Patrik Aspers and Jens Beckert, "Value in Markets," in *The Worth of Goods: Valuation and Pricing in the Economy* (Oxford, 2011), 3–38. That said, the reduction to monetary value has hardly gone uncontested. For a particularly vivid example, see Viviana Zelizer, *Morals and Markets: The Development of Life Insurance in the United States* (New York, 1979).

[14] Karl Marx et al., *Capital: A Critique of Political Economy*, vol. 1 (London, 1992). For a more recent argument along these lines, see Fabian Muniesa, ed., *Capitalization: A Cultural Guide* (Paris, 2017).

[15] Thomas Piketty, *Capital in the Twenty-First Century*, trans. Arthur Goldhammer (Cambridge, Mass., 2014), 571.

thing similar holds true for science as well. Not only do ideas circulate, but they are also accumulated in "centers of calculation" such as museums, libraries, and all of the other institutions that collectively make up the epistemic infrastructure for knowledge production.[16] Hence, scientific knowledge too is subject to both processes of circulation and acts of accumulation, which, in turn, helps to explain the uneven distribution of economic wealth and epistemic power.[17]

However we choose to describe the political economy of modern capitalism, one thing is certain: its roots run far deeper and its reach is far more expansive than the way we do business alone. Rather, capitalism may be likened to a Wittgensteinian "form of life," one that valorizes an impersonal, calculating sort of rationality as the cornerstone of sound judgement. Of course, modern science often makes similar claims for itself also, and it is widely invoked as both a model and litmus test of the right way to reason.[18] Thus, while we deliberately eschew making normative claims about the correct way to demarcate the boundaries of either science or capitalism, we recognize that both categories have been invested with considerable normative power. Rather than engaging in boundary disputes about what truly constitutes science or distinguishes capitalism, this volume therefore includes a number of essays that explicitly ask how precisely these boundaries were policed and enacted in practice (Arunabh Ghosh's piece on "capitalist" vs. "communist" statistics during the mid-twentieth century and Julia Fein's essay on the commodification of specimens in Stalinist Russia perhaps being the clearest examples).[19] Not seeking to ignore normative questions and controversies, we regard precisely these kinds of debates as especially fruitful objects of empirical investigation, because they help to illuminate how the changing relationship between science and capitalism was understood by specific people at particular times in history.[20]

The decision to treat science and capitalism as hotly contested but historically constituted categories also helps to delimit the chronological scope of this volume. Not only do science and capitalism both have a performative dimension, informing people's behavior while shaping the institutions that materially govern our lives, but, in a striking convergence, both also became objects of scholarly contemplation during the long nineteenth century. In each case, this happened as part of an effort to distinguish the right way to produce knowledge and the best way to organize a political economy. For example, whereas a number of words deriving from the Latin *scientia* have been used to characterize experiential knowledge of various kinds for hundreds of years,

[16] On "centers of calculation," see Bruno Latour, *Science in Action* (Cambridge, Mass., 1987). On the infrastructure of knowledge production more broadly, see Lorraine Daston, ed., *Science in the Archives: Pasts, Presents, Futures* (Chicago, 2017); Joanna Radin, *Life on Ice: A History of New Uses for Cold Blood* (Chicago, 2017).

[17] See Jessica Ratcliff, "The Great Data Divergence: Global History of Science within Global Economic History," in *Global Scientific Practice in an Age of Revolutions, 1750–1850*, ed. Patrick Manning and Daniel Rood (Pittsburgh, 2016), 237–54.

[18] For a history of the "epistemic virtue" of objectivity in particular, see Lorraine Daston and Peter Galison, *Objectivity* (New York, 2007).

[19] Arunabh Ghosh, "Lies, Damned Lies, and (Bourgeois) Statistics: Ascertaining Social Fact in Midcentury China and the Soviet Union"; Julia Fein, "'Scientific Crude' for Currency: Prospecting for Specimens in Stalin's Siberia," both in this volume.

[20] For more on this approach, see Thomas Gieryn, "Boundary-Work and the Demarcation of Science from Non-Science," *Amer. Sociol. Rev.* 48 (1983): 781–95; Gieryn, *Cultural Boundaries of Science: Credibility on the Line* (Chicago, 1999).

there was no such thing as the "scientific method" (or the professional scientist, for that matter) until historians and philosophers such as William Whewell began speculating about the best way to generate reliable knowledge. It was thus in a highly prescriptive context that the notion of science as detailed and factual knowledge produced through an objective process of rigorous hypothesis testing arose.[21] The case of "capitalism" is even more clear. Although the word "capital" had long been used to denote assets, money, or commodities more broadly and "capitalist" as anyone who dealt in or otherwise had access to capital, the neologism "capitalism" was deliberately coined by mid-nineteenth-century radicals such as Proudhon and Marx to criticize a form of social organization they viewed as ruthless, unjust, and ultimately unstable. Before long, the word "capitalism" came into much wider use to describe a distinctly modern political economy in which a calculative acquisitiveness informed more and more everyday decision making.[22] To borrow terminology from Hallam Stevens's essay in this volume, both science and capitalism may therefore be said to constitute a performance that takes place on the stage of everyday life: to be a scientist or a capitalist is to perform a particular role in society.[23] And such performances only became possible once an appropriate "script" was available. For example, a large number of observers since Marx have pointed out that to be a capitalist is to be a particularly future-oriented cognitive agent, always projecting oneself into an imagined space where present investments may generate profits or incur losses.[24] Because this mercantile practice of projecting oneself into an imagined future was transferred to the creation of large-scale industrial endeavors during the long nineteenth century, the bulk of the essays that make up this volume concern events and circumstances that range from that period to the present day.

Once science and capitalism became objects of knowledge, their histories were also subjected to intense scrutiny and debate. In another striking convergence, the origins of both were then traced back to sixteenth- and seventeenth-century Europe. As this happened, both categories became means for distinguishing between the ancient and modern, the developed and primitive, the West and the rest, meaning that a fairly provincial script was used to appraise the knowledge-making and wealth-generating performances of people outside nineteenth-century Europe. Given this fraught historiographic terrain, we would have been remiss in ignoring the relationship between

[21] Henry M. Cowles, "The Age of Methods: William Whewell, Charles Peirce, and Scientific Kinds," *Isis* 107 (2016): 722–37; Sydney Ross, "Scientist: The Story of a Word," *Ann. Sci.* 18 (1962): 65–85; Richard Yeo, *Defining Science: William Whewell, Natural Knowledge and Public Debate in Early Victorian Britain* (Cambridge, 2003).

[22] Werner Sombart, *Der moderne Kapitalismus* (Leipzig, 1902); Max Weber, *Die protestantische Ethik, und der Geist des Kapitalismus* (Tübingen, 1905). For a brief history of the word "capitalism," see the introduction to Jürgen Kocka, *Capitalism: A Short History*, trans. Jeremiah Riemer (Princeton, N.J., 2016). See also Kocka and Marcel van der Linden, eds., *Capitalism: The Reemergence of a Historical Concept* (London, 2016).

[23] See Hallam Stevens, "Starting up Biology in China: Performances of Life at BGI," in this volume; Erving Goffman, *The Presentation of Self in Everyday Life* (Garden City, N.Y., 1959); Richard Schechner, *Performance Studies: An Introduction*, 2nd ed. (New York, 2006). For an influential application of performance studies to the practice of science, see Stephen Hilgartner, *Science on Stage: Expert Advice as Public Drama* (Stanford, Calif., 2000).

[24] Jens Beckert, *Imagined Futures: Fictional Expectations and Capitalist Dynamics* (Cambridge, Mass., 2016), 1–2. For a similar kind of analysis, see Jonathan Levy, "Capital as Process and the History of Capitalism," *Bus. Hist. Rev.* 91 (2017): 483–510; Fabian Muniesa, ed., *Capitalization: A Cultural Guide* (Paris, 2017). Finally, see also Cook, *The Pricing of Progress* (cit. n. 9).

knowledge and commerce in other places and time periods. For that reason, we have also made sure to include a number of essays that cover events and developments prior to the nineteenth century and outside of Europe, without thereby seeking to produce an origin story for either science or capitalism. Instead, our aim is to show how epistemic and commercial values—matters of fact and matters of exchange—intersected in other time periods and geographies also. What is more, several of the essays that follow go further and adopt an explicitly transnational focus, showing how knowledge and profits were both generated through encounters and interactions between people from different parts of the globe.

Finally, we would like to zoom out somewhat to discuss the organization of our table of contents as a whole. This volume begins with a wide-ranging historiographical think piece from Harold J. Cook, which we intend as a "companion piece" to this introduction.[25] Next, we have chosen to group the rest of this volume's essays according to the different kinds of entanglements they most clearly address as a way to highlight the centrality of the entanglement concept to our way of thinking. Of course, it goes without saying that each of these essays speaks to more than just one kind of entanglement. Hence, they could have been organized into different clusters as well. Nonetheless, we do hope the organization of our table of contents will prove both interesting and illuminating to readers. The first of these clusters consists of three essays that, in one way or another, discuss the cognitive and manual labor that is required to not only create but also maintain the various kinds of infrastructures that support the entanglement between science and capitalism. The second cluster groups together three essays that all deal with a particular kind of knowledge work (indeed, what may be the most iconic and well-known kind of knowledge work) that has shaped the way economic transactions are carried out: calculation. Third, there is a larger cluster of five essays that all address how the entangled histories of science and capitalism have helped to give rise to new kinds of objects, entities, or relationships in the world, ranging from the Comstock Lode to the accident-prone driver, among several others. Finally, there is another cluster of three essays that all foreground the transnational entanglement between science and capitalism, connecting events, people, and processes in Europe and North America with Asia. We hope the contributions will help strengthen the volume's three principal scholarly interventions, each of which are discussed in more detail next.

DIVERGENCE AND ENTANGLEMENT

Whereas this volume deliberately eschews the question of when science truly began, or how capitalism really got started, others often invoke the rise of modern science to explain the so-called Great Divergence (or "Enrichment") that took place between Europe and the rest of the world during the long nineteenth century. Economic historians drawing upon the work of Douglass C. North, for example, emphasize the significance of formal and informal institutions in creating the conditions for growth and prosperity. Besides the creation of a strong system of private property law, a habit of plain dealing, and a valorization of thrift, these historians often cite a love of learn-

[25] Harold J. Cook, "Sciences and Economies in the Scientific Revolution: Concepts, Materials, and Commensurable Fragments," in this volume.

ing, a unique openness to useful ideas, and a constant desire to devise better ways of getting things done as direct contributions to the creation of modern capitalism.[26] One of the most strident articulations of such an argument is by Joel Mokyr, who attributes the disproportionate economic success of Europe and North America to their exceptional culture. As recently as 2016, Mokyr argued that it was technological innovation that primarily fueled the Great Divergence: "the explosion of technological progress in the West was made possible by cultural changes," he contends, which "affected technology both directly, by changing attitudes toward the natural world, and indirectly, by creating and nurturing institutions that stimulated and supported the accumulation and diffusion of 'useful knowledge.'"[27] For a number of complex and interrelated reasons that included a faith in progress and concomitant irreverence for the wisdom of ancients, these cultural changes took place in Europe. As a result, it was Europe (as well as its former colonies in North America) that diverged from the rest of the world.[28]

The notion that uniquely Western cultural innovations were primarily responsible for the Great Divergence has not escaped criticism, however. For example, economic historians informed by world systems theory have tried to "reorient" both economic and world history to demonstrate the crucial role of non-Western societies in the making of the early modern and modern world.[29] By establishing Asia as one of many centers in the early modern global economy, they effectively provincialize narratives of the European miracle and Western exceptionalism in the history of capitalism.[30] But these histories remain fundamentally comparative in their approach, and one can even detect the specter of civilizational comparisons in some attempts to identify

[26] See, e.g., Stephen H. Haber et al., eds., *Political Institutions and Financial Development* (Stanford, Calif., 2008); Deirdre N. McCloskey, *Bourgeois Equality: How Ideas, Not Capital or Institutions, Enriched the World* (Chicago, 2016); McCloskey, *Bourgeois Dignity: Why Economics Can't Explain the Modern World* (Chicago, 2010); Douglass C. North, "Institutions," *J. Econ. Perspect.* 5 (1991): 97–112; North, "Institutions, Ideology, and Economic Performance," *Cato J.* 11 (1992): 477–96; North, *Understanding the Process of Economic Change* (Princeton, N.J., 2005); North, John Joseph Wallis, and Barry R. Weingast, "A Conceptual Framework for Interpreting Recorded Human History" (working paper, National Bureau of Economic Research, 2006).

[27] Joel Mokyr, *A Culture of Growth: The Origins of the Modern Economy* (Princeton, N.J., 2016), 7.

[28] See also Joel Mokyr, *The Gifts of Athena: The Historical Origins of the Knowledge Economy* (Princeton, N.J., 2002); Margaret C. Jacob and Larry Stewart, *Practical Matter: Newton's Science in the Service of Industry and Empire* (Cambridge, Mass., 2004); Mokyr, *The Enlightened Economy: An Economic History of Britain 1700–1850* (New Haven, Conn., 2009); Jacob, *The First Knowledge Economy: Human Capital and the European Economy, 1750–1850* (Cambridge, 2014). For a recent review of economic and economic-historical literature on this subject, see Cormac Ó Gráda, "Did Science Cause the Industrial Revolution?," *J. Econ. Lit.* 54 (2016): 224–39.

[29] Key scholars associated with this approach include Janet Abu-Lughod, *Before European Hegemony: The World System A.D. 1250–1350* (Oxford, 1991); Andre Gunder Frank, *ReOrient: Global Economy in the Asian Age* (Berkeley and Los Angeles, 1998); Kenneth Pomeranz, *The Great Divergence: China, Europe and the Making of the Modern World Economy* (Princeton, N.J., 2000), among others. For more on the world systems theory approach to global inequality, see Daniel Chirot and Thomas D. Hall, "World-System Theory," *Annu. Rev. Sociol.* 8 (1982): 81–106; Immanuel Maurice Wallerstein, *The Modern World-System*, 4 vols. (Berkeley and Los Angeles, 2011); Wallerstein, *The Essential Wallerstein* (New York, 2000).

[30] Pomeranz, *The Great Divergence* (cit. n. 29) may be the most influential study to make this argument. Others include Roy Bin Wong, *China Transformed: Historical Chang and the Limits of European Experience* (Ithaca, N.Y., 1997). See also Richard von Glahn, *Fountain of Fortune: Money and Monetary Policy in China, 1000–1700* (Berkeley and Los Angeles, 1996), which demonstrates late Ming China's massive appetite for imported silver from the Americas and its far-reaching consequences for the globalization of the early modern world economy.

an Asian age as an alternative to Euro-American versions of capitalism and industrialization.[31] As such, many world systems theorists cannot be said to fully reject the fundamental units of analysis—the "West" and the "rest"—that underlie triumphalist accounts of European exceptionalism. Finally, while there is no denying that many areas of Europe and North America became far richer and more commercially powerful than other parts of the world during the past three or four centuries, narratives of divergence tend to pay less attention to the way global capitalism was continually being made and remade after the point of divergence.[32] Ironically, insofar as they neglect non-Western players as active agents in the global economy after the point of divergence, these narratives fail to account for the dynamic and truly global character of modern capitalism during all periods of its development.

A more recent critique of the Great Divergence argument has been articulated by historians of capitalism who insist that slavery, imperialism, and other means of coercive value extraction must be placed at the center of any narrative about the phenomenal enrichment of Europe and North America between the seventeenth and late nineteenth centuries.[33] In his recent book on the worldwide web of cotton production, for example, Sven Beckert coins the term "war capitalism" to describe the forceful extraction and transfer of wealth from Asia, Africa, and the Americas that took place during Europe's "Age of Exploration." In so doing, he explicitly emphasizes the extent to which violence, coercion, and political power were leveraged to build the material infrastructure upon which free trade ideology has been erected.[34] In this view, the Great Divergence resulted not so much from an explosion in technical know-how as from a willingness on the part of Europe's imperial powers to bring their war-making capacities to bear on extracting and channeling the world's productive resources to fuel their own economic development. Although this line of argument does share some family resemblances with world systems theory, it avoids many of the pitfalls that plague the comparative method by adopting a more truly transnational approach. Moreover, historians of capitalism also depart from world systems theory in their tendency to foreground particular choices made by individual people working in concert to further their interests over structural analyses of how the West came to dominate the rest of the world.

The claim that modern capitalism was built on a foundation of imperial exploitation, military expropriation, and coerced labor offers a welcome corrective to the triumphalist narrative in which technological innovation primarily fueled economic

[31] For a study that seeks to reverse the comparative asymmetry in examining early modern world economic development, see Wong, *China Transformed* (cit. n. 30). For a book that takes extra care to choose comparable units of economic development for purposes of comparing world economic change, see Pomeranz, *The Great Divergence* (cit. n. 29).

[32] Pomeranz, *The Great Divergence* (cit. n. 29), for example, jumps at the end of the study from the point of "divergence" during the eighteenth century to the twenty-first century, when Asian global economic dominance once again seems impossible to deny.

[33] For more on the argument that slavery drove the development of North American capitalism in particular, see, e.g., Edward E. Baptist, *The Half Has Never Been Told: Slavery and the Making of American Capitalism* (New York, 2014); Sven Beckert and Seth Rockman, eds., *Slavery's Capitalism: A New History of American Economic Development* (Philadelphia, 2016); Walter Johnson, *River of Dark Dreams: Slavery and Empire in the Cotton Kingdom* (Cambridge, Mass., 2013); Peter Linebaugh and Marcus Rediker, *The Many-Headed Hydra: Sailors, Slaves, Commoners, and the Hidden History of the Revolutionary Atlantic* (Boston, 2000).

[34] Sven Beckert, *Empire of Cotton: A Global History* (New York, 2014).

growth. But one might nonetheless worry that revisionist accounts go too far in writing the history of science and technology out of the story altogether. Kenneth Pomeranz, for example, makes the counterfactual claim that if China had access to the same resources as the West, the Great Divergence might not have occurred. But this line of reasoning neglects what one scholar refers to as the "human factor," which gave rise to a shift from appreciating coal as a resource upholding livelihood—a principal goal of imperial statecraft during the late Qing Empire—to regarding it as a necessary fuel for survival in an industrial world order.[35] Perhaps even worse is that a failure to address knowledge production leaves the history of capitalism vulnerable to the counterfactual claim that absent the prevalence of coerced labor and imperial expropriation, the Great Divergence would have still taken place.[36]

Ultimately, Great Divergence narratives largely ignore the degree to which science, capitalism, and imperialism all coproduced one another. As a great deal of work in the history of science makes abundantly clear, there is no separating imperial expansion and commercial motives on the one hand from the production of useful knowledge on the other.[37] For example, Harold Cook's influential account of the Dutch East India Company's scientific work convincingly demonstrates that without taking the "activities of commerce, including the trading ventures once called voyages of discovery" into account, it would be very difficult to answer the deceptively obvious question of why such "an enormous amount of personal time and effort, and economic and other resources, come to be devoted to seeking out and acquiring precise and accurate descriptive information about natural things."[38] Thus, even if it were possible to eliminate imperialism from European history conceptually, there is no reason to suspect that the institutions responsible for the growth of both modern science and the production of new technologies would remain unchanged. Given how closely the practice of science and imperial statecraft were bound up with one another, it is impossible to maintain that one, not the other, must be afforded a primary causal role in explaining the Great Divergence.

It is for precisely that reason that we favor the idiom of entanglement, which has previously been put to powerful use by scholars such as Michelle Murphy, who draws upon it to explain the complicated, transnational, and sometimes unsettling inter-

[35] See Shellen Wu, *Empires of Coal: Fueling China's Entry into the Modern World Order, 1860–1920* (Stanford, Calif., 2015).

[36] See, e.g., Deirdre N. McCloskey, "The Industrial Revolution, 1780–1860: A Survey," in *The Economic History of Britain since 1700*, ed. Roderick Floud and Deirdre N. McCloskey (Cambridge, 1981), 242–70. The controversy has become especially fierce in debates about the importance of slavery to the historical development of American capitalism in particular. See, e.g., Alan L. Olmstead, review of *The Half That Has Never Been Told*, by Edward Baptist, *J. Econ. Hist.* 75 (2015): 919–23. For an overview of the recent controversy, see Marc Parry, "Shackles and Dollars," *Chronicle of Higher Education*, 8 December 2016. For a methodological defense of the counterfactual method for apportioning causal power, see Tim De Mey and Erik Weber, "Explanation and Thought Experiments in History," *Hist. & Theory* 42 (2003): 28–38.

[37] See, e.g., Lucile Brockway, *Science and Colonial Expansion: The Role of the British Royal Botanic Gardens* (New York, 1979); Richard Drayton, *Nature's Government: Science, Imperial Britain, and the "Improvement" of the World* (New Haven, Conn., 2000); John Gascoigne, *Science in the Service of Empire: Joseph Banks, the British State and the Uses of Science in the Age of Revolution* (Cambridge, 1998); Lisbet Koerner, *Linnaeus: Nature and Nation* (Cambridge, Mass., 1999); Schiebinger, *Plants and Empire* (cit. n. 3).

[38] Harold John Cook, *Matters of Exchange: Commerce, Medicine, and Science in the Dutch Golden Age* (New Haven, Conn., 2007), 45, 6.

actions between reproductive health efforts, feminist political movements, technoscience, capitalist enterprise, and American imperial ambitions in the 1970s and 1980s.[39] Instead of trying to isolate direct causal influences, the idiom of entanglement highlights complicated circuits, unanticipated trajectories, and feedback loops. We argue that this idiom is particularly useful for exploring the interrelated development of both science and capitalism for at least three key reasons. First, entanglement emphasizes the complexity, contingency, and variety of relationships that make up the science-capitalism nexus. To think of science and capitalism as entangled is to suggest the futility of insisting that any one causal thread should have primacy over another. Instead, doing so highlights the fact that, when viewed in detail in specific contexts, the relationship between science and capitalism is highly convoluted and does not follow any single prescribed path or trajectory. As Murphy notes, such entanglements can be "uneasy," and their ramifications unpredictable. Second, entanglement also offers a way to think about the durability of science and capitalism by showing how they are often reinforced and strengthened through their interaction. As anyone who owns a pair of earbuds will know, tangles are not only complex but also intransigent. They tend not to untangle easily and thus can be hard to undo. Third, entanglement offers a useful way—a useful "topology," as Murphy puts it—through which to think about the geographic spaces and scales across which science and capitalism interact.[40] As the essays in this volume show, science and capitalism have often become wound together in distinctive and powerful ways within specific local settings. But the threads that feed into those local knottings are also part of networks and circuits that far transcend the local and can be truly global. Perhaps most important of all is that entanglement offers a useful idiom for thinking about science and capitalism as part of the same, larger assemblage. Thus, rather than attempting to parse out the relative importance or causal power of each, this volume instead wants to suggest that both derive their considerable power and significance from being so readily, and so often, conjoined.

KNOWLEDGE WORK

If this volume's first scholarly intervention involves a shift from narratives of divergence to ones that foreground entanglement, its second is to inquire into the cognitive labor—what we call "knowledge work"—that generated many of the nodes around which the history of science became so entangled with the history of capitalism. To do so, we want to foreground specific practices more so than general ideas—focusing on the routinized activities that constitute economic life and scientific inquiry, along with the material things through which those activities are conducted, the "know-how" that makes them possible, the institutions and mores that structure them,

[39] Michelle Murphy, *Seizing the Means of Reproduction: Entanglements of Feminism, Health, and Technoscience* (Durham, N.C., 2012). There is also a rich anthropological literature on entanglement, notably as a way to theorize the interactions between different cultural groups (e.g., Western and non-Western groups in colonial encounters) and between human social life and the material world. See Nicholas Thomas, *Entangled Objects: Exchange, Material Culture, and Colonialism in the Pacific* (Cambridge, Mass., 1991); Ian Hodder, "Human-Thing Entanglement: Towards an Integrated Archaeological Perspective," *J. Roy. Anthropol. Inst.* 17 (2011): 154–77; Hodder, *Entangled: An Archaeology of the Relationship between Humans and Things* (Malden, Mass., 2012).

[40] Murphy, *Seizing the Means* (cit. n. 39), 11–21.

the emotions they elicit, and so on.[41] Attention to practice has, of course, been one of the defining methodological trends in the history of science, especially since the 1990s, when scholars turned to the close examination of scientific practice as a way forward out of vexing disputes about the realism versus constructed-ness of scientific theories. Something similar is true for the history of capitalism as well. To no small degree, the history of capitalism and the history of science might therefore be written as histories of gerunds: managing, planning, measuring, calculating, predicting, experimenting, modeling, collecting, classifying, and so on, to name just a few. By studying these technical practices "in action," we can examine how seemingly natural, inevitable, or "black-boxed" aspects of economic order or scientific knowledge were in fact the product of local cultures, personal interests, contested choices, and historical contingencies. In a word, we want to focus attention on the intellectual labor through which science and capitalism were coproduced.

One broad realm of practice—one gerund—that offers an especially good opportunity for collaboration between historians of science and capitalism is, simply put, thinking. Among the most impressive achievements of the historiography of science has been the ability to show that scientific thinking does not simply proceed through individual inspiration or relentless methodicality but is rather diverse, disorderly, and surprising. Older histories of scientific ideas or scientific thought, understood as an accretion of static units of knowledge, have long since given way to an image of scientific thinking as a dynamic social and material process. By comparison, attention to the dynamism and complexity of economic thinking is of a somewhat more recent vintage. Dominant models of capitalism, both Marxist and neoclassical, long left relatively little room for thinking as an open-ended and generative activity and thus downplayed its significance as an object of social and economic analysis. In Marxist analyses, conscious acts of thinking have been seen as the expression of underlying material interests and the class consciousness they beget. In neoclassical models, the presumption is that all individuals, or at least those who move markets, act in ways that maximize their self-interest; what they "think" they are doing is far less important than what their economic choices reveal about their true preferences.

Yet recent scholarship, both in the new history of capitalism and in other fields like economic sociology and anthropology, has put thinking back at the center of capitalist action, showing that economic actors are "subjects" and that capitalism is, in a profound sense, an epistemic system.[42] This effort to reopen the cognitive space of capitalism can be seen, for example, in the efforts of historians of capitalism to recover how foundational concepts in modern economic life—credit, risk, profit, the economy,

[41] For a discussion of the value of studying practice in the history of capitalism, see Kenneth Lipartito, "Connecting the Cultural and the Material in Business History," *Enterprise Soc.* 14 (2013): 686–704. The literature on scientific practice is certainly far too vast to do justice to here. Some especially influential examples include Andrew Pickering, ed., *Science as Practice and Culture* (Chicago, 1992); Robert Kohler, *Lords of the Fly:* Drosophila *Genetics and the Experimental Life* (Chicago, 1994); Pickering, *The Mangle of Practice: Time, Agency, and Science* (Chicago, 1995); Galison, *Image and Logic* (cit. n. 10).

[42] On the importance of seeing economic actors as "not simple decision makers but also *thinking subjects*," see Hirokazu Miyazaki, *Arbitraging Japan: Dreams of Capitalism at the End of Finance* (Berkeley and Los Angeles, 2013), 6. On experimental systems and epistemic things, see Rheinberger, *Toward a History* (cit. n. 10).

and so on—have been forged, refashioned, and made durable.[43] Or in the burgeoning attention historians, including many historians of science, have paid to the details of technical knowledge practices within business enterprises.[44] Or in economic sociology and science and technology studies (STS) research examining the complex assemblages of epistemic devices—economic models, evaluation techniques, calculating instruments—that are needed to allow economic agents to make "rational" choices in market settings.[45] The list could be extended almost indefinitely, and nearly all of the essays in this volume have something to contribute to this conversation in one way or another.

Several of our essays explicitly feature economic and scientific thinking as a form of knowledge work. William Deringer, Martin Giraudeau, and Arunabh Ghosh focus on how calculation, a crucial technical practice common to science and capitalism, serves as a crucial site for the entanglement of both domains.[46] All three challenge the assumption that calculation is simply a mechanical expression of a unitary capitalist or scientific rationality and instead show that calculative practices are creative and contested domains where actors experiment with different epistemologies and explore alternative futures. A key theme in the chapters is the way that the authority of calculators and calculative expertise is coproduced with visions of political-economic order. In his essay, Deringer sheds light on the culture and nature of exchange among British "men of science" between 1660 and 1720. He explains how two computational methods for dealing with annuities were presented as having mathematical ingenuity and financial utility in their promise to streamline many common financial transactions and why both were ultimately rendered obsolete for capitalist practice. Giraudeau characterizes a text written in 1800 by Irénée Du Pont de Nemours, used to raise funds for what was to become the Du Pont Corporation, as similar in kind to

[43] Timothy Mitchell, *Rule of Experts: Egypt, Techno-politics, Modernity* (Berkeley and Los Angeles, 2002), chap. 3; Carl Wennerlind, *Casualties of Credit: The English Financial Revolution, 1620–1720* (Cambridge, Mass., 2011); Jonathan Levy, *Freaks of Fortune: The Emerging World of Capitalism and Risk in America* (Cambridge, Mass., 2012); Arwen Mohun, *Risk: Negotiating Safety in American Society* (Baltimore, 2012); Levy, "Accounting for Profit and the History of Capital," *Crit. Hist. Stud.* 1 (2014): 171–214.

[44] The literature on actuaries is especially extensive. See, e.g., Theodore M. Porter, *Trust in Numbers: The Pursuit of Objectivity in Science and Public Life* (Princeton, N.J., 1995), chap. 5; Timothy Alborn, *Regulated Lives: Life Insurance and British Society, 1800–1914* (Toronto, 2009), chap. 4; Levy, *Freaks of Fortune* (cit. n. 43), chap. 3; Bouk, *How Our Days Became Numbered* (cit. n. 9). Another notable example, among many, is credit scoring: Martha Poon, "Scorecards as Devices for Consumer Credit: The Case of Fair, Isaac & Company Incorporated," in "Market Devices," ed. Fabian Muniesa, Yuval Millo, and Michel Callon, suppl. 2, *Sociol. Rev.* 55 (2007): 284–306; Josh Lauer, "Making the Ledgers Talk: Customer Control and the Origins of Retail Data Mining, 1920–1940," in *The Rise of Marketing and Market Research*, ed. Helmut Berghoff, Philip Scranton, and Uwe Spiekermann (New York, 2012), 153–69.

[45] This literature is vast, but see esp. Daniel Beunza and David Stark, "Tools of the Trade: The Socio-Technology of Arbitrage in a Wall Street Trading Room," *Indust. Corp. Change* 13 (2004): 369–400; Donald MacKenzie, *An Engine, Not a Camera: How Financial Models Shape Markets* (Cambridge, Mass., 2006); Alex Preda, "Socio-Technical Agency in Financial Markets: The Case of the Stock Ticker," *Soc. Stud. Sci.* 36 (2006): 753–82; Muniesa, Millo, and Callon, "Market Devices" (cit. n. 44); Trevor Pinch and Richard Swedberg, eds., *Living in a Material World: Economic Sociology Meets Science and Technology Studies* (Cambridge, Mass., 2008); MacKenzie, *Material Markets: How Economic Agents Are Constructed* (Oxford, 2009); Canay Özden-Schilling, "The Infrastructure of Markets: From Electric Power to Electronic Data," *Econ. Anthropol.* 3 (2016): 68–80.

[46] William Deringer, "Compound Interest Corrected: The Imaginative Mathematics of the Financial Future in Early Modern England"; Martin Giraudeau, "Proving Future Profit: Business Plans as Demonstration Devices"; both in this volume; see also Ghosh, "Lies" (cit. n. 19).

demonstration devices employed to assess profit and loss by natural philosophers at the time. Finally, Ghosh provides us with a Cold War case study that illustrates how and why socialist statistics emerged in the latter half of the twentieth century as a "social science" to serve as a powerful antidote to bourgeois liberal mathematical statistics. In doing so, he denaturalizes universalistic claims of liberal statistics that crucially rested on purportedly "pure" methods of probabilistic thinking in order to identify the ideological concerns behind them, including the desire on the part of capitalist states to increase their control and modernize statecraft, a goal that they shared with socialist nations.

Collecting and accumulation are forms of knowledge work and practice often positively associated with modern scientific inquiry and capitalism. Courtney Fullilove, Julia Fein, and Sarah Milov turn our attention to different forms of collecting objects—biomatter, natural history specimens, and epidemiological data—and consider how such knowledge work functions to make or unmake commodities in particular political economies.[47] Fullilove's essay features contemporary bioprospecting, the collecting of pest-resistant cereal endophytes for capitalist purposes of profit-oriented international gene banks, outlining the emerging characteristics of the twenty-first-century political economy that underlies the practice and informs what biota are treated as a commodity and why, as well as who benefits and loses. Fein's essay examines how material—in this case, "scientific crude," or natural history specimens—is collected for the purposes of turning it into a global commodity in, surprisingly, a communist command economy. Examining the collection of such "crude" from the Siberian periphery during Stalin's first Five-Year Plan (1928–32), Fein specifically illustrates how the socialist state mobilized the collection of such material originally to sell it on the global market until Moscow ultimately shifted its policy toward requisitioning the material away from the Siberian periphery to the center as a form of national heritage. Milov's essay turns our attention to the unexpected transnational significance of data collection on wives of Japanese smokers. She shows how American tobacco that flowed into Japanese markets and lungs served as the basis of Japanese epidemiological data, which, in turn, traveled back to the United States and was used by anti-tobacco grassroots activists as evidence for public smoking bans against the interests of big tobacco.

If calculation, classification, and collection are relatively iconic forms of knowledge work for scientific and capitalist endeavors, other contributors pay attention to ways of knowing and acting that are not usually associated with the two domains. Eugenia Lean and Hallam Stevens draw our attention to the act of copying, conventionally reviled as a problematic practice obstructing innovation and free market dynamics in more hagiographic accounts of science and capitalism.[48] Rethinking the place and value of copying in scientific and capitalist innovation, both Lean and Stevens draw on case studies from China, which has been targeted for engaging in exceptionally unethical copying since the late nineteenth century. By examining early twentieth-century international disputes over alleged Chinese counterfeiting of Bur-

[47] Courtney Fullilove, "Microbiology and the Imperatives of Capital in International Agro-Biodiversity Preservation," in this volume; Fein, "'Scientific Crude'" (cit. n. 19); Sarah Milov, "Smoke Ring: From American Tobacco to Japanese Data," in this volume.

[48] Eugenia Lean, "Making the Chinese Copycat: Trademarks and Recipes in Early Twentieth-Century Global Science and Capitalism," in this volume; Stevens, "Starting" (cit. n. 23).

roughs, Wellcome's popular vanishing cream, Hazeline Snow, Lean shows how copying the product's trademarks and adaptation of its recipes proved crucial in helping Chinese merchants innovate their own products and compete globally in a competitive pharmaceutical market. To stem this rising tide of Chinese manufacturing power, Burroughs, Wellcome and other pharmaceutical companies aggressively promoted an emerging intellectual property (IP) regime, identified Chinese copying as unethical, and pursued alleged copycats. Stevens moves us forward in time to focus on how the Beijing Genomics Institute (BGI), a DNA-sequencing research institute in contemporary Shenzhen, engages in acts of adaptation to establish itself as a highly creative, hybrid corporation that is competitive worldwide. Both contend that copying and innovation have not been mutually exclusive in modern science and capitalism and show that while the Chinese actors they consider are savvy in their acts of adaptation, they have never been singular in their copy work, because copying and adaptation have taken place in all corners of the modern and contemporary world.

The laborious and unglamorous act of maintenance is another form of knowledge work that has often been given short shrift in accounts of science and capitalism that focus on revolutionary leaps in scientific invention and capitalist innovation. Here, two of our contributors—Emily Pawley and David Singerman—focus their attention on the knowledge and drudgery involved in the day-to-day and season-to-season processes of engineering the reproductive capacities of sheep and the technical upkeep of sugar machinery in capitalist industries of the eighteenth and nineteenth centuries.[49] Pawley's account takes us to the fields of mid-eighteenth-century Britain, at a moment when the onset of new forms of "agricultural capitalism" was dramatically transforming the way food was produced and eaten. Her study examines how agricultural experts and practicing farmers built up a new body of natural knowledge about livestock, aimed at producing tender bodies available at all times of the year to meet the increasingly voracious demands of an emerging consumer market in meat. At the center of this biological project were a battery of new techniques, including feeding regimens and the design of natural landscapes, intended to direct and manage animals' sexual desires and reform those occasionally recalcitrant beings into agreeable producers of new animal bodies. Singerman's "Sugar Machines" illustrates the various material parts and paper devices that Scottish engineers utilized in order to maintain and manage temperamental machines that were shipped to far-flung environments in the Caribbean as the production of sugar had become increasingly global by the nineteenth century. If scholars interested in global circulation of commodities and knowledge have focused on the epistemic work that goes into creating standardized, mobile units, Singerman shows that such standardized circuits were only possible because of the maintenance work done by technicians and engineers bearing highly specialized and often tacit forms of technical knowledge that defied standardization.

Given the important contributions historians of science have made to the study of thinking and knowing in practice, the recent surge of attention to the epistemic dimensions of capitalism offers an obvious opportunity for further collaboration. In fact, historians of science and STS scholars have been key contributors to many of

[49] Emily Pawley, "Feeding Desire: Generative Environments, Meat Markets, and the Management of Sheep Intercourse in Great Britain, 1700–1750"; David Singerman, "Sugar Machines and the Fragile Infrastructure of Commodities in the Nineteenth Century," both in this volume.

the key trends listed above—from the study of technical devices in shaping markets and "futures" to agnotology. Yet this shared attention to the epistemic dimensions of capitalism does not imply a desire to see the history of capitalism become a purely intellectual history, or for scholars to neglect the material dimensions of life under capitalism. One of the reasons historians of science can offer an especially useful perspective on economies past is precisely that they are used to reckoning yet another entanglement, namely, the way "epistemic things" and material objects are constantly commingled. One of the most provocative examples is the argument by STS scholars like Michel Callon and Donald MacKenzie that economic models play a crucial role in shaping how markets are constructed, the decisions economic actors can make, and what forms of economic action are rational. At its strongest, this argument about the "performativity" of economic models suggests that certain claims about how the economic world operates, like the Black-Scholes-Merton model for pricing stock options, or "Moore's law" regarding the development of the semiconductor industry, may act as a self-fulfilling prophecy, meaning they have the power to reshape economic phenomena in their own image.[50] While this performativity thesis has generated much debate, it offers a vital reminder that, to borrow a different idiom, economic order is coproduced with economic knowledge—and an invitation to historians of science to help explain such historical entanglements.[51]

Thus, this volume does not only seek to treat knowledge work as a practice, a form of intellectual labor. Our aim is also to lay bare the deep entanglement between words and things, theory and reality, epistemology and ontology. Indeed, several essays in this volume even go so far as to argue that cognitive practices can lead to the creation of new entities and relationships in the world. To highlight this point and make it explicit, we have chosen to group some of the essays into a section on "Entangled Ontologies."[52] For example, Lee Vinsel's essay on auto regulation shows how a group of industrial psychologists created a lucrative niche from which to augment their professional power by arguing that a new technology, the automobile, led to the creation of a new kind of person, the accident-prone driver. Similarly, Victoria Lee tracks the way science as an institution of epistemic authority and the microbe as an object of knowledge coproduced one another in late Meiji Japan. Paul Lucier shows how the involvement of geologists in the practice of what he provocatively calls "Comstock Capitalism" led to the consolidation of a new, material entity: the single, continuous,

[50] Michel Callon, "Introduction: The Embeddedness of Economic Markets in Economics," in *The Laws of the Markets*, ed. Michel Callon (Oxford, 1998), 1–57; MacKenzie, *Engine* (cit. n. 45); Donald MacKenzie, Fabien Muniesa, and Lucia Siu, eds., *Do Economists Make Markets? On the Performativity of Economics* (Princeton, N.J., 2007), esp. the chapter by Callon, "What Does It Mean to Say Economics Is Performative?," 311–57; Peter Miller and Ted O'Leary, "Mediating Instruments and Making Markets: Capital Budgeting, Science and the Economy," *Account. Org. Soc.* 32 (2007): 701–34; Franck Cochoy, Martin Giraudeau, and Liz McFall, "Performativity, Economics, and Politics: An Overview," *J. Cult. Econ.* 3 (2010): 139–46. For a notable critique, see Philip Mirowski and Edward Nik-Khah, "Markets Made Flesh: Performativity, and a Problem in Science Studies, Augmented with Consideration of the FCC Auctions," in MacKenzie, Muniesa, and Siu, *Do Economists Make Markets?*, 190–224.

[51] Sheila Jasanoff, "The Idiom of Co-Production," in *States of Knowledge: The Co-Production of Science and Social Order*, ed. Sheila Jasanoff (London, 2004), 1–12.

[52] Victoria Lee, "The Microbial Production of Expertise in Meiji Japan"; Lee Vinsel, "'Safe Driving Depends on the Man at the Wheel': Psychologists and the Subject of Auto Safety, 1920–55"; Paul Lucier, "Comstock Capitalism: The Law, the Lode, and the Science"; Lukas Rieppel, "Organizing the Marketplace," all in this volume; Fein, "'Scientific Crude'" (cit. n. 19).

extremely valuable, and thus hotly contested "Comstock Lode." Finally, Fein shows how a new kind of commodity—scientific crude—was created in response to Stalin's first Five-Year Plan during the early twentieth century, whereas Lukas Rieppel argues that as a new science of life came into being around the turn of the nineteenth century, organization was refashioned from a feature or property of living beings to a thing in itself—the organism—whose functional integration subsequently came to serve as a model for business organization, especially in the context of large, multidivisional corporate firms. Several other essays in the volume that are not explicitly grouped into this category also discuss the creation of new entities that have emerged with the modern entanglement of science and capitalism. All share the concern of shedding light on how the imperatives of science and capitalism resulted in the articulation of new social roles (the accident-prone driver, the Chinese copycat, the scientist) and objects of knowledge (the microbe, socialist statistics, organization) that, in turn, profoundly shaped concrete material practices as well as political and economic relationships, both local and global.

CIRCUITS OF EXCHANGE

Our third scholarly intervention involves adopting a global approach to the study of the entanglement of science and capitalism from the early modern period to the twenty-first century. We are indebted to developments in both the history of capitalism and the history of science fields. Historians of capitalism have long adopted a global purview, even as some have remained more comparative in their approach.[53] But, as noted above, some recent historians of capitalism are moving beyond comparative methods by paying attention to what were often violent transnational and global relations in the production of commodities like cotton.[54] Similarly, the history of science has increasingly adopted a more global approach in seeking to unravel triumphalist narratives that see the scientific and industrial revolutions as somehow unique to the West, or as automatically desirable and inexorable. If revisionist historians of capitalism have helpfully shed light on how early modern circuits of silver and modern commodities have flowed in multiple directions, historians of science have been particularly effective in attending to the global movement of knowledge. Some have engaged in more theoretically oriented inquiries to conduct a sustained conversation about how fundamentally interconnected the world has been in the making of modern science. Postcolonial scholars of science and medicine, for example, have been among the most critically engaged in articulating the moral imperative to

[53] The comparative approach is especially clear in the "varieties of capitalism" literature. See, e.g., John R. Bowman, *Capitalisms Compared: Welfare, Work, and Business* (Los Angeles, 2014); Barry Stewart Clark, *The Evolution of Economic Systems: Varieties of Capitalism in the Global Economy* (New York, 2016); David Coates, ed., *Varieties of Capitalism, Varieties of Approaches* (New York, 2005); Peter A. Hall and David W. Soskice, eds., *Varieties of Capitalism: The Institutional Foundations of Comparative Advantage* (Oxford, 2001); David Hundt and Jitendra Uttam, *Varieties of Capitalism in Asia: Beyond the Developmental State* (London, 2017); Martha Prevezer, *Varieties of Capitalism in History, Transition and Emergence: New Perspectives on Institutional Development* (New York, 2017).

[54] There have been criticisms of Beckert's important work, however. For a Marxist-inflected critique that demands a more explicit theorization of valuation in the discussion of capitalism, see Aaron G. Jakes and Ahmad Shokr, "Finding Value in *Empire of Cotton*," *Crit. Hist. Stud.* 4 (2017): 107–36. In addition, it has also been noted that Beckert fails to address the ecological implications of cotton emerging as a quintessential commodity in war capitalism.

challenge Eurocentric narratives that posit how modern science emerged singularly in the West and was subsequently exported abroad, rendering it challenging to even distinguish "Western" from "non-Western" science.[55] Warwick Anderson insists on "a critical engagement with the present effects . . . of centuries of 'European expansion'" to decenter "conventional accounts of so-called 'global' technoscience, revealing and complicating the durable dichotomies [of global/local, first world/third world, Western/indigenous, and big science/small science], produced under colonial regimes, which underpin many of its practices and hegemonic claims."[56]

Explorations into the motifs of circulation, movement, and exchange have taken up such a task by providing a powerful framework from which to complicate such durable dichotomies. Because knowledge only becomes recognized as such once it has been widely shared within a community of knowing subjects, circulation is now widely seen to be part and parcel of how knowledge is made, not just an afterthought. The importance of this insight can hardly be overstated, particularly given the methodological space it has opened for writing a global history of science that does not valorize the importance of Europe and North America over the rest of the world. There is also now considerable recognition that knowledge about the natural world emerged through encounters and exchanges between people from all parts of the globe, not just between Europe and the "rest."[57] Finally, related efforts have focused on the way global brokers or mediators facilitated the circulation of knowledge, often in ways that move beyond a strict metropole-colony axis.[58] Rather than privileging Western actors as global agents and dismissing non-Western actors as "local," "indigenous," or somehow particularized,[59] these histories of science emphasize the way go-betweens have often been more "cosmopolitan" than counterparts who remained in the metropole.[60]

[55] For one of the best-known and most controversial proponents of the idea that modern science was disseminated to non-Western parts of the world and that non-Western societies were merely passive receptors of scientific knowledge, see Basalla, "Spread of Western Science" (cit. n. 8).

[56] Warwick Anderson, "Introduction: Postcolonial Technoscience," *Soc. Stud. Sci.* 32 (2002): 643–48, on 644.

[57] Daniela Bleichmar, *Visible Empire: Botanical Expeditions and Visual Culture in the Hispanic Enlightenment* (Chicago, 2012); Antonio Barrera-Osorio, *Experiencing Nature: The Spanish American Empire and the Early Scientific Revolution* (Austin, Tex., 2006); Neil Safier, *Measuring the New World: Enlightenment Science and South America* (Chicago, 2008); Bernard V. Lightman et al., eds., *The Circulation of Knowledge between Britain, India, and China: The Early-Modern World to the Twentieth Century* (Leiden, 2013).

[58] Raj, *Relocating Modern Science* (cit. n. 12); Simon Schaffer, Lissa Roberts, Kapil Raj, and James Delbourgo, eds., *The Brokered World: Go-Betweens and Global Intelligence, 1770–1820* (Sagamore Beach, Mass., 2009).

[59] Some work that falls more squarely in the postcolonial framework of exploring the transmission of knowledge from the metropole to the colony has still tended to treat Western actors as the "global" actor, and the non-Western agents as the "local," and particularized, or "indigenous" agent.

[60] Translators and diplomats, Chinese and Naxi guides to British botanists, and native informers to colonial scientists in early twentieth-century Africa are but a few examples of figures who participated in transnational circuits of knowledge and materials, actively helping to constitute modern science. On translators and diplomats, see Raj, *Relocating Modern Science* (cit. n. 12); Marwa Elshakry, *Reading Darwin in Arabic, 1860–1950* (Chicago, 2013). On Chinese and Naxi guides to imperialist botanists, see Fa-ti Fan, *British Naturalists in Qing China: Science, Empire, and Cultural Encounter* (Cambridge, Mass., 2004); and Erik Mueggler, *The Paper Road: Archive and Experience in the Botanical Exploration of West China and Tibet* (Berkeley and Los Angeles, 2011), respectively. On African informers to colonial naturalists, see Nancy Jacobs, *Birders of Africa: History of a Network* (New Haven, Conn., 2016); Helen Tilley, *Africa as a Living Laboratory: Empire, Development, and the Problem of Scientific Knowledge, 1870–1950* (Chicago, 2011). In this light, Jesuits might be seen not as

With the term "broker" being so intimately associated with economic exchange and circulation, it strikes us that our interest in the entangled history of science and capitalism would similarly benefit by being approached as a story of linkages that emphasizes practices of brokerage and translation, points of convergence, and globally circulating networks of expertise and material, and that eschews the dichotomies of "global/local," "West/rest," and others. Indeed, several of our contributors approach their case studies in this manner, focusing on points of convergence and brokers that have facilitated the global circulation of both science and capitalism in the period spanning the nineteenth to twenty-first centuries. Singerman's "Sugar Machines" looks at one of the quintessential commodities of modern capitalism—sugar—and recovers the key historical actors—a coterie of Glasgow-based engineers—and their ceaseless intellectual and physical labor that went into maintaining the fragile material infrastructure of nineteenth-century transnational sugar production.[61] Ghosh, focusing on a much more recent era, demonstrates how Soviet theoreticians and statisticians were global actors who helped circulate socialist statistics to other parts of the emerging socialist world, including China.[62] This transnational circuit of knowledge producers generated a form of statistics that was not somehow derivative of authentic capitalist statistics but emerged as a calculative culture of the socialist world that made sense within the geopolitical context of the Cold War. In his study of BGI, Stevens sheds light on how this hybrid research center was not merely an imitation of the Western factory, engaged in rampant copying (as Western journalists regularly charge), but functions as a "broker" of sorts. He argues that BGI "performs" *shanzhai*, a do-it-yourself mode of innovation that reflects Shenzhen's biotech and manufacturing culture more generally and relies on the ability to copy and adapt in order to compete effectively on a global scale.[63]

While the global circulation of knowledge as one among many economically valuable commodities offers a highly suggestive way to bring the histories of science and capitalism into dialogue, we also see a number of dangers that emerge from the recent enthusiasm for "knowledge in transit," as James Secord has evocatively described it.[64] Precisely because the motif of circulation is so closely connected to both classical and neoliberal models of the way value is generated, it behooves scholars to question more carefully the conditions in which knowledge is made to travel. Doing so not only means taking seriously those cases in which knowledge and things resolutely stay put, refusing to partake in global circuits of intellectual and economic exchange. It also requires that we take seriously the asymmetries in epistemic, economic, and technical power that shape the way knowledge and commodities alike can be, and have been, mobilized. For that reason, we would be remiss if we were to restrict ourselves to the logic of circulation alone. The way knowledge and other resources are accumulated is decisive as well, producing asymmetries and inequalities with clear

Western missionaries transmitting religion and science from the European metropole to other parts of the world in the early modern period, but as a cosmopolitan cadre of go-betweens who shared information among a variety of empires as they traveled across the globe. On this point, see Laura Hostetler, *Qing Colonial Enterprise: Ethnography and Cartography in Early Modern China* (Chicago, 2001).

[61] Singerman, "Sugar Machines" (cit. n. 49).

[62] Ghosh, "Lies" (cit. n. 19).

[63] Stevens, "Starting" (cit. n. 23).

[64] Secord, "Knowledge in Transit" (cit. n. 12).

consequences for what can be known and who knows it. Given these considerations, we advocate taking our cues from scholars who emphasize the arduous labor, coercion, expropriation, and at times even violence that histories of the way knowledge circulates often belie. In particular, while we agree that a claim must be widely shared in the community to be seen as legitimate, the constitution of that very community is achieved by creating boundaries and erecting barriers to exclude those who are not seen to have a proper place in the group. And while we acknowledge that there can be no doubt that efforts to promote movement and communication play an indispensable role in the history of science and capitalism, these efforts always operate in combination with concordant attempts to control, manage, and, at times, explicitly arrest the movement of objects, ideas, and people.[65]

Several contributions here are explicit in their consideration not only of global circulation but also of instances when movement is obstructed and circulation is arrested, when exclusive communities and boundaries are erected. Lean's essay, for example, explores how the emerging early twentieth-century IP regime of trademark infringement sought to obstruct the circulation of manufacturing knowledge to alleged Chinese copycats in order to improve global market conditions for British pharmaceutical corporations. Fein's essay identifies the process by which Siberian "scientific crude," or natural history specimens, was decommodified in the Soviet Union during the Stalinist period. This was not simply because of the ideological imperatives of a communist command economy, but more because Moscow, which was invested in securing global exports at the time, ultimately came to value this particular material more as national patrimony to be stocked domestically than as an export commodity to be sold abroad. Sarah Milov's contribution sheds light on how the collection of data from Japanese smokers of American tobacco unexpectedly served to underpin a grassroots movement against American big tobacco in the second half of the twentieth century. And Fullilove's piece demonstrates how contemporary international gene banks, which rely on botanic and Linnean classification systems to commoditize biomatter, are now challenged and, indeed, being rendered obsolete by emerging biomatter commodities, such as the fungal endophyte, which feature the genome as the key to their value. By historicizing the instances when the circulation of scientific knowledge or capital has been obstructed, halted, or somehow altered, these essays delve into why global flows and stoppages occurred and what was at stake when they did.[66]

Finally, it should be noted that, even though this volume eschews any "origins" narrative, it does seek to identify the historically specific conditions under which science and capitalism emerged as powerful institutions and ideological regimes by the nineteenth century. In the early modern period, the pursuit of profit in rational, calculative ways, along with empirical ways of knowing the natural world, did not exist solely in the West and could be found in many parts of the world.[67] Yet, even as we

[65] See Warwick Anderson, "Making Global Health History: The Postcolonial Worldliness of Biomedicine," *Soc. Hist. Med.* 27 (2014): 93–113.

[66] Lean, "Making the Chinese Copycat" (cit. n. 48); Fullilove, "Microbiology"; Milov, "Smoke Ring" (both cit. n. 47).

[67] These various forms of empiricism and pursuits of profit did not automatically result in the development of modern science and capitalism and must be studied on their own terms. Revisionist historians of science have unearthed "non-Western" systems of natural, technological, and healing knowledge to demonstrate how rich ways of knowing that appear to mirror the empiricism of Western

recognize the historical existence of multiple forms of empiricism, we are interested in historicizing the specific conditions of the eighteenth and nineteenth centuries—including, crucially, imperial encounters and global exchange—under which specific forms of empiricism did develop to justify and underlie practices of pursuing profit and producing knowledge through extractive and often violent means. A system of capitalist accumulation in which profits were pursued with a calculated acquisitiveness that was predicated upon key social processes (including ways to manage labor and define relations of production, extract material resources, and rearrange geopolitical dynamics) emerged to guarantee ever-cheaper costs of industrial capital. So, too, did conditions that enabled scientific knowledge production that claimed to be value neutral and free of economic and political implication, even as that knowledge was in fact dependent upon extractive approaches to data collection, called for a particular exploitative relationship toward the natural world and was often inextricably linked with capital or state power. Taken together, these developments in science and capitalism came to radically reshape economic and epistemological practice around the world, reconfigure relations of production, labor regimes, and geopolitical dynamics, as well as usher in new ways of understanding and interacting with the natural world.

To better understand the historically specific conditions under which the nineteenth-century coproduction of science and capitalism occurred, we include several European case studies from the early modern period. Again, not seeking the "origins" of science or capitalism, these pieces seek to help decenter any origins narrative from within by refusing a teleological perspective and attending to the historically contingent circumstances that informed economic and epistemological practices—the knowledge work—that did emerge. Some of these practices documented in these essays were retrospectively deemed the "roots" of science and capitalism. Some were not. Deringer's and Giraudeau's contributions, for instance, both recover epistemic practices of calculation that did not inexorably lead to capitalism and science. In showcasing multiple ways that an early modern financier might have calculated futures, Deringer, for example, uncovers methods of the early modern era on their own terms that were rendered obsolete when compound-interest discounting became "black-boxed" in modern capitalist practices. In an investigation into how market demand for lamb during the off-season engendered new breeding techniques and ways to shape the environment, Pawley similarly avoids a straightforward story of rationality or pure market demand for mid-eighteenth-century British agricultural capitalism. Instead, she shows us how this new system engendered a host of actors—botanists, physicians, professional feeders—invested in maintaining a system that rested on

technoscience did not inevitably lead to the rise of modern science, but proved productive, generative, and instrumental for a host of other institutional and ideological purposes. For examples in the case of early modern China, see Benjamin Elman, *On Their Own Terms: Science in China, 1550–1900* (Cambridge, Mass., 2005), who documents how the rise of empirical studies in Qing China, where philological and evidentiary knowledge was applied to knowing the cosmos for purposes of moral cultivation and imperial politics. Far from a "failure" or missed opportunity on the part of the Qing to develop an exact Chinese analogue to European capitalism or Western science, this application of precise knowledge for purposes of moral and political power made perfect sense within the institutional and cultural context of a vibrant early modern empire. For another study that examines the production of scientific—in particular, cartographic and ethnographic—knowledge in the context of the Qing empire in its own right (in contrast to the "empire" of "science and empire" studies, which is implicitly always assumed to be Western imperialism), see Hostetler, *Qing Colonial Enterprise* (cit. n. 60).

what might be seen as an illogical practice of manipulating what were natural rhythms in sheep reproduction. If these three essays examine the eighteenth century, the remaining contributors examine science and capitalism from around the world during the nineteenth and twentieth centuries. Together, the pieces from the modern period until today document how the complex entanglement of science and capitalism was hardly a development that simply moved from the West to the rest after a point of purported divergence. Instead, this entanglement has continuously evolved and remade itself via global and transnational connections *and* obstructions. And it has done so in a variety of locations and in a variety of ways.

Sciences and Economies in the Scientific Revolution:

Concepts, Materials, and Commensurable Fragments

*by Harold J. Cook**

ABSTRACT

Debates about the nature of the so-called Scientific Revolution can be treated as a touchstone describing many of the fundamental changes in the field of the history of science. The establishment of the history and philosophy of science in the second half of the twentieth century occurred at a time when leading academic scientists among the victorious Western allies were intent on keeping the sciences apart from direct political and economic entanglements. In contradistinction to the Marxist-inspired scientists of the 1930s, they sought to raise scientific ideas to the rank of the highest expression of the human spirit, standing alongside Shakespeare and the like. A history of "pure" scientific ideas therefore motivated the field. By the 1970s, attention to ideologies, social relations, and practices began to open up other analytical possibilities. But other analytical approaches began to look for ways to understand mind and body as joined rather than as separate. The material objects to which scientists attend can be made commensurable, flowing like coins across many borders, suggesting that scientific processes are not confined to one branch of human history or one region of the world. The sciences have economies that are larger than moral economies alone.

> I hope too that the truths I set forth . . . will have currency in the world in the same way as money, whose value is no less when it comes from the purse of a peasant than when it comes from a bank.
>
> —Descartes, ca. 1641[1]

Filthy lucre is a dangerous substance. The Midas touch transforms all things in a desire for gold alone. But of course King Midas found that he could not remain alive when even his food and drink turned into the precious metal. Misers like Midas are by definition wretched and miserable. Medieval clerics therefore worried when centers of learning were threatened with being turned into nurseries for lawyers and medical doctors, who seemed simply to wish to monetize knowledge. Knowledge is a good,

* History Department, Box N, Brown University, Providence, RI 02912; harold_cook@brown.edu.

[1] René Descartes, "Search for Truth by Means of the Natural Light," ca. 1641, in *The Philosophical Writings of Descartes*, ed. John Cottingham, Robert Stoothoff, and Dugald Murdoch (Cambridge, 1984), 2:401.

they said, indeed one of the highest, and cannot be bought or sold, only shared with those who put in the effort of study.[2] If knowledge is meant to sustain well-being, it cannot be equated with gold.

It comes as no surprise, then, that when the history and philosophy of science came to be established in the postwar West, long-standing concerns about the corruption of science by material interests were built into its foundations.[3] Part of the strategy was to familiarize the subject for nonscientists by treating it as a branch of philosophy or literature. But since scientists keep referring back to how they interact with natural substances, is its form of knowledge like those others? The English word "knowledge" can be confusing, being inclusive of what other European languages call *kennen* (or *connaissance*: what we know from acquaintance) and *wissen* (or *savoir*: what we know from explanation). On those grounds, scientific knowledge is as much about interacting with tangible things as it is about producing peer-reviewed papers. Perhaps it is therefore a process rather than a literature, a verb rather than a noun. If it is a verb, then the situated activities of the humans who make it happen must be included in the scope of the word, too. Studies of scientists and their assistants, and their instruments and subjects of attention, show that most of the time they work well together, trusting one another, the materials, and the words and signs they produce. But when doubts are raised, they often turn not to authority but to proof. Proof is in turn a word associated with knowing by taste as well as with the process of stamping a mark, as in page proofs or a proof sample of manufactured coins. To obtain proof coins that are all identical in weight, fineness, and appearance—or to consistently extract thyrotropin-releasing hormone (TRH) from organic tissue—requires a remarkably disciplined collection of concepts, skills, procedures, tools, and materials, all mobilized and overseen by humans.[4] Scientific activities therefore mingle minds and bodies, ideas and materials, as do many activities usually termed economic.[5]

As a matter of fact, in Midas's own day, the making of coins and wisdom were located in the same place. Our very word "money" comes from the coins struck from noble metal at the Roman temple to Moneta, otherwise known as Mnemosyne, goddess of memory and mother of the muses. Other temples, too, in many regions of the world, have been sites for the production of money. The Athenian mint, for example, was placed in the most secure precinct of the temple of the wise goddess who watched over her city, her sacral space guaranteeing the quality of the coins it produced. Such standardized tokens of value, made from gold or silver and marked with the sign of a god or

[2] Faye Marie Getz, "The Faculty of Medicine before 1500," in *The History of the University of Oxford*, vol. 2, *Late Medieval Oxford*, ed. J. I. Catto and Ralph Evans (Oxford, 1992), 373–405.

[3] The essay that follows is meant to be indicative of general changes up to the turn of the recent century rather than full and complete. Apologies to the many important authors who are not named explicitly. For other historiographical considerations, see Harold J. Cook, "The History of Medicine and the Scientific Revolution," *Isis* 102 (2011): 102–8; Cook, "Early Modern Science and Monetized Economies: The Co-Production of Commensurable Materials," in *Wissen und Wirtschaft: Expertenkulturen und Märkte vom 13. bis 18. Jahrhundert*, ed. Marian Füssel, Philip Knäble, and Nina Elsemann (Göttingen, 2017), 97–114; Cook, "Problems with the Word Made Flesh: The Great Tradition of the Scientific Revolution in Europe," *J. Early Mod. Hist.* 21 (2017): 394–406.

[4] TRH, or TRF (thyrotropin-releasing factor), figured prominently in Bruno Latour and Steve Woolgar, *Laboratory Life: The Construction of Scientific Facts* (1979; repr., Princeton, N.J., 1986).

[5] Michel Serres, *The Five Senses: A Philosophy of Mingled Bodies*, trans. Margaret Sankey and Peter Cowley (London, 2008); Bruno Latour, *An Inquiry into Modes of Existence: An Anthropology of the Moderns* (Cambridge, Mass., 2013).

potentate, were indeed being used by kings and merchants of the eastern Mediterranean around the time of King Midas of Phrygia, in the late eighth century BCE.[6] The special ability of money to take a regularized physical form arose from careful control of the substance and weight of each instance of the specie coupled with stamping it with a controlled mark, all of which required not only production by knowledgeable artisans but careful supervision by assayers and sacral authorities. In later centuries, a part of the process required sample coins to be randomly selected and placed in a locked box for later testing under the gaze of trusted officials: the "trial of the pix," sharing its name with the box on the altar containing the consecrated Eucharist.[7] Money was therefore a readily transportable, stable substance of more or less identical instances of a kind, produced under conditions of sacred trust and bearing validating information marks. Paper currency in the United States still bears the motto "In God We Trust." Both science and money flow from the temple of Athena, evoking power and mystery.

Yet in the second half of the twentieth century, a long-standing framework holding that science is a field for the development and deployment of explanatory ideas [*wissen*] motivated most historians of science. Sociologists and anthropologists of science upset many assumptions by studying the activities and values of communities of scientists, introducing politics and ideology into the discussion and opening up possibilities for explanations that united minds and hands; economic historians pressed for links between technology and science; historians of medicine saw in market competition some of the reasons for the rise of empiricism. The field of Science and Technology Studies has even more recently developed a strongly materialist orientation. But dualistic privileging of mind over matter continues to echo throughout the literature of the history of science. Since the history and philosophy of science set out to keep ideas far from lucre, including materialism in its frameworks did not come readily. Perhaps Midas's problem, however, was not coveting gold per se but imagining that only one thing has value, causing him to forget that he had to touch many things to remain alive.

MATERIAL DISAPPEARANCES: AESTHETIC MODERNISM AND PURE SCIENCE

When the history and philosophy of science emerged as a field of academic study in its own right in the 1950s and afterward, its practitioners were keenly attuned to the ideological controversies of the decades leading up to the Second World War as well as the Cold War of their own day. They carved out a space meant to explain one of the most hopeful lines of human history: the exciting intellectual endeavor of understanding the universe. That strategy also served to circumvent questions that might arise from the political power of technocratic expertise or from the weighing of recent results of scientific projects, be they the Bomb or eugenics; it also served to turn science into an ally of religion, since both were seeking a knowledge of first things. In the first half of the century, other analyses had been heard loudly, but with fresh concerns

[6] William N. Goetzmann, *Money Changes Everything: How Finance Made Civilization Possible* (Princeton, N.J., 2016), 15–136; for East Asia, Tamara T. Chin, *Savage Exchange: Han Imperialism, Chinese Literary Style, and the Economic Imagination* (Cambridge, Mass., 2014). See also David Graeber, *Debt: The First 5,000 Years* (Brooklyn, 2012).

[7] Christine Desan, *Making Money: Coin, Currency, and the Coming of Capitalism* (Oxford, 2014), 75; Stephen M. Stigler, *Statistics on the Table: The History of Statistical Concepts and Methods* (Cambridge, Mass., 1999), 383–402.

about the relationships between science and politico-economies arising in the middle of the century, the leaders of the field, together with their scientific patrons, set out to build walls that could protect the purity of science.

Given the political importance of science, technology, and medicine (STM), the interests represented within the field come as no surprise. It is worth remembering that the history of science had long been seen as a part of a program meant to improve the human condition by integrating the natural sciences with philosophical and moral studies. Policies for domesticating natural knowledge by making it into a field (or fields) for the exploration of grand ideas rather than technical expertise or irreligion go back millennia and were also voiced loudly by many of the early modern *virtuosi* who hoped to find in the study of Creation a better understanding of God and humanity, or at least protection against charges of atheism. Ideals about the search for truths in nature also promised to expose the authoritarian ambitions of church and state. But any general account of science has faced the special difficulties raised by the technical and linguistic, sometimes mathematical, expertise required in each of the several arenas of natural science as well as considerations for their material consequences for life. Given the association of science with the ability to cause bodily betterment or harm, it was clear that it was not merely associated with words, meanings, and values. The various positions taken by historians of science who aimed at making their subject similar to the history of other kinds of ideas therefore generated considerable friction.

A little over a century ago, European conversations about hopes for further Enlightenment against the forces of reaction were often shaped by positivism. The struggle to root human society in positive knowledge not only held out the hope of freeing intellectuals from the domination of institutionalized religion but provided criteria for hierarchies of talent and for international cooperation. In the United States, Progressives adopted similar ideals in their advocacy for disinterested government against corruption and obscurantism.[8] For instance, the first president of Cornell University, Andrew Dickson White, expressed a positivist faith in his *A History of the Warfare of Science with Theology in Christendom* (1896). He also served as one of the trustees and as a member of the executive board of the Carnegie Foundation for International Peace, which came to the support of one of the founders of the field of the history of science, George Sarton, who emerged from a similar milieu: Francophone anticlericalism and socialist idealism enabled him to imagine how his knowledge of chemistry and mathematics could improve society.[9] Like-minded people agreed that accumulating the facts of nature would progressively improve the human condition materially and morally.

But in the aftermath of the Great War and Bolshevik Revolution, and the growing pressure for independence within the European colonies, both scientific idealism and progressivism struggled. Technoscience had bared its sharpest teeth in the bloodletting of industrialized and chemical warfare, while faith offered comfort in an otherwise cold world. The universalizing hopes offered by Communism continued to share affinities with the materialist aspects of positivism, but other alternatives took up a new term: *kul-*

[8] David A. Hollinger, "Inquiry and Uplift: Late Nineteenth-Century American Academics and the Moral Efficacy of Scientific Practice," in *The Authority of Experts: Studies in History and Theory*, ed. Thomas L. Haskell (Bloomington, Ind., 1984), 142–56.

[9] Arnold Thackray and Robert K. Merton, "On Discipline Building: The Paradoxes of George Sarton," *Isis* 63 (1972): 473–95; Merton, "George Sarton: Episodic Recollections by an Unruly Apprentice," *Isis* 76 (1985): 470–86.

tur. The word had earlier been used as a category of human natural history, but by the late nineteenth century—partly under the influence of comparative studies of literature, myth, folklore, and religion—the word began to be applied to human relationships: culture came to indicate the felt ways of life that bind people together, expressed in meaning and action. A corollary held that human beings interpret the world in ways that begin with conceptual expectations, which are imbibed from cultural upbringing rather than directly from nature. Gestalt psychology and Freudian analysis, art history, linguistics, ethnography, and many other disciplines seized on the primacy of mental worlds. Max Weber offered support in an antipositivist and antimaterialist social science that took relationships of meaning (including religions) and their institutions to be fundamental in establishing social bonds and their patterns. The meaning of Western civilization was being reshaped.[10]

A new aesthetic modernism also encouraged people to imagine that big ideas produced by brilliant minds could locate the forms of beauty in the world.[11] Science was therefore often coupled with the creative and applied arts as an expression of the human spirit. Moreover, the view that radical historical change emerged from conceptual shifts may have been given its most public form by Einstein's theory of relativity of 1915; when in 1919 it was announced that Arthur Eddington had confirmed his theory, *The Times* of London headlined a "scientific revolution" (apparently the first use of the term), and Einstein quickly became a popular hero.[12] The publicity around the new theoretical physics reinforced the decades-old view of science as produced by creative geniuses whose minds alone could discover new worlds, offering a non-Red modernizing revolution.

A "metaphysical" revolution associated with mathematics therefore began to be a common element in accounts of science and its history. H. Floris Cohen has identified this line of interpretation as the Great Tradition of the scientific revolution.[13] It held science to be created by profound minds who were seeking the keys to the universe. Although the polymathic L. J. Henderson helped Sarton obtain a room in Harvard's Widener Library in return for coteaching the second half of a yearlong course in the history of science with him, Sarton's positivistic aims no longer gripped his audiences. He was later described as lecturing with enthusiasm and energy, but mainly to convey information about numerous persons and events rather than illuminating the ideas of a few heroes, as Henderson did.[14] In 1940, the president of Harvard, James Bryant Conant, a chemist, finally promoted Sarton to the rank of tenured professor (although Sarton continued to be paid mainly by the Carnegie Foundation), but he also expressed

[10] Kapil Raj, "Rescuing Science from Civilisation: On Joseph Needham's 'Asiatic Mode of (Knowledge) Production,'" in *The Bright Dark Ages: Comparative and Connective Perspectives*, ed. Arun Bala and Prasenjit Duara (Leiden, 2016), 255–80.

[11] For an attempt at a definition, see Lawrence S. Rainey and Robert Von Hallberg, "Editorial/Introduction," *Modernism/Modernity* 1 (1994): 1–3. See also Thomas P. Hughes, *Human-Built World: How to Think about Technology and Culture* (Chicago, 2004).

[12] Alfred North Whitehead, *Science and the Modern World* (1927; repr., New York, 1967), 13. On the excitement around Einstein, Alastair Sponsel, "Constructing a 'Revolution in Science': The Campaign to Promote a Favourable Reception for the 1919 Solar Eclipse Experiments," *Brit. J. Hist. Sci.* 35 (2002): 439–67. For an American statement of disinterested knowledge, Abraham Flexner, *The Usefulness of Useless Knowledge* (1939; repr., Princeton, N.J., 2017).

[13] H. Floris Cohen, *The Scientific Revolution: A Historiographical Inquiry* (Chicago, 1994).

[14] I. Bernard Cohen, "A Harvard Education," *Isis* 75 (1984): 13–21, on 13.

disappointment that Sarton had not produced the "badly needed" inspirational synthesis for the history of the sciences that Macaulay had done for history per se.[15]

Sarton was nevertheless taking up the new formulas. "The positivistic philosophy has become untenable" he wrote in 1931.[16] In an editorial of 1935 expressing irritation that "the history of medicine has been studied more systematically and by a larger number of scholars than the history of any other branch of science," he proposed that the "core of the history of science should not be the history of medicine but the history of mathematics and mathematical sciences."[17] The new director of the Institute of the History of Medicine at Johns Hopkins University, Dr. Henry E. Sigerist, responded with friendly praise and support for Sarton's project.[18] But he also stressed that the history of medicine was not a branch of the history of science but a field in and of itself, more closely allied with history and economics than with mathematics. He also praised the Soviet Union, which he had recently visited, for its developing scientific culture, while noting that Germany was badly slipping, also "due to economic and political conditions." "Economic and political history form the core of the history of humanity," he argued. After all, "medical history is to a large extent economic history." Sigerist also reminded Sarton that "nobody will forget the part played by the Soviet delegation at the International Congress [of the History of Science and Technology] in London in 1931."[19]

The confrontations in London to which Sigerist referred remain a famous episode in the historiography of science. A delegation of Soviet scientists had gathered considerable public notice by flying in at the last minute (just four years after Charles Lindbergh crossed the Atlantic) to explain how, on the basis of state support for science, the Soviet Union was rapidly improving the material condition of its citizens. Some of their historical papers, most importantly the one by Boris Hessen on economic interests and Newtonianism, helped to inspire historical work by British scientists such has J. D. Bernal, J. B. S. Haldane, and Joseph Needham, and they in turn were sources of inspiration for others who were advancing arguments about science and society.[20] Their work in turn had a powerful effect on historical studies during the next decade, such as the 1938 dissertation of the young American sociologist, Robert K. Merton, completed with Sarton's blessing.[21] Still others who were at the meeting, such as the soon-to-be-powerful historian G. N. Clark, admitted much of Hessen's

[15] Thackray and Merton, "On Discipline Building" (cit. n. 9), 490.

[16] George Sarton, *The History of Science and the New Humanism* (New York, 1931), 175.

[17] George Sarton, "Second Preface to Volume XXIII: The History of Science versus the History of Medicine," *Isis* 23 (1935): 313–20, on 313, 317.

[18] Elizabeth Fee and Theodore M. Brown, *Making Medical History: The Life and Times of Henry E. Sigerist* (Baltimore, 1997).

[19] Henry Sigerist, "The History of Science *and* the History of Medicine," *Bull. Hist. Med.* 4 (1936), 1–13, quotations on 11, 6, 11; see also Sigerist, *Socialized Medicine in the Soviet Union* (New York, 1937).

[20] Boris M. Hessen, "The Social and Economic Roots of Newton's Principia," in *Science at the Cross Roads* (1931; repr., London, 1971); Gary Werskey, *The Visible College: A Collective Biography of British Scientists and Socialists of the 1930s* (London, 1978); Loren Graham, "The Socio-Political Roots of Boris Hessen: Soviet Marxism and the History of Science," *Soc. Stud. Sci.* 15 (1985): 705–22; Pamela O. Long, *Artisan/Practitioners and the Rise of the New Sciences, 1400–1600* (Corvallis, Ore., 2011), 10–29.

[21] Robert K. Merton, *Science, Technology and Society in Seventeenth Century England* (1938; repr., New York, 1970); Merton, "Science and the Economy of Seventeenth Century England," *Sci. & Soc.* 3 (1939): 3–27; Merton, "George Sarton" (cit. n. 9).

case even as he pushed back against Hessen's historical naïveté about the British seventeenth century and argued for the importance of Newton's mind.[22] The materialist argument that science and economy were intertwined clearly remained powerful.

But the ongoing crises in Central Europe pushed most scientists toward mind over matter. Ludwig Fleck, for instance, took up Gestalt psychology as a helpful resource for the position that science could not be equated with work. Fleck tried making a career in a German-language biomedical system that provided few opportunities for a Polish and Jewish laboratory scientist like himself. He turned to Gestalt to understand the scientific establishment, seeing their knowledge claims as emerging from *Denkkollectiven* ("thought styles") that are rooted in the common questions posed by particular communities. The insiders simply "get it," although others might see something entirely different from the evidence at hand. The term he used for the difference was "incommensurability," indicating that the distinctions between each thought collective were unbridgeable.[23]

The Hungarian-born Polanyi brothers, too, resisted fascism but split over the best alternative, as Mary Joe Nye has recently shown. Karl Polanyi is now known chiefly for the book he later published as *The Great Transformation* (1944), which arose from his Christian socialist and Soviet-sympathizing views pursued in "Red Vienna." But his brother, Michael, fundamentally disagreed. Like Fleck, he experienced the chaos in Germany before fleeing to Britain in 1933, where he began to make arguments that shifted attention from material forces and political collectivities to personal skills and contemplation. The "scientist's art of knowing" emerged as a kind of connoisseurship. In other words, for Polanyi the "tacit component" of scientific knowledge depended not only on the interpersonal methods of communication between scientists in the service of a common aim but on the importance of working experientially with materials that have their own inexpressible truths, none of which were comprehended by functionaries.[24] He therefore sided with the Austrian neoclassical economists in arguing that economy is "the most general characteristic of life": that is, it is the form by which natural law regulates human societies. He pressed for setting nature free by separating science from politics. "Defeating the social relations of science movement in Great Britain became one of the essential aims in [Michael] Polanyi's intellectual and political life around 1940."[25]

The interpretation of science as a product of state-sponsored development would soon become powerfully reinforced by the manner in which the warring nations furthered rapid growth in their technosciences during the Second World War. Among the Western allies, however, the politicization of science in both Nazi Germany and the Soviet Union—popularized in works such as Bertolt Brecht's "Life of Galileo" (1938; revised 1945) and Arthur Koestler's *Darkness at Noon* (1940)—deeply disturbed many

[22] G. N. Clark, *Science and Social Welfare in the Age of Newton* (1937; repr., Oxford, 1949).

[23] Ilana Löwy, *Medical Acts and Medical Facts: The Polish Tradition of Practice-Grounded Reflections on Medicine and Science, From Tytus Chalubinski to Ludwik Fleck* (Kraków, 2000), 88–119; Christian Bonah, "'Experimental Rage': The Development of Medical Ethics and the Genesis of Scientific Facts: Ludwig Fleck: An Answer to the Crisis of Modern Medicine in Interwar Germany?," *Soc. Hist. Med.* 15 (2002): 187–207. Fleck's 1935 essay later appeared in English as *Genesis and Development of a Scientific Fact*, ed. Robert K. Merton and Thaddeus J. Trenn (Chicago, 1979).

[24] Michael Polanyi, *Personal Knowledge: Towards a Post-Critical Philosophy* (Chicago, 1958), 70.

[25] Mary Jo Nye, *Michael Polanyi and His Generation: Origins of the Social Construction of Science* (Chicago, 2011), 175, 184.

leaders of science. As the war drew to a close, the dominant view in "the West" held that proper science is the result of inspired minds who are free to pursue knowledge wherever inspiration might lead, for which the term "pure" science was used.

A recent account by Roger Backhouse and Harro Maas of some of the discussions within the U.S. government is revealing about what the new term implied. With the end of the war in sight, President Roosevelt had asked his science adviser, Vannevar Bush, to propose what should be done after victory to further secure the prosperity and safety of the American people, for which Bush established several advisory committees led by prominent scientist-administrators. In the behind-the-scenes debates within the committees, the nature of science was declared to be more intellectual than technical. The debates divided members such as the economist Paul Samuelson and the historian Henry Guerlac from others like the historian of science I. Bernard Cohen. The former two argued that government should prevail over the self-interest of capitalists and continue to plan and support science, the latter that science was an activity of the intellect and should not become politicized. Bush's *Science: The Endless Frontier* (1945) laid out an institutional mechanism that would continue to keep federal dollars flowing into scientific research but would also keep politicians at arm's length.[26] What they termed "pure science" would therefore be a republic of scientists governed by a brotherhood of eminent men who stood apart from economic and political interests.

THE RETURN OF THE POLITICS OF SCIENCE

Within a brief time, however, the arguments about whether science was based on "free thought" or tied to economic and state interests would take on the ideological positioning of the Cold War. The policing of thought soon came to prominence, not only in the Soviet Union. In 1954 even the former head of the project to develop an atomic weapon, J. Robert Oppenheimer, was investigated by the U.S. government and cast aside. Giorgio de Santillana, professor of the history of science at MIT, drew the parallels in the Preface to his *Crime of Galileo* (1955), depicting scientists as defenders of eternal "humanist" verities who were prosecuted by small-minded bureaucrats claiming to be defenders of their polities when they were really acting out of personal interest. But he did not mention the case of a colleague, Dirk Struik, also of MIT. Struik was a mathematician, historian, and self-avowed philosophical Marxist, whose distress about neocolonial attempts to suppress the independence movement in the Dutch East Indies led him to speak out frequently on the subject. In 1951 Struik was indicted under Massachusetts's "Anti-Anarchy Act" for conspiring to overthrow the Commonwealth and the U.S. government. A Struik Defense Committee was established and chaired by George Sarton and included eminent people like Norbert Wiener of MIT, but it did not include Santillana, nor I. B. Cohen of nearby Harvard. Perhaps the case posed clear dangers to a field not yet strongly entrenched. Five years lapsed before it was dropped by the courts.[27]

Others exited the country entirely: the socialist Sigerist returned to Switzerland in 1947; Ludwig Edelstein, a distinguished historian of classical medicine who had first

[26] Roger E. Backhouse and Harro Maas, "A Road Not Taken: Economists, Historians of Science, and the Making of the Bowman Report," *Isis* 108 (2017): 82–106.

[27] Dirk Struik Defense Committee, *Facts Relevant to the Struik Case* (West Newton, Mass., 1952).

found refuge with Sigerist in Baltimore in 1934, resigned from the University of California rather than sign the 1950 loyalty oath and returned to Baltimore; Erwin Ackerknecht, a former Trotskyite who had moved from Baltimore to the new Department for the History of Medicine in Madison in 1947, became very unhappy in middle America and returned to Zürich in 1957. As John McCumber has recently demonstrated, departments of philosophy were also experiencing purges by stealth. In the United States, Continental existentialism and related subjects implying a universe without a Christian God found loud opposition from religiously minded citizens who worried about "Godless Communism." Instead, the new philosophy of science became a backbone for a libertarian and technocratic approach to the study of "reason" as a decision tree of "choices."[28] It began to be difficult to find voices to articulate the relationships between science and worldly interests, at least within academic departments.

Postwar history and philosophy of science therefore stayed clear of politics and economy, and given the clear preference of senior academic scientists for pure science the field focused instead on topics such as metaphysics and epistemology. In France, the first chair of the History and Philosophy of Science was held by the antipositivist Gaston Bachelard, who had studied physics and then turned to historical epistemology [*l'esprit scientifique*]. In Britain, Herbert Butterfield took over chairing the Cambridge Committee on the History of Science, which had previously been headed by Marxist-inflected Needham. Butterfield had published his relativist critique of the Whig interpretation of history in 1931, but when his *Origins of Modern Science, 1300–1800* appeared in 1949, his conversion to progressivist idealism was plain: he treated science as a "way of looking at the universe," a conceptual "system" produced by individual great minds. Winston Churchill himself, it is reported, had been the cause of Butterfield's conversion to the defense of scientific truths.[29]

In the United States, the Gaullist Alexandre Koyré possessed the most influential voice. Koyré stated that science emerged from "the mathematization (geometrization) of nature" and from no other source, echoing what Sarton had been articulating.[30] When Guerlac instead wrote about the interests of Lavoisier "in practical matters, such as street lighting," Koyré responded by labeling the point "un peu Marxiste."[31] One of Koyré's chief advocates and admirers was Charles Gillispie, a founder of the history of science program at Princeton, who had fought in France with the U.S. Army as an artillery officer and wanted to study history for what it expressed about the hopeful aspects of the human spirit.[32] One of Koyré's chief collaborators was I. B. Cohen of Harvard, who wrote of the *Principia*: "In Newton's achievement, we see how science advances by heroic exercises of the imagination, rather than by patient collecting and sorting of myriads of individual facts. Who, after studying Newton's magnificent contribution to thought, could deny that pure science exemplifies the creative accomplishment of the human spirit as its pinnacle?"[33] By the end of the 1950s, even Guerlac de-

[28] John McCumber, *The Philosophy Scare: The Politics of Reason in the Early Cold War* (Chicago, 2016).

[29] A. Rupert Hall, "On Whiggism," *Hist. Sci.* 21 (1983): 45–59.

[30] Alexandre Koyré, *Newtonian Studies* (1965; repr., Chicago, 1968), 6.

[31] Quoted in Nye, *Michael Polanyi* (cit. n. 25), 241.

[32] I remember Gillispie saying this in answer to questions at a session of the History of Science Society annual meeting in the mid-1980s. For a tribute to Koyré, see Charles C. Gillispie, *The Edge of Objectivity: An Essay in the History of Scientific Ideas* (Princeton, N.J., 1960), 523.

[33] I. Bernard Cohen, *The Birth of a New Physics* (1960; repr., New York, 1985), 184.

scribed his two-semester course in the history of science as meant to convey "a sense of science as an enterprise of the mind."[34]

The aesthetic modernist view of science was consolidated in Thomas Kuhn's famous *The Structure of Scientific Revolutions* (1962). Kuhn had been educated in physics but at the suggestion of President Conant had taught in the General Education program Conant had set up at Harvard in 1949 for the instruction of nonscientists, using case studies of important historical texts to elucidate key ideas about the nature of science.[35] But when Kuhn moved to California in the later 1950s and began to converse more intensively with scholars in the humanities and social sciences, he experienced a shock: scholars in those other fields seemed always to be debating fundamental principles, whereas he had seen science as a consensus-building enterprise. Borrowing heavily from the work of Fleck, Polanyi, Jean Piaget, and others, Kuhn came to the conclusion that in the sciences the fundamental questions were uncontroversial within a "paradigm" (Fleck's word), a kind of gestalt thought collective that existed within the community of scientists. Each paradigm expressed a view that compared to its predecessor was incommensurable (also Fleck's word), so that science proceeded by a series of revolutionary steps. Or at least the conceptual "esoteric" sciences operated in that fashion, since inferior fields like medicine, technology, and law responded to "external social need" and so had different causalities. How academic scientists came to agree on paradigms other than through a kind of collective gestalt intuition was never quite resolved in Kuhn's work, which in later years led to criticism about whether he had joined the "relativists." But at the time, Kuhn spoke for the view from the top of the American academic establishment about the superior, disinterested intellectual power of the pure sciences.[36] The scientific cutting edge of Enlightenment radicalism had been domesticated, made into the equivalent of good literature.[37]

The intellectualist view had also opened doors for accounts of religion as a friend rather than an enemy of science. The potentially atheistical materialism of the new science of the early modern era appeared almost nowhere in the new literature. Instead, the "Merton thesis" was redescribed as showing how Puritanism caused science, an interpretation that Merton himself later found puzzling.[38] In its wake, a flood of studies concerned themselves with science and religion (mainly in Protestant varieties).[39] Even leaders in the growing field of development economics shared such opinions. For instance, in 1960 Walter W. Rostow launched his much-reprinted argument on the stages of economic growth that he subtitled a "non-Communist manifesto."

[34] Marshall Clagett, ed., *Critical Problems in the History of Science* (1959; repr., Madison, Wis., 1969), 240.

[35] For his debt to Conant, see Thomas S. Kuhn, preface to *The Copernican Revolution: Planetary Astronomy in the Development of Western Thought* (Cambridge, Mass., 1957).

[36] Thomas S. Kuhn, *The Structure of Scientific Revolutions* (1962; repr., Chicago, 1970); for the esoteric nature of science and the absence of social need from its concerns, 19, 20, 23; Steven Shapin, "The Ivory Tower: The History of a Figure of Speech and Its Cultural Uses," *Brit. J. Hist. Sci.* 45 (2012): 1–27.

[37] Conant praised Kuhn's *Copernican Revolution* for its success in enabling "the scientific tradition to take its place alongside the literary tradition in the culture of the United States"; James Bryant Conant, foreword to Kuhn, *Copernican Revolution* (cit. n. 35), xviii.

[38] See Merton's new preface to *Science, Technology and Society* (cit. n. 21).

[39] See, e.g., the selection made by I. Bernard Cohen, ed., *Puritanism and the Rise of Modern Science: The Merton Thesis* (New Brunswick, N.J., 1990). See also Steven Shapin, "Understanding the Merton Thesis," *Isis* 79 (1988): 594–605.

In his view, the historically first economic "takeoff" had arisen in Britain and its colonial outposts because of religious nonconformity, which needed only "economic and technical" additives to explode into endless growth.[40] The interest continued for many years. For instance, the Marxist-leaning historian of science and medicine Charles Webster wrote about utilitarian sciences at length in his *Great Instauration* (1976), but his chief interpretative line was that of Puritanism and science. In the growth period of the later 1950s and 1960s, therefore, the main aim of academics in the new field was to show the relationships between scientific ideas and other intellectual genres, especially religion and philosophy.

During the 1970s, however, a well-known clash between "internalists" and "externalists" took shape. The former group focused on puzzle solving from within scientific fields, while the latter mainly demonstrated how scientific texts owed their assumptions to other fields of thought. Some externalists, particularly those in touch with sociology (and sometimes with anthropology), also found ways to examine the social foundations of scientific ideas. Some also identified science as an ideological formation, or as deeply affected by ideology, bringing politics into the heart of the argument. The latter were often subjected to denunciation for having Marxist roots.

The views of Michel Foucault were the most visible manifestation of the new mood. His *Madness and Civilization: A History of Insanity in the Age of Reason* had appeared in English in 1967, and in an age of deinstitutionalization the way it questioned the categories that had been imposed on the phenomenon of unreason—which by its very nature remained inarticulable—had an enthusiastic response. (This was also the moment of Thomas Szaz's *The Myth of Mental Illness* of 1961 and Ken Kesey's *One Flew over the Cuckoo's Nest* of 1962.) The English-language publication of Foucault's *The Birth of the Clinic: An Archaeology of Medical Perception* in 1973 had strong effects on the history of medicine. But for the history of science, *The Order of Things: An Archaeology of the Human Sciences*, also published in English in 1973, struck home. It followed the beginnings of the Cultural Revolution in Mao's China, the end of the Prague Spring, the uprisings in Paris, Berkeley, New York, and many other places, the shooting of student protesters at Kent State, and Solzhenitsyn's exposures of the Soviet Gulag, and with his own contemptuous scowl for the French establishment Foucault's famous formula about knowledge/power was heralded by the New Left. Influenced by Bachelard, Georges Canguilhem, and other historically aware philosophers of science, Foucault wrote about the ways in which epistemes conveyed by discourse create disciplines of the self that constrain the possible within thought worlds—which, like Kuhn's paradigms, can suddenly shift shape in revolutions that are incommensurate with what came before.

Although Foucault's writing had enough strangeness to make it both difficult and somewhat mysterious, and although the totality of the repressive thought systems he described were reverse images of progressivism, by writing about the primacy of epistemes he invoked themes familiar to historians of science. The absence of coherent explanations for change—the sharp breaks between layers of civilization, as his archeological metaphor had it—nevertheless posed challenges for those who followed in his wake. With a powerful ability to see the exclusions and repressions of reason but

[40] W. W. Rostow, *The Stages of Economic Growth: A Non-Communist Manifesto* (1960; repr., Cambridge, 1990), 17.

without the empiricists' or structuralists' ability to assess whether claims found in either the sources or the interpretations of them were more or less probable, aestheticism, ideology, or silence presented the chief strategies for taking a position. More generally, Foucaultianism was but one indication of the problem of culture as ideology: too often it meant dominant systems of meaning that governed all those who lived under them when diversity of life, interest, opinion, and affect were also present.[41] Foucault was better at ferreting out the presence of repression than at attending to expressions of carnival or in tracing the tentacles of corporate interest, as pursued by Noam Chomsky, for instance.

The New Left became increasingly excited about intellectual critique, helping to foster a growing interest in the construction of knowledge. While deconstruction might be said to be the early Foucault's forte, related questions about the "credibility" of science began to appear. The growth of the sociology of science certainly contributed to many questions about knowledge claims. In the 1970s, Harry Collins at Bath, and David Edge, Barry Barnes, and David Bloor at Edinburgh began to challenge the fundamental assumptions about how science was like philosophy, with Bloor and Barnes being especially important in developing the "strong programme" of the sociology of scientific knowledge (SSK). SSK aimed to examine the construction of scientific knowledge without presupposing that its claims were accepted either because they were correct or because they harmonized with other ideational forms, opening the subject to considerations of social processes as determinants of truth claims.[42] The approach was brought to the United States through people like Arnold Thackray, founding chair of the Department of the History and Sociology of Science at the University of Pennsylvania (established in 1970).

For early modernists, the most important engagement with the growing interest in social-historical explanations about credibility was Steven Shapin and Simon Schaffer's jointly authored *Leviathan and the Air-Pump* (1986). It investigated the experimental activities hosted by Robert Boyle as a counterweight to the rationalist approaches of Thomas Hobbes and his like, showing how science arose from social consensus rather than from mathematics or philosophy. They explored in detail the practices of Boyle and those around him, showing, for instance, how the operation of instruments such as the air pump depended on the tacit knowledge of persons as much as on formal concepts. On such foundations the authors put the case for science as grounded in the face-to-face politics that built the Restoration Commonwealth on the backs of gentlemanly norms (although they also famously concluded that "Hobbes was right").

Schaffer would continue to develop his interests in the production of scientific knowledge as a form of history that combined minds and hands, individuals and societies, and multiple cultural encounters.[43] But Shapin advanced a powerful claim for the social causes of scientific credibility in his own further study of Boyle, *A Social*

[41] Note the criticism in Carlo Ginzburg, *The Cheese and the Worms: The Cosmos of a Sixteenth-Century Miller*, trans. John Tedeschi and Anne Tedeschi (1976; English trans., Baltimore, 1982), xvii–xviii.

[42] For a review, see Jan Golinski, *Making Natural Knowledge: Constructivism and the History of Science* (Cambridge, 1998).

[43] For early examples, see Simon Schaffer, "Natural Philosophy and Public Spectacle in the Eighteenth Century," *Hist. Sci.* 21 (1983): 1–43; Schaffer, "Scientific Discoveries and the End of Natural Philosophy," *Soc. Stud. Sci.* 16 (1986): 387–420.

History of Truth (1994). It provided a sociological explanation for the world familiar from Conant, Polanyi, Kuhn, and others, analyzing the norms of gentlemen of science and how any appearance of material interests were written out of scientific knowledge claims.

After earning his PhD in Thackray's department at Penn and then moving on to the Science Studies Unit at Edinburgh, Shapin was deeply concerned about how the brotherhood of science agreed on truth claims. It was because of their higher status that gentlemen made the new science visible, despite the importance to their activities of servants and other "invisible technicians." Shapin found the credibility of the statements made by his subjects to arise from mutual trust: they were gentlemen together. He argued that in Boyle's world the content of truths about natural phenomena were determined by trustworthy personal bonds, whereas in our own period "we are obliged to trust in impersonal systems." Further, the judgment of gentlemen, he argued, could be trusted because "the concepts of freedom and trust are codependent," and gentlemen were the only persons who possessed freedom of action both because they were without economic want and because they possessed an honor code based on truth-telling and "perceptual competence." It was above all their "disinterestedness," founded on their "freedom," that made gentlemen "credible" seekers of truth. The Royal Society's import for science therefore lay in its "predominantly gentlemanly . . . membership," with their "factual testimony . . . almost never gainsaid" in public.[44] In other words, for Shapin the production of scientific knowledge was made credible by the social status of gentlemen because they were trusted, which in turn depended on their disinterestedness, which emerged from not being entangled in the wants and interests of economic life. Although he was arguing for sociopolitics as constitutive of science, Shapin identified the social relations that gave rise to something like pure science. In a backhanded way, he implied the effort that went into concealing the material interests of learned gentlemen, turning economy into a moral economy.[45]

MOBILE INFORMATION AND COMMENSURABLE EXCHANGE

Challenging established systems of thought with sociology and ideology certainly had important repercussions. From other directions, however, came a growing interest in embodied culture, which began to open up possibilities for bridging dualism. Some of the urgency came from concerns about identities that arose in part from physical presence. In 1976, a historian of Victorian science obtained a name change, from Walter F. Cannon to Susan Faye Cannon. Certainly, early versions of feminist writing in the history of science made important strides in recovering the personal stories of a hugely marginalized population—as in Margaret Rossiter's *Women Scientists in America* (1982)—on the basis of which some of the structural elements of past discrimination could be identified. Among early modernists, Londa Schiebinger completed a PhD dissertation that provided a foundation for her later *The Mind Has No Sex?* (1989), showing how ideas of scientific objectivity were gendered, keeping women from full participation in the new science despite a few notable (and undervalued) exceptions. By the mid-eighteenth century, she argued, the anatomy of sexual difference, by pointing to smaller

[44] Steven Shapin, *A Social History of Truth: Civility and Science in Seventeenth-Century England* (Chicago, 1994), on xxi, 15, 39, 73–5, 83–4, 122, 124.

[45] See also Lorraine Daston, "The Moral Economy of Science," *Osiris* 10 (1995): 3–24.

heads and wider pelvises in women, further naturalized such prejudices. Some of the literature on childbirth and monsters also brought women's bodies and minds into parts of the historical literature on scientific ideas.[46] Indeed, accounts of how midwives had been attacked by male physicians as part of the patriarchal assault known as the witch craze had shaped one of the early historical arguments of new age feminism, that of Barbara Ehrenreich and Deirdre English.[47] Elsewhere, Margaret Pelling and Charles Webster had begun to look into English archival sources to identify nonelite medical practitioners, many of whom were women.[48] Feminism certainly made it clear that bringing the history of women into the narrative in any significant manner—especially for periods before equal opportunity legislation gained a foothold—would require widening the definition of science to include activities not usually a part of the intellectual gentleman's club.[49]

Feminist critiques of scientific reductionism sometimes also raised the problem of capitalist alienation straight on. For example, Caroline Merchant's book *The Death of Nature: Women, Ecology, and the Scientific Revolution* (1980) made an important intervention by turning the much-lauded mechanical worldview into an example of (male) attack on (female) organic community. She added that a dominating metaphor—nature as feminine—invited the patriarchy to explore ways to control "her" to their advantage, noting that the Scientific Revolution was occurring in the same period as the witch craze, the agricultural revolution, and the first phases of capitalism. More generally, Susan Sontag had already launched her attack on intellectualist aesthetics ("Against Interpretation," 1966), broadening the critique to include reductions of disease as metaphorical states in *Illness as Metaphor* (1978); Donna Haraway, who had earned a PhD in biology with an awareness of how metaphors shaped research agendas, produced an essay known as the *Cyborg Manifesto: Science, Technology, and Socialist-Feminism in the Late Twentieth Century* (first appearing in 1985 but revised and republished in 1991).[50] It explored how embodied feminism required an understanding of the hegemonic methods of networked technoscience. Although the emphasis remained on thought worlds, in Sontag and Haraway one can see a powerful interest in challenging the reign of (male-dominated) conceptual systems in favor of embodied experience, creating openings for fresh views of spirit and matter mingled.

[46] Katharine Park and Lorraine Daston, "Unnatural Conceptions: The Study of Monsters in Sixteenth- and Seventeenth-Century France and England," *Past & Present* 92 (1981): 20–54; Adrian Wilson, "Participant or Patient? Seventeenth Century Childbirth from the Mother's Point of View," in *Patients and Practitioners*, ed. Roy Porter (Cambridge, 1985), 128–44.

[47] Barbara Ehrenreich and Deirdre English, *Witches, Midwives, and Nurses: A History of Women Healers* (Old Westbury, N.Y., 1973).

[48] Margaret Pelling and Charles Webster, "Medical Practitioners," in *Health, Medicine and Mortality in the Sixteenth Century*, ed. Charles Webster (Cambridge, 1979), 165–235; women medical practitioners also figured strongly in Doreen G. Nagy, *Popular Medicine in Seventeenth-Century England* (Bowling Green, Ohio, 1988).

[49] Although racial identities were often front and center in accounts of European empires, it would take some decades before nonwhite ways of knowing began to be included within categories such as the Scientific Revolution or science and Enlightenment.

[50] Susan Sontag, *Against Interpretation* (New York, 1969); Sontag, *Illness as Metaphor* (New York, 1978); Donna Haraway, "Manifesto for Cyborgs: Science, Technology, and Socialist Feminism in the 1980s," *Socialist Rev.* 80 (1985): 65–108, reprinted as "Cyborg Manifesto: Science, Technology, and Socialist-Feminism in the Late Twentieth Century," in her *Simians, Cyborgs, and Women: The Reinvention of Nature* (New York, 1991), 149–82.

A kind of monist ecology also framed many of the writings of Bruno Latour and Steve Woolgar, Michel Callon, and others associated with what came to be called actor-network theory (ANT). They did more than anyone to turn science from a noun into a verb, partly by attending closely to the economic metaphors embedded within scientific work. In *Laboratory Life* (1979), Latour and Woolgar reported on an observational study of how scientists behaved, demystifying science by attending to the quotidian practices of their subjects as they transformed materials into inscriptions. On the basis of the relationships among the researchers and between the researchers and their instruments and subjects, they sought to erase the common dualism of mind and matter, immaterial ideas and material substances, seeking explanations based on process. In doing so, they employed numerous analogies derived from economic discourse, as in discussions of "credit" for one's work, or the "exchange" of information, or the accumulation or deployment of "intellectual capital." And while their study was explicitly about how knowledge is constructed from practices in local settings, they also noted that researchers would come and go, and that those who win agonistic encounters have the most allies, pointing to the construction of scientific networks that reached into the distance.

An important question therefore became pressing: if local and personal relationships provide the foundation for credible knowledge claims, how does knowledge become mobile in either time or space? Shapin, for example, was certainly aware of the question. In 1991, a paper he wrote with Adi Ophir noted that the success of SSK in "displaying the situatedness of knowledge generates its successor problem. How is it, if knowledge is indeed local, that certain forms of it appear global in domain of application? Is the global—or even the widely distributed—character of, for example, much scientific and mathematical knowledge an illusion?" If not, perhaps scientific knowledge claims are distributed (or, to use an older vocabulary, "diffused") because of a "correspondence with reality"; their own preference, however, was to understand them as a reflection of "the success of certain cultures in creating and spreading the very means and contexts of application."[51] One might go on to infer from such formulas that the knowledge claims of disinterested gentlemen had the ability to affect people in many other times and places because of those claims' relationship to methods of power. In other words, the history of colonialism or imperialism provided a common answer to the problem of the extension of the social foundations of scientific knowledge, with many studies showing the powerful interconnections between the rise of European empires and the extension of one or another kinds of sciences.[52] It was certainly a signal reminder that what had once been called transmission was never pure or simple, and never without implications for human domination.

But Latour's *Science in Action* (1987) took another approach, directly evoking networks (as well as nonhuman actants) as enabling some kinds of knowledge to become mobile. In considering knowledge not as a totality but as an assembly of fragments—"bricolage" became another term of art in the era—something like commensurable bits of stable information seemed to be lurking around the edges of scientific activity once more. Latour and other advocates of ANT therefore aimed to account for credi-

[51] Adi Ophir and Steven Shapin, "The Place of Knowledge: A Methodological Survey," *Sci. Context* 4 (1991): 3–22, on 15–16. Also note the astute discussion in Mary Poovey, *A History of the Modern Fact: Problems of Knowledge in the Sciences of Wealth and Society* (Chicago, 1998), 4–27.

[52] For instance, Paolo Palladino and Michael Worboys, "Science and Imperialism," *Isis* 84 (1993): 91–102; David Arnold, *Science, Technology and Medicine in Colonial India* (Cambridge, 2000).

bility and practices not by employing abstract nouns such as society but by rediscovering the networks and materials of particular kinds of concerns that enabled the deployment and mobility of impersonal resources.[53] By the later 1980s the ANT approach was under fierce attack from those who thought that it paid no attention to facts, as well as those who disliked its emphasis on science as practice and its consequent avoidance of theory, even of relativism.[54] It was also disputed by advocates of SSK because it did not focus on the social roots of personal trust or authority, and it was sometimes described as "flat" in not identifying centers of power. But Shapin himself observed that "Latour's inventory of the means by which technoscientific knowledge is extended amounts to a descriptive vocabulary of power as well as of institutionalization."[55]

As Shapin's comment implies, Latour became best known for describing how knowledge claims might move through space, exploring the kinds of objects and information units that can be stabilized and transported. Transcription devices, for instance, might turn events in a rainforest into traces of graphic information on paper, and these inscriptions could be sent on to "centers of calculation" in order to compile numerical results; specimens and artifacts might be boxed up and sent along as physical evidence as well. Such "immutable mobiles" are artfully prepared and decontextualized materials, remaining unchanged on their journeys in space or time or else being capable of exact reconstruction at their destination. Interpretations of what these mobiles mean might change, but the stabilized units of information can themselves move relatively quickly and easily across geographical and cultural boundaries, allowing transformative passages to new worlds of interpretation.[56] Schaffer, Shapin's former coauthor, would also give increasing attention to the importance of calibrated instruments and technique in establishing mobile inscriptions, while some of the practitioners of ANT would later turn to the mechanisms of finance.[57]

Other hints of the return of materialism and economy were beginning to enter from many directions. Keith Thomas's magisterial *Religion and the Decline of Magic* (1971) proposed that the development of instruments to regularize material life, such as the emergence of banking and insurance, were more important than ideas about "reason" in accounting for the rise of faith in laws of nature, while Ian Hacking linked the rise

[53] Michel Callon, "Some Elements of a Sociology of Translation: Domestication of the Scallops and Fishermen of St Brieuc Bay," in *Power, Action and Belief: A New Sociology of Knowledge?*, ed. John Law (London, 1986), 196–229; see also Bruno Latour, "Postmodern? No, Simply *A* modern! Steps towards an Anthropology of Science: Essay Review," *Stud. Hist. Phil. Sci.* 21 (1990): 145–71; Latour, "Why Has Critique Run Out of Steam? From Matters of Fact to Matters of Concern," *Crit. Inq.* 30 (2004): 225–48. On the impersonal, Chandra Mukerji, *Impossible Engineering: Technology and Territoriality on the Canal du Midi* (Princeton, N.J., 2009).

[54] In one well-known instance, the physicists and mathematicians of the Institute for Advanced Study in Princeton vetoed the appointment first of Bruno Latour and then of Norton Wise, who had defended Latour's views. ANT and STS became stalking horses in Paul R. Gross and Norman Levitt, *Higher Superstition: The Academic Left and Its Quarrels with Science* (Baltimore, 1994).

[55] Steven Shapin, "Here and Everywhere: Sociology of Science Knowledge," *Annu. Rev. Sociol.* 21 (1995): 289–321, on 309.

[56] Bruno Latour, *Science in Action: How to Follow Scientists and Engineers through Society* (Cambridge, Mass., 1987); more recently, Peter Howlett and Mary S. Morgan, eds., *How Well Do Facts Travel? The Dissemination of Reliable Knowledge* (Cambridge, 2011).

[57] For instance, Simon Schaffer, "Golden Means: Assay Instruments and the Geography of Precision in the Guinea Trade," in *Instruments, Travel and Science: Itineraries of Precision from the Seventeenth to the Twentieth Century*, ed. Marie-Noële Bourguet, Christian Licoppe, and H. Otto Sibum (London, 2002), 20–50; Michel Callon, Yuval Millo, and Fabian Muniesa, *Market Devices* (Malden, Mass., 2007); Donald MacKenzie, *Material Markets: How Economic Agents Are Constructed* (Oxford, 2009).

of mathematical probability and statistics to methods of actuarial insurance.[58] Ecology framed the work of Alfred Crosby on *The Columbian Exchange* (1975), while postcolonialism affected Michael Adas's understanding of Western self-regard as an outgrowth of technological power.[59] Chandra Mukerji depicted the Scientific Revolution as a manifestation of the growing commercial materialism of the period, while Cook, Roy Porter, and others wrote about the medical marketplace.[60] Pamela Smith interpreted the work of one of the best-known chemists and projectors in Germany in the years before Leibniz, Johann Joachim Becher, as fundamentally framed by Cameralism; a few years later, Lisbet Koerner did the same for Carl Linnaeus.[61] A former student of Guerlac, Margaret Jacob, pointed to "urban and mercantile elites" who aimed to exploit as well as explain nature, and she and Larry Stewart would argue that popular Newtonianism prepared the way for the Industrial Revolution.[62]

The nature of social relationships was also contested. For a century, studies of culture had mainly concentrated on trying to parse other systems of thought so as to more or less fully comprehend them; but the daily activities of getting on with life raise multiple instances of human interactions in which goods or information are exchanged in brief encounters among virtual strangers. Sociolinguists noticed the importance of "weak" rather than "strong" ties in systems of communication, just as the language and literature studies of Mary Louise Pratt pointed to the simplified pidgin languages employed in contact zones, and historians of science like Susan Leigh Star and James Griesemer commented on the phenomenon of "shallow" or "thin" information exchanges.[63]

From their side, economic historians were also asking fresh questions about credibility. They opened up the black box of technological determinism and found in it a complex web of information networks and knowledge claims. Joel Mokyr, for example, had begun to write about "technological creativity" in relation to the Industrial Revolution, and in his *The Gifts of Athena: Historical Origins of the Knowledge Economy* (2002), he took up some aspects of Jacob's argument about Newtonianism to make the case for the importance of "propositional" knowledge. Since then, a great many economic historians have joined in the study of how different kinds of knowledge claims, many closely associated with STM, are implicated in different kinds of

[58] Ian Hacking, *The Emergence of Probability: A Philosophical Study of Early Ideas about Probability, Induction and Statistical Inference* (Cambridge, 1975); see also Lorraine Daston, *Classical Probability in the Enlightenment* (Princeton, N.J., 1988).

[59] Michael Adas, *Machines as the Measure of Men: Science, Technology, and Ideologies of Western Dominance* (Ithaca, N.Y., 1989).

[60] Chandra Mukerji, *From Graven Images: Patterns of Modern Materialism* (New York, 1983); Mark S. R. Jenner and Patrick Wallis, introduction to *Medicine and the Market in England and Its Colonies, c. 1450–c. 1850* (Houndsmills, Basingstoke, 2007).

[61] Pamela Smith, *The Business of Alchemy: Science and Culture in the Holy Roman Empire* (Princeton, N.J., 1994); Lisbet Koerner, *Linnaeus: Nature and Nation* (Cambridge, Mass., 1999).

[62] Margaret C. Jacob, *The Cultural Meaning of the Scientific Revolution* (Philadelphia, 1988); Jacob, *Scientific Culture and the Making of the Industrial West* (New York, 1997); Larry Stewart, "Other Centres of Calculation, or, Where the Royal Society Didn't Count: Commerce, Coffee-Houses and Natural Philosophy in Early Modern London," *Brit. J. Hist. Sci.* 32 (1999): 133–53.

[63] Mark S. Granovetter, "The Strength of Weak Ties," *Amer. J. Sociol.* 78 (1973): 1360–80; Harold J. Cook and David S. Lux, "Closed Circles or Open Networks? Communicating at a Distance during the Scientific Revolution," *Hist. Sci.* 36 (1998): 179–211; Susan Leigh Star and James Griesemer, "Institutional Ecology, 'Translations' and Boundary Objects: Amateurs and Professionals in Berkeley's Museum of Vertebrate Zoology, 1907–39," *Soc. Stud. Sci.* 19 (1989): 387–420; Mary Louise Pratt, *Imperial Eyes: Travel Writing and Transculturation*, 2nd ed. (London, 1992); Peter Galison, "Trading with the Enemy," in *Trading Zones and Interactional Expertise: Creating New Kinds of Collaboration*, ed. Michael Gorman (Cambridge, Mass., 2010), 25–52; and esp. Theodore M. Porter, "Thin Description: Surface and Depth in Science and Science Studies," *Osiris* 27 (2012): 209–26.

commerce, capitalism, and other economic forms.[64] Attention to the study of the generation and deployment of information in colonial settings also helped to shift discourses of imperialism while opening up possibilities to explore the sharing of information between different groups on the basis of common interests in natural kinds, including but not limited to commodities.[65] Such considerations led Ian Hacking to write, "There is perfect commensurability, and no indeterminacy of translation, in those boring domains of 'observations' that we share with all people as people."[66] By the early 2000s, for some historians, the appearance of structures for conveying information and substances within knowledge economies had become a clear matter of concern. Some kinds of commensurability could span the world even as the nature of knowledge itself became more fragmented.

CONCLUSIONS

Over the course of the last two decades, conversations between historians of science and authors of other genres have deepened. Real-world and variable entanglements rather than pure and free minds have come to dominate explanatory strategies. The material concerns exemplified in scientific practices have been fundamental in many recent historical accounts, including those written by historians earlier concerned with epistemology, helping to focus on questions of "how" things happened rather than "why."[67] Contingency rather than determinism has a prominent place at the table, but the constraints of working with material things clearly set limits on the possible. In some cases, such as the projects of Pamela Smith, Lawrence Principe, or William Newman, historians have even begun to work in laboratories in order to explore what might be added to extant descriptions of past practices by aiming to achieve the recorded results.[68]

Equally important, opening the gates in the wall of the garden of science has allowed all kinds of fresh faces to appear in historical accounts of it. The study of Iberian, Germanic, Dutch, Nordic, Slavic, Ottoman, and other European regions—and other ethnicities such as Jews and Muslims—has complicated older narratives about the rise of science along an Italian-French-English axis. But Europe, too, has been provincialized. The game of comparing civilizations in order to explain Western exceptionalism

[64] Harold J. Cook, "Sharing the Truth of Things: Mistrust, Commerce, and Scientific Information in the 17th Century," in *"Eigennutz" und "gute Ordnung," Ökonomisierungen der Welt im 17. Jahrhundert*, Wolfenbütteler Arbeiten zur Barockforschung vol. 54, ed. Sandra Richter and Guillaume Garner (Wiesbaden, 2016), 273–291.

[65] C. A. Bayly, *Empire and Information: Intelligence Gathering and Social Communication in India, 1780–1870* (Cambridge, 1996); Bernard S. Cohen, *Colonialism and Its Forms of Knowledge: The British in India* (Princeton, N.J., 1996); Richard Drayton, *Nature's Government: Science, Imperial Britain and the "Improvement" of the World* (New Haven, Conn., 2000); Kapil Raj, *Relocating Modern Science: Circulation and the Construction of Scientific Knowledge in South Asia and Europe, 17th–19th Centuries* (Delhi, 2006); Harold J. Cook, *Matters of Exchange: Commerce, Medicine and Science in the Dutch Golden Age* (New Haven, Conn., 2007).

[66] Ian Hacking, *Historical Ontology* (Cambridge, Mass., 2002), 171–2; see also G. E. R. Lloyd, *Cognitive Variations: Reflections on the Unity and Diversity of the Human Mind* (Oxford, 2007).

[67] For instance, Hans-Jörg Rheinberger, *An Epistemology of the Concrete: Twentieth-Century Histories of Life* (Durham, N.C., 2010).

[68] For more about Pamela Smith's work, see http://www.makingandknowing.org (accessed 7 July 2017). One of Principe's experiments can be found at https://www.youtube.com/watch?v=2vS4aPQI80M (accessed 27 January 2018). For more about William Newman's work, see http://io9.gizmodo.com/5910577/incredible-videos-recreate-isaac-newtons-experiments-with-alchemy (accessed 7 July 2017). See also Pamela H. Smith, Amy R. W. Meyers, and Harold J. Cook, eds., *Ways of Making and Knowing: The Material Culture of Empirical Knowledge* (Ann Arbor, Mich., 2014).

lost strength after the projects of William McNeill and Fernand Braudel, who pioneered world and regional histories rooted in material culture, with further steps taken in Immanuel Wallerstein's world systems theory and more recently with the emergence of global history. The latter category has been led by economic historians who recognized the fundamental importance of China to the early modern world, which undermined both Marxist and Weberian frameworks for understanding the shape of history.[69] Connected histories are another important new genre, growing out of subaltern studies and microhistory, treating intermingled questions of power and sociability in ways that establish multiple links among particular persons and places rather than reconstructing thick lines emanating from colonial or imperial centers. Mestizo, mélange, and other mixtures; translation; and intermediaries and information brokers (and spies) are now of much interest to historians of science as well as everyone else.[70] For early modernists who are interested in natural knowledge, such connections and minglings have brought the Americas and Africa into the frame, which now loom as large as South and East Asia; even Southeast Asia is on the agenda, and Central Asia and the Pacific archipelagos cannot be far behind. Outside the walled garden of carefully tended, named hybrid flowers, a luxuriant ecology of emergent and interdependent forms has not ceased to establish its own varieties. From the global and connected viewpoint, no single group can be said to have invented or discovered the assembly of durable procedures that is modern science—although with recent pushbacks in favor of the eminence of the West's legacy, it is too soon to know where such paths will lead.

One productive line through the intertwined fabric of de-essentialized mind and matter can be found by attending once more to the mutual production of the activities we call science and commerce, latterly capitalism. Commercial materialism and philosophical materialism may even have been coproduced. For instance, in one thought-provoking recent account, David Graeber claims that monetization gave body to spirit, shifting the inventive abstractions of human thought earthward toward considerations about the productions of art and nature. "Everywhere we see the military-coinage-slavery complex emerge, we also see the birth of materialist philosophies," he writes.[71] The persons most attentive to the assessment of objects and their descriptions were merchants, who relied on material commensurabilities and proofs to make exchange and accumulation possible. Immutable mobiles could cross countless borders before being consumed or transformed according to local custom. Merchants everywhere also shared the disciplines that allowed their own communicative interactions and their sometimes coercive relationships with other kinds of people. We can say that economies and sciences go together like body and mind.

As for the European Scientific Revolution: where merchants gained power, so did the kinds of commensurable material knowledge they most valued. In a growing num-

[69] R. Bin Wong, *China Transformed: Historical Change and the Limits of European Experience* (Ithaca, N.Y., 1997); Andre Gunder Frank, *Reorient: Global Economy in the Asian Age* (Berkeley and Los Angeles, 1999); Kenneth Pomeranz, *The Great Divergence: China, Europe, and the Making of the Modern World Economy* (Princeton, N.J., 2000).

[70] See, e.g., Sanjay Subrahmanyam, "Connected Histories: Notes towards a Reconfiguration of Early Modern Eurasia," *Mod. Asian Stud.* 31 (1997): 735–62; Serge Gruzinski, *The Mestizo Mind: The Intellectual Dynamics of Colonization and Globalization*, trans. Deke Dusinberre (New York, 2002); Mario Biagioli, *Galileo's Instruments of Credit: Telescopes, Images, Secrecy* (Chicago, 2006); Simon Schaffer, Lissa Roberts, Kapil Raj, and James Delbourgo, eds., *The Brokered World: Go-Betweens and Global Intelligence, 1770–1820* (Sagamore Beach, Mass., 2009).

[71] Graeber, *Debt* (cit. n. 6), 244, 248.

ber of places, merchants governed the cities that depended on their activities; they also managed to engineer city-states and republics. Even where they could not gain sovereignty, however, they became indispensable to the rulers of principalities and kingdoms by lending them the ready money required for the exercise of power. Because of the stipulations required of personal and national debtors, governments agreed to abide by contractual rules.[72] (That did not happen everywhere: in the powerful imperial system of China, for instance, the government lent money rather than borrowed it, keeping the merchants in check.[73]) By the later seventeenth century, a class of rentiers can be identified whose wealth derived chiefly from investment instruments—the word "capitalist" first appears in the 1620s as a designation for the purposes of taxation of the wealthiest owners of property[74]—who added modeling and the channeling of energy to their interest in descriptions and calculations of material and space. As merchants and capitalists began to make robust claims on the future of nations, the kind of knowledge processes they had long valued grew in prominence, opening doors to what would come to be called modern science.

Perhaps, then, the lessons to be drawn from the unwise Midas are not what we thought they were. If we look again at Ovid's version of the story, the tale begins not with Midas but with Dionysus, the often-smiling god of life-giving irrational ecstasy. Dionysus became aware that his foster father, Silenus, had disappeared while drunk, and that some Phrygians had discovered him and brought him to their king, who entertained him graciously. As a reward, Dionysus offered a wish to Midas, who in turn chose the golden touch that became a curse. The story therefore begins with the affection of a pupil for his teacher, people meeting strangers, and respectful relationships leading to exchanges of favors. But Ovid does not stop here. Midas returned to Dionysus and begged for relief, and obtaining his second wish was allowed to wash away his deadly touch. Midas subsequently gave up the desire for power and wealth, joining the followers of Pan. Yet his story does not end here, either: when asked to judge the quality of a music contest, Midas voted for his patron out of loyalty rather than the truth, which led Apollo to embarrass him by turning his ears into those of an ass.[75] In other words, relationships and judgments drive the several events in the account of the misfortunes of the eager but shortsighted Midas, demonstrating that questions of goodness and quality cannot be separated from the real situations that give rise to them. Knowledge cannot be measured in gold. It is joined to life and its interactions, and to judging the real truth.

The need to keep substance and ideation together in the same body therefore not only generates creative interpretative frictions but also opens one of the most exciting points of contact between the history of science and other fields. Guarding ourselves against the corruptions and debasements of filthy lucre remains critical. But like transubstantiation,[76] the transformative possibilities of understanding the realms of spirit and matter as joined rather than separate are immanent in the productions of science. Movements as mysterious as those continue to command attention.

[72] See, e.g., Avner Greif, *Institutions and the Path to the Modern Economy: Lessons from Medieval Trade* (Cambridge, 2006).

[73] For a recent discussion, see Goetzmann, *Money Changes Everything* (cit. n. 6), 141–99.

[74] Marjolein C. 't Hart, *The Making of a Bourgeois State: War, Politics and Finance during the Dutch Revolt* (Manchester, 1993), 122–4.

[75] Ovid's account is in *Metamorphoses*, 11.94–216.

[76] I thank Mario Biagioli for using this word in discussion during the meeting held in preparation for this volume of *Osiris*.

ENTANGLED INFRASTRUCTURES

Feeding Desire:
Generative Environments, Meat Markets, and the Management of Sheep Intercourse in Great Britain, 1700–1750

*by Emily Pawley**

ABSTRACT

As a system of profit based on reproduction, growth, and eating, animal husbandry offers an ideal place to examine how capitalism shapes knowledge of bodies. Recent work on the history of breeding demonstrates this, showing how new markets in "blood" helped define new theories of heredity and race. This essay expands on this literature by examining eighteenth-century British efforts to control a different aspect of animal reproduction: desire. Spurred by changing meat markets in out-of-season lamb and expanding property structures that created sex-segregated herds, shepherds, farmers, and agricultural writers worked to provoke the seasonally dependent desires of ewes by feeding them aphrodisiac foods, changing the ways that sex was staged, and creating landscapes of "artificial" grass timed to help ewes escape the constraints of the seasons. Their efforts draw our attention to a broader range of bodily experts, from physicians, to professional feeders, to Linnaean botanists, who were interested in the ways that landscapes could be made to shape bodies. The essay suggests that these forms of environmental control, which still undergird capitalist farming, have left significant modern traces on both knowledge and landscapes and offer a rich and relatively untapped source of bodily knowledge.

William Ellis knew a surprising number of aphrodisiacs for sheep, including, but not limited to, "blades of onion and garlick," "the ladyfinger grass, the Tyne Grass, and the Honeysuckle Grass," "Half a Pint of good *October*, mellow, silky Beer," and, more bleakly, two dogs to "fright and run [ewes] about the ground."[1] Ellis was particularly proud of this last point. "If I was to write a full Detail of all the valuable Services this one Secret, that I have here exposed publickly, will do the World, I might write a Volume on it," he declared. Indeed, the inclusion of sheep aphrodisiacs was crucial to the appeal of Ellis's book, *A Compleat System of Experienced Improvements, Made on Sheep, Grass-lambs, and House Lambs* (1749). On the cover, he promised to explain "How to make an hundred Ewes take Ram in an Hour's Time, either by artificial, or by

* Department of History, Dickinson College, P.O. Box 1773, Carlisle, PA 17013; pawleye@dickinson.edu.

[1] William Ellis, *A Compleat System of Experienced Improvements, Made on Sheep, Grass-Lambs, and House-Lambs* (London, 1749), 296–9; here and throughout, emphasis in the original.

natural means, at any Time of the Year." The most prolific English author on agricultural topics during the 1740s, Ellis was not an isolated or eccentric voice; for example, Pehr (or Peter) Kalm, Linnaeus's Finnish protégé who stopped in Britain on his way to North America, would visit him as part of a broader Linnaean project to gather and exploit profitable landscapes.[2] Ellis was also far from alone in suggesting remedies to excite desire in sheep: he referred frequently both to related printed works and to the practices of "the most accurate Grasiers, Farmers, Sheep-Dealers, and Shepherds of England," which he had gathered as part of an informational tour around Great Britain. This market for aphrodisiac knowledge reminds us that, in agriculture, sex is fundamental to production. It also reminds us that sex is more than a straightforward transmission of information.

It is information transfer or, less anachronistically, the rise of ideas of heredity that has made Ellis's period and place, eighteenth-century Britain, significant to scholarship on the intersections of natural knowledge and capitalism. In particular, a substantial and still growing group of historians has shown how new ideas of "inheritance," that is, of physical similarity passed down over time from parent to child, emerged in part from new markets in "blood" that took shape during the eighteenth and nineteenth centuries. "Improved" horses, cattle, and sheep, sold or rented as sires, and valued for their ability to stamp their offspring with their good qualities, acted as highly publicized models of transmissible bodily change, models that would have wide-ranging effects on racial, evolutionary, and ultimately eugenic and genetic theory.[3] At the same time, the herdbooks, bodily standards, ideas of breeding, and actual animals developed in this period still powerfully shape the millions of animals whose lives constitute modern systems of meat, dairy, and wool production.[4] Since the commodification of "blood" in the eighteenth and nineteenth centuries is an obvious precursor to the rapid commodification of genetic material toward the end of the twentieth, the interest in se-

[2] Pehr Kalm, *Kalm's Account of His Visit to England on His Way to America*, trans. Joseph Lucas (London, 1892).

[3] See, e.g., Harriet Ritvo, "Possessing Mother Nature: Genetic Capital in Eighteenth-Century Britain," in *Noble Cows and Hybrid Zebras: Essays on Animals and History* (Charlottesville, Va., 2010), 157–74; Margaret Derry, *Masterminding Nature: The Breeding of Animals, 1750–2010* (Toronto, 2015); Derry, *Bred for Perfection: Shorthorn Cattle, Collies, and Arabian Horses* (Baltimore, 2003); Nicholas Russell, *Like Engend'ring Like: Heredity and Animal Breeding in Early Modern England* (Cambridge, 2007); James A. Secord, "Darwin and the Breeders: A Social History," in *The Darwinian Heritage*, ed. David Kohn (Princeton, N.J., 1985), 519–42; Theodore James Varno, "The Nature of Tomorrow: Inbreeding in Industrial Agriculture and Evolutionary Thought, in Britain and the United States, 1859–1925" (PhD diss., Univ. of California, Berkeley, 2011); Staffan Müller-Wille and Hans-Jörg Rheinberger, "Heredity—The Formation of an Epistemic Space," and Roger J. Wood, "The Sheep Breeders' View of Heredity Before and After 1800," both in Müller-Wille and Rheinberger, *Heredity Produced: At the Crossroads of Biology, Politics and Culture, 1500–1870* (Cambridge, Mass., 2007), 3–35, 230–2; B. Matz, "Crafting Heredity: The Art and Science of Livestock Breeding in the United States and Germany, 1860–1914" (PhD diss., Yale Univ., 2011); Daniel J. Kevles, "New Blood, New Fruits: Protections for Breeders and Originators, 1789–1930," in *Making and Unmaking Intellectual Property*, ed. Mario Biagioli, Peter Jaszi, and Martha Woodmansee (Chicago, 2011), 253–69; Müller-Wille and Rheinberger, *A Cultural History of Heredity* (Chicago, 2012); Emily Pawley, "The Point of Perfection: Cattle Portraiture, Bloodlines, and the Meaning of Breeding, 1760–1860," *J. Early Repub.* 36 (2016): 37–72; Sarah Franklin, *Dolly Mixtures: The Remaking of Genealogy* (Durham, N.C., 2007).

[4] Rebecca Jane Houghton Woods, "The Herds Shot round the World: Native Breeds and the British Empire, 1800–1900" (PhD diss., Massachusetts Inst. of Technology, 2011); Woods, "From Colonial Animal to Imperial Edible: Building an Empire of Sheep in New Zealand, ca. 1880–1900," *Comp. Stud. South Asia, Africa, Middle East* 35 (2015): 117–36.

lective breeding has been amplified by another rising theme in the history of the life sciences, the study of "biocapital."[5]

Scholars of genes and blood have shown us how invisible features of heredity became a new sort of commons, rapidly privatized and speculated on, with enormous scientific, social, and environmental consequences. Like a great deal of history of science focusing on capitalism, however, it is primarily concerned with the moment of commodification. By focusing largely on the novel commodities of heredity, which have emerged relatively recently, it implies that the connection between bodies and markets is new—that "bio" and "capital" are to be found together for the first time in pureblood pedigrees or patented genes.

However, in mid-eighteenth-century Britain, as now, graziers, farmers, and shepherds were not primarily struggling toward an understanding of the principles of inheritance. Rather, they were developing ways to produce vigorous, tender, heavy bodies at the right times for an increasingly sophisticated consumer market in meat. Meat has been a commodity for longer than there have been laboratories or corporations; we can see its long-term imbrication with money and value in the term "chattel" (which comes from the same root as "cattle").[6] It would be ridiculous to suggest that meat stopped changing as a bodily product once it had been commodified. Following meat rather than genes or blood leads us to different genres of bodily knowledge, those concerning not commodification but changes in production. Producing vigor, tenderness, weight, and timing was dependent on a range of intimate interventions extending beyond the selection of sires and dams.

To get at the changing forms of knowledge that shaped processes of meat production, we need to look at a different scale of life, not the hidden capabilities of "blood" or DNA, but at living animals and plants, at the fields or pastures in which they lived, and the social institutions and kinds of knowledge that created those fields.[7] Those kinds of knowledge encompass a much wider array of subjects, addressing what we might now call studies of instinct, nutrition, and appetite in the animals themselves,

[5] E.g., Stefan Helmreich, "Species of Biocapital," *Sci. Cult.* 17 (2008): 463–78; Kaushik Sunder Rajan, *Biocapital: The Constitution of Postgenomic Life* (Durham, N.C., 2006); Rajan, ed., *Lively Capital: Biotechnologies, Ethics, and Governance in Global Markets* (Durham, N.C., 2012); Nikolas Rose, *The Politics of Life Itself: Biomedicine, Power, and Subjectivity in the Twenty-First Century* (Princeton, N.J., 2007). For similar material on the commodification of novel living beings and, to a certain extent, genetic material, see, e.g., Londa Schiebinger and Claudia Swan, eds., *Colonial Botany: Science, Commerce, and Politics in the Early Modern World* (Philadelphia, 2005); Harold J. Cook, *Matters of Exchange: Commerce, Medicine, and Science in the Dutch Golden Age* (New Haven, Conn., 2007); Vandana Shiva, *Biopiracy: The Plunder of Nature and Knowledge* (Berkeley and Los Angeles, 2016); Jack Kloppenberg, *First the Seed: The Political Economy of Plant Biotechnology* (Madison, Wis., 2005). See also Courtney Fullilove, "Microbiology and the Imperatives of Capital in International Agro-Biodiversity Preservation," in this volume.

[6] *Oxford English Dictionary*, s.v. "cattle," http://www.oed.com/view/Entry/29037?redirectedFrom=cattle (accessed 22 February 2018).

[7] Of course, production and commodification are not fully separate kinds; as Victoria Lee's study shows, often processes of production help constitute both species and commodity category. It is only because Japanese brewers worked so hard to stabilize their product that the microbes could be isolated and imagined as unchanging species. See Lee, "The Microbial Production of Expertise in Meiji Japan," in this volume. Historians of chemistry have paid more attention to process; see, e.g., Ursula Klein and Wolfgang Lefevre, *Materials in Eighteenth-Century Science: A Historical Ontology* (Cambridge, Mass., 2007). For works examining the value of process and agricultural knowledge, see Judith Carney, *Black Rice: The African Origins of Rice Cultivation in the Americas* (Cambridge, Mass., 2001); Courtney Fullilove, *The Profit of the Earth: The Global Seeds of American Agriculture* (Chicago, 2016).

as well as the botanical, climatic, and hydrological knowledge needed to shape pastures.

Ellis's sheep aphrodisiacs, and the complex system of bodily production that they reveal, can help us illuminate these forms of knowledge. As meat markets expanded and meat-producing landscapes were privatized, meat animals' sexual encounters, indeed their lives as a whole, were increasingly stage-managed.[8] As a result, the sometimes-recalcitrant desires of animals became the subject of lively speculation and experimentation. This essay shows how knowledge about animal bodies shaped and was shaped by forms of expertise not only in selection, but in feeding and in the creation of specialized landscapes of food. In doing so, it links a broader array of bodily experts, from shepherds, graziers, and butchers, to doctors and pornographers, Linnaean botanists, members of the Royal Society, and authors of the growing genre of agricultural texts. These forms of knowledge, like breeder knowledge, have left durable and fundamental traces in the present day. Even as twenty-first-century meat producers trade in the biocapital of genes, they continue to engineer living spaces, behaviors, and desires in the animals they seek to control, transforming landscapes and organisms and, in doing so, making knowledge.

* * *

Writing in mid-eighteenth-century Britain, Ellis played a role in a period crucial both to histories of agricultural capitalism and to histories of capitalism as a whole—not least because of the continuing controversies surrounding the "agricultural revolution." According to the oldest school in this debate, the increases in food production that enabled the Industrial Revolution emerged during the second half of the eighteenth century, as a result of a top-down set of shifts: the new orientation toward profit among landed elites, the accelerated transformation of commonly held land into enclosed private property, the rise of new forms of crop rotation, and of course the development of new forms of animal breeding based on selection. From the perspective of this classic school, Ellis's book (1749) appeared right before the real action started. However, from the perspective of the competing school, Ellis appears not as prologue to agricultural change but as one culmination of it. According to this Marxian school, the heroic late eighteenth-century innovators praised by the classic view often profited from earlier waves of bottom-up change—innovations in agricultural practice and property regimes stretching back into the sixteenth century.[9]

Contemporary accounts of the 1740s certainly give us a picture of a rural landscape in transition, a bewildering mosaic of old and new property regimes with widely differing forms of production and relationships to the market. Throughout England, but

[8] For reproductive control, see Jason Hribal, "'Animals Are Part of the Working Class': A Challenge to Labor History," *Labor Hist.* 44 (2003): 435–53.

[9] For the original British Agricultural Revolution literature, see Baron Rowland Edmund Prothero Ernle, *English Farming, Past and Present* (London, 1912), 176–89; J. D. Chambers and G. E. Mingay, *The Agricultural Revolution, 1750–1880* (London, 1966), 66–9. For responses, see E. Kerridge, "The Agricultural Revolution Reconsidered," *Agr. Hist.* 43 (1969): 463–76; Robert C. Allen, "Tracking the Agricultural Revolution in England," *Econ. Hist. Rev.* 52 (1999): 209–235; Mark Overton, "Re-Establishing the English Agricultural Revolution," *Agr. Hist. Rev.* 44 (1996): 1–20; S. Todd Lowry, "The Agricultural Foundation of the Seventeenth-Century English Oeconomy," *Hist. Polit. Econ.* 35 (2003): 74–100; Mauro Ambrosoli, *The Wild and the Sown: Botany and Agriculture in Western Europe, 1350–1850* (Cambridge, 1997), 363.

particularly in the Midlands, "commoners" still grazed their animals on common pasturage and plowed long strips of open, unfenced fields. They made decisions about crop rotations, fallows, and grazing levels collectively, working through a sophisticated system of interlocking customary rights, and they managed a range of organisms—cattle, sheep, barley, geese, and so forth—in part for sale in distant markets, but also for local consumption.[10] Sometimes adjoining commons and open fields, however, were tracts of enclosed land, where conditions differed sharply. Though the great age of parliamentary enclosures had not yet begun, general and private enclosures organized by local rights holders had already transformed perhaps 22 percent of English land during the seventeenth century, a process that accelerated in the first decades of the eighteenth century.[11]

Nowhere were the effects of enclosure clearer than in the counties around London. Some parts of Middlesex, for example, were entirely geographically specialized, focusing on a single crop: grass. Well-off tenant farmers from that area supplied the horses and cattle of London with hay; on the way home, they filled their wagons with street manure, purchased for a fee, that sometimes allowed them an astonishing three crops a year. Hay profits in turn paid high rents and the wages of migrant laborers: Irish men who came for the summer and returned to Ireland in the winter. Landscapes like these displayed the regularly agreed upon features of agricultural capitalism: the rise of distant markets and decline of local subsistence cycles; the rise of cash and decline of barter; the development of new forms of calculating and recording profits, and of private property markets in land as well as the destruction of commons rights and transhumance; the geographic specialization of labor, that is, the development of regions specializing mainly in corn, cattle, apples, or sugar; the rise of monocrop production; and the emergence of characteristic new exploitative labor forms.[12]

Though commons and enclosures operated at opposite ends of a spectrum of land tenure, they were linked by flows of meat and money. Throughout the eighteenth century, hundreds of thousands of British cattle and sheep moved through commercially determined migratory stages, starting on distant pastures and commons and ending as chops, shanks, and roasts in London markets. Each phase of their life might take place in a different county, managed by a different specialist. Animals born and raised as far away as Scotland and Wales were walked in droves toward the counties around London, where they passed from drovers to graziers. Graziers were professional fatteners—maintaining rich pastures and developing concoctions of fodder crops to fatten animals for their final destination: Smithfield Market. Smithfield butchers, in turn, stored animals in even more intensive feeding landscapes, turning them out into a close ring of manicured pastures, or confining them to fattening pens where they might eat Middlesex hay, keeping in condition while they awaited slaughter. The wealthiest eaters

[10] J. M. Neeson, *Commoners: Common Right, Enclosure, and Social Change in England, 1700–1820* (Cambridge, 1996).

[11] Mark Overton, *Agricultural Revolution in England: The Transformation of the Agrarian Economy* (Cambridge, 1996), 148; Tom Williamson, *The Transformation of Rural England: Farming and the Landscape, 1700–1870* (Exeter, 2002).

[12] Immanuel Wallerstein, *The Modern World-System I: Capitalist Agriculture and the Origins of the European World-Economy in the Sixteenth Century* (Berkeley and Los Angeles, 2011); Jason W. Moore "The Capitalocene, Part I: On the Nature and Origins of Our Ecological Crisis," *J. Peasant Stud.* 44 (2017): 594–630; Sven Beckert and Seth Rockman, eds., *Slavery's Capitalism: A New History of American Economic Development* (Philadelphia, 2016); Allan Kulikoff, "The Transition to Capitalism in Rural America," *William Mary Quart.* (1989): 120–44.

also enjoyed meat from more specialized and intimate local subsystems. Essex, for example, had become famous for its "white veal," produced by a combination of confinement, heroic bleeding, and powdered chalk added to the calves' feed.[13] Money for such luxuries came in part from even farther-flung systems of agricultural profit: the plantation-style agriculture then biting deeper frontiers into Ireland and the Americas, and the imperial Spanish transatlantic slave-trading contracts, founded on sugar profits, that the British had won thirty years before.[14]

This market in meat helped support a related market in texts, one that Ellis had managed successfully to enter long before his treatise on sheep. When Linnaeus's disciple Pehr Kalm arrived in Britain in 1748, he set off to visit Ellis in Little Gaddesden, Hertfordshire. Baron Bjelke of the Royal Academy of Sciences of Stockholm had commissioned Kalm to find Ellis because of his "beautiful books on Rural Economy."[15] Ellis had already written at least eight, publishing in both London and Dublin. Arriving in Little Gaddesden, Kalm was disgusted to find a man transparently profiting from his writing rather than his farming. While Ellis's fields stood neglected, Kalm complained, the manuscripts that would become his treatise on sheep lay on a nearby table, awaiting "the printer who would pay most for them."[16] Indeed, Ellis's neighbors "maintained that Mr. Ellis's principal occupation consists in writing books, and selling to gentlemen the ploughs and implements which he has lauded therein."[17]

Ellis might have been the most prolific author benefiting from this market, but his books did not monopolize it. Ellis's competitor in the matter of sheep aphrodisiacs, Richard Bradley, Professor of Botany at Cambridge, had issued a similar stream of books in the 1720s and 1730s, which were still circulating in the 1740s.[18] Both Ellis and Bradley had relied in turn on a broader husbandry literature, which supplied copy, probably inadvertently, for pasted-together collections of extracts like the *Dictionarium Rusticum* or the later *Compleat Grazier*.[19] Both Ellis and Bradley drew some level of authority from their references to this textual tradition and their familiarity with classical authors. They drew even more from their promise to record the experiences of actual specialists from across the system of meat production. William Ellis claimed that his treatise derived from the knowledge of the "most accurate Grasiers, Farmers, Sheep-Dealers, and Shepherds of England."[20] According to Kalm, indeed, Ellis's neighbors, though scornful of his writing and his farming, often employed the techniques and species that Ellis described and occasionally claimed as his own inventions. Here too Ellis was part of a larger eighteenth-century pattern, an epistemological elevation

[13] Robert Trow-Smith, *A History of British Livestock Husbandry, 1700–1900* (New York, 2013).
[14] See, e.g., Jason W. Moore, "Sugar and the Expansion of the Early Modern World-Economy: Commodity Frontiers, Ecological Transformation, and Industrialization," *Rev. Fernand Braudel Center* 23 (2000): 409–33; Bertie Mandelblatt, "A Transatlantic Commodity: Irish Salt Beef in the French Atlantic World," *Hist. Workshop J.* 63 (2007): 18–47.
[15] Kalm, *Visit to England* (cit. n. 2), 188.
[16] Ibid., 189.
[17] Ibid., 191.
[18] S. M. Walters, *The Shaping of Cambridge Botany: A Short History of Whole-Plant Botany in Cambridge from the Time of Ray into the Present Century* (Cambridge, 1981), 15–30.
[19] Edward Lisle, *Observations in Husbandry*, vol. 2 (London, 1757); Nathan Bailey, ed., *Dictionarium Rusticum, Urbanicum et Botanicum*, 3rd ed. (London, 1726); Richard Bradley, *The Gentleman and Farmer's Guide for the Increase and Improvement of Cattle*, 2nd ed. (London, 1732); Edward Whitaker, *Compleat Grazier*, 3rd ed. (London, 1767).
[20] Ellis, *Compleat System* (cit. n. 1), cover.

of practical knowledge that led to the recording and codification of knowledge from many rural people, shepherds in particular.[21]

Ellis's turn to treatises specializing in sheep was perhaps strategic, following a market shift from beef to mutton. The significance of both tended to seesaw back and forth, since, eighteenth-century authors pointed out, great outbreaks of cattle disease accompanied drought, whereas sheep epidemics followed heavy rains. As Ellis wrote, the seesaw had just tipped sharply toward mutton. Starting in 1745, a thirteen-year outbreak of hoof-and-mouth disease devastated herds and, as the state enforced a quarantine, partially paralyzed the beef trade from infected regions. Whereas 86,787 cattle had passed through Smithfield in 1739, by 1748 only 67,681 did so. Likely in response, the number of sheep sold at Smithfield over roughly the same period increased by almost 200,000: from 468,120 in 1743 to 656,340 in 1750. This increase in sheep population had struggled against disease epidemics as well—sheep had died of famine in 1740 and had caught the rot in enormous numbers in 1745.[22] (This may help explain another of the selling points on the cover of William Ellis's book: "How to make the most Profit of Rotten Sheeps Carcasses, or those that die by Accident.")[23] It was in these same years, as the market for mutton expanded, that new groups of breeders would struggle to develop an animal notable for meat, rather than wool, eventually producing animals that were famously round-bodied and short-legged, better built to fatten than to walk. As practices of selection shifted, however, so too did other practices needed to produce such bodies.

* * *

The French physician Nicolas Venette was reasonably certain that brute animals were less lecherous than men. "Man . . . the most lascivious of creatures" was "disposed for the delights of Love at every hour, and in every Season," as Venette pointed out in *The Mysteries of Conjugal Love Reveal'd*, a text popular in the late seventeenth and early eighteenth centuries. Indeed, a considerable number of conjugal love's mysteries seemed to surround the choice of time for reproductive sex. Was early morning better than early evening? How many times per night were advisable? By contrast, Venette observed, "most other Creatures wait for certain periods of time, in order to Copulate."[24] This assertion of relative animal passionlessness may ring oddly against ideas of animals as lusty beings, widespread even at the time. However, for eighteenth-century Europeans, who lived among many animals, the timing of animal desire was an everyday annoyance—even in cities, mares had to be kept from stallions when in heat. (The prevalence of geldings among city horses was, among other things, a necessity for the smooth flow of traffic.)

For those responsible for producing animal bodies, the potential absence of animal passion was even more problematic. In part this had to do with the functionality attributed to desire in the early eighteenth century. In animal husbandry texts, as in texts on

[21] See Vladimir Jankovic, *Reading the Skies: A Cultural History of English Weather, 1650–1820* (Manchester, 1988), 134.
[22] Thomas Southcliffe Ashton, *An Economic History of England: The Eighteenth Century* (1955; repr., New York, 2013), 53–4.
[23] Ellis, *Compleat System* (cit. n. 1), cover.
[24] Nicolas Venette, *The Mysteries of Conjugal Love* (London, 1707), 8.

human sex, the level of desire was itself expected to be a significant productive force. "Lustiness" throughout both literatures might refer to both desire and health, which could in turn be equated with bodily heat. Thus, for example, the 1707 *Husbandman's Instructor* suggested that ewes younger than two years old had not "contracted a sufficient heat to produce Lambs strong and lusty enough to continue their Health."[25] Older ewes, having contracted more heat, were all the "more easily entreated to the rut."[26] Husbandry and human reproduction texts described the level of desire felt by both partners as vital to the health and form of the calf, lamb, or child.[27]

However, seasonality also made animals' passions difficult to channel profitably. This was particularly true in the case of ewes, which were generally recognized as more seasonally rigid than cows, mares, or women. Like deer, most ewes in temperate climates only come into season annually, a period eighteenth-century texts called becoming "blythesome" or "coming into blossom." In keeping with eighteenth-century ideas about human reproduction, blossom was cast as female desire and also referred to as "bringing [ewes] to lust."[28] Left alone, most eighteenth-century English ewes blossomed between July and September but varied individually and from season to season. This uneven timing could threaten the production of lambs: those born too early froze or caught the rot; those born too late might be too small to navigate the furrowed fields they were turned into in May. As regimes of animal production shifted in the late seventeenth and early eighteenth centuries, the uncertainty of ewe desire became increasingly problematic.

Most notably, encounters between male and female sheep were changing. Mid-seventeenth-century manuscripts described almost constant interaction between the sexes; "tups" (rams) and ewes had run together on commons and sheepwalks throughout the year until the autumn. Then, tups too young to breed and riggons (males with testicles in their back ridges who could not be castrated) might be separated from the rest in a makeshift enclosure. The remaining males would be expected to manage their own encounters mostly by fighting over the ewes—care was taken here to keep the horned rams from killing the hornless ones, which were seen as producing more easily birthed lambs.[29] In the early eighteenth century, Edward Lisle described further layers of potential control—sharp sticks were tied to the tails of one-year-old ewes, considered too young to produce healthy lambs, to keep the rams from mounting them, and thick-wooled ewes would be shaved in the back to keep rams from bouncing off, preventing an otherwise inexplicable epidemic of sterility.

As enclosures expanded, however, it became possible to constrain sexual encounters more rigidly, using a more capital-intensive contraceptive technology: the fence. With sufficient fencing, rams could be kept entirely separated from ewes until speci-

[25] A. S., *Husbandman's Instructor or, Countryman's Guide* (London, [1707?]), 53.

[26] Bradley, *Gentleman and Farmer's Guide* (cit. n. 19), 20.

[27] Roy Porter and Lesley Hall, *The Facts of Life: The Creation of Sexual Knowledge in Britain, 1650–1950* (New Haven, Conn., 1995); Müller-Wille and Rheinberger, "Heredity" (cit. n. 3). On female desire and fertility, see Thomas Walter Laqueur, *Making Sex: Body and Gender from the Greeks to Freud* (Cambridge, 1992); Karen Harvey, "The Century of Sex? Gender, Bodies, and Sexuality in the Long Eighteenth Century," *Hist. J.* 45 (2002): 899–916; Laura Gowing, "Women's Bodies and the Making of Sex in Seventeenth-Century England," *Signs* 37 (2012): 813–22; Robin Ganev, "Milkmaids, Ploughmen, and Sex in Eighteenth-Century Britain," *J. Hist. Sexual.* 16 (2007): 40–67.

[28] On seasonality, menses, and desire in women and animals, see Venette, *Mysteries* (cit. n. 24), 101.

[29] Henry Best, *Rural Economy in Yorkshire in 1641* (Durham, 1857), 1–2, 28.

fied moments. The *Husbandman's Instructor*, for example, suggested that to produce lambs for breeding rather than for meat, two-year-old ewes should be separated from the flock and should "receive the *Ram* in a warm close pasture."[30] These encounters were even more choreographed when ram rental became regionally common. Rented for a season or a single leap, "a tup" or "a ram" could refer both to an animal and to a sometimes-costly sex act. This was part of a broader commodification of sexual services in animal husbandry. In enclosed parishes, parish bulls gave way to leased bulls, meaning that poor families could often no longer afford to keep a cow for home dairying—keeping her pregnant was prohibitively expensive.[31]

More staged encounters could pose some problems—when rams were left alone with two-year-old ewes, for example, both rams and ewes were often reluctant. The *Husbandman's Instructor*, noting that the "*Ram* rather covets the old than the young *Ews* [*sic*] because they are easier wooed," suggested that both be given "Blades of Onions and Garlic . . . that eating them they may stir up Desire, and render them both the willinger to a compliance."[32] Rams also needed special feeding and special treatment to make it through their assigned quotas of ewes—the size of these quotas and their effects on vitality and stamina were much debated.[33] The commodification of tupping only increased demand for rams who could provide service on demand.

While these questions absorbed authors, other questions surrounding female desire excited even greater attention. Separation of the sexes narrowed the window in which ewes could come to blossom, leaving many unproductive. At the same time, expanding luxury economies of meat made the timing of ewe desire matter even more. At a guinea apiece, "house lambs," which were born in the late autumn or winter and reared indoors, were worth a lot of trouble. "Some Time ago, Lamb was a Rarity at *Christmas*," wrote Bradley, the Cambridge botanist, in 1729, "but now, unless it be two Months in the Year, some of our Farmers have got the Knack of bringing their sheep to blossom every Month."[34] Bradley's friend Sir John D'Oyley produced a hundred house lambs a year. Ellis, writing sixteen years later, promised "the newest Methods of suckling House-Lambs, in the greatest perfection."[35] House lamb flesh was analogous in tenderness and refinement to that of the bloodless Essex veal calves: "it is certain the Flesh of an housed Lamb is much more delightful to nice Palates, than that of grass Lamb," wrote Bradley, "so where Luxury prevails the most, such Dainties will fetch the most Money."[36] In its out-of-season luxury, house lamb resembled the hothouse fruits and flowers produced under glass in Chelsea. Making tender flesh out of season, however, meant shifting the seasonality of desire.

* * *

Given the increasing emphasis on precisely timed or out-of-season blossom, it is perhaps not surprising that the husbandry literature described a great variety of aphrodi-

[30] A. S., *Husbandman's Instructor* (cit. n. 25), 53.
[31] Elaine S. Tan, "'The Bull Is Half the Herd': Property Rights and Enclosures in England, 1750–1850," *Explor. Econ. Hist.* 39 (2002) 470–89.
[32] A. S., *Husbandman's Instructor* (cit. n. 25), 53.
[33] Lisle, *Observations* (cit. n. 19), 159.
[34] Bradley, *Gentleman and Farmer's Guide* (cit. n. 19), 19.
[35] Ellis, *Compleat System* (cit. n. 1), cover.
[36] Bradley, *Gentleman and Farmer's Guide* (cit. n. 19), 23.

siacs and desire-provoking practices. Some were physical interventions. Authors debated the value of exposing ewes to rams suddenly after a long period of deprivation or weakening their resistance by chasing them with dogs. Later texts suggest that "teaser rams"—cheaper rams fitted with aprons and introduced to the paddock to interest the ewes and to identify those in blossom before the arrival of the expensive tup—first appeared in this period.[37] It is also clear that the practice extended beyond what some authors were willing to recommend. "There are some Drugs may be given to Sheep to force them to couple, besides the Herbs I have mentioned," Bradley remarked, "but they weaken the Sheep when the forcing is over, and therefore I hold them unprofitable."[38]

However, more frequently than dogs or teaser rams, the same collection of stimulating plants, Bradley's "Herbs," appeared over and over again, including sharp-tasting or spicy plants, like garlic, turnip, or onion leaves, or saltwater or salt-marsh grass, part of a broader regimen of shepherds' herbal remedies.[39] Texts on human desire from this period help place these aphrodisiac foods into a broader therapeutic context. Discussing the emergence of desire in fourteen- or fifteen-year-old girls, for example, the 1704 edition of *Aristotle's Masterpiece*, the most widely circulated English-language manual on human reproduction, argued for the value of similar flavors: "if [girls] eat salt, sharp Things, Spices, & c. whereby the Body becomes Still more and more heated, then the Inclination and Proneness to *Venerial Embraces* is very great."[40] The same turnips and onions recommended for sheep, indeed, were specifically mentioned as aphrodisiacs in both *Aristotle's Master-Piece* and Venette's *Conjugal Mysteries*.[41] Herbals referred to a similar list of potentially therapeutic plants, assuming readers' knowledge of the causal link between culinary heat and desire. In fact, as recent scholarship has shown, concern with fertility in the latter half of the seventeenth century had helped legitimize a wide range of aphrodisiacs. Arguments for the aphrodisiac and thus fertility-promoting capacities of "heating," spicy, or salty foods for both men and women were a constant element of popular medical texts, midwives' manuals, apothecaries' advertisements, pornographic pamphlets, and popular ballads.[42] The various versions of *Aristotle's Master-Piece* and *Conjugal Mysteries Reveal'd* were inheritors of these older arguments.

This same body of texts can help to illuminate seemingly more prosaic but equally common advice, that to bring ewes to blossom on schedule one should also feed them more and better food. Blossom, observed Bradley, is "evidently brought about by increasing the Richness of [ewes'] Food sometime before we would have them couple; for Richness of Food increases the Vigour of the Body."[43] Bradley's friend, the house lamb specialist Sir John D'Oyley, managed his monthly blossoms mainly through this method, which "depended chiefly upon the assorting the Sheep in different Pastures,

[37] William Youatt, *Sheep: Their Breeds, Management, and Diseases, to Which Is Added, The Mountain Shepherd's Manual* (London, 1837), 3, 493.

[38] Bradley, *Gentleman and Farmer's Guide* (cit. n. 19), 21.

[39] Ibid., 48–64.

[40] *Aristotle's Master-Piece: Or the Secrets of Generation Display'd in All the Parts Thereof* (London, 1704), 4.

[41] Ibid., 8; Venette, *Mysteries* (cit. n. 24).

[42] Jennifer Evans, *Aphrodisiacs, Fertility and Medicine in Early Modern England* (Woodbridge, Suffolk, 2014).

[43] Bradley, *Gentleman and Farmer's Guide* (cit. n. 19), 18.

and the Richest, or what produced the most nourishing Food."[44] Ellis elaborated on Bradley's theory of the value of rich food directly. "Had Mr. *Bradley* known the Virtues of the Lady-finger Grass, the Tyne Grass, and the Honeysuckle Grass," Ellis wrote, "he would undoubtedly have recommended the Sowing of these most excellent sorts for this Purpose."[45] Seventeenth- and eighteenth-century texts on human reproduction elaborated on the centrality of rich food to desire and fertility alike: "such as are subject to Barrenness," *Aristotle's Master-Piece* suggested, "should eat such Meats only as tend to render them fruitful," specifically, "all Meats of good Juice, that nourish well, and make the Body lively and full of Sap."[46] Just as spicy foods might increase the body's heat, rich foods were commonly described as helping create blood, from which human seed (sometimes male and female, sometimes only male) might be elaborated.[47]

If calls for rich fodder had their echoes in pornographic and medical texts, they also had roots in forms of knowledge and practice that were specific to husbandry, and that had already proved key to major transformations in British food production, landscapes, and agricultural knowledge. During the first decades of the eighteenth century, feeding rather than selection was the aspect of animal production that occupied the most attention and debate. It was the techniques of graziers, indeed, that the Linnaean disciple Pehr Kalm hoped to investigate. Arriving in England, Kalm was astonished by the "fatness" and "delicious taste" of English meat, which he attributed to "the excellent pasture, which consists of such nourishing and sweet-scented kinds of hay as there are in this country, where the cultivation of meadows has been brought to such high perfection."[48] Walking the countryside around London and Hertfordshire accompanied by obliging members of the Royal Society, Kalm inspected meadows, pastures, and haystacks, commenting incessantly on their astonishing exuberance, their earliness compared to Swedish meadows, and the practices used to create them. Indeed, graziers were among the most visible and commercially oriented specialists in animal husbandry. Occupying the land closest to London and most densely dotted with country seats of the urban rich and the aristocracy, they were usually richer than the farmers who supplied them with store animals and were likely to participate in the cycles of moneylending that were already common in rural England before 1750, both as borrowers and as lenders.[49] This wealth allowed some of them to lever themselves into a new class. Visiting a grazing area in the Vale of Aylesbury in the 1720s, Daniel Defoe (himself a tallow chandler, butcher's son, and cattle breeder) observed, "all the Gentleman here are Grasiers, tho' all the Grasiers are not Gentlemen."[50]

Coming as we do from a system of food production where nutriment can be generally supplied to animals regardless of place—think of chicken factories supplied with corn from global markets—it is worth remembering the extreme complexity inherent

[44] Ibid., 19.
[45] Ellis, *Compleat System* (cit. n. 1), 296.
[46] *Aristotle's Master-Piece* (cit. n. 40), 7.
[47] Evans, *Aphrodisiacs* (cit. n. 42), 100; Jennifer Evans, "'Gentle Purges Corrected with Hot Spices, Whether They Work or Not, Do Vehemently Provoke Venery': Menstrual Provocation and Procreation in Early Modern England," *Soc. Hist. Med.* 25 (2012): 2–19.
[48] Kalm, *Visit to England* (cit. n. 2), 15.
[49] B. A. Holderness, "Credit in English Rural Society before the Nineteenth Century, with Special Reference to the Period 1650–1720," *Agr. Hist. Rev.* 24 (1976): 97–109.
[50] Paula R. Backscheider, "Defoe, Daniel (1660?–1731)," in *Oxford Dictionary of National Biography* (Oxford, 2004; online ed., 2008), http://www.oxforddnb.com/view/article/7421 (accessed 5 May 2016); Daniel Defoe, *A Tour through the Whole Island of Great Britain*, vol. 2 (London, 1748), 216.

in the timed supply of "rich food" in more seasonally dependent and locally limited markets. In early eighteenth-century England, where forage and fodder shortages periodically devastated herds and flocks, the flow of food was far from certain. Bradley's writing about the optimal timing of blossom gives a sense of both the variety of rural landscapes and the complexity of managing them. In a section about unforced tupping, he explained that the time to bring rams and ewes together depended largely on the lands available and the speed with which they might produce grass in the spring. Farmers with "good Winter Pasture for Sheep that springs early in the year" might allow rams and ewes to live together all year. But if they had "only grass in common," which he assumed to be poorer and less reliable, then they should be kept from each other until July. The poorer the land, the later the date of encounter: "If the Farmer has only a Run of Sheep upon a common field amongst the Arables, then tis time enough about Michaelmas [September]." But in "mountainous and rocky lands, which have no Pastures or common fields . . . 'twill be Time enough to bring the Rams and Ewes together about *St. Simon* and *St. Jude's* Day, which is towards the end of October."[51]

These different spaces—rich pasturage, grass in common, arable lands, and mountainous land—were not natural kinds. Neither was the period of scarcity that constituted "winter." Around Britain, new techniques of landscape modification that depended on manipulating grass were developing and spreading. It is these transformations, indeed, that historians arguing for an older, more practitioner-centered agricultural revolution have emphasized. For example, the "good winter pasture for sheep that springs early in the year," which Bradley saw as necessary to support early-blossoming ewes, was increasingly being supplied by a capital-intensive system for controlling the growth of grass that had spread rapidly in the early eighteenth century. "Water meadows" or "floating meadows" were artificial spaces, watered in sloped country by small redirected brooks or on carefully flattened land near rivers by networks of trenches.[52] Covered with an inch of water for most of the winter, and drained to breathe at intervals, the grass was defended against damaging frosts; in the summer it was comparatively immune to drought and fed by river or hill sediment. Such meadows could quadruple the hay crop of unfloated meadows, ensuring a steady supply of fodder to promote ewe blossom through the winter, and could produce spring grass ready for the bite of a newborn lamb as early as March. It is in the context of these sorts of manipulations that we should view the technique of D'Oyley's method, which required the creation and maintenance of "pastures of different Kinds, some of higher Food than others; but chiefly the high Lands, short in Grass."

If the timing of the grass that enabled blossom required one form of expertise, so too did the choice of species that accounted for its richness. While the word "grass" may now evoke an indistinguishable mass of lawn green, for eighteenth-century rural Europeans, it was populated by a series of distinct entities assigned remarkably different qualities. Mindful of this, Kalm and his companions painstakingly picked through

[51] Richard Bradley, *The Gentleman and Farmer's Guide for the Increase and Improvement of Cattle*, 1st ed. (London, 1729), 22.

[52] A "meadow" is a place where hay is cut. A "pasture" is a place where animals are sent to graze. Confusingly, animals are sometimes set to graze on meadows, which are still meadows because hay is cut there. Eric Kerridge, "The Sheepfold in Wiltshire and the Floating of the Watermeadows," *Econ. Hist. Rev.* 6 (1954): 282–9; J. H. Bettey, "The Development of Water Meadows in Dorset during the Seventeenth Century," *Agr. Hist. Rev.* 25 (1977): 37–43.

haystacks and took great pains, on their walks, to ascertain the species growing in meadows, making notes for an intended "Oeconomico-Botanical" treatise on British meadows, part of a broader Linnaean project to spread productive landscapes.[53] To learn the specific functions of species, they also interviewed meadow owners, from whom Kalm learned, for example, that "ray grass," when planted with clover, combated clover's tendency to inflate cattle suddenly to fatal proportions.[54]

The species on Kalm's lists had reached their meadows and haystacks through local exchange and international botanical networks and markets. Ellis's aphrodisiac lady-finger grass and honeysuckle grass were entering a crowded field of recently arrived, commercially traded "artificial grasses." Sometimes known as "French" or "sown" grasses, artificial grasses had been spreading slowly in Great Britain since at least the mid-seventeenth century. Mostly imported from France or the Low Countries, clover, sainfoin, and other artificial grasses had been the objects of British natural philosophical speculation as early as the 1640s, when Samuel Hartlib and the members of the Hartlib Circle had experimented with them.[55] By the late seventeenth century, a network of seedsmen were regularly funneling grass and fodder crop seeds to English farmers from the Netherlands; by the early eighteenth, seedsmen had developed cheaper, local sources of supply (though clover was still imported from Flanders, where mechanized threshing methods provided seedsmen with a competitive advantage).[56] Enclosures, which during the low grain prices of the late seventeenth and early eighteenth centuries often resulted in the conversion of arable land to grazing land, expanded these markets; to accelerate their transformation, formerly ploughed lands could be sown with a mixture of hayloft sweepings and imported artificial grass seed.[57] As Ellis published, purchases of new grasses were accelerating further, perhaps in response to the fodder crisis of 1740.[58]

Ellis clearly hoped to join these networks of exchange and to profit from the expectation of novelty. He sold the seeds of his grasses directly to his readers, claiming, "I am their first Discoverer, and the two first sorts are sold by none but myself in England." With them he sold, for one guinea, the recipe for "Mr. Livings's Manure," which, applied to the grasses, would "not only cause Ewes to have a Propensity to take Ram, but [would] assuredly produce the sweetest Butter, Cheese, and Flesh."[59] On making inquiries, Kalm cast some doubt on Ellis's claims to discovery, showing that Ellis's "lady's finger grass," or *Lotus corniculatus*, was used by his scornful neighbors in Hertfordshire and known there under the same folk name.[60] Here as elsewhere perhaps Ellis capitalized on the unevenness of British agricultural practice—converting the knowledge and techniques of his neighbors into "discoveries," a practice that would only accelerate in the latter half of the century.

[53] Kalm, *Visit to England* (cit. n. 2), 30, 58.

[54] Ibid., 152.

[55] Joan Thirsk, *Agrarian History of England and Wales*, vol. 3 (Cambridge, 1967), 555. For a comprehensive history of artificial grasses, see Ambrosoli, *The Wild and the Sown* (cit. n. 9).

[56] Thirsk, *Agrarian History* (cit. n. 55), 529–31.

[57] Overton, *Agricultural Revolution* (cit. n. 11), 35–7; Allen, "Tracking" (cit. n. 9), 32. On artificial grass seed trade in Britain in the eighteenth century, see Ambrosoli, *The Wild and the Sown* (cit. n. 9), 262–96.

[58] Thirsk, *Agrarian History* (cit. n. 55), 117.

[59] Ellis, *Compleat System* (cit. n. 1), 297.

[60] Kalm, *Visit to England* (cit. n. 2), 225.

* * *

If it seems jarring that a story that started with sheep aphrodisiacs has morphed gradually into disquisitions on fencing and the seasonality and composition of grass (even aphrodisiac honeysuckle grass), it is worth recalling a point about early eighteenth-century bodily knowledge that has been important to historians of both racial theory and breeding: both humans and domesticated animals were expected to derive their physical qualities from their environments. Indeed, sheep were considered particularly sensitive to place; the *Dictionarium Rusticum* noted, typically, "Tis observable, that fat Pastures breed straight and tall *Sheep*; but Hills and Short Pastures broad and square ones."[61] Such ideas are often categorized as "environmental determinism" and, in the context of the rise of the idea of inheritance, as obsolete racial and breeding theories, supplanted by later theories of stable blood.

Such a characterization, I would argue, is misleading. First, the label "environmental determinism" implicitly turns breeds and races into natural kinds, the consequences of fixed environments. Tracing meat production processes, however, reminds us that agricultural environments were being deliberately and profoundly altered to change the bodies they contained. "Fat pastures" were products of knowledge that spread through texts, markets, and formal and informal networks of botanical exchange. They were shaped by forms of expertise in nourishment and herbal medicine that the focus on selection makes invisible. The story of aphrodisiacs gives us an entryway into these forms of expertise, the pathways of plants, practices, and ideas on which they depended, and the ideas about human bodies that they drew on and influenced.

Second, while theories of environmental influence on inherited characteristics certainly lost power across the nineteenth and twentieth centuries, the story of meat production shows that the structuring of environments to reshape bodies has remained a primary focus of knowledge making in animal husbandry. It occupies the vast majority of the everyday work, attention, and knowledge of farmers and agricultural laborers. Later textual records tend to confirm this; even as selective breeding has taken pride of place in the first chapters of manuals of animal husbandry, the manipulation of environments—confining animals or relaxing their constraints, modifying their foods, changing their access to light, minimizing their exposure to disease, and orchestrating their encounters with each other—has continued to make up their bulk.

Like eighteenth-century breeding knowledge, these forms of environmental knowledge have had material consequences. Just as breeders created cattle and horses whose descendants dominate modern industrial populations, the environments developed by eighteenth-century feeders have left enormous, if not immediately obvious, physical traces. We can see the outlines of the fences and pastures of the eighteenth century very clearly in, for example, the sheep-centered landscapes of New Zealand. Walking contemporary fields, Pehr Kalm would catalog artificial grasses in temperate climates around the globe. Near my home in Pennsylvania, Ellis's honeysuckle grass is on the list of invasive species, and alfalfa, spread in the eighteenth century as "lucerne," is a cash crop.[62] Because in the twenty-first century domesticated mammals literally and vastly

[61] Bailey, *Dictionarium Rusticum* (cit. n. 19), "Shepherd's Observations."

[62] Recent work on the history of grass includes Tom Brooking and Eric Pawson, "Silences of Grass: Retrieving the Role of Pasture Plants in the Development of New Zealand and the British Empire,"

outweigh wild ones, the spread of the landscapes and techniques developed to feed them is not a trivial historical development.[63]

The long links that bind eighteenth-century forms of environmental manipulation to the bright lights, pens, incubating chambers, vaccines, and ventilation shafts of modern feeding operations have, for the most part, not been traced. Perhaps this is because of the naturalness we retrospectively attribute to eighteenth-century pastures. Some connections are not difficult to see: debates about the timing of sheep ovulation, for example, still weigh the relative importance of feeding, photoperiodicity, and the "ram effect" on ewe seasonality. Others require more drawing out, though we can perhaps begin to see, in the close pastures and warm house-lamb pens of the eighteenth century, ancestors of, say, the gestation crates of modern pig production. We can certainly see elaborate interventions in intercourse in those same crates, where sows are now sexually stimulated by drugs and human workers to make them more receptive to artificial insemination, and where antibestiality laws must carefully avoid outlawing what have become standard practices of meat production.[64]

If we are looking for the effects of such techniques of bodily control on what Nikolas Rose, studying biocapital, has called "moral economies of life and self," we have farther to go. Certainly to find these connections we should look further back than the late twentieth or early twenty-first centuries, periods in which farm animals have become invisible rather than ubiquitous, obscured by declining agricultural populations in industrialized states, by the shifting of many meat-production operations indoors, and by the secrecy measures, most visibly the "Ag-Gag" laws enacted in the United States, that meat producers use to conceal even their most standard practices. That accounts of meat production are often called "exposures" helps us gauge the extent to which animals have been withdrawn from everyday life and conversation. When the United States and Western Europe still maintained large agricultural populations, domesticated animals were public bodies, open to discussion and generally recognized manipulation, which could be used to confirm or develop both vernacular and learned ideas about human bodies.

J. Imperial Commonwealth Hist. 35 (2007): 417–35; Jeremy Vetter, "Capitalizing on Grass: The Science of Agrostology and the Sustainability of Ranching in the American West," *Sci. Cult.* 19 (2010): 483–507; Maura Capps, "Fleets of Fodder: The Ecological Orchestration of Agrarian Improvement in New South Wales and the Cape of Good Hope, 1780–1830," *J. Brit. Stud.* 56 (2017): 532–56; Albert G. Way, "'A Cosmopolitan Weed of the World': Following Bermudagrass," *Agr. Hist.* 88 (2014): 354–67; H. J. D. Rosa and M. J. Bryant, "The 'Ram Effect' as a Way of Modifying the Reproductive Activity in the Ewe," *Small Ruminant Res.* 45 (2002): 1–16; Graeme B. Martin and Hiroya Kadokawa, "'Clean, Green and Ethical' Animal Production Case Study: Reproductive Efficiency in Small Ruminants," *J. Reproduction Develop.* 52 (2006): 145–52; M. Balasse and A. Tresset, "Environmental Constraints on the Reproductive Activity of Domestic Sheep and Cattle: What Latitude for the Herder?," *Anthropozoologica* 42 (2007): 71–88.

[63] Vaclav Smil, *The Earth's Biosphere: Evolution, Dynamics, and Change* (Cambridge, Mass., 2003), 186.

[64] See the articles in Sarah Wilmot, ed., "Between the Farm and the Clinic: Agriculture and Reproductive Technology in the Twentieth Century," special issue, *Stud. Hist. Phil. Biol. Biomed. Sci.* 38 (2007), esp. Wilmot, "From 'Public Service' to Artificial Insemination: Animal Breeding Science and Reproductive Research in Early Twentieth-Century Britain," 411–41; Gabriel N. Rosenberg, "A Race Suicide among the Hogs: The Biopolitics of Pork in the United States, 1865–1930," *Amer. Quart.* 68 (2016): 49–73, on 60–4; Alexander David Blanchette, "Conceiving Porkopolis: The Production of Life on the American 'Factory' Farm" (PhD diss., Univ. of Chicago, 2013); Jeannette Vaught, "Animal Sex Work," *Platypus: The Castac Blog*, 15 June 2016, http://blog.castac.org/2016/06/animal-sex-work/.

If following inheritance reminds us to acknowledge the knowledge of breeders, following the story of desire and environmental manipulation in the eighteenth century reminds us to look for other potential sources of bodily expertise: the nutritional and behavioral knowledge of graziers and shepherds, for example, or the anatomical knowledge of butchers. We might look for their effects, not only in botany or theories of sexual reproduction, but also in continuing theories of instinct and discipline, desire, vigor, seasonality, and growth. We might examine the effect of widespread animal feeding regimens on ideas of human fatness and nourishment, bodily growth, and industry. We might examine how ideas about embodied masculinity were shaped by the twinned practices of stud rental and normalized castration.

Paying attention to these processes of production will perhaps help us see a different face of capitalist knowledge making. The production of biocapital, like the practices of taxonomy and commodification as a whole, requires a process of abstraction, the rendering of living organisms as names, codes, and exchangeable goods. Historians of science, unsurprisingly, have been quick to see the significance of abstraction. However, the intensified creation, feeding, and slaughter of bodies pushed agricultural workers and agricultural experts not only into complex calculations of profits and newer forms of quantification, not only into markets for commodified blood, but also into more and more intense forms of manipulation. It was not distancing, for example, when eighteenth-century shepherds, encouraged by the demand for lamb, established new bonds between orphaned lambs and bereft ewes by strapping the strange lamb into the flayed skin of its predecessor, based on a profit-oriented understanding of "maternal affection." As a continuing practice in sheep husbandry, it is not distancing now, either.[65] As we continue to track the knowledge that comes from the distancing effects of capitalist farming, the knowledge emerging from its often grisly intimacies requires our attention as well.

[65] Edward O. Price, *Principles and Applications of Domestic Animal Behavior* (Cambridge, 2008), 150.

Sugar Machines and the Fragile Infrastructure of Commodities in the Nineteenth Century

*by David Singerman**

ABSTRACT

This essay uses sugar machinery to explore the fragile infrastructure that allowed global commodity traffic to emerge. In the nineteenth century, the cane sugar industry transformed the Caribbean, the Hawaiian Islands, and much of the rest of the tropical world. Observers then and now tied sugar's revolutionary power to the invention and spread of advanced mechanical technologies. Yet the origins and lives of those machines themselves have remained obscure. The superficially effortless circulation of standardized material goods like sugar depended on carefully cultivated systems for managing people, paper, objects, and knowledge—and such things could not be standardized so easily.

INTRODUCTION

Sugar is both a central commodity in the history of capitalism and a model for students of that history.[1] As with any natural substance, understanding how capitalism transforms sugar into a commodity means understanding the technologies that make such commodification possible. Among commodities, however, sugar is distinctive in that sophisticated machinery has been its particularly close companion for hundreds of years.

The recognizable shape of the early modern Caribbean plantation emerged during the "sugar revolution" of the eighteenth century, when Europe turned whole islands

* Corcoran Department of History and Program in American Studies, Nau Hall–South Lawn, University of Virginia, Charlottesville, VA 22904; ds2ax@virginia.edu.

A long list of generous archivists made this work possible. Particular thanks to Alma Topen, Rachael Egan, William Bill, Claire Daniel, Colin Vernall, Emma Yan, Claire Patterson, and George Gardner at the University of Glasgow; to Dr. Anne Cameron at the University of Strathclyde; to Susannah Waters at the Glasgow School of Art; and to Juan Carlos Román at the Archivo General de Puerto Rico. I am grateful also to the staffs of the Mitchell Library in Glasgow, the University of Glasgow Centre for Business History, the Instituto de Estudios del Caribe in San Juan, and the Massachusetts Historical Society in Boston. Humberto García Muñiz, Ray Stokes, and Frances Robertson were especially supportive. And of course thanks to William Deringer, Lukas Rieppel, and Eugenia Lean for inviting this essay, and to the editors, editorial staff, and reviewers for *Osiris*.

[1] Sidney Mintz, *Sweetness and Power: The Place of Sugar in Modern History* (New York, 1985).

into sugar factories in order to slake its own addiction.[2] Planters designed their properties around the fact that the juice of the sugarcane, once cut, begins to ferment within a day. So plantations became what Sidney Mintz described as a "synthesis of field and factory" in which industry worked to agricultural rhythms and agriculture arranged itself to efficiently supply machines.[3] At the center of the plantation, no more than a day's carriage from the field, were the mill and the boiling-house, and inside them the plantation's large and costly technological objects: rollers, furnaces, cauldrons, troughs, and clay molds for drying the wet sugar.

From the beginning of the seventeenth century, sugar planters and allied natural philosophers tried hard to portray the process of sugar manufacture as mechanical rather than artful, driven by European ingenuity rather than enslaved people's know-how. In fact, as historians have long emphasized, and as contemporaries grumblingly acknowledged, making sugar was a complex art that required all the skill and craft of enslaved people, not merely their labor.[4] More recently, historians such as Eric Otremba and Daniel Rood have demonstrated that slaves were frequently responsible for improvements to machinery—improvements for which planters and savants inevitably claimed credit.[5]

In the nineteenth century, machinery was again key to a second sugar revolution that reconfigured the relationships between field and factory and between labor and capital. Whereas in 1800 all the sugar mills of the world were powered by water, wind, or animals, over the following decades European and North American firms began to produce mills driven by steam. With steam power came other machines, like centrifuges and sealed vacuum chambers, many of which were adaptations of devices already in use by beet-sugar producers in Europe. By 1860, steam drove 70 percent of the mills in Cuba, the best-capitalized and most technically advanced sugar producer.[6] Over the following decades, plantation estates consolidated into larger agricultural units built around huge factories, and new factories were constructed in undeveloped regions of the Caribbean and across the world. Sugar, purer and more consistent than ever before, glutted the world market. By the end of the century, slavery had been overthrown or abolished, replaced (at least de jure) by other systems of labor.[7]

This second revolution has, like the first, been analyzed in largely technological and deterministic terms. The intrinsic economies of scale of these new technologies,

[2] B. W. Higman, "The Sugar Revolution," *Econ. Hist. Rev.* 53 (2000): 213–36, on 213.

[3] Mintz, *Sweetness and Power* (cit. n. 1), 47.

[4] See, e.g., Richard S. Dunn, *Sugar and Slaves: The Rise of the Planter Class in the English West Indies, 1624–1713* (Chapel Hill, N.C., 1972); Manuel Moreno Fraginals, *The Sugarmill: The Socioeconomic Complex of Sugar in Cuba, 1760–1860*, trans. Cedric Belfrage (New York, 1976).

[5] Eric Otremba, "Inventing Ingenios: Experimental Philosophy and the Secret Sugar-Makers of the Seventeenth-Century Atlantic," *Hist. & Tech.* 28 (2012): 119–47; Daniel B. Rood, *The Reinvention of Atlantic Slavery: Technology, Labor, Race, and Capitalism in the Greater Caribbean*, 1st ed. (New York, 2017).

[6] Jonathan Curry-Machado, *Cuban Sugar Industry: Transnational Networks and Engineering Migrants in Mid-Nineteenth Century Cuba* (New York, 2011).

[7] The Cuban historian Manuel Moreno Fraginals argued that the island's planter elite abolished slavery in the late nineteenth century because slavery as a system, and slaves as a class, were incompatible with those planters' desires for mechanized and technically sophisticated production. Moreno Fraginals, *The Sugarmill* (cit. n. 4), 112; Moreno Fraginals, "Plantations in the Caribbean: Cuba, Puerto Rico, and the Dominican Republic in the Late Nineteenth Century," in *Between Slavery and Free Labor: The Spanish-Speaking Caribbean in the Nineteenth Century*, ed. Moreno Fraginals, Frank Moya Pons, and Stanley L. Engerman (Baltimore, 1985), 3–21.

combined with their higher cost, led to the "rationalization" of sugar production in the form of "central factories" that could take advantage of those efficiencies by processing much more cane from a much larger area.[8] From the seventeenth century to the twentieth, therefore, machines have been central to historians' understanding of sugar economies. Yet even as historians have made so much of the new technologies of sugar production, they have remained largely incurious about the machines themselves or who made them. This essay begins to tell the story of sugar machines in order to expose the fragile infrastructure upon which familiar Atlantic commodity networks were built.

The industrialization of sugar production affords an opportunity to incorporate the workshops of empire into the empire of sugar. Such workshops might be in England, or the United States, or France.[9] But mostly they were in Glasgow, Scotland. The city of Glasgow rose to mercantile prominence in the eighteenth century with the West India trade, and it became the heavy-engineering center of the nineteenth century after those merchants invested in mining, cotton, railroads, and shipbuilding.[10] And through 1914 Glasgow on its own produced 80 percent of the world's sugar manufacturing equipment.[11] Mirrlees Watson, perhaps the world's largest maker, dispatched over 2,000 mills from its 800-strong Glasgow workshop, "one of the most splendid and completely equipped engineering establishments" in Britain by the 1890s.[12] Yet for all its contemporary fame in the sugar economy—in 1878, the British consul in San Juan reported that "such well-known names as Tait and Mirelees [*sic*] and Buchanan are to be seen on scores of sugar plantations"—Glasgow has remained largely absent from historians' analyses of that economy.[13] When the city's manufacturers are mentioned, it is usually just to use the city's engineering reputation to add heft to claims about the technical sophistication of a new sugar factory.[14]

[8] Dale Tomich, "Commodity Frontiers, Spatial Economy, and Technological Innovation in the Caribbean Sugar Industry, 1783–1878," in *The Caribbean and the Atlantic World Economy: Circuits of Trade, Money and Knowledge, 1650–1914*, ed. Adrian Leonard and David Pretel, Cambridge Imperial and Post-Colonial Studies (New York, 2015), 184–216; J. H. Galloway, *The Sugar Cane Industry: An Historical Geography from Its Origins to 1914* (Cambridge, 1989), 139.

[9] Nadia Fernández-de-Pinedo and David Pretel, "Circuits of Knowledge: Foreign Technology and Transnational Expertise in Nineteenth-Century Cuba," in Leonard and Pretel, *The Caribbean* (cit. n. 8), 263–89, table 12.1, on 269.

[10] Michael S. Moss and John R. Hume, *Workshop of the British Empire: Engineering and Shipbuilding in the West of Scotland* (London, 1977), 3–5; T. M. Devine, "The Golden Age of Tobacco," in *Glasgow*, ed. Devine and Gordon Jackson (Manchester, 1995), 139–83; Richard Pares, "A London West India Merchant House 1740–69," in *The Historian's Business, and Other Essays*, ed. R. A. Humphreys and Elisabeth Humphreys (Oxford, 1961), 198–226, on 210.

[11] Fernández-de-Pinedo and Pretel ("Circuits of Knowledge" [cit. n. 9], 269) show that the large majority of sugar equipment came from Britain between 1844 and 1857, although during that period slightly more (264 vs. 233) grinding mills came from the United States.

[12] John Mayer, *Notices of Some of the Principal Manufacturers of the West of Scotland* (Glasgow, 1876), 66, 118.

[13] "Report by Consul Bidwell on the Trade and Commerce of the Island of Porto Rico for the Year 1878," in *Puerto Rico en La Mirada Extranjera: La Correspondencia de Los Cónsules Norteamericanos, Franceses E Ingleses, 1869–1900*, ed. Gervasio L. García Rodríguez and Emma Aurora Dávila Cox (Río Piedras, Puerto Rico, 2005), 165; Andrés Ramos Mattei, "Technical Innovations and Social Change in the Sugar Industry of Puerto Rico," in Moreno Fraginals, Moya Pons, and Engerman, *Between Slavery and Free Labor* (cit. n. 7), 158–78.

[14] The index for J. H. Galloway's *Sugar Cane Industry* does not even include an entry for "Glasgow," and while that for Noël Deerr's standard and encyclopedic *History of Sugar* does mention the names of specific machinery producers, it includes only an entry for Glasgow "as refining centre." Galloway, *Sugar Cane Industry* (cit. n. 8), 261; Noël Deerr, *The History of Sugar*, 2 vols. (London, 1949), 2:610.

For historians of science and capitalism, however, Glasgow's engineering workshops provide an opportunity to study just what forms of knowledge it takes to produce "standard" commodities on a global scale. It would be reductive to declare the relationship between sugar, on the one hand, and the machinery for making it, on the other, to be a case for the history of technology rather than the history of science. Historians of science have long accepted that factories and plantations, as well as a wide variety of spaces of exchange, are zones where knowledge can be produced just as much as in a laboratory.[15] Those who build the infrastructure for knowledge production in the marketplace merit no less scrutiny than do makers of scientific instruments. This is especially true in the sugar market for two reasons. First, as this essay shows, building industrial machines for sugar depended on networks and skills of information management that were just as complex as any instrument builder's. Second, the commodity status of nineteenth-century sugar depended more heavily on modern science than any comparable natural object. Yet the more sugar seemed to become immutable and mobile, the more that machines for making it became customized and unruly. The circulation of standardized natural goods—superficially effortless—in fact depended entirely on deeper currents of people, objects, and knowledge that could not be standardized so easily.

Over the course of the nineteenth century, sugar acquired its modern reputation as a nearly ideal commodity. The Cuban sociologist Fernando Ortiz could write in 1940, for instance, of the "equal chemical and economic standing of all the sugars of the world, which, if they are pure, sweeten, nourish, and are worth the same."[16] Mechanically processed, sugar seemed to effortlessly transgress social and even natural boundaries: it "is born brown and whitens itself; at first it is a syrupy mulatto and in this state pleases the common taste; then it is bleached and refined until it can pass for white [and] travel all over the world."[17] But just like the act of passing, there was nothing easy about the global travel of pure sugar, nor anything natural about the categories that shaped it. Instead, sugar's commodification required constant labor on the part of factory workers, chemists, and engineers, just as its production had required the skills of slaves not long before.[18] Even so, the labor that produced sugar from plants was subjected to the de-skilling and standardizing pressures of capital, pressures akin to those facing workers in late eighteenth-century shipyards or twentieth-century manufacturing.[19]

James Secord and Kapil Raj, among others, have called for histories of science in transit, lest an emphasis on the specific local cultures of scientific knowledge can distract from the ultimate objective of understanding how knowledge moves beyond that culture.[20] Efforts to standardize practices, especially in the nineteenth century, fre-

[15] Harold J. Cook, *Matters of Exchange: Commerce, Medicine, and Science in the Dutch Golden Age* (New Haven, Conn., 2007).

[16] Fernando Ortiz, *Cuban Counterpoint: Tobacco and Sugar*, trans. Harriet De Onís (Durham, N.C., 1995), 24–5.

[17] Ibid., 9.

[18] David Roth Singerman, "The Limits of Chemical Control in the Caribbean Sugar Factory," *Radic. Hist. Rev.* 2017, no. 127 (2017): 39–61.

[19] William J. Ashworth, "'System of Terror': Samuel Bentham, Accountability and Dockyard Reform during the Napoleonic Wars," *Soc. Hist.* 23 (1998): 63–79; David F. Noble, *Forces of Production: A Social History of Industrial Automation*, 1st ed. (New York, 1984), chap. 1.

[20] James A. Secord, "Knowledge in Transit," *Isis* 95 (2004): 654–72; Kapil Raj, *Relocating Modern Science: Circulation and the Construction of Knowledge in South Asia and Europe, 1650–1900* (New York, 2007).

quently involved the circulation of standard objects. Precision devices only travel fitfully and after much social work has smoothed their way, but that travel is what makes it possible to claim that knowledge is universal.[21] And likewise, historians have supposed that the distribution of standard objects and devices has made the easy travel of commodities and other artifacts possible. Those standards sometimes take exemplary form, such as jars of sugar or samples of cotton, and sometimes the form of inspection procedures and instruments.[22] The "highly sophisticated machinery" that filled central factories in the late nineteenth century made possible the efforts of factory owners to standardize their workers' labor.[23]

Like Courtney Fullilove's plant genes, however, these sugar machines were fundamental to a standardized commodity precisely because they resisted standardization themselves.[24] Moreover, factory owners' efforts to de-skill labor depended on that idiosyncrasy, on machines tailored by Glaswegian manufacturers to their sugar factory's exact requirements. Well into the twentieth century, as this essay shows, sugar machinery was designed and maintained through individualized systems of design, irregular channels of communication, and the ability to recontextualize people and things. Glasgow's sugar machinery should join uranium and financial derivatives in reminding us that human craft, skill, and labor are impossible to eliminate in the infrastructures of commodity production. They can only be shuffled out of sight.[25]

PEOPLE

Sometime around 1800, one of Glasgow's West India sugar merchants placed an order for a sugar mill with a flax millwright named James Cook. By 1805, Cook had so many orders that he was forced to move his workshop to a larger site on the south bank of the Clyde, and within two decades Cook's works were among the city's leading industrial operations.[26] So many engineers were trained there, wrote one industry observer a century later, that Cook's works became known as "The College." His firm anchored a cluster of sugar-engineering works on the south side of the city.[27]

These firms, like the others that would dominate Glasgow's sugar-machinery business, all began as joint partnerships. Some partners contributed capital, others engineering expertise, and others familiarity with or connections to the West India merchant class.[28] In the mid-1860s, for instance, the young engineer Duncan Stewart was

[21] Marie-Noëlle Bourguet, Christian Licoppe, and Heinz Otto Sibum, "Introduction," in *Instruments, Travel and Science: Itineraries of Precision from the Seventeenth to the Twentieth Century* (London, 2002), 1–19, esp. 3–9; see also M. Norton Wise, ed., *The Values of Precision* (Princeton, N.J., 1995); Joseph O'Connell, "Metrology: The Creation of Universality by the Circulation of Particulars," *Soc. Stud. Sci.* 23 (1993): 129–73.

[22] Michel Callon, Cécile Méadel, and Vololona Rabeharisoa, "The Economy of Qualities," *Econ. & Soc.* 31 (2002): 194–217, esp. 198–9; Ken Alder, "Making Things the Same: Representation, Tolerance, and the End of the Ancien Regime in France," *Soc. Stud. Sci.* 28 (1998): 499–545.

[23] Moreno Fraginals, "Plantations" (cit. n. 7), 8.

[24] Courtney Fullilove, "Microbiology and the Imperatives of Capital in International Agro-Biodiversity Preservation Projects," in this volume.

[25] Gabrielle Hecht, *Being Nuclear: Africans and the Global Uranium Trade* (Cambridge, Mass., 2012); Vincent Antonin Lépinay, *Codes of Finance: Engineering Derivatives in a Global Bank* (Princeton, N.J., 2011).

[26] Robert Harvey, *Early Days of Engineering in Glasgow* (Glasgow, 1919), 8.

[27] Michael Pacione, *Glasgow: The Socio-spatial Development of the City*, World Cities Series (Chichester, 1995), 75–7; Harvey, *Early Days* (cit. n. 26), 9.

[28] S. G. Checkland and Anthony Slaven, eds., *Dictionary of Scottish Business Biography, 1860–1960* (Aberdeen, 1986), 188–9.

offered the position of chief engineer by a Glasgow merchant firm that owned seven Demerara estates.[29] Although he had recently started his own London Road Ironworks, Stewart accepted the appointment in order to "master the process of making sugar from the cane," which he could not do just by observing Scottish refineries. When he returned to Glasgow, a biographer wrote later, he brought "wider technical knowledge which proved to him of greater worth than the mere possession of capital."[30]

The "technical knowledge" that Stewart brought back from his years overseas incorporated the whole complex of sugar production, human as well as mechanical. His reputation was built on a series of patents for hydraulic rams that regulated the pressure a mill's rollers applied to the cane.[31] Early modern sugar mills were notorious for maiming or killing human beings, but in a steam mill it was far less likely that a body part of a worker would get caught. Instead, Stewart set to work on his hydraulic cylinders to prevent damage from an overload of cane. Just as Lee Vinsel shows elsewhere in this volume how the auto industry disclaimed responsibility for protecting against "accident-prone" drivers,[32] so too the management of sugar factories invariably blamed a broken roller on the "injudicious and unskilful [*sic*]" workers. An "accident," in sugar engineering parlance, switched from meaning an injury to a human to an injury to a machine. This explains why Stewart's patented hydraulics were called a "safety device": not because they prevented a person from being milled but because they ensured the safety of the capital investment in machinery.[33]

The engineers who traveled back and forth between the Caribbean and Britain played crucial roles as go-betweens, mediating the relationship between sugar estates and the firms that manufactured their machines.[34] Cheaper and faster oceanic travel meant these engineers could easily spend the grinding season supervising a factory in the Caribbean, and the rest of the year elsewhere, sometimes in the cane fields of Louisiana, which harvested their cane on a different schedule, or in beet-sugar factories. Boosters of Glasgow industry argued that these engineers' voyages provided unusually direct experience with the processes of sugar making: "By their long connection with the trade, and especially from the circumstance that principals of several of the firms engaged in it have travelled in the sugar-growing countries of both hemispheres, they have come face to face with the planters upon their estates, and acquired an exact acquaintance with their wants." Through such careful management of foreign customers, the sugar-machinery business "like the sugar plantations themselves . . . has

[29] T. M. Devine, "An Eighteenth-Century Business Élite: Glasgow-West India Merchants, C. 1750–1815," *Scot. Hist. Rev.* 57 (1978): 40–67.

[30] "The Implement and Machinery Review," 1 September 1886, in "Newspaper cuttings," TD185/5, Records of Messrs. Duncan Stewart & Co., Mitchell Library, Glasgow.

[31] "The story of Duncan Stewarts [sic], and their association with the British beet sugar industry," address given by Mr. K. S. Arnold at the General Managers' Annual Conference of the British Sugar Corporation at Peterborough, 1 June 1955, University of Glasgow Archives & Special Collections, Records of Duncan Stewart & Co., Ltd., GB 248 UGD 052/1/4/3, University of Glasgow Archives & Special Collections (repository no. GB 248), Glasgow (hereafter cited as "UGA").

[32] Lee Vinsel, "'Safe Driving Depends on the Man at the Wheel': Psychologists and the Subject of Auto Safety, 1920–55," in this volume.

[33] A. J. Wallis-Tayler, *Sugar Machinery: A Descriptive Treatise Devoted to the Machinery and Processes Used in the Manufacture of Cane and Beet Sugars* (London, 1895), 36.

[34] For the entrepreneurial dispersal of British engineers overseas, though neglecting the kinds of networks that produced the movement of engineers in sugar, see R. A. Buchanan, "The Diaspora of British Engineering," *Tech. & Cult.* 27 (1986): 501–24.

been anxiously, intelligently, and enterprisingly cultivated."[35] One engineer, for instance, "began his life of wandering" immediately upon joining Duncan Stewart's company as an apprentice in 1876, a journey that took him to the company's clients in Argentina, Barbados, and Cuba, as well as to its displays at international exhibitions.[36] Most of the machinery makers in Glasgow's other engineering industries did not need to travel such distances to see their products in action.

Planters came to rely on traveling engineers as crucial intermediaries with the firms from whom they wished to order new equipment or replacement parts. Purchasing agents who spent most of their time in Britain were better placed to communicate with engineering firms, but too far from the plantation to know what needed ordering. By contrast, itinerant engineers were uniquely positioned to understand how sugar factories actually worked.[37] While many such travelers were clearly employed by a particular British firm, plenty of others exploited oceanic distance. They might, for instance, maintain a consulting position for a Glasgow firm while still working for one or more sugar estates. Not infrequently an engineer would earn a commission from the plantation for ordering a machine from a firm and then be hired by the same firm to install that very machine.

W. H. Ross, an engineer who placed orders for many Cuban estates and was a confidant to their American owners, traveled back and forth between customers in Havana and manufacturers in New York, Glasgow, and Liverpool.[38] For instance, Edwin Atkins, the Bostonian owner of several Cuban factories, asked Ross's "opinion of the approximate value of a new mill del[ivere]d at Cienfuegos . . . made by Tate Mirless & Watson [*sic*]" and left much up to Ross himself to decide.[39] Shortly after he had purchased a plantation called Soledad, Atkins wrote Ross "giving order for a mill for regrinding on Soledad, you will note that we leave the matter in its detail pretty much to your judgement [*sic*], as I am sure you will take [care to] give us only the best; and your long experience in these matters make[s] your judgment of value." What made Ross valuable was not only his experience with sugar in general but with Atkins's new plantation in particular: "You know the kind of grinding we aim at doing, and trust you will govern your work in accordance."[40] To maintain himself in Mirrlees Watson's good graces, Ross asked Atkins for data about the performance of the sugar boiling equipment, which Atkins allowed Ross to share with the manufacturer.[41] Those who set up shop in Britain, trying to manage Caribbean affairs remotely, did not have access either to orders or to the currency of Atlantic information; as a result, they frequently found themselves without clients.[42]

In 1883, the Manchester merchants N. P. Nathan's & Sons purchased a full sugar-production arrangement from Duncan Stewart for a site in the Canary Islands. That

[35] Mayer, *Principal Manufacturers* (cit. n. 12), 116.

[36] *Beardmore News*, November 1923, Records of Duncan Stewart & Co., Ltd., UGD 052/1/4/2, UGA.

[37] Curry-Machado, *Cuban Sugar Industry* (cit. n. 6), chap. 3.

[38] Mirrlees Watson Order Book, 1891, Job no. 386, Records of Mirrlees Watson Co., Ltd., sugar machinery manufacturers, Glasgow, Scotland, UGD 118/2/4/10, UGA. On W. H. Ross, see Fernández-de-Pinedo and Pretel, "Circuits of Knowledge" (cit. n. 9), 273.

[39] Edwin Atkins to W. H. Ross Esq., 22 August 1885, II.6.287, Atkins Family Papers, Massachusetts Historical Society, Boston (hereafter cited as "MHS").

[40] Atkins to Ross, 17 April 1888, II.7.394, Atkins Family Papers, MHS.

[41] Atkins to Ross, n.d., II.8.260, and 14 August 1889, II.8.287–88, Atkins Family Papers, MHS.

[42] Curry-Machado, *Cuban Sugar Industry* (cit. n. 6), 87–90.

price included everything needed for a factory: a horizontal engine with double gearing, a sugar mill with hydraulics, a fifty-foot cane carrier, clarifiers, filter presses, a triple-effect evaporator, vacuum pan, four centrifugals, and molasses tanks, plus piping, mounting, staging, spares, and tools for erecting, all of which cost £7,304.[43] After the parts had been produced in the works, the entire machine was assembled next door in the erecting shop, tested as much as possible without any sugar, and photographed. Then it was disassembled, packed, and shipped, along with erecting tracings and photographs (fig. 1).[44] For the manufacturer, those photographs served as insurance against future claims and, along with the tracings, were used to reconstruct the machines after delivery. Unlike the iconic TEA (transversely excited atmospheric) lasers and air pumps of science studies, these reconstructions of sugar machines were attempts to put together a fixed set of existing parts, not build copies from scratch based on two-dimensional representations or textual descriptions.[45] But clearly the engineers at Duncan Stewart believed paper insufficient, and the tacit skills of knowledgeable people necessary. One of the firm's engineers, Robert Gilbert, accompanied the Nathan's order to the Canaries to supervise.

Although he was Duncan Stewart's eyes and hands, Gilbert was legally an independent contractor. In November 1883, he signed a contract directly with Nathan's, witnessed by the manager of Duncan Stewart and the company clerk. The main task for which he had been hired, at £5 a week plus "suitable Board and Lodging," was to supervise "the erection of and fitting and putting into working order, of Cane Crushing and Evaporating Machinery."[46] Once in Las Palmas, Gilbert was responsible for the machinery, including its unloading from the ship.

Individuals such as Gilbert, capable of supervising the construction of a whole sugar factory, were in short supply. The terms of his contract as a traveling erecting engineer emphasize the difficulty of the task. He was to obey orders from Nathan's, their manager in the Canaries, or their attorneys. At the same time, Gilbert was also obliged to "devote his whole time and attention to and employ his whole art and skill in the said erection and fitting up of the said machinery, and in getting the same in to proper and satisfactory working order."[47] Gilbert could not rely on quick or frequent communication with those who had designed the machines, and at the same time, his erstwhile employers were stifled in their efforts to understand what was going on.

Contracted erecting engineers sent as agents of a firm did not necessarily keep their principals apprised of their activities. In July 1884 Gilbert was still in the Canaries, and the personnel of Duncan Stewart were not even sure of how the work was progressing. He had not kept in close touch. The manager of the works in Glasgow asked Gilbert to "kindly write to us on receipt of this and let us know when you expect to come home." Not only did they want to know how his mission was going, but they needed him back in Glasgow so they could ship him off again—this time on a mission

[43] Duncan Stewart Cost Book 1881–1901, p. 18, UGD 052/1/1/1, UGA.

[44] It is not clear when firms began using progress photographs; vol. 1 of A. & W. Smith's photograph albums is dated 1907. UGD 118/1/5/1, UGA.

[45] Harry Collins, "The TEA Set: Tacit Knowledge and Scientific Networks," *Sci. Stud.* 4 (1974): 165–85; Steven Shapin and Simon Schaffer, *Leviathan and the Air-Pump: Hobbes, Boyle, and the Experimental Life* (Princeton, N.J., 1985), 229.

[46] "Copy minute of agreement between N. P. Nathan's Sons and Robert Gilbert (Nov. 1883)—Forbes and Bryson, writers, Glasgow," UGD 052/1/7/2, UGA.

[47] Ibid.

Figure 1. *A. & W. Smith centrifugals, 1915, with note affixed: "this photograph should be given to the engineer who is to erect the machines." Records of Mirrlees Watson Co., Ltd. (cit. n. 38), UGD 118/1/51, UGA.*

to Brazil, accompanying "a large Sugar Plant which will be ready for shipment early next month."[48] He missed that trip. He was still in the Canaries in mid-September, and the sugar factory was still not finished, awaiting replacement parts from Duncan Stewart. "We have sent as much of the details as you have ordered," the company wrote, "as we possibly could in the short time they gave us before the ship sailed." The tyranny of distance worked both ways, frustrating both Gilbert's attempts to get his parts and Duncan Stewart's attempts to recall him.[49]

A year and a half after he had left Glasgow, in May 1885, Gilbert was still in the Canaries and still waiting for parts. They had been shipped, at least, just in time to be installed for the next grinding season. "We hope therefore all will reach you in good time so that you may be able to start the factory to [crush] the greater part of the crop satisfactorily," the firm wrote. Despite the delays and frustration, however, the presence of an engineer like Gilbert could serve a firm's interests. While twiddling his

[48] David. H. Andrew to Robert Gilbert, 7 July 1884, UGD 052/1/7/4, UGA.

[49] Duncan Stewart & Co. to Robert Gilbert, 15 September 1884, to Arucas, Canary Islands, UGD 052/1/7/5, UGA. On the "tyranny of distance," see Jim Endersby, "A Garden Enclosed: Botanical Barter in Sydney, 1818–39," *Brit. J. Hist. Sci.* 33 (2000): 313–34.

thumbs in the Canaries, Gilbert had been in contact with other planters and was on the verge of securing orders for two additional factories.[50]

Such orders were capital expenses that plantations hoped to undertake only once or twice a century. A glance at company order books shows that plantation owners usually ordered just one mill-and-engine combination at a given time, and rarely more than two. Properly maintained, such a pair might still match the efficiency of new machines fifty years later.[51] One of Mirrlees Watson's first mills, shipped in 1850, was not broken up until 1927.[52] Even more impressively, while in St. Lucia in 1973, Prince Charles stumbled across a rusting Mirrlees mill that was over a century old.[53] The bulk of the work of the engineering firms was not on new machines but on extensions, additions, and repairs to old ones. As the next section shows, the phenomenal cost and mass of these machines shaped both the plans of the sugar producer and the information demanded by the engineering firm.

PAPER

While mills lasted for decades in the Caribbean, their paper representations were carefully guarded back in Glasgow, where engineering companies were reluctant to part with original paper plans long after they had been turned into metal.[54] The transatlantic history of the equipment that filled the industrial sugar factory is thus not only a story of attenuated communication but also a story of difficult maintenance at a distance.[55] Historians and other scholars are increasingly interested in studying maintenance and disrepair as ordinary states of affairs, rather than ones that are extraordinary or "pathological."[56] Sugar machinery spent little time at factory spec, as the hot, noisy, and humid factory placed immense demands on people and metal. During the six-month grinding season, a mill was run almost around the clock. Every day brought more things to lubricate and to tighten; every week or two everything had to be thoroughly inspected. At the end of each season the machines were partly dismantled and packed away to protect them from the humidity, a curious combination of robustness and impermanence.[57] In such an environment, machines broke often, and when they broke they were far from the tools and expertise that had made them. Historians of the sugar industry have largely paid attention to when devices were invented or, more usefully, when they were widely adopted. But for most of their lives sugar machines were "old, existing things" rather than new ones, and they ought to be analyzed as such.[58]

[50] Duncan Stewart & Co. to Robert Gilbert, 21 May 1885, to Arucas, Canary Islands, UGD 052/1/7/6, UGA.

[51] Sugar Department Letterbooks, Mirrlees Watson & Co., Records of Mirrlees Watson Co., Ltd., UGD 062 1/3/1, UGA.

[52] Mirrlees Watson Mill and Krajewskis Order Book, UGD 118/2/4/37, UGA.

[53] "Mirrlees, Tait & Watson, Eglinton Works, Cook St., Engineers," AGN 479, Mitchell Library, Glasgow.

[54] Moss and Hume, *Workshop* (cit. n. 10), 162.

[55] David Edgerton, *The Shock of the Old: Technology and Global History since 1900* (New York, 2006), 77. On the perils of managing seventeenth-century sugar plantations at a distance, see Dunn, *Sugar and Slaves* (cit. n. 4), 200–201.

[56] Simon Schaffer, "Easily Cracked: Scientific Instruments in States of Disrepair," *Isis* 102 (2011): 706–17, on 707.

[57] Curry-Machado, *Cuban Sugar Industry* (cit. n. 6), 80–2, 94–5.

[58] Andrew Russell and Lee Vinsel, "Innovation Is Overvalued: Maintenance Often Matters More," *Aeon*, https://aeon.co/essays/innovation-is-overvalued-maintenance-often-matters-more (accessed 16 September 2017).

In order to be able to fulfill requests for replacement parts, engineering firms made it their business to know the fate of their machines many decades after original manufacture. The annotations in the margins of the Mirrlees Watson mill order book testify not only to the kind of information the firm wanted but also to the process by which that information was obtained (fig. 2). For example, mill number 360 was originally purchased for the Caledonia estate in Cuba in March 1859. A marginal note shows that in 1886 the mill was transferred to San Isidro, and whoever recorded that move took care to note the letter that carried it: number 9367 in 1904, fifty-five years after the machine was first bought. Or mill number 449, a horizontal mill sent to the Armonia estate in Cuba in June 1861. In 1903, the company received word that the mill unit had been transferred to La Reglita on the island. But the next year, they received conflicting information: perhaps it was only the accompanying engine, number 365, that had been moved. Other entries note the company's own modifications, like the one for mill 723: built in 1869 for Demerara as a three- or four-roller mill, it was expanded in 1889 to five rollers, "reverted" to three some years later, and finally sold to Bradford estate.[59]

These were fragile systems for acquiring knowledge. In December 1857, the Cuban owner of the Soledad plantation purchased a mill with thirty-inch-wide rollers from Mirrlees Watson, number 257. A few years later, the San Antonio estate purchased mill number 445, which sometime in the next two decades was sold to Soledad. In 1882, Edwin Atkins came into possession of Soledad by foreclosure and found its Mirrlees apparatus to be older than he wanted. "We replaced it with more modern machinery," he recalled, but "it was still in good condition and we sold it to another estate."[60] His new mill had significantly larger rollers—thirty-eight inches in diameter, the largest that Mirrlees then fabricated. The property of the estate when Atkins took it over also included mill 1406, which Mirrlees had just shipped there in February 1883. After activity in this secondary market, it might take Mirrlees Watson years to find out, despite yearly contact between Atkins and Mirrlees's agent, if they found out at all. The last time Mirrlees's books record mills 257 and 445, for example, they are at Soledad. If Atkins sold them, the engineering firm never knew.[61]

Rarely could sugar factories upgrade their entire production lines at once, so they cobbled together pieces of others. Sugar factories were agglomerations of machines from different makes as well as different vintages, so Glaswegian firms needed to know how their apparatus were being fitted together, not just who owned what. In 1907, the purchasing agent Victor Mendoza ordered mill number 1745 for Cuba's Central Mercedita. This was a standard three-roller unit, which the company recorded was "making with mills 1405, 1712 & 1713 a 12 roller train." The position of the unit in the "train" was important to know because it helped predict pressure and wear patterns and thus the likelihood of the unit's breakage. Similarly, Dos Hermanos in Cuba took delivery of mill 1326 in 1882 and mill 1555 ten years later. Each mill came with an engine and gearing, but in 1909 the company received word, via Mendoza, that the engines and mills had been switched, so that the later engine was driving the earlier

[59] Mirrlees Watson Mill and Krajewskis Books no. 1 (1841–1912) and no. 2 (1883–1964), UGD 118/2/4/37 and 118/2/4/38, UGA.

[60] Edwin Farnsworth Atkins, *Sixty Years in Cuba: Reminiscences of Edwin F. Atkins* (Cambridge, Mass., 1926), 93.

[61] Mirrlees Watson Mill & Krajewskis Book no. 1 (cit. n. 59).

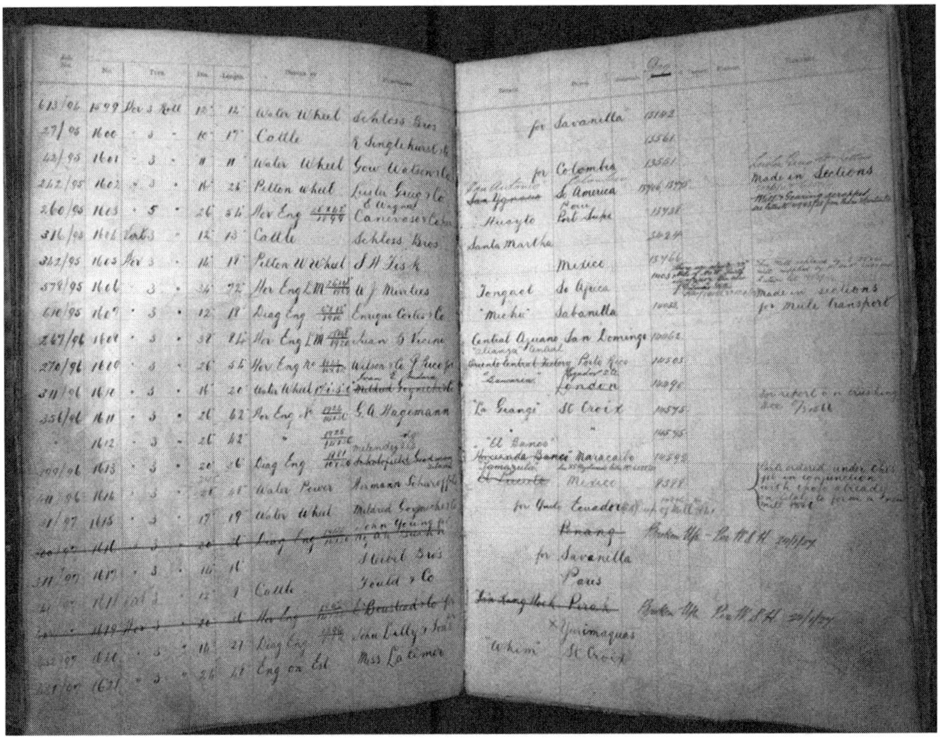

Figure 2. *Mirrlees Watson mill order book, showing annotations where new information was received. Records of Mirrlees Watson Co., Ltd. (cit. n. 38), UGD 118/2/4/37, UGA.*

mill. The earlier mill's rollers were both longer and wider than the newer ones and thus would experience different pressures and volumes of cane, feature different patterns of wear, demand different amounts of power, and have different life spans.[62]

Engineering firms maintained their relationships with overseas clients by exchanging drawings and tracings. Conversely, Caribbean engineers used Glasgow firms' practices of record-keeping to test their reliability as manufacturers and credibility as suppliers of parts (fig. 3). In 1838, the locomotive engineer Peter McOnie, having fallen on hard times, wrote to a friend in the West Indies seeking work on an estate but was advised to set up a repair shop in Scotland instead.[63] When this friend needed spares for his Cook-made mill, he needed to have "sent home such complete sketches with all sizes carefully marked" before McOnie could produce working patterns for the forge.[64] Later, when the plantation needed a new mill, the friend "made careful drawings of the set he required and sent them home to McOnie, who started at once to make working drawings from the sketches." The estate's owners were skeptical of McOnie's ability to complete the order and only signed the deal "on Mr. McOnie explaining that he had working drawings already made to sketches from D. Cook's engine and sugar mill in Trinidad."[65]

[62] Mirrlees Watson Mill & Krajewskis Book no. 2 (cit. n. 59).
[63] Moss and Hume, *Workshop* (cit. n. 10), 31.
[64] Robert Harvey, "The History of the Sugar Machinery Industry in Glasgow," *Int. Sugar J.* 19 (1917): 57–61, 112–7, on 112.
[65] Ibid., 113.

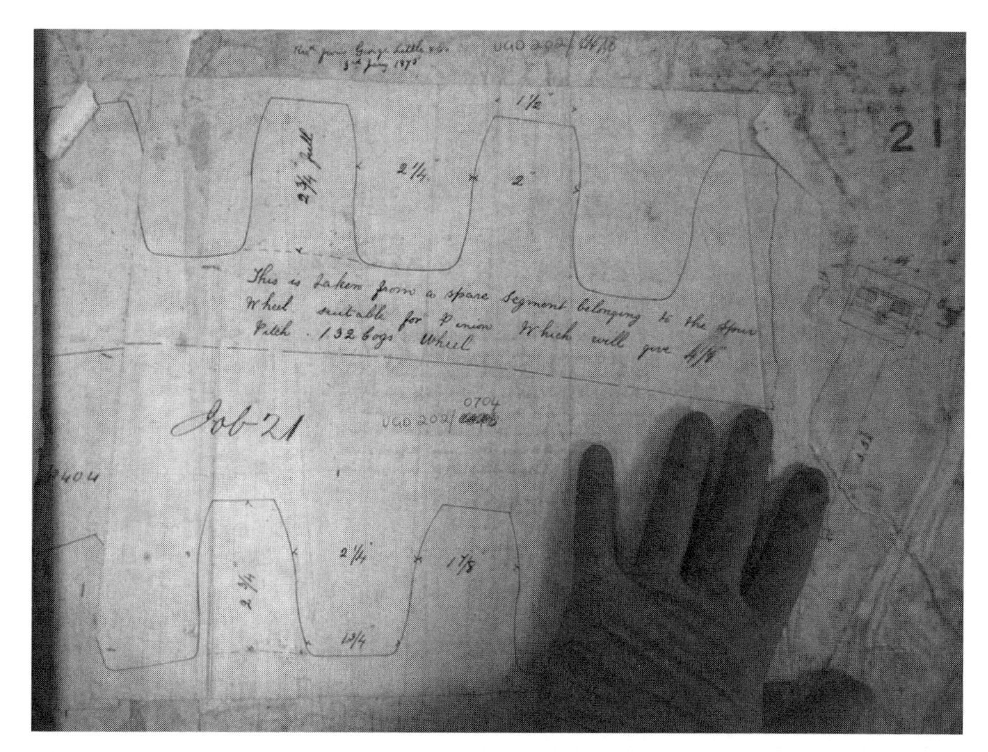

Figure 3. *Full-size tracing of gear teeth of a sugar mill spur wheel sent to Mirrlees Watson, with hand for scale. Records of Mirrlees Watson Co., Ltd. (cit. n. 38), UGD 202/0704, UGA.*

This was true not only for new partnerships but for long-standing ones as well. An engineer in Puerto Rico, at the end of the 1864 sugar season, sent Mirrlees Tait—Watson joined in 1868[66]—a request for two new mill cheeks on behalf of the estate of Señors Patxot and Polidura in Mayaguez, "as they have split the two this last crop across by the centre brass." The correspondent even told them where to look in their own books. "Knowing that *you have always your plans and models of the machines at hand in case any accident such as above should happen*, for that reason I only send you the number of mill 374, and year 1859, *thinking that is quite enough*." He attached only a "small sketch" to show where the original pieces had failed. But he also wanted the firm's guarantee that the new cheek would be functionally identical, "finished so that the same brasses will suit when necessary."[67] A few weeks later the firm received another letter on the same matter from a different correspondent, identifying the estate as "Ysabel for which you have executed before this several orders."[68] According to the company's books, Ysabel's mill was still being driven by a water wheel, and this client had required an unusually large number of drawings.[69] The engineer Thomas Dodd, writing in 1859 for Estate Florida near Ponce, on the southern

[66] Moss and Hume, *Workshop* (cit. n. 10), 33.
[67] Drawing from C. A. Hasche in Mayaguez, dated 8 July 1864, UGD 202/2430, UGA; emphasis added.
[68] Ibid.
[69] Mirrlees Watson Mill & Krajewskis Book no. 2 (cit. n. 59). See also Descriptive Drawings Index, "Horizontal Sugar Mills," p. 502, UGD 118/2/7/12, UGA.

coast of Puerto Rico, enclosed a drawing of the links of a driving chain of which he needed twenty-four feet. "I forgot to take the diameter of the pitch wheels to give the curve of the links," he wrote, "*but you will have it in your dimension Books*."[70] Dodd, apparently satisfied, was still ordering from Mirrlees in 1891.[71]

As a result of their importance, draftsmen could be powerful figures in a sugar engineering firm, and drawing sugar machines required knowledge that counted as specialized even within the world of engineers. The ascent of Robert Harvey, who began as the best turner at James Cook's lathes, emphasizes these points. In the early 1830s, Harvey's skill as a portraitist led to a successful trial as a draftsman. After a year, Harvey left in search of a raise and was hired by Neilson, a general engineering works that had received an order from Cuba for an engine and mill. Neilson's manager, William Tait, told the partners that he could not build a sugar mill "as it was all strange to him," so he sought out Harvey for advice. The expertise acquired from working with Harvey later brought Tait into business with James Mirrlees, forming the largest and longest-lasting of the Glasgow sugar partnerships, while Harvey returned to his old firm as managing partner in the 1850s.[72]

In the 1830s, when Harvey moved into the drafting office, self-trained draftsmen took techniques from many exemplars and sources, from drawing manuals to magazines. A decade or so later, however, Harvey began to teach mechanical drawing to supplement his income: first at home in the evenings, then in the Mechanics' Institute, and at the Government School of Design, which became the Glasgow School of Art.[73] The demand for these courses testified to the attraction of the potentially more "gentlemanly" nature of the work performed by draftsmen compared to even skilled factory work—as, indeed, did Harvey's willingness to take an initial pay cut.[74] At the same time, however, educational sites like the Mechanics' Institute, and Anderson's Institution, from which it had split, helped render draftsmen the "invisible technicians" of engineering workshops, including those in sugar.[75]

By the latter decades of the century, sugar engineering firms, like those in other lines of manufacture, boasted large drawing offices that included men of widely varying skills, training, background, and status.[76] In organization and discipline, the drafting office of a firm came to resemble the works itself.[77] Some draftsmen were themselves designers of machinery, while others worked to translate those designs into shapes that the works could manufacture. The largest group of draftsmen worked as tracers, skilled

[70] Chain link drawing from Thomas Dodd, 16 September 1859, Records of Mirrlees Watson Co., Ltd. (cit. n. 38), UGD 202/1555, UGA; emphasis added.

[71] Mirrlees Watson Order Book, 1891 (cit. n. 38), Job. no. 304.

[72] Harvey, "History" (cit. n. 64), 59.

[73] Harvey, *Early Days* (cit. n. 26), 11.

[74] Peter Jeffrey Booker, *A History of Engineering Drawing* (London, 1963), 134.

[75] See Steven Shapin, "The Invisible Technician," *Amer. Scient.* 77 (1989): 554–63; Larry Stewart, "Assistants to Enlightenment: William Lewis, Alexander Chisholm and Invisible Technicians in the Industrial Revolution," *Notes Rec. Roy. Soc. Lond.* 62 (2008): 17–29; George Emmerson, *Engineering Education: A Social History* (Newton Abbot, England, 1973), 91–100, 184–5. See also Steven Shapin and Barry Barnes, "Science, Nature, and Control: Interpreting Mechanics' Institutes," *Soc. Stud. Sci.* 7 (1977): 31–74; Maxine Berg, *The Machinery Question and the Making of Political Economy, 1815–1848* (New York, 1980), 147–55.

[76] See John Laidlaw letters, T-HB 72, Mitchell Library, Glasgow. For apprentices at Duncan Stewart and their salaries, see TD 158/11, Apprenticeship Book, Mitchell Library.

[77] Frances Robertson, "Manufacturing the Visual Economy in Nineteenth-Century Britain" (paper presented at the annual meeting of the International Society for Cultural History, Lunéville, France, 2 July 2012).

at making many reproductions of the same image. The more complex process of manufacturing itself now demanded multiple copies of drawings where just one had previously been necessary, and what counted as an adequate copy depended on the person for whom it was being produced, their place in the workshop, and what information they needed from a drawing.[78] Finally, some drawings needed to impress customers; as Frances Robertson points out, "Superhumanly neat inscriptions on paper functioned as a promise to deliver the goods in the material world."[79]

Outside the drafting office, the patternmakers were responsible for translating draftsmen's designs into wooden forms. Like its drawings, a firm's patterns were its most crucial assets, and the ability of a company to classify, store, retrieve, and reuse drawings and patterns was key to its fortunes.[80] So patterns, like the machines they represented, were built to last. In constructing a pattern by gluing multiple cross-grained layers of wood, the patternmaker had to consider how useful the pattern might be in the future and build it accordingly: cheaply if it was to be discarded, or solidly to last, with layers of shellac to protect it from warpage and precisely designed joints that would not alter with age.[81] When the partnership that James Cook founded finally dissolved, the managing partner sold the works but "bought all the patterns and drawings."[82]

The Puerto Rico engineers Robert Bennett and Thomas Dodd sent many orders to the Mirrlees firm in the late 1850s and early 1860s. "As I am not exactly acquainted with your patterns" for juice clarifiers, wrote Bennett in 1857, "I have sent you a sketch of the position they are to be placed with regard to the engine and you can make any other alterations that may be required to fit your patterns."[83] They wrote again on the eve of the 1861 grinding season to order a new engine, cane carrier, and shafts for Estate Vista-Alegre. "Cane carrier sides to be fitted to your mill No. 116," but "if the patterns have been altered since [the mill was made] let us know & we will send you a sketch, as we have not had time to do so."[84] The next year they ordered new cheeks for the mill but trusted the company's records less, sending them the "shape of cheek . . . taken of casting." They added that "the drawing you have corresponds with these dimensions so you can make the gudgeons [to] your drawing."[85] Neither drawings nor patterns, however, dictated the construction of sugar machines. The cooperative processes of design and construction meant that shop-floor workers had to use their judgment to figure out how to translate paper to wood and then to iron and steel.[86]

[78] Moss and Hume, *Workshop* (cit. n. 10), 160; Booker, *History* (cit. n. 74), 133.

[79] Frances Robertson, "Delineating a Rational Profession: Engineers and Draughtsmen as 'Visual Technicians' in Early Nineteenth Century Britain" (paper presented at the Three Societies Meeting, Philadelphia, 11 July 2012).

[80] "Inventory and Valuation of Machinery Plant, and Tools at Eglinton Engine Works, Cook Street, Glasgow, made by Messrs John Turnbull Jr. & Sons, Consulting Engineers, 18 Blythswood Sq. Glasgow, 20th February 1897," UGD 118/1/7/1, UGA.

[81] Sarah Fayen Scarlett, "The Craft of Industrial Patternmaking," *J. Mod. Craft* 4 (2011): 27–48.

[82] Harvey, *Early Days* (cit. n. 26), 60–1.

[83] Drawing dated 24 June 1857, UGD 202/0044, UGA.

[84] Drawing dated 24 December 1861, UGD 202/2148, UGA.

[85] Drawing dated 25 May 1862, UGD 202/1890, UGA.

[86] David McGee, "From Craftsmanship to Draftsmanship: Naval Architecture and the Three Traditions of Early Modern Design," *Tech. & Cult.* 40 (1999): 209–36; Jonathan Zeitlin, "Between Flexibility and Mass Production: Strategic Ambiguity and Selective Adaptation in the British Engineering Industry, 1830–1914," in *World of Possibilities: Flexibility and Mass Production in Western Industrialization*, ed. Charles F. Sabel and Zeitlin (New York, 1997), 241–72, on 247.

Glasgow companies even used their drawing offices' competence as a selling point, by advertising how little information they needed to issue a spare. "To request spare parts it is sufficient to give us the number of the centrifuge, which is inscribed in the vertical axis of each one," boasted Watson, Laidlaw & Co. in a brochure sent to the Spanish Caribbean in the 1920s and retained by a Puerto Rican factory. "We hold designs and plans of every centrifugal made by us, from the beginning of our firm in 1870 until the present, in order to be able to respond to any request for spares."[87] The drawing and pattern offices, no less than sugar factories or the shop floor, expose the constant process of maintenance and repair that underlay the surface of mechanically standardized sugar—a process that relied completely on the judgment, collaboration, and physical labor of human beings.

Glasgow's firms sought information about their products, and therefore about the state of sugar manufacture more generally, from across the Caribbean and the sugar-producing world. As such it may be tempting to think of the city's sugar workshops as something like Latourian centers of calculation, "mobilizing" and "accumulating" the world on paper through fixed inscriptions that travel easily.[88] In reality, however, these pieces of paper were severely recontextualized when they traveled, to borrow a term from the anthropologist Matthew Hull.[89] A tracing of a gear wheel meant one thing in Puerto Rico, next to the broken machine from which it was drawn. It meant quite another in Glasgow, amidst a company's ledgers and dimension books, patterns from the wheel's construction, and the metallurgical skills of its makers. Moreover, Glasgow's manufacturers were less central or coordinating than they were vulnerable, and their business depended completely on what they could learn from distant plantations.

Through the nineteenth century, sugar plantations were unquestionably where knowledge about sugar production would be generated. A century of production and maintenance by Glasgow's manufacturers enabled an industrial transformation in those plantations and in how sugar was made. At the turn of the twentieth century, that transformation began to enable claims that the Scottish city itself might generate knowledge about making sugar, just as the city's own importance began to waver.

MODEL FACTORIES

By the beginning of the twentieth century, the newest sugar factories in the world were many times larger than their predecessors of a few decades earlier. Glasgow firms were frequently called upon to design and construct these new, technologically sophisticated "central factories" from the ground up. The Harvey Engineering Company produced fourteen complete factories between 1905 and 1909 alone.[90]

[87] "Tenemos diseños y plantillas de todas la centrífugas fabricadas por nosotros desde la fundación de nuestra casa en 1870 hasta la fecha, así que podemos atender á cualquier pedido de repuestos." Watson, Laidlaw, & Co. catalog, "Centrifugas," in Colección Central Mercedita, caja "Católogos Comerciales 1920s–1940s," Archivo General de Puerto Rico, San Juan.

[88] Bruno Latour, *Science in Action: How to Follow Scientists and Engineers through Society* (Cambridge, Mass., 1987), 232–7.

[89] Matthew Hull, *Government of Paper: The Materiality of Bureaucracy in Urban Pakistan* (Berkeley and Los Angeles, 2012), 23–4.

[90] Harvey Engineering Company advertisement at the end of Llewellyn Jones and Frederic I. Scard, *The Manufacture of Cane Sugar* (London, 1909).

In the fall 1911 term, the Glasgow and West of Scotland Technical College (the descendant of the Mechanics' Institute) introduced a new post of lecturer in sugar manufacture, financed by "firms and individuals interested in this industry," including representatives of Glasgow's wealthy community of West India traders, estate owners, and commission merchants.[91] More than 75 percent of the financing for the lecturer's salary, however, was given by the city's sugar engineering firms, including the Harvey Engineering Company, Mirrlees Watson, A. & W. Smith, and Duncan Stewart.[92]

The centerpiece of the sugar school was a remarkable model of a sugar factory.[93] Mirrlees Watson provided a vacuum pan, Duncan Stewart delivered crystallizers, and Watson Laidlaw furnished its specialty centrifuges.[94] The college's 1913 annual report claimed that "the equipment for the demonstration of all the important processes in the treatment of sugar juice is now complete."[95] The first lecturer, a former West Indian factory chemist named Thomas Heriot, wrote in the *International Sugar Journal* in 1916 that the college had constructed a "complete factory in miniature" (fig. 4), one in which "every essential feature of the factory plant is reproduced on a smaller scale."[96]

Yet that emphatic sense of completion excluded crucial elements of the sugar factory. Most obviously, there was no possibility of a miniature mill for grinding sugarcane, which could not be brought to Glasgow before its juice hopelessly degraded. Similarly, neither the workers who would populate a real factory nor the ability to manage them were part of Heriot's miniature model. Commanding a workforce—"driving" them, in Heriot's word—"requires no technical skill, and may be better performed by others," rather than by the trained chemist or chemical engineer.[97] He acknowledged the importance of "administration" but felt it could be delegated to "native foremen" without much trouble.[98] And, finally, the machines were not subject to cycles of assembly and disassembly or heat and moisture that confronted actual factories in the tropics.

The Royal Technical College's model was derived from the products of machinery firms. Like the famous Phillips machine that modeled hydraulic Keynesianism, Heriot's miniatures represented the operations of a sugar factory as procedures subject to fine control.[99] With the help of the manufacturers and instrument makers, Heriot had

[91] Glasgow & West of Scotland Technical College (GWSTC) annual report 116th session (1912), p. 20, OE/4, Archives and Special Collections, University of Strathclyde Library. On the Audubon Sugar School in Baton Rouge, see John Alfred Heitmann, *The Modernization of the Louisiana Sugar Industry, 1830–1910* (Baton Rouge, La., 1987), chap. 10.

[92] GWSTC annual report 116th session (cit. n. 91), p. 64, University of Strathclyde Archives; Thomas H. P. Heriot, "Technical Training for the Sugar Industry," *Int. Sugar J.* 28 (1916): 173–9, n. 1. The salary for 1911 is found in OE/6/1/1, GWSTC staff card index, University of Strathclyde Archives.

[93] For models in general, see Soraya de Chadarevian and Nick Hopwood, eds., *Models: The Third Dimension of Science* (Stanford, Calif., 2004).

[94] This information comes from the GWSTC annual reports for 1912 and 1913; I was unable to find entries for these donations in the relevant books of the various companies.

[95] GWSTC annual report 117th session (1913), p. 24, OE/4, University of Strathclyde Archives.

[96] Heriot, "Technical Training" (cit. n. 92), 175–6.

[97] Thomas Hawkins Percy Heriot, *Science in Sugar Production: An Introduction to Methods of Chemical Control* (Altrincham, 1907), 7.

[98] Heriot, "Technical Training" (cit. n. 92), 178.

[99] Mary S. Morgan and Marcel Boumans, "Secrets Hidden by Two-Dimensionality: The Economy as a Hydraulic Machine," in de Chadarevian and Hopwood, *Models* (cit. n. 93), 369–401, esp. 392.

Figure 4. *Evaporator, School of Sugar Manufacture, Glasgow and West of Scotland Technical College, c. 1930, OP4/145, Archives and Special Collections, University of Strathclyde Library.*

installed "conveniences for exact experimental work which are lacking in the sugar factory," which allowed him to educate students "*in a more direct manner than by practical experience.*"[100] He intended that an enrollee in the sugar manufacturing course would learn "all essential principles outside the factory, by means of lectures and laboratory experiments, so that, when he first enters the factory, he understands

[100] Heriot, "Technical Training" (cit. n. 92), 173; emphasis added.

what he sees, and needs no other instructor than his own eyes and intelligence."[101] In other words, Heriot claimed that his school was in fact superior to any actual factory for learning about sugar production.

And students came, first from the engineering companies themselves. The evening course was packed with almost sixty employees of the local sugar machinery works. They heard lectures that followed sugar from its agricultural beginnings as "raw material" through milling, diffusion, clarification, concentration to syrup, crystallization, drying, and packing, all before "chemical control of manufacture." Within a few years, the school added a new laboratory course that included "analysis of sugars, juices, &c."[102]

The course began as a way to teach local sugar-machinery-making students how factories worked, but it became a model of a factory that taught colonials how their factories were supposed to work. Over the next decade, the college reported that its enrollment shifted to include many more West Indian students, who came in the hope "that their work [here] will do much to increase the knowledge of modern methods of sugar production and manufacture *which they have come to this country to acquire*."[103] For a hundred years, the infrastructure of sugar production had been characterized by just the opposite flows of people, paper, and knowledge. Glasgow's own success at industrializing the production of sugar had led to institutions like the "factory in miniature" at the city's engineering college.

It also set the stage for Glasgow's undoing, as rival centers of sugar-production expertise could emerge. A 1901 survey of "Local Industries of Glasgow and the West of Scotland" painted a gloomy picture of the future of the sugar-machinery industry. The British West Indian sugar market had run dry, though this was partly compensated by sales to independent Latin American nations, and to imperial possessions in South Africa, India, and Queensland. The real concern for the Glasgow sugar machinery makers was being shut out of two of the largest and richest sugar producers in the world. A quarter century earlier, after the Kingdom of Hawai'i signed a reciprocity treaty with Washington in 1876, Glasgow had supplied the islands with their first "complete" sugar factory. But the archipelago, "once a good market, is now practically closed to British manufactures." Fortunately, Cuba was "showing signs of revival, and may yet prove a fairly good market," according to the survey, "if it is not turned into an American preserve in a similar way to Honolulu."[104]

Far more than a "preserve," the Hawaiian capital would soon become home to the most advanced sugar-machinery firm in the world. The Honolulu Iron Works had been founded in 1853 by David Weston, who sold foreign rights to his patent for a self-stabilizing centrifuge to Mirrlees Watson. Like so many of Glasgow's firms, the Iron Works began by repairing others' machinery. By the 1880s, however, its work was

[101] Thomas H. P. Heriot, "The Sugar Industry after the War," *Proc. Roy. Phil. Soc. Glasgow* 49 (1918): 31–44, on 31.

[102] E10/2/6, "Guide to Evening Classes in Science and Technology," 1912–3, 29, 55, University of Strathclyde Archives.

[103] GWSTC annual report 116th session (cit. n. 91), 20; E10/2/2, Glasgow & West of Scotland Technical College Prospectus of Day Classes 1911–2, University of Strathclyde Archives; emphasis added.

[104] William G. Hall, "The Future of Sugar Is in the Orient," *Trans Pacific* (September 1921), 57–60, on 59; Henry Dyer, "Mechanical Engineering," in *Local Industries of Glasgow and the West of Scotland*, ed. Angus McLean (Glasgow, 1901), 35–91, on 64.

in high demand, "especially since the great distance from manufacturers in America and Europe made it extremely necessary to obtain what was needed right at home."[105] Honolulu Iron Works profited from the creation of the Hawaiian Islands as a model sugar production zone, free from the historical labor constraints that had plagued planters in the Caribbean. In this sense it was a scaled-up version of Heriot's "complete factory in miniature," complete not despite its lack of skilled labor but because of it.

The status of Glasgow as the world's dominant heavy-engineering city diminished in realms beyond sugar machines. Following the boom of the Great War, when many sugar-machinery makers retooled for armaments, the crop of 1920 brought the highest sugar prices in a century. Mountains of unfilled orders rose in the sales offices of sugar engineering companies, and so did profits. But those lucrative harvests were temporary, merely filling the vacuum left by European beet sugar. When the beet crop returned in 1921, prices collapsed.

Moreover, in return for tariff reciprocity with sugar producers, American protectionism had led to an increase in the relative cost of Scottish machinery. New factories were constructed in Cuba and the Hawaiian archipelago, demanding new machinery, because sugar from those islands could now enter the United States tax-free. But now American machinery faced no taxes upon entering the islands either. British manufacturers hoped this tariff surcharge would not entirely outweigh their firms' long-standing relationships with engineers and owners. But as Cuba's central factories failed following the 1921 collapse, they often fell into the hands of American creditors, who strongly preferred to buy from domestic engineering firms. That year, only a single order of centrifugals arrived in Scotland. Glasgow newspapers now reported on their city's sugar firms in different tones: they spoke of "moderate activity," business that was "fairly active," with "work not so plentiful as in the years past." Some firms received no new contracts until 1923.[106]

By the 1920s, by contrast, the Honolulu Iron Works were producing factories for Cuba, the Dominican Republic, Formosa, and the Philippines. The firm boasted not only a branch in New York City but a full office in Havana, including engineering staff—a far more substantial presence than the consulting engineers on whom Glasgow firms had historically relied, and one that gave customers the possibility of far closer coordination. Technology, expertise, and individuals could move through multiple overlapping circuits of the sugar economy. If it remains strange to think of sugar expertise in sugar-free Glasgow, it should be no less discordant to our received geographies of technology, knowledge, and capitalism to find a prestigious ironworks in the Hawaiian Islands—an archipelago entirely lacking in useful ores.[107]

In 1921, the manager of the Honolulu Iron Works published an article in a trade journal called *The Trans Pacific*, and with industry as well as agriculture in mind, titled it "The Future of Sugar Is in the Orient."[108] That same year, the board of A. & W. Smith dispatched the aptly named Martin Ironside to Cuba "for the purpose of

[105] Hall, "Future of Sugar" (cit. n. 104), 59.

[106] D. Bradley, "Fletcher & Stewart Ltd: A Business History" (MPhil thesis, Univ. of Nottingham, 1972), 118, 122, 211, UGD 118/11/1/48, UGA.

[107] "Ancient Hawaiians: Plenty of Oars but No Ores," U.S. Geological Survey Hawaiian Volcano Observatory, 24 September 2009, https://volcanoes.usgs.gov/observatories/hvo/hvo_volcano_watch.html?vwid=359 (accessed 17 April 2018).

[108] Hall, "Future of Sugar" (cit. n. 104).

gaining knowledge of the most modern practice" in sugar making and in machinery. They judged that "such a visit would be advantageous" to the firm and, perhaps acknowledging the usefulness of Honolulu Iron Works' Havana office, "a similar visit on the part of a member of the Works Staff would be even more so." Ironside returned six months later, having toured not just "the Sugar Factories of Cuba" but also Trinidad, Louisiana, and elsewhere in the United States.[109] He may well have visited Louisiana's new and "radical" Adeline Sugar Factory, which had been "designed in its entirety by the Honolulu Iron Works Company of New York"—a company name that concisely expresses the geographic inversions of expertise in the early twentieth-century sugar economy.[110]

CONCLUSION

The geography of the cane sugar economy placed singular demands on the companies in Glasgow that built machines for that economy in Glasgow. Machines for milling grain into flour or for spinning cotton into thread, to name a few other creations of Glasgow industries, only had to travel as far as other British cities. Sugar machines, by contrast, spent their lives in tropical factories, where they could work close to the plant itself and to its agriculture. They had to be durable yet flexible, customized yet interoperable. They had to function for decades without intervention from their manufacturers yet be immediately recognizable to those manufacturers when they were called upon to make repairs. They also had to provide the infrastructure for the precise chemical and human operations that turned the sugarcane plant into sugar—a substance that, by the twentieth century, appeared to approximate the ideal commodity as closely as any natural object could.

That superficial appearance of an efficient commodity market belied the difficult and constant work deep below. Sugar plantations and factories famously required the adaptive skills of the people who worked within them. So too, this essay has shown, did the workshops that built the machines on which those plantations and factories depended. Glasgow's sugar engineering firms depended on peregrinations of specific people to carry information and knowledge that paper could not convey and to reassemble machines in ways that were not self-evident. They nonetheless relied on movements of idiosyncratic inscriptions—from tracings made directly from broken machines to patterns within the works itself—to build new machines and learn the status of their old ones. Meanwhile, on the other side of the Atlantic, clients used a firm's management of those inscriptions to test its ability to supply real products when they would be needed most.

Despite their distance from the sources of sugar and of knowledge about it, Glasgow's firms successfully managed these fragile networks for much of the nineteenth century. Their success meant they played a crucial role in industrializing sugar production in the Caribbean and across the world. That industrialization, however, and the reconfiguration of sugar making around modern science, had unexpected consequences for both Glasgow and the tropics. First, if sugar production were now a matter of machines, then it became plausible for someone like Heriot to claim that a model

[109] A. & W. Smith Minute Book no. 2, 24 October 1921 and 3 May 1922, UGD 118 1/1/1, UGA.
[110] "The Adeline Sugar Factory," *American Sugar Industry and Beet Sugar Gazette* 14 (1912): 36.

factory was complete if it included nothing but machines. Second, plantations in the Hawaiian Islands were constructed as a model of what fully industrialized sugar production might look like. Helped by tariffs and proximity, the Honolulu Iron Works became the leading manufacturer not only in its own archipelago but for much of the world.

The exclusion of Glasgow from histories of the sugar world has led to neglect of the close connections between the standardization of commodities and the variety of tools and techniques that make commodification possible. Precisely because of the distance and fragility of sugar's networks, and because of the immense reaches of time and space that sugar machines needed to cross, such machines make those connections visible.

Starting up Biology in China:
Performances of Life at BGI

*by Hallam Stevens**

ABSTRACT

BGI (*hua da ji ying*; 华大基因; "China Great Gene") counts among the world's largest and wealthiest institutions for biomedical research. Located in Shenzhen, the new megacity in southern China, BGI is now a critical site for understanding the relationship between biomedicine and the economic development of China. This essay uses performance studies and the notion of *shanzhai* ("copycatting") to understanding how this laboratory poses a challenge to traditional modes of understanding technoscience. This marks an attempt to understand BGI, its work, and its workers on their own terms, or at least on local terms. Just as *shanzhai* challenges our notions of originality, BGI's hybridity challenges our notions of where and how scientific knowledge is produced. Performing not merely as a "laboratory," but also, and at the same time, as a "factory," and a "company," BGI is an unfamiliar kind of hybrid scientific-industrial-commercial-governmental-philanthropic space that draws its repertoire from its very particular regional, national, and local-urban circumstances.

BGI (*hua da ji ying* ; 华大基因 ; "China Great Gene") counts among the world's largest and wealthiest institutions for biomedical research. Located in Shenzhen, the new megacity in southern China, BGI is now a critical site for understanding the relationship between biomedicine and the economic development of China. As Winnie Wong has suggested, the rise of Chinese science, especially biomedicine, "requires us to consider configurations of 'Asian' biotechnology and capital that as yet remain uncaptured by US-centric descriptions."[1] What do such configurations look like? And how can we understand them on their own terms?

* History Programme, Nanyang Technological University, 14 Nanyang Drive, 05-07, Singapore 637332; hstevens@ntu.edu.sg.

For helping to arrange visits to BGI, I would like to acknowledge the invaluable assistance of Bicheng Yang, Scott Edmunds, Aizhu Wang, Chao Chen, and Huanming Yang. I am also grateful to other BGI employees who agreed to be interviewed. During 2014, the first round of fieldwork for this project was undertaken with Dr. Eddie Paterson (Department of Culture and Communications, University of Melbourne). During the summer of 2017, I was hosted as a visiting scholar at the Center for Special Economic Zone Research at Shenzhen University. For welcoming me there and assisting with my research activities, I would like to thank Professor Tao Yitao, Professor Yuan Yiming, and Lai Zehua. This work was supported under a Tier 1 grant from the Ministry of Education (Singapore), "Performances of Life: The Political, Social, and Economic Contexts of Biomedicine in Singapore, China, and Japan" (RG 56/13).

[1] Winnie Wong, "Speculative Authorship in the City of Fakes," *Curr. Anthropol.* 58 (2017): S103–S112, on S109. Wong is here echoing Aihwa Ong's and others' call for attention to the new modes of biotechnology in Asia. See Ong, *Fungible Life: Experiment in the Asian City of Life* (Durham,

What we read about BGI, in the Western press at least, revolves around an almost irresistible analogy: the factory. "Now, as the world's scientists focus with increasing intensity on transforming the genetic codes of every living creature into information that can treat and ultimately prevent disease, Shenzhen is home to a different kind of factory: B.G.I., formerly Beijing Genomics Institute, the world's largest genetic-research center."[2] The "factory" represents Western fascination with China's rapid development, Western fears about China's rise (taking away manufacturing jobs, Dickensian treatment of workers), and an implicit derogation of China as a place of "copying" and imitation. The ubiquitous "Made in China" label serves as a constant reminder of the massive presence and productivity of China's factories.

Instead of starting with the "factory," this essay draws on a set of tools from performance studies. Building on the work of Erving Goffman and Victor Turner, since the 1970s, performance studies has deployed a range of methods for examining practices beyond the theater, including rituals, political speeches, sports, and other cultural events.[3] This work has drawn attention to the fact that performances cannot be understood in isolation from their broader contexts and—borrowing from gender studies and queer theory—it has also highlighted the "performativity" of everyday acts—building identity through doing and speaking.[4] In science and technology studies, the notion of performance (and performativity) has been mobilized to analyze the performance of bodies, to draw attention to action (rather than words or text), and to explore the ways in which scientific theories and models can generate or shape elements of reality.[5] Following Latour, others have examined how scientists "stage" or "direct" their findings to scientific or public audiences.[6]

Here I will be less concerned with self-conscious performances to a given audience (a conference presentation or a courtroom, for instance) and instead focus on the performative aspects of everyday speech and work, not just within the lab, but outward to the world. The setting, the costumes of workers, speeches, posters, advertisements, machines, and so on are all aspects of a "performance" that communicates the laboratory's intentions, ideals, and hopes. Turning to BGI's performances opens up several

N.C., 2016); Ong and Nancy Chen, eds., *Asian Biotech: Ethics and Communities of Fate* (Durham, N.C., 2011); Kaushik Sunder Rajan, *Biocapital: The Constitution of Postgenomic Life* (Durham, N.C., 2006).

[2] Michael Specter, "The Gene Factory," *New Yorker*, 6 January 2014, http://www.newyorker.com/magazine/2014/01/06/the-gene-factory.

[3] Victor Turner, *The Anthropology of Performance* (Cambridge, Mass., 1986); Erving Goffmann, *The Presentation of Self in Everyday Life* (New York, 1959). See also Turner and Richard Schechner, *Between Anthropology and Performance* (New York, 1983); Schechner, *Performance Studies: An Introduction*, 2nd ed. (New York, 2006).

[4] On performativity, see esp. John L. Austin, *How to Do Things with Words*, ed. James O. Urmston and Marina Sbisà (Oxford, 1962); Judith Butler, "Performative Acts and Gender Constitution: An Essay in Phenomenology and Feminist Theory," *Theatre J.* 40 (1988): 519–31.

[5] Hannah Landecker, "Microcinematography and the History of Science and Film," *Isis* 97 (2006): 121–32; Natasha Myers and Joseph Dumit, "Haptic Creativity and the Mid-Embodiments of Experimental Life," in *A Companion to the Anthropology of the Body and Embodiment*, ed. Frances E. Mascia-Lees (Chichester, 2011), 239–61; Karen Barad, "Posthumanist Performativity: Toward an Understanding of How Matter Comes to Matter," *Signs* 28 (2003): 801–31.

[6] Bruno Latour, *Science in Action: How to Follow Scientists and Engineers through Society* (Cambridge, Mass., 1987); Stephen Hilgartner, *Science on Stage: Expert Advice as Public Drama* (Stanford, Calif., 2000).

modes of analysis. First, BGI's geographical location, appearance, architecture, and relationship to Shenzhen become evidence of its self-presentation. Second, it allows us to attend to BGI's own speech about itself (its slogans, advertisements, media appearances, etc.) as evidence of its efforts to create a specific kind of work and vision. Third, performance calls attention to the importance of BGI's attempts to create certain kinds of institutional spaces and certain kinds of workers. The "performativity" of laboratory speech and acts demonstrates how BGI and its employees are actively attempting to shape a future, not only for the lab itself, but for China and Chinese citizens.

I refer to these performances as "performing *shanzhai* (山寨)." The notion of *shanzhai* (literally "mountain stronghold") has, as we shall explore in more detail, a complicated set of meanings. Although at first taken to denote low-quality "copycat" goods produced in Shenzhen, *shanzhai* has gradually become associated with notions of creativity, innovation, and Chinese national pride. In performing *shanzhai*, then, BGI is enacting a particular vision of China's future: one rooted in the copy, but ultimately generative of surprising, new, and original elements. *Shanzhai* creates a vision in which China is seen to be ultimately able to surpass its rivals through a particular combination of imitation and innovation.

BGI begins as a place of the copy. It has built its worldwide reputation largely on DNA sequencing. This work involves extracting DNA molecules from living things (humans, animals, plants, viruses, cancer cells, etc.) and processing them such that they can be digitally "read out" as a series of letters (A, G, T, or C) that comprise the genetic code. DNA molecules are composed of two complementary strands: they are fundamentally structured as a double helix. This structure allows organisms to make near-identical copies of their genes that can be passed on to all the cells in a body and to the next generation. It is precisely this property—its "copyability"—that also allows biologists to read, edit, and manipulate DNA. But BGI is not engaged in mere copying. The success of its work—institutional, social, and scientific—relies on subtle recombinations and hybridizations that allow it to make better copies. This is copying with a (helical) twist.

By understanding BGI in terms of "performing *shanzhai*," we can see how this laboratory poses a challenge to traditional modes of understanding technoscience. This marks an attempt to understand BGI, its work, and its workers on their own terms, or at least on local terms. Just as *shanzhai* challenges our notions of originality, BGI's hybridity challenges our notions of where and how scientific knowledge is produced. Not merely performing as a "laboratory," but also, and at the same time, as a "factory," and a "company," BGI is an unfamiliar kind of hybrid scientific-industrial-commercial-governmental-philanthropic space that draws its repertoire from its very particular regional, national, and local-urban circumstances.

This account is based on ethnographic fieldwork at BGI and in Shenzhen between 2014 and 2017. This included interviewing scientists, visiting the laboratory (the headquarters in Yantian, the China National Gene Bank in Dapeng, and the labs in Hong Kong), and attending conferences, workshops, talks, and other activities at the lab. This work occurred during several visits to Shenzhen, including an extended visit (two months) in 2017. Although BGI has permitted significant access to its sites, it is in part a private (and more recently, public) company, and access is thus necessarily restricted and incomplete in some respects. As such, the ethnography here is supplemented by media sources and interviews; for the purposes of this work, these form a valuable set of primary sources that provide a wider range of information than could be obtained by ethnographic work alone.[7]

I begin with a section describing the context in which BGI has developed, examining the history of Shenzhen itself. The second part examines the various meanings of "copycatting" and *shanzhai* in Shenzhen's factories and electronics markets. The third and central part of the text turns to BGI's work in more detail, examining BGI's various performances: of laboratory, of factory, and of company. I conclude by reflecting on how BGI's performances of *shanzhai* are not only critical for its own work, but also for understanding how biomedicine, and even technoscience more broadly, is developing in China.

SETTING THE STAGE

BGI was founded in 1999 in order to coordinate China's contribution to the Human Genome Project. Although their eventual 1 percent contribution brought the lab considerable attention, by the middle of the 2000s they found their funding dwindling and began to search for a new home. After a brief stint in Hangzhou, BGI moved to Shenzhen in 2007. Why did BGI's leaders choose Shenzhen? What kind of place was it, and what did BGI hope to achieve by moving there?

The city is now a metropolitan area of over 20 million people in Guangdong Province. Most widely known for its electronics manufacturing industry, this region is now sometimes called the "Silicon Valley of China."[8] In 1980, Shenzhen became China's first special economic zone (SEZ). The SEZ was part of Deng Xiaoping's attempt to open China up to foreign investment. To achieve this, the Shenzhen government adopted a range of policies, including creating special tax incentives for foreign investment, granting greater independence in international trade, encouraging development sponsored by foreign capital, supporting Sino-foreign joint ventures and partnerships, promoting the manufacture of products for export, and allowing market forces freer rein.[9]

This opening up meant, at first at least, largely garment manufacturing. During the 1980s and 1990s, thousands of small and medium-sized factories making shoes, clothing, and toys opened their doors. Many were owned by Hong Kong or Taiwanese-based investors often connected through family networks—overseas Chinese supplied capital to their relatives in the SEZ, who could gain access to cheap land and cheap sources of labor.[10] These factories attracted migrants from the Chinese countryside, coming to Shenzhen (and other cities in southern China such as Dongguan and Guangzhou) to escape farm and village life and make their economic futures. Almost all of these migrant workers were young, and the majority of them were women.[11]

[7] In some cases, too, I have cited media sources where they corroborate evidence from ethnographic fieldwork and interviews.

[8] This phrase recurs across a variety of media and popular outlets. See, e.g., Tom Whitwell, "Inside Shenzhen: China's Silicon Valley," *Guardian*, 13 June 2014, http://www.theguardian.com/cities/2014/jun/13/inside-shenzen-china-silicon-valley-tech-nirvana-pearl-river.

[9] Yue-man Yueng, Joanna Lee, and Gordon Kee, eds., "China's Special Economic Zones at Thirty," special issue, *Eurasian Geogr. Econ.* 50, no. 2 (2009).

[10] Alan Smart, "Flexible Accumulation across the Hong Kong Border: Petty Capitalists as Pioneers of Globalized Accumulation," *Urban Anthropol. Stud. Cult. Systems World Econ. Develop.* 28 (1999): 373–406.

[11] Leslie T. Chang, *Factory Girls: From Village to City in a Changing China* (New York, 2009); Pun Ngai, *Made in China: Women Factory Workers in a Global Workplace* (Durham, N.C., 2005).

This migration has had several kinds of impacts on Shenzhen. First, it transformed the area from a large number of villages on the Pearl River delta (inhabited by about 300,000 people) into a massive, sprawling metropolis. In the 1990s and 2000s, Shenzhen was one of the world's fastest growing cities, and it is now China's third largest after Beijing and Shanghai. This population growth was matched by growth in infrastructure.[12] In the 1980s and early 1990s, Shenzhen built more skyscrapers than any other Chinese city, including some of the world's tallest.[13]

The overwhelmingly migrant population lends the city a unique vibe.[14] Young people who have come from all over China make Shenzhen a dynamic and increasingly cosmopolitan city. The antiquated Chinese household registration system (*hukou*) means that many of the residents remain "unofficial"—unregistered, uncountable, and often unable to obtain health care, get a driver's license, or send their children to school.[15] The result is a high degree of transience, impermanence, and uncertainty among much of the population. Migrants have come to make their fortunes in a place where, often, they know no one and the rules of the game have to be made up as they go along. There is money to be made, but the risks involved for individuals are high.[16]

The location of Shenzhen is also critical to its flavor: the city is separated from the Hong Kong "Special Autonomous Region" by the Shenzhen River. There are several checkpoints in Shenzhen through which tourists, workers, and businesspeople can enter directly into the New Territories. The proximity of Shenzhen to Hong Kong has no doubt contributed to its success.[17] Hong Kong has long been an international city where foreigners have established commercial relationships and feel comfortable visiting and doing business. From there, extending connections a few miles into the "mainland" has proved relatively easy; bosses from Hong Kong can easily supervise operations across the border. The Shenzhen–Hong Kong border, then, provides a kind of "window" from the People's Republic of China to the rest of the world.[18]

Shenzhen's geographic location has also influenced its political climate. Close to Hong Kong, but far from the central government in Beijing, the city has also become a place notable for the loosening of central authority. As the Beijing government searched for ways to promote economic development, Shenzhen "emerged out of a period of illicit (and often outright politically unapproved) experimentation."[19] More recently, in 2010, the "small government, big society" political experiment initiated by Premier Wen Jiabao allowed local government to scale back its influence in civil society. This has led to the flourishing of nongovernmental organizations (NGOs), many

[12] On the early development of Shenzhen, see Ezra Vogel, *One Step Ahead in China: Guangdong under Reform* (Cambridge, Mass., 1989), esp. 125–60.

[13] Wendell Cox, "The Evolving Urban Form: Shenzhen," *New Geography*, 25 May 2012, http://www.newgeography.com/content/002862-the-evolving-urban-form-shenzhen.

[14] See Xiangming Chen and Tomas de'Medici, "The 'Instant City' Coming of Age: Production of Spaces in China's Shenzhen Special Economic Zone," *Urban Geogr.* 31 (2013): 1141–7, http://dx.doi.org/10.2747/0272-3638.31.8.1141.

[15] Ngai, *Made in China* (cit. n. 11), 5.

[16] Although, by developed-world standards, factory wages are very low. See Kaxton Siu, "The Working and Living Conditions of Garment Workers in China and Vietnam," in *Chinese Workers in Comparative Perspective*, ed. Anita Chan (Ithaca, N.Y., 2015), 105–31; tables on 109.

[17] Joseph Y. S. Cheng and Guo Shiping, eds., *Shenzhen and Hong Kong: Competitiveness and Cooperation in Technology* (Hong Kong, 2008), 199.

[18] Ibid.

[19] Mary Ann O'Donnell, Winnie Wong, and Jonathan Bach, "Introduction: Experiments, Exceptions, and Extensions," in *Learning from Shenzhen: China's Post-Mao Experiments from Special Zone to Model City* (Chicago, 2017), 1–19, on 3.

of which are funded by the city government to help deal with social problems such as mental health, migrant worker education, or factory suicides.[20] Such changes proceed slowly and cautiously, but they have opened up possibilities for new kinds of organizations playing new kinds of social roles in China.

This history suggests that Shenzhen itself is *shanzhai*, a copy. Borrowing from Hong Kong, China's leaders set out to emulate the financial success of that city. But of course, Shenzhen could never become an exact copy of Hong Kong; for one thing, political liberalization was not supposed to go along with economic freedoms. Indeed, as Jun Zhang has argued, the Chinese Communist Party also used the more authoritarian city-state of Singapore as a model for Shenzhen. As such, Shenzhen aimed to reproduce aspects of both cities, recombining and tweaking them to make something new.[21]

SHANZHAI

Since the 1980s, Shenzhen's economy has gradually shifted away from garments and toward electronics and information technologies. Beginning in the 1990s, the Shenzhen municipal authorities adopted the policy of "utilizing high-technology industry as the 'dragon head' or pacemaker in the construction of an international metropolis."[22] Between 1991 and 2002, the value of "high-tech" industries (including computer hardware, telecommunications, microelectronics, mechanical and electrical integration, new materials and new energy, and biotechnology) in Shenzhen grew from 8.1 percent of the city's total economic output to over 35 percent of total output. In 2002, revenue from high-tech products had increased to over RMB 80 billion, with over RMB 30 billion contributed by the telecommunications giants Huawei (华为) and ZTE (中兴).[23]

Shenzhen's high-tech industries have been particularly visible in the West because of Foxconn, the company contracted by Apple to assemble iPhones and iPads. Foxconn is the trade name of Hon Hai Precision Industry Co., founded in Taiwan in 1974, opening its first plant in the Longhua region of Shenzhen in 1988. The Longhua Science and Technology Park is a "walled campus" that employs more than a quarter of a million people and covers just over one square mile. About one-quarter of the factory workers live inside the walls of "Foxconn City." Apart from fifteen factories, the complex encloses worker dorms, a swimming pool, its own fire brigade, its own television network, grocery stores, bank, restaurants, bookstores, and hospital. Foxconn not only makes iPhones for Apple but also includes among its customers Acer, Amazon, Cisco Blackberry, Dell, Google, Nintendo, Microsoft, Motorola, Sony, Toshiba, Nokia, and Chinese companies such as Huawei and Xiaomi (小米).

However, Shenzhen's electronics industry is hardly confined to Foxconn. As well as other foreign firms from Taiwan and Hong Kong and some of the largest electronics and telecommunications companies in China (Dingoo [丁果], Hasee [神舟], Netac,

[20] Karla W. Simon, *Civil Society in China: The Legal Framework from Ancient Times to the "New Reform Era"* (New York, 2013). "Small government, big society" is part of a broader project of reform initiated by the central government in 2004.

[21] Jun Zhang, "From Hong Kong's Capitalist Fundamentals to Singapore's Authoritarian Governance: The Policy Mobility of Neo-Liberalising Shenzhen, China," *Urban Stud.* 49 (2012): 2853–71.

[22] Cheng and Guo, *Shenzhen and Hong Kong* (cit. n. 17), v.

[23] Ibid., 2–5. Huawei was founded in Shenzhen in 1987 to produce phone switches, and it is now the largest telecommunications manufacturer in the world.

Skyworth [创维], and Coolpad [酷派] as well as the Internet giant Tencent [腾讯]),[24] Shenzhen (and the surrounding cities and towns) hosts a massive network of small to medium-sized factories and suppliers, mostly geared toward the production of specific products (digital cameras, phone cases, tablet screens, etc.).

This cluster of high-tech electronic manufacturing has made Shenzhen into a center of hardware innovation. This production is driven by the large numbers of young people flocking to the city in search of work or moving in and out of the large number of companies. The best evidence of this can be found at the Huaqiangbei electronics markets, by some accounts the largest in the world. These markets sell not merely finished products but also a range of electronic components necessary to assemble almost any device. As one Western journalist explains, "In Shenzhen, you have everything you need to turn a sketch on a napkin into 100,000 smartwatches, bike lights, or drones. . . . The mood here feels like the pre-boom Internet of the late 1990s: a lot of excitement and a few big deals."[25]

The markets, located near downtown Shenzhen, stretch for several blocks. Different shopping mall–sized buildings specialize in different kinds of items: one in power cables and power supplies, the one adjacent to that in mobile phone screens, and the one across the road in LED lights. Vendors are densely packed into floors and floors of small booths, sitting behind benches piled high with microchips or coaxial cables or digital cameras next to ubiquitous flasks of tea. A diverse crowd of shoppers roams the spaces; many carry long lists of components required for building their own devices. Huaqiangbei is a complete ecosystem for making and selling electronics for the global marketplace.

It is this culture that has given rise to the local hardware hacking scene now known as *shanzhai*. This term is sometimes translated as "copycat electronics."[26] And indeed, there is a great deal of copying of many of the international and Chinese electronics brands, especially mobile phones.[27] However, blatant copying has also given way to tweaking, improvement, and innovation. Although *shanzhai* violates intellectual property laws and flouts safety testing, it has developed into a sharing, collaborative economy. Design files for smartphones, smartwatches, and tablets are shared online, and these devices can be assembled from the components on sale at Huaqiangbei. Many such phones and other devices include features for specialized markets, such as dual SIM cards (for those crossing the Hong Kong border frequently) or a built-in compass (for Muslims needing to know the direction of Mecca). Such devices are often priced to sell to customers in the developing world (either within or outside China).

It is easy to dismiss the activities of Huaqiangbei and *shanzhai* as mere copying, or as a kind of piggybacking on Western (especially Silicon Valley) innovation. But there is something more complicated and more interesting occurring here. As Bunnie Huang

[24] Tencent's business includes e-commerce, e-payments, multiplayer online games, and instant messenger services (QQ) as well as China's most popular social media platform, WeChat. See http://www.tencent.com/en-us/index.shtml (accessed 14 August 2017).

[25] Whitwell, "Inside Shenzhen" (cit. n. 8).

[26] The word also has connotations of peasant work equivalent to the English word "tinker." The word also sounds like "Shenzhen."

[27] Some estimate that *shanzhai* now makes up 25 percent of the global market for mobile phones. Silvia Lindtner, Anna Greenspan, and David Li, "Shanzhai: China's Collaborative Electronics-Design Ecosystem," *Atlantic*, 18 May 2014, http://www.theatlantic.com/technology/archive/2014/05/chinas-mass-production-system/370898/.

notes in his guide to Huaqiangbei, "'fake' is not an all-or-nothing concept." Huang goes on to list various ways in which products at the market may be "fakes": genuine parts with old date codes, factory remainders sold as genuine, parts relabeled with a more reputable brand, subassemblies of authentic quality but usually not allowed to be sold separately, lower spec labeled as higher, recycled or refurbished components, preproduction prototypes, and so on.[28] *Shanzhai* functions precisely in and because of these ambiguous middle zones.[29]

The meanings of *shanzhai* are multiple, complex, and even contradictory. Some Chinese dismiss such goods as low class, low taste, and representative of the Chinese inability to be original. For brand-name electronics manufacturers, *shanzhai* has cut into market share and profits.[30] For these more established companies—both Chinese and foreign—*shanzhai* poses a threat and is seen as a scourge. "Counterfeiting" remains illegal in China and, at least for large parts of the establishment, continues to be a derivative, "low," and sometimes dangerous activity.

On the other hand, especially among the young, the unique kitsch of *shanzhai* phones expresses a hip, cool, artistic, and rebellious vibe.[31] In the Chinese novel *Shui Hu Zhuan* (水浒传; The Water Margin), the rebels fighting against the Song dynasty government are described as residing in *shanzhai*.[32] Such outlaws or rebels, like Robin Hood in the West, are also heroes. With respect to electronics, the term has also enjoyed a rehabilitation, having come to be associated not only with mimicry and fakery but also with originality, innovation, and creativity. *Shanzhai* has been embraced, especially by Chinese youth, as a unique Chinese mode of making and doing that will be central to Chinese technological and economic development. Through *shanzhai*, some claim, China will be able to challenge the dominance of global brands such as Samsung and Apple, developing their own products and brands.

In other words, *shanzhai* has become a political tool and even an expression of "a new wave of democracy."[33] For reformers within China, this new mode of expression

[28] Andrew "Bunnie" Huang, *The Essential Guide to Electronics in Shenzhen* (Singapore, 2015), 15–18.

[29] It is interesting to compare here *shanzhai* with "Dafen Artists' Village." At Dafen, on the outskirts of Shenzhen, artists produce a large fraction of the world's reproduction oil paintings. Although this is portrayed by Western media as "Van Gogh from the Sweatshop," Wong argues that "the repetitive work of painting in Dafen was never simply the antithesis of originality, authenticity, or creativity, nor is 'copying' always the same and simple activity. . . . I found little reason to presume that any Dafen painter may *not* be considered an 'artist' working within the same spectrum of aesthetic concerns and limitations as any contemporary artist." Winnie Won Yin Wong, *Van Gogh on Demand: China and the Readymade* (Chicago, 2013), 25. See also Martin Paetsch, "China's Art Factories: Van Gogh from the Sweatshop," *Spiegel Online*, 23 August 2006, http://www.spiegel.de/international/china-s-art-factories-van-gogh-from-the-sweatshop-a-433134.html; James Fallows, "Workshop of the World: Fine Arts Division," *Atlantic*, 19 December 2007, http://www.theatlantic.com/technology/archive/2007/12/workshop-of-the-world-fine-arts-division/7859/.

[30] Yi-Chieh Jessica Lin, *Fake Stuff: China and the Rise of Counterfeit Goods* (New York, 2011), 18–9.

[31] Candy Yang and Lisa Li, "Decoding Shan Zhai Ji," *China Youthology* 17 (17 November 2008), http://chinayouthology.com/insights/27. See also Andrew Chubb, "China's Shanzhai Culture: 'Grab-ism' and the Politics of Hybridity," *J. Contemp. China* 24 (2015): 260–79; Michael Keane and Elaine Jing Zhou, "Renegades on the Frontier of Innovation: The Shanzhai Grassroots Communities of Shenzhen in China's Creative Economy," *Eurasian Geogr. Econ.* 53 (2012): 216–30.

[32] S. Lindtner, A. Greenspan, and D. Li, "Designed in Shenzhen: Shanzhai Manufacturers and Maker Entrepreneurs," in *Proceedings of the Fifth Decennial Aarhus Conference on Critical Alternatives*, Aarhus, Denmark, 17–21 August 2015, 85–96.

[33] Lin, *Fake Stuff* (cit. n. 30), 61.

might even signal a new mode of resistance to the central government.[34] But the form has also been at least partially embraced by the government as a true expression of socialist creativity. *Shanzhai* manufacturers are, after all, producing "working-class" information technology products suitable—in price and functionality—for the average (relativity poor) Chinese citizen. Binjie Liu, a Communist Party official from the National Copyright Administration, argued that "*shanzhai* shows the cultural creativity of the common people." There are numerous examples of state-sanctioned and state-led "copying" in various forms.[35]

Within the last decade, the line between *shanzhai* and "legitimate" has also become increasingly blurred. Some formerly *shanzhai* brands have transformed into legitimate companies (e.g., K-touch/Tianyu [天语]).[36] The concept of *shanzhai* allows electronics makers to be simultaneously "guerilla" and legitimate, "cool" and establishment, creative and imitative, and rebellious and orthodox.

PERFORMING *SHANZHAI*

The remainder of this essay describes how BGI can be understood in *shanzhai* terms. I do this by considering various "performances," each related to a different aspect of the term. My aim is neither to celebrate BGI as an innovator or entrepreneur in biomedicine nor to portray it as a factory or "copycat" institution. Rather, understanding BGI as "performing *shanzhai*" affords us a sharper view of what it might mean and represent to China, to Shenzhen, and to its own employees.

Performing the Laboratory

BGI's most foregrounded performance is that of a laboratory, a scientific space. In science studies, laboratories have been described as places that exert careful control over inward and outward flows of people and information; such control affords labs particular authority over knowledge of the natural world.[37] Part of this authority derives from claims to "openness": in the ideal, the work of a lab (if not the lab itself) is open for others to inspect or verify. Although they may not be public spaces, the knowledge labs produce is usually published in "open" academic journals. Reflecting the "communalist" ethos of science, even laboratories belonging to private companies often share their findings.[38] Although individual examples do not always live up to this model, laboratories usually perform "openness" in various ways.

[34] For example, *shanzhai* cultural events, broadcast on TV, have threatened the state's monopolistic control over cultural content. Ibid., 63.

[35] On the rehabilitation of "imitation" by the central Chinese government, see Winnie Wong, "Shenzhen's Model Bohemia and the Creative China Dream," in O'Donnell, Wong, and Bach, *Learning from Shenzhen* (cit. n. 19), 193–212.

[36] Cara Wallis and Jack Linchuan Qiu, "Shanzhaiji and the Transformation of the Local Mediascape in Shenzhen," in *Mapping Media in China: Region, Province, Locality*, ed. Wanning Sun and Jenny Chio (New York, 2012), 109–25.

[37] See, e.g., Steven Shapin, "The House of Experiment in Seventeenth Century England," *Isis* 79 (1988): 373–404; Adi Ophir and Steven Shapin, "The Place of Knowledge: A Methodological Survey," *Sci. Context* 4 (1991): 3–21. On the control of information within science, and esp. in genomics, see Stephen Hilgartner, "Selective Flows of Knowledge in Technoscientific Interaction: Information Control in Genome Research," *Brit. J. Hist. Sci.* 45 (2012): 267–80.

[38] Robert Merton, "The Normative Structure of Science," in *The Sociology of Science: Theoretical and Empirical Investigations* (Chicago, 1973), 267–80.

The manufacturing economy of Shenzhen is also reliant on openness. It is the sharing of knowledge and designs that facilitates "Shenzhen speed." The rapid design, prototyping, and production processes are enabled by cooperation between suppliers, designers, and factory owners. The *shanzhai* economy, then, is one in which designs, hardware, and software are shared (often in violation of intellectual property rules). This, of course, fits very well with a "hacker" or Western "open-source" ethos, and for this reason it has attracted "makers" from around the world to Shenzhen.[39] But in Shenzhen, philosophical motivations are overshadowed by the practicalities of getting new products to market cheaply and rapidly; such "open" approaches are an indigenous development, not a Western import. There is a certain playfulness to this spirit, too—the emphasis on speed and newness often leads to fun and quirky tweaks on phones, wearables, and other electronic devices.

BGI also performs such openness and playfulness. Like other labs, access to BGI is carefully controlled. Journalists and other visitors (like myself) are usually welcomed (on one day, a film crew from *Vice* was shooting inside the lab), but access to space and people is choreographed by a small "communications" department. Within BGI's headquarters and at the China National Gene Bank (CNGB), a visitor finds exactly what one would expect to find in a lab: large, white-walled rooms, laboratory benches scattered with pipettes, and white-coated workers tending to samples or to large arrays of machinery. Significantly, this work takes place largely behind glass—internal windows along the corridors of the headquarters provide a view into the laboratory spaces proper. At CNGB, too, large glass windows provide a vista from the public spaces onto row upon row of DNA-sequencing machines. The laboratory is "on view," open for employees and visitors to see, and simultaneously inaccessible.

But BGI signals its openness in other ways too. A critical part of the organization is BGI's journal, *Gigascience*. The journal has been a leader in promoting open sharing of scientific data.[40] Like others in the "open science" and "open access" movements, the *Gigascience* team believe that traditional scientific publishing models are inadequate. In particular, it suffers from the problem that many scientific journal articles are behind paywalls and remain accessible only to an elite handful of academics. Even if the content of scientific articles is available, often the data on which the findings are based remain unavailable (this means no one can reuse the data, nor can the findings be adequately verified or replicated). With BGI's funding, *Gigascience* has established itself as a leader in promoting open access to their journal articles and open access to data. Most significantly, it was one of the first journals to "publish" data sets.

An important part of *Gigascience*'s mission is ensuring that all BGI-generated data become available to the scientific community. Both *Gigascience* and BGI stress the importance of this openness: widely distributed data will allow the worldwide scientific community to take the utmost advantage of BGI's work and to potentially build on their work to generate further advances in biomedicine and health. Editors at *Gigascience* told me how making BGI's rice genome data public had allowed it to be used by agricultural scientists around the world who were working closely with farmers to

[39] Silvia Lindtner, "Hackerspaces and the Internet of Things in China: How Makers Are Reinventing Industrial Production, Innovation, and the Self," in "Political Contestation in Chinese Digital Spaces," ed. Guobin Yang, special issue, *China Inform.* 28 (2014): 145–67.

[40] Hallam Stevens, "Gigascience: A Science Journal That Provides Full Data Sets," *SmartData Collective*, 13 August 2014, http://www.smartdatacollective.com/hstevens/221646/gigascience.

improve rice varieties.[41] *Gigascience*'s and BGI's commitment to open data and open access plays a key role in BGI's publicity. On one visit to BGI, I received a *Gigascience* branded bottle opener with "Open Everything" stenciled on the side.

BGI has also developed products, such as "Genebook Tipsy," that tap into the playful hacker spirit of *shanzhai*. "Tipsy"—packaged to look like a hardback book—is actually a "spit kit" (fig. 1). The customer fills the enclosed vial with his or her saliva and sends it to BGI for genotyping; BGI will report whether the customer has mutations in genes associated with alcohol metabolism, including the "flush reaction." This is supposed to be lighthearted, but BGI intends to use any profits from the sale of such kits to fund other health research projects. This mixture of fun, novel technology, openness, and serious science suggest how BGI "performs *shanzhai*" through its appearance, its work, and its products.

The performance of openness is critical to BGI's image as a laboratory. But BGI's openness derives only partly from traditional scientific values (such as "communalism"). Rather, the kinds of "openness" articulated by *Gigascience* and BGI's leaders owe more to the *shanzhai* spirit of hackers and sharers. By articulating a radical or total openness, BGI offers a critique of existing (Western) models of scientific practice and publication. Implicit in BGI's work is the idea that its science, like *shanzhai* electronics, offers a chance for China to catch up to or surpass the West. Wide sharing of data, especially data from rice or other food crops, is depicted as a "socialist" science that can bring benefits to ordinary people.

Performing the Factory

It is factories that actually produce the *shanzhai* products found in markets in Shenzhen and elsewhere. Such goods may be produced in "ghost shifts" of the factories that produce "real" products, or, more likely, in other small-scale factories around Shenzhen.[42] To what extent does BGI perform the work of a "factory"? Factories (like Foxconn City) are often extremely closed spaces. Like laboratories, they tightly control flows of people, knowledge, and materials, although their aims are very different. In the production of goods for a profit, factories perform speed, efficiency, and worker discipline.

There are some immediate similarities between BGI and the factories of the Pearl River Delta. Like Foxconn and other factories, BGI has grown rapidly and scaled up quickly to move into a leading position in genomics. The lab, too, has attracted young workers from all over China who mostly come to live in a campus-like environment (although BGI's is not nearly as closed as Foxconn's). Staff eat together and sleep in dorms or apartments near the lab. BGI also looks rather like a factory. It is located in an industrial zone, and the eight-floor main building used to be a shoe factory. BGI headquarters lie east of the center of Shenzhen itself—Yantian is located about twenty-five minutes by car from downtown. The area, surrounded by steep mountains, is dominated by crisscrossing freeways and a large port complex.

[41] Interview with *Gigascience* employee A by Hallam Stevens, 19 July 2014, Hong Kong. All interviews were conducted in confidentiality, and the names of interviewees are withheld by mutual agreement.

[42] Factory bosses may pay workers for "off the book" extra shifts to produce surplus inventory not ordered by the customer; thus "fake" goods may actually be produced during "ghost shifts" at the same factory in which "real" goods are produced. See Huang, *Essential Guide* (cit. n. 28), 16.

Figure 1. *"Genebook Tipsy." A "spit kit" sold by BGI. Purchasers can send their saliva to BGI in the enclosed vial, and BGI will test it for the genetic mutation associated with the so-called Asian flush reaction. (Photograph by Scott Edmunds; used by permission.)*

Like Shenzhen's factories, location is important; BGI's proximity to Hong Kong (where it also has laboratories and offices) provides an important conduit to the outside world. BGI also benefits from the kinds of subsidies and tax breaks that have attracted other factories to Shenzhen's SEZ. BGI was enticed to Shenzhen by a US $12.8 million grant from Shenzhen's municipal government as well as free rent for three years.[43]

Perhaps most important, BGI's work also performs "production." The lab has been, and continues to be, focused on high-throughput aspects of biology, especially DNA and RNA sequencing. The process of sequencing involves extracting DNA from living cells and "reading out" its genetic code. "Reading" a molecule—even a large one like DNA or RNA—requires reproducing it millions of times in order to produce a large enough signal to be detected by a sequencing machine. These machines usually make

[43] Christina Larson, "Inside China's Genome Factory," *MIT Technology Review*, 11 February 2013, https://www.technologyreview.com/s/511051/inside-chinas-genome-factory/.

use of biology's own tools, repurposing enzymes such as DNA polymerase to perform the copying required for sequencing. Despite the involvement of machines, large-scale sequencing requires significant amounts of repetitive labor: extracting DNA, preparing samples, loading machines, and analyzing the data. Large-scale DNA sequencing deploys factory-like labor to enable such molecular duplications.

Many kinds of biological and biomedical work now involve DNA or RNA sequencing. For example, distinguishing different types of cancer cells, identifying different pathogenic bacteria, understanding the differences between varieties of plants (such as rice), and identifying genetic diseases in human embryos all involve sequencing. Such sequencing work can be "outsourced" by sending cells or DNA off-site. Like other outsourcing, large-scale, highly automated DNA sequencing offers economies of scale. It is this "production" sequencing work that BGI's reputation is largely built on—it advertises itself as such and makes money by selling large-scale sequencing services. Customers send their cells or samples to BGI, and BGI sends back thoroughly analyzed, high-quality DNA (or RNA) data. For BGI, DNA sequencing has become its commodity product.

In some cases, BGI is happy to receive credit instead of cash. If scientists external to BGI are conducting high-profile work that requires DNA or RNA sequencing, BGI may do the sequencing in exchange for coauthorship on the resulting scientific papers. As Winnie Wong has argued, this has led to the perception that BGI is a "factory" for scientific journal articles, especially those in highly ranked publications such as *Science* and *Nature*. Significantly, the framed covers of these publications are prominently displayed in the "museum" at BGI headquarters alongside BGI's sequencing machines. Some within the scientific community perceive this as an illegitimate "copy-catting" of "real" scientific work. In other words, this gaming of the academic publication system has gained BGI a reputation for "faking" science.[44]

BGI has certainly done a lot of DNA sequencing, including participating in the Three Million Genomes Project, the Rice 10,000 Genome Project, the Earth Microbiome Project, the 1000 Genome Project, and the 10,000 UK Genome Project.[45] But BGI has also expanded to all kinds of other domains including proteomics and cell biology, cloning, bioinformatics and big data analysis, biobanking, reproductive health, pharmacogenomics, cancer therapy, metabolism and cardiovascular research, geriatrics, agricultural biotechnology and breeding, low-carbon economy research, marine organism breeding, epidemic surveillance, SARS research, and women's health. In fact, it is hard to think of any area of medical or agricultural research in which BGI is not now involved. Nevertheless, much of this work is also based on DNA and RNA sequencing; that is, these other domains are explored through sequencing more and more DNA, faster and faster.[46] Speed, efficiency, and consistent, high-quality output are highly valued.

But in sequencing DNA, BGI is not only performing imitative work. For example, BGI scientists also spoke to me about their development of a noninvasive prenatal

[44] Wong, "Speculative Authorship" (cit. n. 1).

[45] "BGI Unveils Significant New Global Research Collaborations at the 6th International Conference on Genomics," press release, 16 November 2011, http://www.genomics.cn/en/news/show_news?nid=98952.

[46] E.g., BGI's primary work on cancer involves developing cheaper, faster, less invasive, and more sensitive tests that can be performed by detecting tumor DNA in a patient's blood via sequencing.

genetic screening program conducted in Shenzhen (called NIFTY).[47] The lab worked with local public health authorities in Shenzhen to develop low-cost tests that could be covered under social insurance schemes. Although prenatal tests for many genetic diseases already existed, these tests were expensive or invasive, or they involved a lengthy wait for results. BGI's work involved developing versions of the test that could be performed quickly on sequencing machines from a routine maternal blood test. These platforms utilized existing sequencing technologies and deployed existing knowledge of genetic diseases. However, development of the tests involved putting these technologies and knowledge to work in novel ways: speeding up the sequencing process, lowering the cost, and making them accessible to more people.[48] Such work is both novel and valuable in multiple ways: economically valuable for BGI's owners, scientifically valuable for BGI's workers, and biomedically valuable for Shenzhen's citizens. It enacts a form of "socialist" science—a science specifically directed toward the people.

BGI places the DNA double helix at the center of its branding. A 20-foot-high mural near the entrance to the laboratory depicts an Orwellian hand gripping (crushing?) a helix (fig. 2). The helix motif appears on T-shirts, coffee cups, and stationery. Part of BGI's performance, then, is one of copying—copying of DNA, copying of models of scientific publication, and copying the model of the factory. Like *shanzhai*, BGI depends—historically, economically, and institutionally—on the copy. But like the mobile phones at Huaqiangbei, BGI's copies are not mere copies, because they carry meaning beyond that of "imitation" or "fake." Like *shanzhai* phones, BGI's products are produced in a hybrid space: BGI recombines elements of factories and laboratories, open and closed, automation and innovation, production and invention. It values *both* traditional scientific production (publication) and the kinds of speeding up, efficiency, and control over space and people that are associated with factories. Many commercial laboratories in China and elsewhere combine science with for-profit enterprise. But "performing *shanzhai*" draws attention to the unique ways in which BGI combines imitation, automation, speed, socialism, and commodity production with more traditional forms of scientific knowledge making.

Performing the Start-Up

If factories form the foundation of Shenzhen's *shanzhai* economy, then the superstructure consists of small-scale electronics entrepreneurs. The markets at Huaqiangbei are populated by thousands of small entrepreneurs trying out new designs and building new devices. The innovativeness of these individuals stems partly from their ability to circumvent the rules, recombining, reappropriating, and hybridizing without borders. It is therefore partly a renegade set of practices. It exists largely in grey zones and grey markets; its institutional loci are multiple and ambiguous. It is effective partly because of its semilegitimacy. BGI mimics such performances. Borrowing from both local models (*shanzhai*) and foreign ones (Silicon Valley), BGI acts out the model of a kind of "start-up." In Silicon Valley, start-ups offer disruptive innovations

[47] Interview with BGI employee B by Hallam Stevens, 17 July 2014, Shenzhen.

[48] See, e.g., Yan Xu, Xuchao Li, Hui-juan Ge, Bing Xiao, Yan-Yan Zhang, Xiao-Min Ying, Xiao-Yu Pan, et al., "Haplotype-Based Approach for Noninvasive Prenatal Tests of Duchenne Muscular Dystrophy Using Cell-Free Fetal DNA in Maternal Plasma," *Genet. Med.* 17 (2015): 889–96.

Figure 2. *"My Life in My Hands." Displayed on a 20-foot-high billboard adjacent to the entrance to BGI. (Photograph by Hallam Stevens.)*

that aim to remake entire industries through information technology. Start-ups have also become zones of organizational experimentation, refiguring the work and lives of their employees in novel ways. In Shenzhen, such technological and cultural disruption intersects with the "renegade" performances of *shanzhai*.

BGI certainly has a Silicon Valley–style garage start-up myth. In September 1999, BGI was founded as a "dysfunctional adjunct" to the Chinese Academy of Sciences Institute of Genetics. Participating in the Human Genome Project was a radical idea in China, and BGI remained a small enterprise with fifty employees, limited capital, and a few sequencing machines. China's slow start in genetics and biotechnology meant that BGI was largely on its own.[49] In the early days, the operation was crowded into an apartment in Beijing: "Their furniture consisted of the cardboard shipping boxes that had contained their new equipment."[50] Later, the company moved to a warehouse near the Beijing airport.[51] These oft-repeated stories form an important part of the ethos of BGI: Huanming Yang and the other founders were mavericks who defied normal (Chinese) scientific practices, taking risks that ultimately paid off.[52]

[49] Laurence Schneider, *Biology and Revolution in Twentieth-Century China* (Lanham, Md., 2005), chap. 9.

[50] Specter, "The Gene Factory" (cit. n. 2).

[51] See BGI's website, http://www.genomics.cn/en/navigation/show_navigation?nid=2946 (accessed 14 August 2017).

[52] BGI's founders were all educated outside China, returning in the 1990s as partial outsiders.

BGI took a further risk by divorcing itself from the Chinese Academy of Sciences and moving away from Beijing. This shift has allowed BGI to portray itself as "renegade," eschewing what BGI leaders perceived as the slow-paced science and moribund bureaucracy of the capital and moving to the "frontier" of the SEZ. The move to Shenzhen became a way for BGI to free itself of political constraints and government influence. In the words of one investor speaking to the *Financial Times*: "Shenzhen is as far from Beijing as you can get."[53] In statements such as this, BGI attempts to distance itself from the Beijing government and from the more established centers of Chinese science.

This distance has given BGI the opportunity not only to engage in new and different kinds of scientific work (namely, genomics) but also to construct different (and particularly more capital-oriented) ways of practicing science. After its move to Shenzhen, BGI was labeled "the first citizen-managed, non-profit research institution in China"—a new type of institution.[54] "We represent a new model of an international Chinese organization," Wang Jun (BGI's former chief executive officer) has said, acknowledging possible tensions between business and basic research. "If we are too commercial, we lose sight of the future, . . . but if we are only thinking of the future, that isn't suitable either."[55] BGI's leaders recognize the need to adopt and combine multiple roles—academic, industrial, commercial, and nonprofit.

BGI's funding and revenue model is also similar to that of a start-up. Although BGI has made some money selling products and services, it has also succeeded in establishing lucrative research partnerships and contracts with corporations, governments, and NGOs. BGI receives funding from universities, pharmaceutical companies, the Chinese central government, local governments (such as Shenzhen's), international organizations such as the Bill and Melinda Gates Foundation, and even the Silicon Valley venture capital firm Sequoia Capital.[56]

In many ways, BGI's organization of people and work performs the start-up image. Workers report little factory-like discipline within the organization. Most of the staff are free to come, go, and work according to flexible schedules. For those working with high-throughput experiments, their days revolve largely around the rhythms and cycles of the machines themselves (the "runs" of a DNA-sequencing machine, for example). For bioinformatics staff, it is not unusual to find them coding late into the night and sleeping through the mornings or napping at their desks during the day. A sort of antihierarchy is enforced: business suits are forbidden at headquarters. Flexible teams are convened to work on specific projects that need not be initiated or led by scientists with PhDs; individuals are judged by their ability to work together to produce high-quality publications or technologies by specified deadlines.

[53] Henny Sender, "Chinese Innovation: BGI's Code for Success," *Financial Times*, 17 February 2015, https://next.ft.com/content/9c2407f4-b5d9-11e4-a577-00144feab7de.

[54] The original source of this phrase is unclear, but it has been widely circulated.

[55] Quoted in Sender, "Chinese Innovation" (cit. n. 53).

[56] BGI has received significant funding from the state-operated China Development Bank ("BGI to receive $1.5B in 'Collaborative Funds' Over 10 Years from China Development Bank," 29 January 2010, http://www.genomics.cn/en/news/show_news?nid=98698). The Gates Foundation and BGI have an ongoing collaboration in the fields of global health and agricultural development ("BGI and the Bill and Melinda Gates Foundation Sign Memorandum of Understanding on Collaboration for Global Health and Agricultural Development," 25 September 2012, http://www.genomics.cn/en/news/show_news?nid=99229). This has included funding for the 10K International Rice Project (http://www.genomics.cn/en/navigation/show_navigation?nid=2679 [accessed 14 August 2017]). On Sequoia's investment, see Sender, "Chinese Innovation" (cit. n. 53).

The BGI staff I spoke with emphasized the "openness" and "flatness" of their organization.[57] They stressed that bosses were accessible and that people worked and talked across projects. While these claims may have been exaggerated, the kinds of spaces and interactions I observed seemed to support this notion of flatness. The internal laboratory spaces are characterized by open plan offices, frequent face-to-face interaction, and ample space for informal gatherings.

BGI intervenes in its employees' lives in an attempt to create a specific institutional culture. Social lives are dominated by the lab. Extracurricular activities (including various team-building exercises and a thriving basketball league) form a central part of the day-to-day life of the staff (fig. 3). During their time off, workers organize English conversation classes or hike up the steep mountains behind the lab: "we work together and we play together," one staff member told me.[58] A senior member of the lab stressed how important this was for young people in a society where many had come from single-child families and were not necessarily used to playing well with others. BGI's employees present themselves as community-minded team members who are enthusiastic about their work and mission.

Posters on the walls of the lab exhorted workers to take part in myriad activities, not only as a means of fostering teamwork but also as a means of staying healthy. Contributions to global biomedicine required, one poster noted, "first making yourself healthy." Between meals, employees could find healthy snack foods distributed at various "bars" located on each floor. The fruit and other items are provided at a nominal cost (staff were supposed to pay by putting money in a box next to the food). The elevator in the eight-story headquarters building had been programmed to stop only at floors 1, 5, and 8 in order to encourage employees to take the stairs. Even more dramatic were posters encouraging the use of a monitoring device called "GeneBook" that would feed data about employees' blood pressure, cholesterol, and other health information into an app (fig. 4). Such monitoring was again in the name of keeping employees healthy.

Google is famous for providing, at its Googleplex in California, a sort of work-play environment complete with free food; table tennis, pool, and foosball tables; video games; gyms; and a variety of party and entertainment options in a spectacular campus-like environment designed to attract the best young (mostly American) talent.[59] BGI is not (yet) quite like the Googleplex.[60] But it is deliberately cultivating some of the same characteristics: the nonhierarchical organization, the integration of living and working, the intensive socialization, and the attention to health and play.

But beyond this Silicon Valley model, BGI also draws more concretely on the entrepreneurial culture that pervades Shenzhen. Although Huaqiangbei and *shanzhai* are centered on electronics, not biomedicine, the high-tech buzz of the city is important for attracting talented, technically minded young people to Shenzhen. As BGI expands its business into sequencing machines, diagnostic kits, and bioinformatics, there is increasing crossover with the computational and electronic domains. In 2015, BGI's former executive director, Jun Wang, left BGI to form *iCarbonX* (碳云智能). The com-

[57] Interview with BGI employee C by Hallam Stevens, 28 June 2017, Shenzhen.
[58] Interview with BGI employee D by Hallam Stevens, 17 July 2014, Shenzhen.
[59] Steven Levy, *In the Plex: How Google Thinks, Works, and Shapes Our Lives* (New York, 2011).
[60] In the near future it will certainly look a little more like Google: a new purpose-built "campus" for BGI will be completed in 2018.

Figure 3. *"2014 BGI Basketball League." Employees at the laboratory engage in a range of recreational activities, including a basketball league. (Photograph by Hallam Stevens.)*

pany plans to combine genomics, proteomics, metabolomics, and microbiomics with artificial intelligence in order to contribute to drug development, cosmetics, nutrition, diagnostics, and health profiling. One of its major funders is TenCent.[61]

[61] Eva Xiao, "iCarbonX Becomes China's First Biotech Unicorn," *technode*, 14 April 2016, http://technode.com/2016/04/14/icarbonx-becomes-first-chinas-biotech-unicorn-1-billion-rmb-series/; Nick Paul Taylor, "Ex-BGI CEO Startup Hits $1B Valuation in 6 months," *Fierce Biotech*, 15 April 2016, http://www.fiercebiotech.com/it/ex-bgi-ceo%E2%80%99s-startup-hits-1b-valuation-6-months.

Figure 4. *"BGI Manufacturing's Genebook." Poster displayed inside the laboratory. These devices are made and promoted by BGI to monitor their employees' health. The second line reads, "BGI employee welfare!" (Photograph by Hallam Stevens.)*

This kind of work fits naturally into a city that is now considered the world capital of open hardware. This became particularly evident after BGI's acquisition of Complete Genomics (based in Mountain View, Calif.). When BGI attempted to design its own sequencing machine based on Complete's technologies, it was able to recruit a crack team of hardware hackers and engineers to work in Shenzhen. Drawing on local expertise, the availability of raw materials, and rapid prototyping capabilities of the city, the team was able to produce the BGISeq 500 in six months.[62] Such "disruption" of the DNA-sequencing industry emerged from a literal combination of Silicon Valley and Shenzhen know-how. Significantly, so far, the machine is only for sale in China because some of its technology remains in patent disputes in the United States and Europe.[63]

A start-up model infuses the ideals and the mission of BGI, too. This is a vision that comes from the top. In the Danish documentary *DNA Dreams*, BGI Chairman Huanming Yang is captured giving a rousing speech at the 2011 International Conference on Genomics: "I have a dream," he begins, invoking a famous renegade. If Martin Luther King Jr.'s dream was one of racial justice and harmony, Yang presents his goals as no less socially and politically important. For BGI, the dream is a techno-humanist vision of solving the world's health and social problems through biomedicine. This vision inspires and animates many of those I met at BGI. Of course, BGI provided a stable (even prestigious) job with good pay, but many employees saw their work as part of a larger, and important, project. This ambition has socialist (improving health care), nationalist (securing China's economic future), and global (exporting its success) dimensions.

Silicon Valley's dream, deeply rooted in the history of the Internet and the counterculture, revolves around the transformation of society through information and communication technology.[64] In particular, it sees technology as the means for spreading economic and political freedoms. Self-styled "disruptors" present themselves as renegades, railing against conventional wisdom and practice. BGI's rhetoric serves a similar purpose—it presents the company and its employees as outsiders, rebels, and heroes, battling against tradition, convention, and bureaucracy. This has a particular salience in Shenzhen, a city known not only for *shanzhai*, but also for its unorthodox capitalist and political experimentation. By mixing together elements of the laboratory, the factory, and the renegade start-up, BGI produces an institution that is simultaneously imitative and innovative, "Californian" and Chinese, capitalist and socialist, open and closed.

LIFE AS COPY

A visit to Shenzhen is not complete without a stop at its main tourist attraction—"Window of the World." Here, locals and tourists can see the Eiffel Tower (354 feet high!), the Pyramids, the Taj Mahal, the Acropolis, the Coliseum, Angkor Wat, Mount Fuji, the Sydney Opera House, Mount Rushmore, and the White House all in one place. The theme park is Shenzhen's attempt to capture the wonders of the world

[62] BGI had another team working concurrently at Complete's headquarters in California. They designed a much larger sequencing machine called "Revelocity," of which only a handful have been produced. See Dave Yuzuki, "BGISeq-500 debuts at the International Congress of Genomics 10," *Weblog*, 24 October 2015, http://www.yuzuki.org/bgiseq-500-debut-at-the-international-congress-of-genomics-10/.

[63] Keith Robison, "BGI Launches BGISeq 500," *Weblog*, 27 October 2015, http://omicsomics.blogspot.nl/2015/10/this-weekend-brought-formal-launch-of.html.

[64] For the history of these ideas see Fred Turner, *From Counterculture to Cyberculture: Stewart Brand, The Whole Earth Network, and the Rise of Digital Utopianism* (Chicago, 2006). For a critique, see Richard Barbrook and Andy Cameron, "The Californian Ideology," *Sci. Cult.* 6 (1996): 44–72.

for China. This theme park reminds us of the purpose of Shenzhen as an SEZ: to bring the world to China and China to the world. Each of Window's elements is, of course, a "copy," but bringing these elements together in one place amounts to something different and something more. Ultimately, BGI's project should be understood on this scale and in this context—it is an attempt not merely to "copy" but to make something new by taking and combining elements from around the world. It is an attempt to dominate the world through replication and recombination.

"Performing *shanzhai*" suggests a variety of ways for understanding BGI and its multiple meanings for China, Shenzhen, and its own employees. It suggests how BGI can be understood as at once a space of copying/faking *and* a space of innovation, invention, re-creation, and openness. It suggests how it is drawing on Shenzhen's history, culture, and social environment, as well as on those of China more broadly. Here, the notion of performance—attending to space, costumes, buildings, setting, slogans, and publicity—sheds light on what BGI is attempting to be and do on its own terms. This framework seeks to avoid either celebrating or denigrating BGI's work, seeing it neither as "fake" or "copycat" nor as a performance of science in the ways with which we are most familiar.

One corollary of this argument is that thinking about BGI as a factory only captures a small slice of what is interesting, important, and unique about it. BGI-as-factory underestimates its possibilities for transforming how we think about and do biological work. BGI draws on many kinds of models—start-up, NGO, industrial lab. Labeling BGI a "factory" is ultimately rooted in contemporary Western notions of China as a zone of "copycatting." This label mischaracterizes what is actually occurring in *shanzhai*. In particular, it misses the innovative and creative aspects of "copying" work. The kind of labor that is performed may actually pose a challenge to contemporary categories of "originality."[65] As Eugenia Lean shows, China has long been portrayed as a space of dangerous copying.[66] In the early twentieth century, too, this reflected Western anxieties about the limits of capitalist and colonial power in an era when intellectual property rules were very much in flux. But in this context, too, "copying" was not slavish imitation. "Masterful copying" provided opportunities for innovation, invention, and creative rearrangement that had "virtuous" qualities. In Shenzhen, *shanzhai*, and BGI we see how "copying" is recast as variously "virtuous": anticapitalist, nationalist, or altruist.

In her book about Dolly, Sarah Franklin attempts to understand why the 1996 cloning of a sheep elicited such wide public excitement. Why do we fear the copy? Ultimately, Franklin suggests, we fear our own individuality and uniqueness may be called into question.[67] Copying threatens some (Western, liberal) basic beliefs about human nature. No doubt such notions are also sharpened by western fears about loss of global economic power, manufacturing industries, and jobs. But this denigration of the copy is culturally and historically located. Historians agree that the dominant form of "making" before the middle of the nineteenth century in the West was the copycat; Philip Scranton has argued that American manufacturing between the Civil War and World War I consisted of small, flexible, specialists working in an ecology

[65] Wong, *Van Gogh* (cit. n. 29), 24–5.

[66] Eugenia Lean, "The Making of a Chinese Copycat: Trademarks and Recipes in Early Twentieth-Century Global Science and Capitalism," in this volume.

[67] Sarah Franklin, *Dolly Mixtures*: *The Remaking of Genealogy* (Durham, N.C., 2007).

that seems similar in many respects to *shanzhai*.[68] And indeed in the West, in debates about music, movies, and especially software, the copy is being rehabilitated: remix is increasingly a celebrated and legitimate (even virtuous) form of copying.[69] Copying—especially in the form of open source and open access—is a form of political resistance and a pushback on corporate power. This "hacktivist" spirit is partly what animates *shanzhai* and draws people to Shenzhen. BGI also participates in this movement, explicitly embracing the rhetoric and practices of open source and open access. It, too, seeks forms of "copying" that are virtuous: innovative, nationalistic, socialistic. In this sense, BGI is part of a broader movement that is forcing us to reevaluate our moral indictment of the copy: not mere copying, but rather copying that is playful, copying as productive scaling up, copying that includes novelty, copying for social good, copying as sharing. The DNA molecule is a symbol of both life and the copy; copying is necessary for life and for pursuing "life itself" in Shenzhen.

Beginning in California in the 1970s, the biotechnology industry reconfigured bioscience work in various ways. The modes of work that the biotech industry engendered were deeply influenced by the culture of Silicon Valley and Stanford University, even as biotech spread outward from there.[70] Labels like "commercialization" and "industrialization" do not capture the complexity of these transformations. Rather, historians and sociologists have turned to terms such as "biocapital," "nature, enterprised up," "biovalue," and "bioeconomics" to characterize the mixing of life science with regimes of production, consumption, capital, and ownership.[71] But for BGI, and China, as the life sciences develop, such categories seem increasingly inadequate or incomplete. *Shanzhai*—with its connotations of the copy, the renegade, the socialist worker, and the entrepreneur—presents a new category that more fully reflects the kinds of work, ideals, and hopes that are being enacted at BGI. This is biomedicine performed through imitation, industrial scaling up, rampant capitalist ambition and speculation, socialist spirit, and nationalism. BGI is remaking "life itself" in the image of Shenzhen.

[68] Philip Scranton, *Endless Novelty: Specialty Production and American Industrialization, 1865–1925* (Princeton, N.J., 1997); Hillel Schwartz, *The Culture of the Copy: Striking Likenesses, Unreasonable Facsimiles* (New York, 1998).

[69] Lawrence Lessig, *Free Culture: The Nature and Future of Creativity* (New York, 2004).

[70] Eric J. Vettel, *Biotech: The Countercultural Origins of an Industry* (Philadelphia, 2006).

[71] Stefan Helmreich, "Species of Biocapital," *Sci. Cult.* 17 (2008): 463–78.

ENTANGLED CALCULATIONS

Compound Interest Corrected:
The Imaginative Mathematics of the Financial Future in Early Modern England

*by William Deringer**

ABSTRACT

What is money in the future worth today? In the seventeenth century, questions about the "present value" of future wealth became matters of practical concern, as businesspeople and governments deployed future-oriented financial technologies like mortgages, bonds, and annuities. Those questions also attracted the attention of mathematicians. This essay examines the excursions two English mathematicians, the indefatigable mathematical gossip John Collins (1625–83) and the lesser-known Thomas Watkins (fl. 1710s–20s), made into the mathematics of financial time. In capitalist practice today, present-value problems are invariably dealt with using a single technique, compound-interest discounting, which has become deeply embedded in commercial, governmental, and legal infrastructures. Yet, for early modern thinkers, the question of how best to calculate the financial future was an open question. Both Collins and Watkins explored imaginative alternatives to what would become the compound-interest orthodoxy. With help from his network of correspondents, Collins explored simple-interest discounting, which provoked thorny mathematical questions about harmonic series and hyperbolic curves; Watkins crafted multiple mathematical techniques for "correcting" the compound-interest approach to the financial future. Though both projects proved abortive, examining those forgone futures enables us to examine the development of a key element of capitalistic rationality before it became "black-boxed."

Among the earliest surviving pieces of correspondence written by a young Isaac Newton are a series of three letters written in January and February 1670 to John Collins, an accountant, sometime government clerk, and mathematics instructor who was also the era's most enthusiastic dealer in mathematical gossip.[1] The two had met in

* Program in Science, Technology, and Society, Massachusetts Institute of Technology, 77 Massachusetts Avenue, E51-188, Cambridge, MA 02139; deringer@mit.edu.

I would like to thank all of the participants in the June 2016 *Osiris* workshop for their thoughts on an early draft of this essay. I would especially like to thank my coeditors, Lukas Rieppel and Eugenia Lean; the *Osiris* general editors, Suman Seth and Patrick McCray; and two anonymous reviewers for their thoughtful comments on the manuscript. I would also like to thank Katherine Marshall and the staff at the Royal Society for assistance with images.

[1] Isaac Newton to John Collins, 19 January 1669/70, MS Add. 9597/2/18/1, Cambridge University Library (hereafter cited as "Camb. UL"); Newton to Collins, 6 February 1669/70, MS Add. 9597/2/18/2, Camb. UL; Newton to Collins, 18 February 1669/70, MS Add. 9597/2/18/5, Camb. UL. These

 0369-7827/11/2018-0006$10.00

November or December of the previous year, and Collins, clearly impressed by his new associate's talents, passed along some problems in algebra he was eager to see solved. These letters have piqued the interest of Newton scholars and historians of mathematics, both for their mathematical content—they contained creative work on infinite series and on approximating solutions to equations and initiated a dialogue that ultimately inspired Newton's *Universal Arithmetick*—and because those letters marked one of the first times the publicity-shy Newton circulated mathematical work to a larger community.[2] But one crucial feature of those letters has not garnered much notice: they were talking about finance.

In particular, the two were discussing mathematical problems that arose concerning annuities, a kind of financial instrument that promises to pay the owner a recurring income over some future period. The question at hand in Newton's second letter was indicative. Consider an annuity that promised the owner £100 every year for thirty-one years. Imagine that the owner of that annuity had purchased it by paying £1,200 up front. What interest rate, assuming compound interest, was the owner effectively making on the original £1,200 laid out? Using logarithms, Newton constructed a formula to estimate the answer within a few pence—7.43 percent in the particular case in question. Newton's first letter had involved calculating the "present value" of an annuity using a technique called simple-interest discounting, an even more vexing problem closely related to finding the area under a hyperbola (on which more to come). These questions were not merely of academic interest. They had significant commercial utility, as finding more accurate and efficient computational methods for dealing with annuities promised to streamline many common financial transactions. But they also hold a deeper interest for historians of science and capitalism, one that goes beyond either their mathematical ingenuity or financial utility. Within them lay profound, unsettled questions about economic value and future time that would have serious consequences for economic practice and culture in ensuing centuries.

The exchange between Newton and Collins was hardly an isolated occurrence, either. During the period from roughly 1660 to 1720, several British "men of science" turned their attention to what could be called the mathematics of financial time. These included several members of the new Royal Society of London, ranging from relatively obscure figures, like Collins (1625–83) and Thomas Watkins (fl. 1710s–20s)—the two main characters in this essay—to far more eminent ones, like Edmund Halley, James Gregory, and of course Newton.[3] Those mathematicians became interested in an array of problems related to debt, interest, and time-dependent financial transactions like annuities and mortgages. These problems arose in numerous areas of commercial life, from loans to land transactions to the dealings that took place in London's "Exchange Alley." They were increasingly a matter of political concern as well, as Britain's government came to rely increasingly on long-term public debt to

---

and other letters written by Newton are available online through the *Newton Project*, http://www.newtonproject.ox.ac.uk/ (accessed 16 April 2018).

[2] Richard Westfall, *Never at Rest: A Biography of Isaac Newton* (Cambridge, 1980), 222–6; Helena Pycior, *Symbols, Impossible Numbers, and Geometric Entanglements: British Algebra through the Commentaries on Newton's "Universal Arithmetick"* (Cambridge, 1997), 177.

[3] For Halley's contribution, see "An Estimate of the Degrees of the Mortality of Mankind, Drawn from Curious Tables of the Births and Funerals at the City of Breslau; With an Attempt to Ascertain the Price of Annuities upon Lives," *Phil. Trans.* 17 (1693): 596–610.

fund state activities, especially after 1688.[4] Implicit in all of these practical mathematical problems was the fundamental question of how to understand—and how to calculate—the relationship between the economic present and the economic future.

From a historical perspective, there is nothing especially remarkable about elite mathematical thinkers exploring questions that arose in commercial settings—one can find countless examples, from al-Khwārizmī (ca. 780–850) to Leonardo of Pisa ("Fibonacci," ca. 1170–1250) to Benoit Mandelbrot (1924–2010). What is most remarkable about the financial-mathematical excursions embarked upon by these Royal Society calculators was the questions they explored and the possibilities they imagined, and what their mathematical explorations can tell us about the economic moment in which they lived. In short, they were using mathematics to explore the economic future—indeed, to imagine an array of different possible economic futures, many of which ultimately did not come to pass. In his epistolary exchange with Newton, and many other mathematicians besides, John Collins was actually experimenting with two very different approaches to putting a price on the future. In the beginning of the next century, Thomas Watkins published a paper laying out some half dozen different approaches. The shape of the financial future was yet to be determined; for adventurous mathematical thinkers like Collins and Watkins, it was a research question.

It would not remain such an open question for long. By about 1800, most of the problems about value and time that these early modern mathematicians were exploring would no longer seem especially problematic. By that point, a single dominant paradigm had emerged for dealing with all such temporal calculations, an approach we can call compound-interest discounting. That technique used the exponential logic of compound interest as the essential means of translating between the present and the future. In the ensuing centuries, that compound-interest technique would be inserted into countless economic practices: the valuation of financial instruments (bonds, stocks, futures), real estate transactions, insurance, retirement planning, firms' capital budgeting decisions, economic modeling, governmental cost-benefit analysis—even climate-change planning. Nearly any specialized domain of economic activity that involves a temporal element relies on compound-interest discounting in some way.[5]

To use a classic formulation from science and technology studies (STS), compound-interest discounting would become a quintessential black box, so deeply embedded in

[4] For an introduction to practical financial calculations and financial texts in the period, see Natasha Glaisyer, "Calculating Credibility: Print Culture, Trust and Economic Figures in Early Eighteenth-Century England," *Econ. Hist. Rev.* 60 (2007): 685–711. On the broader financial context in the period, see P. G. M. Dickson, *The Financial Revolution in England: A Study in the Development of Public Credit* (London, 1967); John Brewer, *The Sinews of Power: War, Money, and the English State, 1688–1783* (London, 1989); Anne L. Murphy, *The Origins of English Financial Markets: Investment and Speculation before the South Sea Bubble* (Cambridge, 2009); Carl Wennerlind, *Casualties of Credit: The English Financial Revolution, 1620–1720* (Cambridge, Mass., 2011).

[5] There are a variety of technical labels for such discounting techniques in modern business and accounting practices, notably "net present value" (NPV) and "discounted cash flow" (DCF). For overviews of the history and significance of exponential discounting techniques, see R. H. Parker, "Discounted Cash Flow in Historical Perspective," *J. Account. Res.* 6 (1968): 58–71; Scott P. Dulman, "The Development of Discounted Cash Flow Techniques in U.S. Industry," *Bus. Hist. Rev.* 63 (1989): 555–87; Liliana Doganova, "Décompter le Futur: La Formule des Flux Actualisés et le Manager-Investisseur," *Soc. Contemp.* 93 (2014): 67–87. On climate change and discounting, see David Weisbach and Cass R. Sunstein, "Climate Change and Discounting the Future: A Guide for the Perplexed," *Yale Law Policy Rev.* 27 (2009): 433–57.

capitalist infrastructures that those practitioners who used it would never question it and most lay people would never know of its critical importance.[6] In the late seventeenth and early eighteenth centuries, though, things were far less settled. The black box had not yet been closed. Alongside compound-interest discounting, Royal Society calculators explored an array of substantially different approaches to the mathematics of financial time. Many of the mathematical excursions carried out by the likes of John Collins and Thomas Watkins did not necessarily get very far and left only limited marks on the development of mathematical research or financial practice. They make for an admittedly abstruse topic for historical research—even within fields like the history of mathematics and the history of accounting, which have high thresholds for technical indulgence. But, I want to suggest, these kinds of technical sources are also remarkably useful sources for studying the history of capitalism. By digging into the technical details of these exploratory calculations, we can glimpse early modern people wrestling with profound questions about value, time, and the future while those questions were still up for grabs.[7] By allowing us to think alongside earlier actors for whom the economic future might have unfolded very differently, these calculations help us to denaturalize an essential piece of capitalism's code.

One of the most striking trends in recent historical and social scientific scholarship on both capitalism and science and technology has been a recognition that "the future matters"—that is, that people's attitudes, aspirations, and images of future time have a profound impact on life in the present.[8] In a sweeping new book, economic sociologist Jens Beckert has argued that capitalism is effectively defined by a particular, forward-looking temporality: "Capitalism is a system in which actors—be they firms, entrepreneurs, investors, employees, or consumers—orient their activities toward a future they perceive as open and uncertain." In his account, people are pulled into economic action—buying, investing, learning, working, consuming—by their imagination, by their sense that acting will produce some (better) expected future. This "temporal disposition" is "crucial to understanding . . . how capitalism diverges from the economic orders that preceded it."[9] Scholars in STS have made a similar observation, stressing that "imaginaries" of the future play a critical role in motivating and organizing scientific and technical enterprise.[10] Scholars of modern science and modern capitalism observe that, in both domains, the future can become a crucial site of

[6] On "black boxes," see the classic statement by Bruno Latour, *Science in Action: How to Follow Scientists and Engineers through Society* (Cambridge, Mass., 1987).

[7] For further discussion of "financial quantification conventions" and capitalism, see Eve Chiapello and Christian Walter, "The Three Ages of Financial Quantification: A Conventionalist Approach to the Financiers' Metrology," *Hist. Soc. Res./Hist. Sozialforsch.* 41 (2016): 155–77, esp. 161.

[8] On recent historiographical interest in the "future," see Susan Friend Harding and Daniel Rosenberg, *Histories of the Future* (Durham, N.C., 2005); David C. Engerman, "Introduction: Histories of the Future and the Futures of History," *Amer. Hist. Rev.* 117 (2012): 1402–10, and other papers in the forum. Note also the workshop "Histories of the Future," convened at Princeton University by Fred Gibbs, Erika Lorraine Milam, and Joanna Radin, 6–7 February 2015. Papers from that event can be found at histscifi.com (accessed 16 April 2018).

[9] Jens Beckert, *Imagined Futures: Fictional Expectations and Capitalist Dynamics* (Cambridge, Mass., 2016), 1–2.

[10] George E. Marcus, ed., *Technoscientific Imaginaries: Conversations, Profiles, and Memoirs* (Chicago, 1995); Sheila Jasanoff and Sang-Hyun Kim, "Containing the Atom: Sociotechnical Imaginaries and Nuclear Power in the United States and South Korea," *Minerva* 47 (2009): 119–46; Jasanoff and Kim, eds., *Dreamscapes of Modernity: Sociotechnical Imaginaries and the Fabrication of Power* (Chicago, 2015).

political conflict in the present as well. In his study of scientific and technical "visioneers" in the 1970s and 1980s, Patrick McCray writes that "creating visions of the future and the technologies that might help shape it is a political act as well as an exercise of imagination. But," McCray observes, "the future is not a neutral space."[11] As Beckert explains, "to have power means: My expectations count!"[12]

This suggests that, in reconstructing the development of both capitalism and science, historians ought to attend closely to the practices of future construction and to changing orientations toward the future in the past. In particular, we ought to identify and examine moments when some aspect of the relationship between present and future came into question, and to examine the practices and interests through which those temporal contests were conducted. I contend that the quirky computational excursions of our Royal Society mathematicians—like the demonstrative business plan of Irénée Du Pont de Nemours, as described by Martin Giraudeau—constitute just such an archive for reconstructing contests over the future in the past.[13]

Early modern mathematicians like Collins and Watkins were not simply using calculation to make assessments about specific financial futures—to assess the value of a specific annuity or project the growth of a specific debt—but rather to explore an array of different kinds of futures, to imagine financial time itself in different ways. To be clear, those early modern mathematicians were not exactly concerned with predicting or speculating what the future might be. Indeed, outright claims to predict the future—particularly those that relied on mathematical calculations—were looked upon with suspicion by many during this period, seen to evoke the worst excesses of astrology.[14] Rather, Collins, Watkins, and their mathematical colleagues were investigating what we might call the mechanics of the economic future, exploring different ways that future might unfold and how, according to those different mechanical models, the future rebounded upon the present. This different mode of futurity may seem unfamiliar to many readers, at once both stranger and less spectacular than the more fantastical "imaginaries" and "visioneering" that have attracted recent scholarly attention. Instead of projecting forward from the present to the future, they were dealing with calculations that traveled backward from the future to now, giving that future a value in the present moment. (Theirs was not the future tense, but the future perfect.) Yet in investigating these alternative future-to-present mechanics, they were forced to engage deep questions about future time itself—the trajectories along which it proceeded, the uncertainties it entailed, the ties of obligation that bound it to the present.

If futurity offers one crucial point of convergence for scholars of science and scholars of capitalism, so too does calculation. Within economic sociology, in particular,

[11] W. Patrick McCray, *The Visioneers: How a Group of Elite Scientists Pursued Space Colonies, Nanotechnologies, and a Limitless Future* (Princeton, N.J., 2013), 16.

[12] Beckert, *Imagined Futures* (cit. n. 9), 80.

[13] Martin Giraudeau, "Proving Future Profit: Business Plans as Demonstration Devices," in this volume.

[14] On changing ideas about prediction in the early modern period, and suspicions of calculative predictions in particular, see Edward M. Jennings, "The Consequences of Prediction," *Stud. Voltaire 18th Cent.* 153 (1976): 1131–50; Will Slauter, "Forward-Looking Statements: News and Speculation in the Age of the American Revolution," *J. Mod. Hist.* 81 (2009): 759–92; Benjamin Wardhaugh, *Poor Robin's Prophecies: A Curious Almanac, and the Everyday Mathematics of Georgian Britain* (Oxford, 2012). I discuss these themes further in Deringer, *Calculated Values: Finance, Politics, and the Quantitative Age* (Cambridge, Mass., 2018), chap. 6.

there has been a flurry of new thinking about how best to conceptualize the role that calculations play in ordering economic life. The most provocative assertion has been the claim that economic calculations are performative, having the power to change economic circumstances and—in exceptional cases—to create the conditions of their own truth.[15] Yet this claim has attracted a wide array of critiques. In his *Imagined Futures*, Beckert offers a particularly creative riposte to this performativity thesis, arguing that capitalism's fictional futures are ultimately narrated, not calculated, and suggesting that economic calculations are ultimately just "props" that assist actors in this narrative process.[16]

The story told here suggests that calculations may also serve a very different, overlooked function in the dramas of economic life: calculations can be the very medium through which economic imagination happens. To modern eyes, the notion that calculation might be imaginative, or imagination calculative, may feel peculiar, as the two are often made out to be essentially antithetical. But those two modes of thinking have not always seemed so distant from one another. For early modern Europeans, they seemed rather closer in spirit, both essential capacities of an intelligent, reasonable, and indeed scientific mind. The work of Lorraine Daston suggests that it was in the late eighteenth and especially the nineteenth centuries when the two began to take on starkly different characters: calculation, once "allied with the higher mental faculties of speculative reason and moral judgment," came to seem mechanical, laborious, and uninspiring, albeit objective; imagination, once seen to be "essential to philosophy and science—the pursuits of reason," became a subjective threat to the impersonal objectivity expected of scientists.[17]

This essay reveals that, for early modern thinkers confronting the new timescapes opened up by their ongoing "Financial Revolution," calculation was crucially imaginative. It shows that broader political-economic visions are forged and fought out through the details of calculative practice—as can also be seen elsewhere in this volume, in Giraudeau's study of novel techniques for "proving financial profit" in France at the dawn of the nineteenth century, and in Arunabh Ghosh's account of Soviet and Chinese scientists' quest to formulate a "socialist, scientific, and correct" mode of statistical practice that corrected the ills of "bourgeois statistics."[18] Calculations themselves can be evocative, generating alternative scenarios, evoking new futures, playing out contests over different values. As anthropologist Jane Guyer and others have recently observed, numbers can constitute an "inventive frontier."[19] In seeking to un-

[15] On the performativity thesis, see Michel Callon, "Introduction: The Embeddedness of Economic Markets in Economics," in *The Laws of the Markets* (Oxford, 1998), 1–57; Donald MacKenzie, *An Engine, Not a Camera: How Financial Models Shape Markets* (Cambridge, Mass., 2006); MacKenzie, Fabien Muniesa, and Lucia Siu, eds., *Do Economists Make Markets? On the Performativity of Economics* (Princeton, N.J., 2007), esp. the essays by Callon, "What Does It Mean to Say Economics Is Performative?," 311–57, and Philip Mirowski and Edward Nik-Khah, "Markets Made Flesh: Performativity, and a Problem in Science Studies, Augmented with Consideration of the FCC Auctions," 190–224 (for a notable critique of performativity); Franck Cochoy, Martin Giraudeau, and Liz McFall, "Performativity, Economics, and Politics: An Overview," *J. Cult. Econ.* 3 (2010): 139–46.

[16] Beckert, *Imagined Futures* (cit. n. 9), 68.

[17] Lorraine Daston, "Enlightenment Calculations," *Crit. Inq.* 21 (1994): 182–202, on 185; Daston, "Fear and Loathing of the Imagination in Science," *Daedelus* 134 (2005): 16–30, on 20.

[18] Giraudeau, "Proving Future Profit" (cit. n. 13); Arunabh Ghosh, "Lies, Damned Lies, and (Bourgeois) Statistics: Ascertaining Social Fact in Midcentury China and the Soviet Union," in this volume.

[19] Jane Guyer et al., "Introduction: Number as Inventive Frontier," *Anthropol. Theory* 10 (2010): 36–61.

derstand how capitalism's future unfolded in the past, scholars would be well served to look at such calculations. This is a task for which historians of science seem especially well equipped.

COMPOUND INTEREST AND THE EXPONENTIAL FUTURE

At the center of nearly all financial practices—borrowing, lending, mortgaging, investing, insuring—are translations of economic value across time, between the present and various points in the future. One of the essential objectives of financial calculation is, therefore, to organize these temporal translations and explain how value varies with time. A simple example is the calculation of "future value," namely, how much a given sum of money will be worth in the future. What will £100 today be worth in twenty years? Arguably even more fundamental to financial practice, though, is the reverse maneuver: moving from future to present instead of present to future. Imagine you will receive £100 in twenty years; what is that £100 worth today? What is the present value of that future sum? From these two basic problems—future value and present value—arise myriad more complicated cases. An especially important one is calculating the present value of a stream of regular payments made in the future, like an annuity that pays £100 per year every year for twenty years. Such streams of future payments arise in many situations, like the regular interest payments made on a bond or loan, the premium payments made on a life-insurance policy, or the rents tenants pay to their landlords. It is easy to concoct many more complex scenarios. What is the present value of a stream of payments that continues indefinitely (a "perpetuity")? How do you assess the value of an income stream that does not start paying until future time, like the value of a landed estate that will not be inherited for a decade?

Over the last three centuries, there has emerged one paradigmatic approach to answering all of these questions. It relies on the logic of compound interest. While there is a certain risk in beginning any historical account at the end point, it is worth laying out the basics of compound-interest discounting because it helps to put into perspective the alternative, imaginative options that mathematicians were exploring in the early modern period. Under the modern compound-interest paradigm, calculating future value involves calculating how much interest will accrue on your original principal, assuming that interest is paid at some fixed rate and that it compounds every year—namely, in each given year you earn interest not only on the principal but also on whatever interest has accrued. This assumes the future value of property grows exponentially. At 5 percent, that £100 will grow to a future value of £265 in twenty years, and to £13,150 in one hundred years. To calculate present values according to this method, simply reverse this procedure. The present value of £100 in twenty years is however much you would have to save today to have £100 in twenty years, assuming you earn compound interest in the interim. Again assuming 5 percent, the present value of £100 twenty years out is £37.69; one hundred years out, £0.76. Using compound-interest discounting, the standard formulas for future value (*FV*) and present value (*PV*) of a principal sum p are as follows, where r is the interest rate, and t the time elapsed:

$$FV(p) = p\ (1+r)^t \qquad (1),$$

$$PV(p) = \frac{p}{(1+r)^t} \qquad (2).$$

Compound-interest discounting has some felicitous mathematical features, and some very peculiar ones. The exponential properties of compound interest are quite useful when dealing with streams of future payments. Consider, for example, an annuity that pays £100 every year for twenty years. Determining the present value of that entire annuity requires summing the present value of £100 at sequential points in the future—£100 in one year plus £100 in two years . . . plus £100 in twenty years. If compound-interest discounting is used, this sum produces a geometric series, which is easy to handle mathematically.[20] Assuming 5 percent interest, the annuity of £100 per year for twenty years is worth £1,246. This is the amount you would need to invest today to have £100 in one year (£95.24), plus the amount you would need to invest today to have £100 in two years (£90.70), plus the amount you would need to have £100 in three years (£86.38), and so on, all the way until the amount you would need to invest today to have £100 in twenty years (£37.69): £1,246 in total. A similar annuity for fifty years has a present value of £1,826; for one hundred years, the present value is £1,985. This same method can be applied even if the income stream is a perpetuity, which pays out forever. According to compound-interest discounting, a perpetuity that pays out £100 per year forever has a finite value: £2,000 (assuming 5 percent interest). This is not all that large a number, especially given that a comparable annuity for 100 years is worth almost as much, £1,985—meaning that all of the payments that come after year 100 are worth a grand total of £15 today. In a sense, the logic of compound interest places a remarkably low value on the distant future. Because even very small sums are envisioned to grow incredibly vast in future centuries, that means that even incredibly vast sums in the future—when discounted back to the present—amount to very small sums today.

The history of compound-interest discounting in the European world is often said to begin with the Pisan mathematician Fibonacci. In his 1202 *Liber Abaci* ("Book of the Abacus"), Fibonacci explained how to calculate future values and present values by comparing the accumulation of interest to the way a merchant compounded profits on sequential trips between trading posts.[21] Compound-interest techniques were reproduced and elaborated by Italian and French mathematicians over the ensuing centuries. In 1582, Flemish mathematician Simon Stevin published practical tables for calculating present-value problems by compound-interest discounting.[22] Those techniques soon made their way to England, where multiple texts on compound-interest mathematics appeared in the early seventeenth century. Texts like Richard Witt's *Arithmeticall Questions* (1613) and William Purser's *Compound Interest and Annuities* (1634) stressed that compound-interest discounting could be used in an array of

[20] See table 1.

[21] William N. Goetzmann, "Fibonacci and the Financial Revolution," in *The Origins of Value: The Financial Innovations That Created Modern Capital Markets*, ed. William N. Goetzmann and K. Geert Rouwenhorst (Oxford, 2005), 123–44.

[22] On the early history of compound interest calculations, see Frank Swetz, *Capitalism and Arithmetic: The New Math of the 15th Century* (La Salle, Ill., 1987); G. W. Smith, "A Brief History of Interest Calculations," *J. Indust. Eng.* 18 (1967): 569–74; Michael E. Scorgie, "Evolution of the Application of Present Value to Valuation of Non-Monetary Resources," *Accounting Bus. Res.* 26 (1996): 237–48; Geoffrey Poitras, *The Early History of Financial Economics, 1478–1776: From Commercial Arithmetic to Life Annuities and Joint Stocks* (Cheltenham, 2000), chap. 5.

applications, including valuing landed property.[23] Therefore, the technical foundations of compound-interest calculation were well established by the middle of the seventeenth century, when our story begins. But those same calculations had not yet become authoritative in the way they later would. It is all too easy to tell a progressive, linear narrative about the triumph of compound-interest thinking that runs from Fibonacci to the present. The story was not so straightforward.

JOHN COLLINS AND HIS CORRESPONDENTS EXPLORE DIFFERENT FUTURES, CA. 1665–85

In seventeenth-century England, the most intrepid explorer of mathematical futures was John Collins. Born to a poor minister near Oxford in 1625, Collins was orphaned early and then parlayed a limited grammar school education into an apprenticeship to a bookseller, a position as clerk in the royal kitchen, and then seven years as purser on a merchant ship trading to Venice. Upon returning to London in 1649, Collins set himself up as a mathematics instructor and subsequently pursued a variety of unstable clerical positions within the inchoate administrative state.[24] His greatest passion was for mathematical learning. His interests were remarkably ecumenical, ranging from basic arithmetical pedagogy, to practical concerns like navigation and barrel gauging, to esoteric questions about logarithmic spirals and the "quadrature" of curves. Beginning in the early 1650s, Collins published a collection of texts on practical mathematical topics, including accounting, "dialing" (the mathematics of sundials), and surveying.[25] He also became Restoration England's most prominent mathematical "intelligencer," dubbed by contemporaries as the "English Mersenne."[26] He built and maintained an extensive network of mathematical correspondents, spanning the British Isles and the Continent, which included John Wallis, Isaac Barrow, and John Flamsteed—not to mention Newton and Leibniz.[27] (His correspondence with the latter two was later used by Newton's Royal Society champions as evidence of Newton's priority in inventing calculus.) Collins actively steered correspondents toward prob-

[23] R[ichard] W[itt], *Arithmeticall Questions: Touching the Buying or Exchange of Annuities . . .* (London, 1613); William Purser, *Compound Interest and Annuities. The Grounds and Proportions Thereof, with Tables to Divers Rates Most Usuall . . .* (London, 1634). See also C. G. Lewin, "An Early Book on Compound Interest: Richard Witt's *Arithmeticall Questions*," *J. Inst. Actuaries* 96 (1970): 121–32; Lewin, "Compound Interest in the Eighteenth Century," *J. Inst. Actuaries* 108 (1981): 423–42, esp. 423–8.

[24] On Collins's biography, see D. T. Whiteside, "Collins, John," in *Complete Dictionary of Scientific Biography*, vol. 3 (Detroit, 2008), 348–9; William Letwin, *The Origins of Scientific Economics: English Economic Thought, 1660–1776* (London, 1963), chap. 4; Pycior, *Symbols* (cit. n. 2), chap. 3.

[25] Representative works include John Collins, *Introduction to Merchants-Accompts: Containing Seven Distinct Questions or Accompts*, new ed. (1652; London, 1675); Collins, *Geometricall Dyalling: or, Dyalling Performed by a Line of Chords Onely, or by the Plain Scale* (London, 1659); Collins, *The Doctrine Decimal Arithmetick, Simple Interest, &c.* (London, 1685).

[26] For a comparison to Mersenne, see Edward Sherburne, *The Sphere of Marcus Manilius: Made an English Poem . . .* (London, 1675), 116–7.

[27] Much of Collins's correspondence is collected in Stephen P. Rigaud, ed., *Correspondence of Scientific Men of the Seventeenth Century*, 2 vols. (Oxford, 1841). On Collins's and related mathematical networks, see Jacqueline Stedall, "Tracing Mathematical Networks in Seventeenth-Century England," in *The Oxford Handbook of the History of Mathematics*, ed. Eleanor Robson and Jacqueline Stedall (Oxford, 2009), 133–52. On scientific correspondence networks during the early modern period, see David S. Lux and Harold J. Cook, "Closed Circles or Open Networks? Communicating at a Distance during the Scientific Revolution," *Hist. Sci.* 36 (1998): 179–211; Justin Grosslight, "Small Skills, Big Networks: Marin Mersenne as Mathematical Intelligencer," *Hist. Sci.* 51 (2013): 337–74.

lems and encouraged members of their network to publicize their work, often helping them to navigate the complexities of the mathematical publishing trade. Collins was recognized for his contributions to mathematics in 1667, when he was elected to the Royal Society of London, the first fellow to be absolved of paying dues.[28]

Collins made his career at the junction of learning and commerce, and the mathematics of financial time was a crucial thread uniting his varied enterprises. Collins wrote a collection of texts on financial mathematics himself and also urged members of his mathematical network to investigate questions that he had identified as having financial applications. One of his greatest interests, for example, was the mathematics of calculation itself—how to make certain common numerical calculations easier and the theoretical possibilities of certain calculative tools. This was evident in his interest in logarithms, developed earlier in the century by British mathematicians John Napier and Henry Briggs. As a practical matter, his posthumously published textbook, *The Doctrine Decimal Arithmetick, Simple Interest, &c. As Also of Compound Interest and Annuities* (1685), was one of the earliest texts to make clear that many financial calculations, particularly those related to compound interest, could be understood as applications of the mathematics of logarithms and thus solved using log tables.[29] Collins also had a keen interest in exploring the theoretical possibilities of what could be done using numerical tables. In the late 1660s and early 1670s, for example, Collins "pestered" the likes of Isaac Newton and James Gregory about whether it might be feasible to approximate the solution to any algebraic equation using just a small selection of tables.[30]

At times, Collins posed financial problems directly to his mathematical correspondents. We have already seen how Collins had communicated with Isaac Newton about the problem of "yield approximation," a question he also posed to James Gregory, Regius Professor of Mathematics at St. Andrews.[31] Collins's interest in the yield approximation problem may have been piqued by one of his lesser-known correspondents, a tobacco cutter named Michael Dary. An amateur mathematician and "philomath" like Collins, Dary had a great interest in applying mathematical research to practical computational problems. Dary's 1677 *Interest Epitomized, both Compound and Simple* was among the era's most mathematically sophisticated financial texts. It used "symbolical" techniques and drew upon algebraic research by John Wallis to put the mathematics of interest on a sounder mathematical footing. Dary ultimately de-

[28] Michael Hunter, *The Royal Society and Its Fellows, 1660–1700: The Morphology of a Scientific Institution* (Chalfont St. Giles, Buckinghamshire, England, 1982) 8, 77.

[29] Collins, *Doctrine Decimal Arithmetick* (cit. n. 25), 8–14. See also "Collins/John," in "Birch's Biographical Notes," British Library Add. MS 4,221, fols. 331–9, esp. fols. 333–4, 336. On logarithms and practical financial calculations, see Timothy Alborn, *Regulated Lives: Life Insurance and British Society, 1800–1914* (Toronto, 2009), 121–7.

[30] H. W. Turnbull, *The Correspondence of Isaac Newton*, vol. 1, 1661–1675 (Cambridge, 1959), 5, 23n13; James Gregory to Collins, 23 November 1670, in Turnbull, *Correspondence of Isaac Newton*, 45; Gregory to Collins, 15 February 1671, in Turnbull, *Correspondence of Isaac Newton*, 61; Collins to Francis Vernon, n.d. (1671?), Letter LIX, in Rigaud, *Correspondence of Scientific Men* (cit. n. 27), 1:155–8, on 156.

[31] Gregory to Collins, 2 April 1674, Letter CCXIV, in Rigaud, *Correspondence of Scientific Men* (cit. n. 27), 2:255–6. On the history of this problem, see Augustus de Morgan, "On the Determination of the Rate of Interest of an Annuity," *Assurance Magazine, and Journal of the Institute of Actuaries* 8 (1859): 61–7; Gabriel A. Hawawini and Ashok Vora, "Yield Approximations: A Historical Perspective," *J. Finance* 37 (1982): 145–56.

Table 1. *Comparison of Formulas for Present and Future Values according to Compound and Simple Interest, Where* p *Is the Principal,* r *Is the Interest Rate, and* t *Is Elapsed Time*

	Compound interest	Simple interest
Future value of a sum p	$p(1+r)^t$	$p(1+rt)$
Present value of a single sum p	$\frac{p}{(1+r)^t}$	$\frac{p}{1+rt}$
Present value of an *annuity* paying p annually for n years	$\Sigma_{t=1}^{n}\frac{p}{(1+r)^t} = \frac{p}{r}\left(\frac{(1+r)^n-1}{(1+r)^n}\right)$	$\Sigma_{t=1}^{n}\frac{p}{(1+rt)} = ?$

veloped his own numerical technique for addressing the yield approximation problem, which he worked into his 1677 textbook.[32]

Such explorations into logarithmic calculation and annuity yields fell mostly within a compound-interest paradigm. But not all of the mathematical explorations by Collins and his collaborators did. They put at least as much effort into a significantly different approach: simple-interest discounting. Like the compound-interest approach, simple-interest discounting translates across time periods by assuming that steady interest is earned as time passes. Where the simple-interest approach differs concerns how that interest is earned. Simple interest does not assume "interest upon interest" like compound interest does. If £100 is set aside and earns 5 percent simple interest, the interest paid is just £5 every year, regardless of how much interest has already accrued. Simple interest grows linearly with time, not exponentially. In many ways, it is just as plausible to create a mathematics of time based on simple interest as compound interest. (See table 1.) Computing the present value of a single sum using the simple-interest approach is quite straightforward. The present value of £100 in twenty years is whatever you have to save today to have £100 in twenty years, assuming simple interest. Simple interest discounting seems to have been relatively common in commercial transactions in the seventeenth century, particularly for individual sums and short time periods. Contemporary guides to commercial calculations, like the popular *Webster's Tables* by William Webster, often foregrounded simple-interest techniques.[33]

The simple-interest approach to financial time posed one especially vexing mathematical problem, though: calculating the present value of annuities. As discussed, compound-interest reasoning made calculating the present value of annuities easy to manage because of the geometric series that arose. When simple-interest discounting was applied, though, the calculations proved much less tractable. In 1677, Dary cautioned that calculating present value according to simple interest "doth not hold but only in the payment of Single Sums, for when many equal payments are made at equal times it faltereth."[34] Mathematically, adding up a stream of present values calculated using simple-interest discounting produced a harmonic series, or what con-

[32] Michael Dary, *Interest Epitomized, both Compound and Simple* (London, 1677), [ii], 1, 8. On Dary, see Stedall, "Tracing Mathematical Networks" (cit. n. 27), 148.

[33] See, e.g., William Webster, *Webster's Tables for Simple Interest Direct, at 10, 8, 7, and 6 p. per Centum . . . For Simple Interest to Rebate, at 8 l. per Centum . . .* , The Second Edition, with Very Large Additions (London, 1629).

[34] Dary, *Interest Epitomized* (cit. n. 32), [iii].

temporaries termed a "musical" series.[35] Unlike geometric series, whose sums could be easily computed using a single algebraic formula, early modern mathematicians were at a loss when it came to calculating the sum of a harmonic series—short of laboriously calculating each term and adding them all up. What is more, harmonic series behaved very differently than their geometric siblings. Most dramatically, harmonic series with an infinite number of terms never seemed to converge to a definite value, as geometric series did.[36] This had profound implications for the valuation of perpetual streams of income, like land rents. If harmonic series always had infinite values, then perpetuities had infinite value according to the simple-interest approach.

Some seventeenth-century financial mathematicians were not willing to let the possibilities of simple-interest discounting go unexplored. For John Collins, finding a way to calculate the sum of harmonic series became a personal campaign. In 1668, for example, Collins wrote to Henry Oldenburg asking for the secretary's help in locating a recent book by Erasmus Bartholin, which Collins thought "perchance handles the Musicall Progression" and could thus help him "enlarge my Paper of interest."[37] Not long thereafter, in January 1670, Collins posed the question of "how to find the aggregate of a series of fractions, whose numerators are the same, and their denominators in arithmetical progression" in his first exchange with Isaac Newton. Newton realized the problem had a broader mathematical appeal because it "much resembles the squaring of the hyperbola."[38] In other words, the problem of simple-interest discounting was closely related to that of finding the area underneath hyperbolic curves, a foundational problem in what would soon be recognizable as integral calculus.[39] Collins queried other members of his network for insights on the harmonic series as well, including James Gregory.[40] Collins explained his ongoing quest most clearly in a letter (ca. 1671) to Francis Vernon, a fellow "intelligencer," laying out in very deliberate terms how the "adding of a musicall progression" related to calculating the "present worth" of an annuity, as well as to finding the "area of the hyperbola."[41]

Despite his extensive solicitations, Collins struggled to make progress on the harmonic series problem. But he never seems to have given up on it. His posthumous *Doctrine of Decimal Arithmetick, Simple Interest, &c.*, published in 1685, began with a discussion of how to calculate future and present-value problems using simple interest, before acknowledging that work still had to be done to figure out how to de-

[35] A harmonic series contains fractions with fixed numerators, whose denominators form an arithmetic sequence, e.g., $(\frac{1}{1} + \frac{1}{2} + \frac{1}{3} + \frac{1}{4})$ or $(\frac{3}{4} + \frac{3}{7} + \frac{3}{10} + \frac{3}{13} + \cdots)$.

[36] On the history of research on the harmonic series, see William Dunham, "The Bernoullis and the Harmonic Series (1689)," in *Journey through Genius: The Great Theorems of Mathematics* (New York, 1990), chap. 8.

[37] Collins to Henry Oldenburg, [? December] 1668, in *The Correspondence of Henry Oldenburg*, vol. 5, 1668–9, ed. and trans. A. Rupert Hall and Marie Boas Hall (Madison, Wis., 1968), 212.

[38] Newton to Collins, 19 January 1669/70 (cit. n. 1); Collins to Newton, 13 July 1670, Letter CCXXX, in Rigaud, *Correspondence of Scientific Men* (cit. n. 27), 2:301–3; Newton to Collins, 20 July 1671, MS Add. 9597/2/18/11, Camb. UL.

[39] Summing the harmonic series involved finding closed solutions to infinite series of the form $\Sigma_n \frac{1}{a+bn}$, closely related to the problem of finding the area under hyperbolic curves of the form $y = \frac{1}{a+bx}$.

[40] James Gregory to Collins, 2 April 1674 (cit. n. 31), 2:2556, and 26 May 1675, Letter CCXVI, in Rigaud, *Correspondence of Scientific Men* (cit. n. 27), 2:259–62.

[41] Collins to Francis Vernon, n.d. [1671], Letter LIX, in Rigaud, *Correspondence of Scientific Men* (cit. n. 27), 1:155–8, on 155.

termine the present value of "an Annuity, at Simple Interest." Collins was optimistic that a solution was close at hand, though, possibly in a new text entitled "Arithmetical Quadratures," by the Bolognese mathematician Pietro Mengoli (1626–86). This was wishful thinking, it seems, as Collins admitted that Mengoli's was "a Book I never saw."[42]

What is most significant in Collins's quixotic harmonic quest to find an effective mathematics of simple interest is not the progress he made—minimal at best—or how that quest became entangled with more elevated mathematical pursuits like integral calculus. Rather, most remarkable is what it reveals about the imaginative possibilities entertained in early modern financial thinking. Collins and his collaborators did not live in an economic world in which there was only one correct way to calculate the value of the future. They knew the basics of compound interest and its many extensions and wanted to explore its potential even further, yet they also imagined that there might be room for substantially different methods. This is not to say that these mathematicians looked upon all futures with equal favor. It is clear that Collins and others had some doubts about the validity and applicability of simple-interest discounting in certain scenarios, particularly when time horizons got long.[43] At least once, in a letter to Vernon, Collins commented on the "absurdity of simple interest," in reference to his calculation that showed that a £100 annuity for 100 years was worth a hefty £3,200 at 6 percent interest, nearly double its value using compound-interest discounting (£1,662).

For Collins, the desire to extend simple-interest techniques may have been influenced by the contemporary legal environment. One commentator on Collins's achievements specifically noted his investigations into mathematical questions related to accounting for mortgages, observing that England's equity courts only allowed simple (not compound) interest to accrue to creditors under certain conditions in cases related to disputed mortgages.[44] Such legal peculiarities reflected the complex place that charging financial interest had long held in English law. Due to the abhorrence of "usury" in the Christian tradition (and many other faith traditions), lending money at interest had been outlawed in England until 1545. From 1545 to 1552, and then from 1571 onward, this blanket prohibition was replaced by laws prohibiting lending above a certain maximum rate: first 10 percent, lowered to 8 percent in the 1620s, 6 percent by 1660, and 5 percent in 1714.[45] While lending at interest was widely practiced in seventeenth-century England, and many financial transactions—including those relating to annuities, bills of exchange, or real property—effectively involved the charging of interest in various indirect or disguised ways, there were still situations

[42] Collins, *Doctrine Decimal Arithmetick* (cit. n. 25), 7–8.

[43] Dary was especially harsh on simple-interest techniques, though perhaps because he favored the computational advantages afforded by the compound-interest approach; see Dary, *Interest Epitomized* (cit. n. 32), [iii]–[iv].

[44] "Collins/John," in "Birch's Biographical Notes" (cit. n. 29), fol. 336.

[45] On the history of usury laws, and contemporary attitudes thereon during the period, see H. J. Habakkuk, "The Long-Term Rate of Interest and the Price of Land in the Seventeenth Century," *Econ. Hist. Rev.*, n.s., 5 (1952): 26–45; Tim Keirn and Frank T. Melton, "Thomas Manley and the Rate-of-Interest Debate, 1668–1673," *J. Brit. Stud.* 29 (1990): 147–73; Peter Temin and Hans-Joachim Voth, "Interest Rate Restrictions in a Natural Experiment: Loan Allocations and the Change in the Usury Laws in 1714," *Econ. J.* 118 (2008): 743–58; Hugh Rockoff, "Prodigals and Projectors: An Economic History of Usury Laws in the United States from Colonial Times to 1900," in *Human Capital and Institutions: A Long-Run View*, ed. David Eltis, Frank D. Lewis, and Kenneth L. Sokoloff (Cambridge, 2009), 285–323.

in which the specter of usury reared its head. This seems to have been particularly true in legal settings, where courts were on guard against practices that might be tantamount to usury (such as letting compound interest penalties rack up on a borrower who was delinquent in his mortgage payments). In such cases, compound interest appears to have been viewed with special wariness, compared to simple interest. One hypothesis is thus that the dubious status of compound interest in English law was a key factor driving Collins's preoccupation with exploring the mathematics of simple-interest discounting. Whatever the source of Collins's, and others', interest, though, and whichever way his own temporal preferences leaned, there is no doubt that he took very seriously the possibility of a very different kind of financial future—one in which the future was not devalued quite so heavily.

THOMAS WATKINS CALCULATES SIX DIFFERENT FUTURES, 1715

For John Collins and his correspondents in the later seventeenth century, their adventures into alternative futures remained exploratory, producing an array of minor insights and small problems, but not programmatic statements or transformative new techniques. A generation later, another commercially minded mathematician, or mathematically minded man of commerce, named Thomas Watkins would make a bolder statement. Most of what can be gleaned about Watkins's life emerges from records of his membership in the Royal Society of London. He was first elected in 1714, alongside the likes of J. T. Desaguliers and Pierre Varignon. One year later he produced a lone article in the society's *Philosophical Transactions*. A 1718 register of the society's membership listed Watkins as "Gent." and gave his address as St. Martin's Church-Court in the Strand. That same register suggests he played at least a modest part in the intellectual life of the society in the later 1710s. The register's author, Thomas Clark, employed a Greek letter code to indicate that certain fellows were the "most proper and able Persons to be consulted upon" any of ten fields—"Astronomy" (β), "Chymistry" (η), and so forth. Most fellows received no mark; President Newton got four (α, β, γ, δ). Watkins merited an α, for "Natural Philosophy and Mathematicks." The register notes that Watkins was also a member of the society's Council, alongside Newton, Edmund Halley, and future President Hans Sloane.[46]

As to Watkins's life outside the society, details are scarce, though like Collins it is clear he traveled between the worlds of business, publishing, and government service, as well as "Natural Philosophy and Mathematicks." In 1717, he published a practical pamphlet on financial mathematics, and that same year was appointed to a governmental post responsible for administering various lottery debts.[47] Around 1724, Watkins and a collaborator published a short text describing "a scheme for making provision for the wives, children, or other relations" of army and naval officers.[48] Based on his writings, he clearly had intricate knowledge of many areas of fi-

[46] [Thomas Clark], *A List of the Royal Society of London; Instituted by His Majesty King Charles II. For the Advancement of Natural Knowledge . . .* (London, 1718), 8–9, 13–4.

[47] Entry for 6 December 1717, *Calendar of Treasury Books: Volume 31, 1717*, ed. William A. Shaw and F. H. Slingsby (London, 1960), http://www.british-history.ac.uk/cal-treasury-books/vol31/pp698-716 (accessed 22 May 2016).

[48] Thomas Watkins and Dan. Combes, *An Abstract of a Scheme for Making Provision for the Wives, Children, or Other Relations of Such Officers. . .* ([London?], [1724]). As listed in the English Short Title Catalogue, estc.bl.uk (accessed 22 May 2016).

nancial practice and some acquaintance with legal matters (though the absence of the professional designation "esq." suggests that he was not a lawyer).

Watkins's first, and only, publication in the Royal Society's *Philosophical Transactions* came in 1715: "Rules for Correcting the Usual Methods of Computing Amounts and Present Values, by Compound as well as Simple Interest; and of Stating Interest Accounts," signed "Thomas Watkins, Gent. F. R. S."[49] Superficially, the paper looked like many mathematical papers that occupied the *Philosophical Transactions*, with dense algebraic calculations in paragraph form, punctuated by the occasional table. Behind this formal style was a practical agenda. Watkins thought the standard financial calculations his contemporaries used were wrong because they relied on unrealistic assumptions and oversimplified mathematics. His goal was to correct those flawed techniques by offering more sophisticated models. Watkins broke down his proposed financial-computational reforms into three sections: "Compound-Interest," "Simple-Interest," and "Interest Accounts." The first two sections are particularly telling.

Watkins opened the first section, "Of Compound-Interest," with a bold assertion:

> The Supposition whereon the Method of computing by Compound Interest is founded; *viz.* That all Interest Money, Rents, &c. are or may be constantly receiv'd, and put out again at Interest, the Moment they become due, without any Charge, or Trouble, being impracticable; therefore all Computations by this method . . . must needs be erroneous.[50]

In other words, Watkins believed that in most cases the logic of compound interest was an unrealistic way to think about value and time because it assumed the financial world worked much more smoothly than it actually did. In particular, those calculations implicitly assumed that investors would always be able to find good places to invest and reinvest their money, both in the present and the future, without any extra cost, delay, or uncertainty. This was hardly a trivial assumption. Capital markets were still in their relative adolescence. Real estate, either by purchasing land or extending mortgages, was a steady investment, but opportunities were not necessarily easy to find. Government bonds were still a new, somewhat unstandardized, and (many hoped) impermanent mode of investment; it would be the 1750s before Britons could count on government "Consol" bonds as a long-term, low-risk way to invest their money. In 1715, it invariably took time and effort to put money to work.

Watkins contended that this oversight had serious consequences. For one, it meant that usual compound interest calculations tended to overstate the future value to which a present sum would grow over time. (Watkins frequently called the future value "the *Amount* of a Sum of Money.")[51] It was simply too optimistic to pick a standard interest rate and assume that money would grow consistently at that rate. Conversely, this meant that the existing compound interest calculations understated the present value of future sums. Watkins's solution was to try to rebuild the apparatus of compound-interest calculation in a way that recognized the messiness of contemporary financial life. His first move was to introduce two new variables into existing compound-interest models. The first represented the annual "Charge and Trouble of

[49] Thomas Watkins, "Rules for Correcting the Usual Methods of Computing Amounts and Present Values . . . ," *Phil. Trans.* 29 (1715): 111–27.
[50] Ibid., 111.
[51] Ibid.; emphasis added.

the Management" of one pound, which Watkins called c. The second variable, t, captured the wasted time "spent in receiving and putting [money] out again at Interest."[52] Algebraically, dealing with the first of these two issues was straightforward: Watkins deducted the charge for management "Trouble" from the interest rate to be received in a way that did not fundamentally change the structure of the calculations. (The rate r became $r - cr = r(1 - c)$, which Watkins renamed dr.) Dealing with the issue of wasted time was far more complicated. Watkins embarked on well over a page of intricate algebraic manipulations designed to build the inevitable time delays into the standard compound-interest calculations. As he did so, he defined a slew of new variables so that the symbolic bulk of his ever-growing formula would not get out of hand (d replaced $1 - c$, e replaced $1 + dtr$, a Gothic-style r replaced $\frac{dr}{e}$, etc.).

Watkins's algebra is exceedingly difficult to follow—at least for myself, and I suspect for many of Watkins's contemporaries. Yet Watkins seemed unconcerned that his calculations might appear inaccessible to readers. He thought that the challenge facing his computational reforms lay elsewhere. "The only difficulty that remains, is the right assuming the Quantities c and t," he explained, "the impossibility of doing which with perfect Exactness, I suppose to be the reason why neither this, nor any Method of Correction to the like purpose, has yet been taken notice of."[53] Such comments were indicative of the striking optimism that pervaded Watkins's paper. He seemed sincere in his hope that his arcane computational labors might change how his contemporaries went about their financial lives.

Watkins did not stop at applying his revised calculations to single sums. The most pivotal calculations were those involving annuities. He concluded his section "Of Compound-Interest" with a numerical "Specimen" of how his "corrected" methods could be applied to the valuation of annuities. He assumed c, the charge for management "Trouble," to be 6 percent, and he estimated t, the average amount of time lost in receiving and reinvesting sums, at four and a half months, or roughly three-eighths of a year. He produced a series of numerical tables giving the results of his revised interest calculations, showing both the future values ("Amounts") and present values of a £1 per year annuity. He surveyed a range of different interest rates (from 5 to 10 percent) and time horizons (for five to 100 years, plus perpetuities). In all of the tables he included a column showing the difference between the standard method of calculating these figures using compound interest and his own corrected method (fig. 1).[54]

Watkins's discussion "Of Simple-Interest," the subject of the second section of his paper, was even blunter. The author wasted no time pointing out the fundamental "absurdity of the Supposition, on which the usual Method of computing present Values by Simple-Interest, is founded." The key problem was that such simple-interest discounting assumed that any money received in the future, for example, from rents or an annuity, was always immediately reinvested ("put out again at Interest"), but interest earned on those sums was not, instead being allowed to "lie dead during the

[52] Ibid., 112.
[53] Ibid., 114.
[54] Strangely, in analyzing annuities, Watkins's corrected method gave lower values than traditional compound interest calculations for both future values ("Amounts") and present values. This is odd given that he argued that, in general, traditional methods of compound interest undervalued the present value of single quantities. Watkins's explanation why, and his mathematical execution, is extremely difficult to follow on this point.

Amounts of 1 *l.* at 5 *per Cent.* computed 6 ſeveral Ways.

Years.	1	2	3	4	5	6
	Simple Int.	*Id. for Bonds*	*Sim. Int. cor.*	*Id. by the red. rat.*	*Co. Int. cor.*	*Comp Int.*
	$prn + p = m$	Id. till $prn = p$	$\frac{1}{2} \times r + 1 \times prn + p = m$	$\frac{1}{2} \times r + 1 \times pr\,r + p = m$	$\overline{1+r}^n \times p = m$	$\overline{1+R}^N \times p = m$
5	1,25	1,25	1,27813	1,25501	1,25655	1,27628
10	1,5	1,5	1,61875	1,56338	1,57892	1,62889
20	2,	2,	2,48750	2,34021	2,49300	2,65330
30	2,5	2,	3,60625	3,33048	3,93625	4,32194
40	3,	2,	4.97500	4,53419	6,21504	7,04000
60	4,	2,	8,46250	7,58194	15,49408	18,67919
80	5,	2,	12,95000	11,48345	38,62672	49,56144
100	6,	2,	18,43750	16,23874	96,29634	131,50126
	Amounts of 1 l. *at* 6 per Cent. *by the ſame Theorems.*					
5	1,3	1,3	1,3405	1,31063	1,31328	1,33822
10	1,6	1,6	1,7710	1,69758	1,72471	1,79084
20	2,2	2,	2,9020	2,70048	2,97463	3,20713
30	2,8	2,	4,3930	4,00870	5,13039	5,74349
40	3,4	2,	6,2440	5,62223	8,84844	10,28572
60	4,6	2,	11,0260	9,76525	26,32086	32,98769
80	5,8	2,	17,2480	15,12954	78,29488	105,79599
100	7,	2,	24,9100	21,71510	232,89852	339,30208
	Amounts of 1 l. *at* 10 per Cent. *by the ſame Theorems.*					
5	1,5	1,5	1,6125	1,54749	1,55970	1,61051
10	2,	2,	2,4750	2,30157	2,43268	2,59374
20	3,	2,	4,9500	4,42951	5,91793	6,72750
30	4,	2,	8,4250	7,38381	14,39643	17,44940
40	5,	2,	12,9000	11,16449	35,02190	45,25925
60	7,	2,	24,8500	21,20493	207,25717	304,48165
80	9,	2,	40,8000	34,55084	1225.5335	2048,4003
100	11,	2,	60,7500	51,20222	7258,5398	13780,6127

Figure 1. *Image of Thomas Watkins's table, showing "Amounts of 1l. . . . computed six several Ways." In Watkins, "Rules" (cit. n. 49), 120. Reproduced with kind permission of the Royal Society.*

whole Term."[55] Yet he acknowledged that certain circumstances demanded a simple-interest approach, notably for legal reasons. British courts took an array of different views on the legality of financial interest—when it could be charged or demanded, how much, whether it could be compounded—based on the circumstances involved. For example, Watkins noted, the courts did not allow "Interest upon Interest" in disputes over debts; if a debtor failed to pay interest on a debt on schedule, the creditor was not allowed to penalize the debtor further by demanding interest on that unpaid interest. Watkins suggested that this was intended to "curb the Exorbitant Avarice of Usurers." (As noted above, for both moral and political-economic reasons, Britain's

[55] Watkins, "Rules" (cit. n. 49), 118.

Table 2. *The Future Value ("Amount") of £1 in 100 Years, 5 Percent Interest Rate, according to Watkins's Six Different Methods*

Column	Method	"Amount" (£)
1	Simple interest	6.00
2	Simple interest for bonds	2.00
3	Simple interest corrected	18.44
4	Simple interest corrected at the reduced rate	16.24
5	Compound interest corrected	96.30
6	Compound interest	131.50

Parliament had long imposed a maximum allowable interest rate, recently reduced from 6 to 5 percent in 1714.) Similarly, landlords could not charge tenants compounding interest on unpaid rents.[56]

Because of such legal vagaries, Watkins admitted that there might be situations in which a "corrected" approach to simple-interest discounting would be useful. He particularly felt that there were many situations in which it was logical, and legally permissible, to assume that interest would be earned upon interest one time but would not continue to compound beyond that—where interest-upon-interest was allowed, but interest-upon-interest-upon-interest was not. Watkins fashioned a formula calculating present values based on this intermediate approach, which he termed "Simple Interest corrected." Building on his work in the previous section, Watkins showed that it was also possible to further revise the "corrected" simple-interest formula by making the same adjustments he had made to the compound-interest formula, factoring in the cost of management and reinvestment delays. He dubbed this, his third new method, "Simple Interest corrected by the reduced rate."

Watkins summarized his models in two numerical tables that showed the results of his three new methods alongside traditional approaches. His first table (fig. 1) showed how to calculate the future value ("Amount") of £1 at 5, 6, and 10 percent interest rates, for terms ranging from five to 100 years, using six different methods: (1) standard simple interest; (2) simple interest for "Debts due by Bond" (a peculiar case where law limited the amount of total interest that could be charged to double the original principal); (3) Watkins's simple interest corrected; (4) his simple interest corrected "by the reduced rate"; (5) his compound interest corrected; and (6) standard compound interest. For example, in 100 years, £1 earning 5 percent interest would reach different "Amounts" depending on which of Watkins's six methods was used (table 2). Subsequently, a smaller table on the following page displayed the present value of a £1 annual annuity calculated according to simple interest corrected, compound interest corrected, and standard compound interest.

Watkins's remarkable tables mapped an array of different financial futures, ranging from the relatively limited growth entailed by simple interest to the explosive, exponential potential of compound interest. Watkins was most interested in exploring the possible futures that fell in between these two idealized extremes, the futures most likely to arise in the messy business of real life. Watkins did not feel the need to settle on one, superior calculation. He wanted to give options and even left open the pos-

[56] Ibid., 118–9.

sibility that other formulas might need to be derived for new cases. A subsequent paragraph explained to readers the different scenarios in which it might make sense to use each of the six columns. For example, he explained that simple interest corrected (col. 3) "answers a Case of a Security or joynt Obligor that has duly and constantly paid the Interest, and at last the principal Sum of a Debt, from which he has a Counter-Bond from the principal Debtor."[57]

Watkins's goal was not to obscure the intricacies of contemporary finance but to reveal them for the public using mathematics. He was fighting against his contemporaries' entrenched tendency to oversimplify and idealize. In closing, he urged readers not to object to his new "Rules and Theorems . . . merely because they tend to produce some Alteration in the present practice." Watkins looked forward to a future in which the expanding power of mathematics permitted a proliferation of more sophisticated calculations, each more sensitive to commercial realities.[58]

BEFORE INTEREST BECAME UNINTERESTING

As with John Collins's quest for a harmonic future based on simple interest, Thomas Watkins's attempt to create multiple calculating regimes did not seem to go very far. There seem to be no further references to Watkins's article in later editions of the *Philosophical Transactions*, and I know of no examples of Watkins's baroque formulas being put into use. What, then, do we make of these abortive excursions into alternative financial futures? First and foremost, they demonstrate that compound-interest discounting was not an exclusive, uncontested, or inevitable way to think about the economic relationship of present and future. However indisputable and rational and natural that calculation seems in current practice, it was not always so. The question of how that black box got closed is a story that lies beyond the scope of this essay, but there is much to be learned by looking at what that box looked like before it was closed—by looking at the scattered computational components that were developed, tested, and thrown away. We learn that early modern thinkers had good reasons to believe that there was a place for simple interest, and that they were concerned that compound-interest models dangerously abstracted from the realities of commercial life. We also learn that early modern calculators, Watkins especially, were entertaining methods of financial calculation that were significantly more complicated mathematically than those that would eventually come to prevail—a reminder that the development of capitalism's logic cannot be told as a monotonic story about increasing "sophistication."

The example of these early Royal Society mathematicians demonstrates why, methodologically, historians of economic life can learn much from archives of scientific and technical practice. Scientific experimentation and technological design are often full of imagination and contestation, possibility and surprise. They are full of problems. They are excellent places to look for things, like the temporal logic of capitalism, that were problematic in the past but came to seem far less so with time. Our story of misbegotten quests to calculate alternative futures points the way toward new avenues for research in early modern economic practice and culture. Watkins's remarkable six-part taxonomy of different financial calculating regimes, for example,

[57] Ibid., 121.
[58] Ibid., 127.

offers a veritable map of different economic value systems at work in early eighteenth-century Britain, each operating with slightly different attitudes about value and time, fairness and obligation.

This essay has primarily told a story about a historical change that never occurred, about futures searched for but never quite found. To what degree can we say that the likes of Collins and Watkins were more directly part of the processes that shaped the economic logics that ultimately would prevail? As a coda, I will close with one suggestive story about Thomas Watkins. His alternative interest calculations did get mentioned in at least one other place, at least obliquely: his own next publishing effort, a 1717 *Table of Redemption*, dedicated to the House of Commons and signed "T. W. F. R. S." Like many such pamphlets of its time, Watkins's text was predominantly filled with tables to facilitate interest calculations. The entries in his tables were not monetary values, though, but times—specifically, the time it would take to pay down a fixed amount of debt (£100), depending on the interest rate the debt carried, the annual funds available for debt repayment, and the frequency with which payments were made. (For example, Watkins's table showed that, given a £100 debt bearing 4½ percent interest, and payments of £9 annually made half-yearly, it would take fifteen years, six months, and twenty-eight days to repay the debt entirely.)[59]

Watkins had a very specific application in mind for his tables. He wanted them to be used by "Parliamentary, or other Committees" who could use them to think through different fiscal policies "without adjourning themselves, or retarding their Conferences, for tedious Calculations."[60] There was good reason to think demand for such a political tool would be high. Public debt was a tense issue at the time, as Parliament had racked up more than £45 million of debt in the course of two costly wars with France since 1688. There was widespread discussion about how best to handle these mounting burdens, and an array of inventive and sometimes outlandish plans for how to do so. Watkins was concerned about the proliferation of such "chimerical Projects of paying Debts without Mony" and hoped he could use his mathematical knowledge to help the nation's Parliamentary representatives navigate through them.[61]

What made these "redemption" calculations tricky was that, as the debt was paid down, interest continued to be paid, but that interest varied as the balance declined. Mathematically, computing how long it would take to redeem certain debts was a variant of the present-value problem. So which of his six different approaches to interest-rate calculations did Watkins use? The answer: plain compound interest. For all his talk about the need for corrected methods in the *Philosophical Transactions*, he fell back upon the uncorrected methods when it came time to apply his expertise to public problems. To Watkins's credit, he did give an extensive explanation in the introduction to his *Table* explaining why he thought that particular method made sense for these specific circumstances—in other words, why "column 6"?[62] But few readers probably realized that Watkins was even making such a choice.

[59] "T. W. F. R. S." [Thomas Watkins], *A Table of Redemption. Shewing at one View in What Time the Principal and Interest of any Debt from Three to Six per Cent. May Be Discharged . . .* (London, 1717).

[60] Ibid., [3].

[61] Ibid., 5.

[62] Ibid., [3].

Here, perhaps, we can begin to see an even more fundamental way in which the histories of science and capitalism were entangled: namely, through the coproduction of forms of scientific expertise and objects of economic knowledge (and economic value). This is a theme that runs throughout several of the essays in this volume: for example, in Paul Lucier's study of the interactions between geological experts and corporate mining interests around the Comstock "lode" and in Victoria Lee's examination of the entwined emergence of fermentation science and industrial brewing in Meiji Japan.[63] More specifically, the story told here is one of how scientific actors, in seeking to exercise influence over new fields of knowledge opened up by transformations in economic life, ended up pushing those fields in unanticipated directions, perhaps at odds with their own scientific judgments or those of their peers. In that sense, it parallels the story told by Lee Vinsel about psychologists in the early automobile age, who, in trying to make a place for themselves in the new field of auto safety through their expertise in mental testing, ended up reifying a vision of the "accident prone driver" that few in the psychological community endorsed.[64]

Just as he was experimenting with how to build new logics of financial calculation, Thomas Watkins was also experimenting with how to fashion his own role as a purveyor of knowledge, particularly with regard to a broader, interested "public." The same was true for John Collins. Both were intensely aware that calculating the relationship between present and future was a messy enterprise, fraught with ambiguities and imperfect assumptions. Yet what they presented to the broader public were predominantly simplified, ready-to-use calculations, based primarily on the logic of compound interest. They took on the task of interpreting the complications of the economic future for their fellow Britons, smoothing out the rough computational and conceptual edges. They left the hard problems to be solved behind closed doors by mathematicians. In attempting to fashion their own authority, they helped to make the very problems they were trying to solve seem less problematic than they knew them to be. They may have inadvertently helped to foreclose the very alternative futures they were investigating.

[63] Paul Lucier, "Comstock Capitalism: The Law, the Lode, and the Science"; Victoria Lee, "The Microbial Production of Expertise in Meiji Japan," both in this volume.

[64] Lee Vinsel, "'Safe Driving Depends on the Man at the Wheel': Psychologists and the Subject of Auto Safety, 1920–55," in this volume.

Proving Future Profit:
Business Plans as Demonstration Devices

*by Martin Giraudeau**

ABSTRACT

This essay is a study of the "Project for the establishment of a war and hunting gunpowder manufactory in the United States," written by Irénée Du Pont de Nemours in 1800, in order to raise funds from potential investors for what was to become the DuPont Corporation. It shows that the "Project" is best understood as a demonstration device akin to those used by natural philosophers at the time. This investment proposal relied on demonstration techniques similar to those of the report of an experiment, textually gathering a crowd of virtual witnesses to address their objections, submitting the proposed manufactory to a number of manipulations to assess its reactions under various circumstances, and relying on a specific experimental device—a profit-and-loss account—that made it possible to produce compelling quantitative results. The essay shows the originality of these techniques at a time when profit-and-loss calculations were uncommon in business practice. It explains that these calculations were intended to ameliorate the entrepreneur's problematic credit situation, which required a specific demonstrative effort, and points to scientific demonstration practices in Physiocratic political economy, and in Lavoisier's chemistry, as likely influences.

Irénée Du Pont de Nemours, born in 1771, landed in the United States on 1 January 1800, along with his father, the Physiocratic economist Pierre Samuel Du Pont de Nemours, his older brother Victor Marie, and his wife and children, and soon settled in a house near New York City, in Bergen Point, New Jersey, which would be their base in the United States.[1] They were following the path of many other elite émigrés who had fled revolutionary France in the 1790s, but they were not as wealthy as some of their predecessors

* Centre de Sociologie des Organisations, Sciences Po, 27, rue Saint Guillaume, 75337 Paris Cedex 07, France; martin.giraudeau@sciencespo.fr.

I am most grateful to the three editors of this volume, William Deringer, Eugenia Lean, and Lukas Rieppel, for their invitation to contribute, their insightful comments, and their kind patience. The *Osiris* editors and two anonymous reviewers also helped me to substantially improve the essay. For reading earlier versions and sharing their subsequent questions and suggestions, I warmly thank Mario Biagioli, Alex Csiszar, Stephanie Dick, Frédéric Graber, Bruno Latour, Vincent Lépinay, Nadia Matringe, Peter Miller, Michael Power, Claude Rosental, Rita Samiolo, Zsuzsanna Vargha, François Vatin, Christine Zabel, Michael Zakim, and the participants in the June 2016 workshop on Science and Capitalism at the Heyman Center for the Humanities at Columbia University.

[1] On the Du Pont family and its arrival in the United States, see Denise Aimé-Azam, *Du Pont de Nemours, honnête homme* (Paris, 1934); Pierre Jolly, *Du Pont de Nemours, soldat de la liberté* (Paris, 1956); Ambrose Saricks, *Pierre Samuel du Pont de Nemours* (Lawrence, Kans., 1965); Marc Bouloiseau, *Les Dupont de Nemours (1788–1799): Bourgeoisie et révolution* (Paris, 1972).

and would therefore need to find ways of earning their living there.[2] For that purpose, they conceived a number of different plans between the fall of 1797 and the fall of 1800, most of them of an agricultural or commercial nature.[3] Among these plans was an industrial one, however, designed by Irénée and for which he wrote, in the fall of 1800, his "Project for the establishment of a war and hunting gunpowder manufactory in the United States."[4] It is this investment proposal, which initiated the founding of what is now called the DuPont Corporation, that I propose to study here.

Lacking capital for his undertaking, Irénée was compelled to seek funds from other parties, and this was the purpose of the document, which he immediately sent to potential investors in Europe. The rhetoric of this specific project was relatively unusual compared with that of other investment proposals of the time. Lacking personal credit, Irénée could not convince his audience through traditional arguments, for instance, by putting forward his reputation, experience, and connections. Instead, he tried to convince his readers of the certain profitability of the proposed manufactory. To do so, he presented a series of accounting simulations, made on a profit-and-loss account, in which expected revenues and expenses were estimated in order to calculate the resulting profits. A variety of hypotheses were considered—different production technologies could be used, gunpowder prices could vary, and so forth—and their impact on profit levels calculated, so as to show that the manufactory would remain sufficiently profitable under all reasonably imaginable circumstances. The author of the "Project" claimed in the document that he had preemptively answered all possible "objections," and that it was thus "proven that a gunpowder manufactory established in America would have sufficient advantages."

I argue below, in the first section, that Du Pont's "Project" must be considered as a demonstration device akin to the ones used in natural philosophy at the time. A careful internal study of its contents indeed reveals that it was designed in many ways like a late eighteenth-century literary and quantitative report of an experiment. The object of experimentation was the proposed manufactory, the experimental device the profit-and-loss account, and the results a series of comparable profit numbers. The "Project" was a "literary technology" allowing the "virtual witnessing" of a business experiment by its readers, as well as a quantitative technology facilitating the removal of the author's subjective self from his experiment.[5] As such, it could silence possible objections and thus hopefully allow its author to gain credit—that is, financial credit, in the form of investments—from the chosen audience. This homology between Du Pont's "Project" and experimental reports is interesting in that it reveals an unusually early instance of

[2] Du Pont's father, who was born into a bourgeois family (his own father was a watchmaker), owned a small rural estate south of Paris, but he had mostly earned his living as an officer of the royal administration under the ancien régime, and then as the owner of a print shop, run by Irénée, from 1791 to 1797. Pierre Samuel's political opposition to the Directoire and to the rise of Napoléon, who were in his view corrupting the ideals of the Constitution of 1791, led to the ransacking of the print shop during the coup of 18 Fructidor, Year 5 (4 September 1797), cutting the family's revenue stream and prompting them to move to the United States.

[3] P. S. Du Pont, folder entitled "*Catalogue des mémoires*," n.d. (1800), Group 2D (box 33), Winterthur MSS, Eleutherian Mills Historical Library, Hagley Museum and Library, Wilmington, Del. (hereafter cited as "EMHL, HML").

[4] E. I. Du Pont, "Projet d'établissement d'une fabrique de Poudre de Guerre et de Chasse dans les États-Unis d'Amérique" [Project for the establishment of a war and hunting gunpowder manufactory in the United States], n.d. (1800), Group 3B (box 11), Longwood MSS, EMHL, HML.

[5] On literary technologies, see Steven Shapin, "Pump and Circumstance: Robert Boyle's Literary Technology," *Soc. Stud. Sci.* 14 (1984): 481–520. On quantitative technologies, see Theodore M. Porter, *Trust in Numbers: The Pursuit of Objectivity in Science and Public Life* (Princeton, N.J., 1995).

the use, in the world of business, of a form of demonstration often thought to pertain primarily, at that time, to the world of science.

Further, the homology gives us an insight into the historical connections between scientific and business demonstration practices, which I explore in the second section by situating the device in its social contexts. If some motivations for Du Pont's use of such a demonstration device can be found in the business context of his time, I indeed show that the influence of contemporary scientific practices also needs to be considered. Specifically, I suggest that Du Pont may have inherited the experimental approach to business developed, relatedly, by Physiocratic economists in farming and by Antoine-Laurent Lavoisier in gunpowder manufacturing. The Physiocrats, and especially his father, had developed accounting simulation techniques similar to that used by Du Pont in his "Project," and at the Régie des Poudres Lavoisier had designed a large-scale economic experiment in the wild. The Régie's gunpowder manufactories were the object of this experiment, and the profit-and-loss accounts were the experimental device. The Du Pont case is one example of the subtle entanglement of science and capitalism at the beginning of the late modern period.

These observations, drawn from the analysis of the Du Pont "Project" and of its origins, appear intriguing when related to the lessons from prior literature. Economic historians have indeed emphasized the role of classically social factors, such as personal status, reputation, and networks, in the mechanisms of personal credit formation for entrepreneurs. In early modern England, credit was ensconced in an economy of obligation, where the rules of gentlemanly civility forbade seeking profit for itself, or at least openly: one was otherwise at risk of being accused of profiteering.[6] Although these rules could be gamed by pretending to make projects for the common good, and they had lost some of their force by the end of the early modern period, the role of social status, reputation, and networks in the formation of personal credit remained crucial in the eighteenth century, on both sides of the Atlantic world.[7] If there were, in this context, devices used to demonstrate creditworthiness, prior scholarship suggests that their role was primarily to provide evidence of the personal value of the entrepreneur, be it his moral value and social reputation (letters of recommendation) or his financial worth (personal accounts in the hands of notaries).[8]

Some historians of science, although well aware of the role of other types of demonstration devices (less centered on the person) in the formation of credit, have reached the same conclusion as economic historians when looking at entrepreneurs. Focusing on the late twentieth century, Claude Rosental has examined how they use live "demos" of technology rather than written documents to publicly display their "proofs of concept" and thus gain credit with potential investors and other stakeholders, so as to eventually "capitalize on science."[9] Steven Shapin has similarly described how today's venture

[6] Natasha Glaisyer, *The Culture of Commerce in England, 1660–1720* (Rochester, N.Y., 2006); Craig Muldrew, *The Economy of Obligation: The Culture of Credit and Social Relations in Early Modern England* (London, 1998).

[7] Pierre Gervais, "Mercantile Credit and Trading Rings in the Eighteenth Century," *Annales* 67 (2012): 693–730.

[8] Philip T. Hoffman, Gilles Postel-Vinay, and Jean-Laurent Rosenthal, *Priceless Markets: The Political Economy of Credit in Paris, 1660–1870* (Chicago, 2000).

[9] Claude Rosental, *Les capitalistes de la science: enquête sur les démonstrateurs de la Silicon Valley et de la Nasa* (Paris, 2007); Rosental, "Toward a Sociology of Public Demonstrations," *Sociol. Theory* 31 (2013): 343–65.

capitalists are persuaded to invest by entrepreneurs, previously recommended to them by trustworthy allies, through the live presentation of their business projects.[10] These studies have, of course, acknowledged that entrepreneurial "demos" are regularly accompanied by visual PowerPoint presentations and the circulation of written "plan" documents. But for the most part, such formal devices were considered to have little influence on the interactions between entrepreneurs and investors. What mattered instead were the speech acts, bodily performance, and interpersonal networks of the participants. "Without the advantages of familiarity," Shapin wrote, "the satisfaction of formal criteria means almost nothing."[11] In the case of business ventures, and especially of the most speculative ones, financial credit would therefore be an "almost" exclusively intersubjective issue.[12]

These studies in economic history and the history of science have had the great benefit of showing the relevance of a focus on demonstration practices in business, but also of challenging the long-standing Weberian notion that "it is one of the fundamental characteristics of an individualistic capitalistic economy that it is rationalized on the basis of rigorous calculation, directed with foresight and caution toward the economic success which is sought in sharp contrast to the hand-to-mouth existence of the peasant, and to the privileged traditionalism of the guild craftsman and of the adventurers' capitalism, oriented to the exploitation of political opportunities and irrational speculation."[13] Modern entrepreneurs are not the rational calculators and cautious planners of future profits presented by Weber, as accounting historians, excessively focused for so long on the rise of double-entry bookkeeping, now also admit.[14]

Yet modern entrepreneurs do calculate and plan, and increasingly so, it seems. Early modern entrepreneurs applied for patents through written presentations of their projects.[15]

[10] Steven Shapin, *The Scientific Life: A Moral History of a Late Modern Vocation* (Chicago, 2008), 209–68.

[11] Ibid., 288.

[12] To be sure, scholars have acknowledged the critical importance of demonstration devices in certain economic domains, like the financial models and consumer credit scores that have come to play such a powerful role in mass financial markets in the twentieth century. See, e.g., Donald MacKenzie, *An Engine, Not a Camera: How Financial Models Shape Markets* (Cambridge, Mass., 2006); Martha Poon, "From New Deal Institutions to Capital Markets: Commercial Consumer Risk Scores and the Making of Subprime Mortgage Finance," *Account. Org. Soc.* 34 (2009): 654–74. But in matters of entrepreneurship, scholarship suggests that human interactions and networks have always had much more influence than formal devices like business plans in the establishment of credit. Symptomatically, Bruno Latour, after having long emphasized the role of material devices in interaction, commented on the hypothetical case of two entrepreneurs: "Yes, of course, they have drawn up a 'business plan.' . . . Still, these calculations do not support any particular course of action. . . . They refine judgments concerning the intersection of scripts, but they are ultimately incapable of untangling the passions needed to launch the same scripts"; Latour, *An Inquiry into Modes of Existence* (Cambridge, Mass., 2013), 427.

[13] Max Weber, *The Protestant Ethic and the Spirit of Capitalism* (London, 1930), 75.

[14] For exceptions, see B. G. Carruthers and W. N. Espeland, "Accounting for Rationality: Double-Entry Bookkeeping and the Rhetoric of Economic Rationality," *Amer. J. Sociol.* 97 (1991): 31–69; Basil S. Yamey, "The 'Particular Gain or Loss upon Each Article We Deal In': An Aspect of Mercantile Accounting, 1300–1800," *Account. Bus. Financ. Hist.* 10 (2000): 1–12; Jonathan Levy, "Accounting for Profit and the History of Capital," *Crit. Hist. Stud.* 1 (2014): 171–214; Pierre Gervais, "Why Profit and Loss Didn't Matter: The Historicized Rationality of Early Modern Merchant Accounting," in *Merchants and Profit in the Age of Commerce, 1680–1830*, ed. Dominique Margairaz, Yannick Lemarchand, and Pierre Gervais (London, 2015), 33–52.

[15] See, e.g., Liliane Hilaire-Pérez, *L'invention technique au siècle des Lumières* (Paris, 2000); Koji Yamamoto, *Taming Capitalism before Its Triumph: Public Service, Distrust, and "Projecting" in Early Modern England* (Oxford, 2018).

They issued printed subscription prospectuses in the press to invite curious readers to financially support their projects.[16] They circulated proposals privately across investor networks, promising dividends and capital gains. For late twentieth- and early twenty-first-century entrepreneurs, creating a "business plan" has become a standard and almost unavoidable requirement of starting a business.[17] It is therefore essential to account for these practices, by looking at the material forms and uses of the actual calculations and plans they produce, as well as at their historical origins and evolution across the modern period. The fact that some of these practices appear to be homologous, and even historically related, to scientific demonstration practices does not make them more rational, nor thereby confirm the Weberian hypothesis, but it prevents us from simply dismissing it either. Entrepreneurs do calculate and plan and have been influenced in doing so by scientific practices. The point is therefore to account for the historically shifting forms and roles of entrepreneurial plans in practice.

The Du Pont case is a particularly appropriate place at which to begin such an inquiry. First, the available archival sources on the case are exceptional, starting with the fact that they include the "Project for the establishment of a war and hunting gunpowder manufactory in the United States." Written entrepreneurial plans, which lacked any legal value, were not always preserved and, when they were, they remain hard to find, hidden as they generally are in archives' boxes tagged as "miscellaneous." The Du Pont archives also have the advantage of including countless other documents from the family over decades from the mid-eighteenth to the late twentieth century, including correspondence, business records, and numerous notes and memoirs. These resources allow for thick description and explain to a great extent why the Du Pont Corporation has become for historians of business the equivalent of the *Drosophila* fly for geneticists: a model organism through which many questions can be thoroughly explored.[18]

But Du Pont's "Project" of 1800 is also interesting for less practical reasons. It is particularly intriguing in its reliance on profit-and-loss accounting as its core instrument. Even in established businesses, the use of profit-and-loss accounts was a recent development at the time.[19] An entrepreneur like Josiah Wedgwood was exceptional in developing thorough profit-and-loss calculations for each product line, for internal management purposes, as early as 1772.[20] The preparation of such accounts only became a common practice in England in the mid-nineteenth century, and it is in the twentieth century that

[16] Marie Thébaud-Sorger, *L'Aérostation au temps des Lumières* (Rennes, 2009).

[17] Benson Honig and Tomas Karlsson, "Institutional Forces and the Written Business Plan," *J. Manage.* 30 (2004): 29–48; Liliana Doganova and Marie Eyquem-Renault, "What Do Business Models Do? Innovation Devices in Technology Entrepreneurship," *Res. Policy* 38 (2009): 1559–70; Martin Giraudeau, "Imagining the (Future) Business: How to Make Firms with Plans," in *Imagining Business: Performative Imagery in Business and Beyond*, ed. François-Régis Puyou, Paolo Quattrone, Christine Mclean, and Nigel Thrift (London, 2011), 213–29; Giraudeau, "Remembering the Future: Entrepreneurship Guidebooks in the US, from Meditation to Method (1945–1975)," *Foucault Stud.* 13 (2012): 40–66.

[18] See, e.g., Alfred D. Chandler and Stephen Salsbury, *Pierre S. Du Pont and the Making of the Modern Corporation* (New York, 1971); Joanne Yates, *Control through Communication: The Rise of System in American Management* (Baltimore, 1989); Naomi R. Lamoreaux and Kenneth Lee Sokoloff, *Intermediaries in the U.S. Market for Technology, 1870–1920* (Cambridge, Mass., 2002); Pap Ndiaye, *Nylon and Bombs: DuPont and the March of Modern America* (Baltimore, 2007).

[19] Yamey, "The 'Particular Gain or Loss'" (cit. n. 14).

[20] Neil McKendrick, "Josiah Wedgwood and Cost Accounting in the Industrial Revolution," *Econ. Hist. Rev.* 23 (1970): 45–67; Anthony Hopwood, "Accounting Calculation and the Shifting Sphere of the Economic," *Eur. Account. Rev.* 1 (1992): 125–43.

it was fully institutionalized, starting in the United States.[21] Du Pont's "Project" thus comes at an important transitional moment in the history of accounting, and of capitalism more generally. Finally, it was influenced by a number of other important transitions, in the era of so-called revolutions. Irénée Du Pont was a student of Lavoisier, a central figure of the Scientific Revolution. The Du Pont family immigrated to the United States in an attempt to escape the French Revolution. And the manufacture of gunpowder was an early American industrial venture during the Industrial Revolution. Standing at various major crossroads, the case allows us to explore openly the connections between business and other fields of activity.

A DEMONSTRATION DEVICE

In this section, I proceed to show that Du Pont's "Project" was akin to scientific demonstration devices of its time. To do so, I first familiarize the reader with the structure and contents of the document, before reviewing the characteristics it shared with the report of an experiment.

The "Project" was a ten-page manuscript in French, which combined text and numeric tables. It was explicitly aimed at proving that the proposed manufactory would generate sufficient profits to make an investment in it worthwhile. For that purpose, right after stating in its introduction the U.S. government's need for gunpowder produced in the United States (rather than imported from England) and the amount of capital required for setting up the manufactory ($36,000), the document announced a forthcoming calculation: "We will calculate in detail the annual cost and production of the establishment in order to give an idea of the profit for which it would be reasonable to hope." Before proceeding with this calculation, however, the narrator explained the assumptions on which it would rest: it would be for a manufactory relying on an old production process "used in France before the revolution" and selling its gunpowder at the same market price as the lower-quality gunpowder sold at the time in the United States. An accounting table followed, inserted within the text: it was an annual profit-and-loss account, which stated the expected revenues (volumes times prices), before listing and summing up the operating expenses in raw materials (saltpeter, sulfur, and charcoal), employee wages (foreman, skilled workers, and laborers), fabrication losses, office expenses, and machine maintenance. The bottom line was for the "remaining earnings," after subtraction of the expenses from the revenues. Earnings were estimated at $19,000, an amount immediately converted into francs for European readers.

This first table was followed immediately by another one of the exact same structure, but based on different assumptions: this time, the reader was briefly told, the improved production processes that had been developed in France during the revolutionary wars, and whose superior efficiency "[could] not suffer any objection," would be used. The profits amounted in that case to $32,000, allowing the narrator a first intermediary conclusion: "it is undoubtedly difficult to find another enterprise that would yield such high earnings with such limited capital and such secure markets." The exact same procedure

[21] Levy, "Accounting for Profit" (cit. n. 14), 173; Stewart Jones and Max Aiken, "The Significance of the Profit and Loss Account in Nineteenth-Century Britain: A Reassessment," *Abacus* 30 (1994): 196–230; Dale Buckmaster and Scott Jones, "From Balance Sheet to Income Statement: A Study of a Transition in Accounting Thought in the USA, 1926–1936," *Account. Audit. Accountabil. J.* 10 (1997): 198–211.

was repeated twice after that, with different assumptions, this time bearing not on production technologies (the new calculations were based, conservatively, on the "old" ones) but on market conditions. Should the price of gunpowder go down, the manufactory would yield profits of $12,376; on the contrary, should it go up a bit, profits would reach $21,696. Each of these results was followed by reassuring comments on other possible assumptions. The lower amount of $12,376 was an absolute minimum given that the manufactory would actually use newer and more economically efficient production processes, that it would produce higher-quality gunpowder than its existing competitors, and that it could also sell "elite" gunpowder on the side, at even higher prices. As for the higher amount of $21,696, it was not affected by possible increases in the price of saltpeter, the main raw material needed, because it would not be in the interest of saltpeter producers to raise their prices, and in any case there were alternative supply sources, the U.S. government having large reserves of saltpeter at hand.

All possible objections had thus been answered, the narrator suggested, which allowed him to wittily conclude, in an address to the potential investors who would be reading it: "One objection remains, we barely have the funds necessary for this establishment, and to this there is no answer. This objection would be just as strong if the new factory could yield earnings of one million instead of one hundred thousand francs per year."

The "Project" was a "literary technology" that relied on similar demonstrative techniques as reports of experiments of the time.[22] Like those reports, it indeed created a dialogue with an audience of "virtual witnesses," who were invited to observe the effects of consecutive manipulations on the proposed object of experimentation, a gunpowder manufactory. The narrator of the "Project" engaged with these spectators during the show, by incorporating their "objections," in direct speech, into his demonstration.[23] For instance, after the first two accounting tables had been presented, the narrator transitioned to the next calculation in the following way: "However, one may say, present prices, on which these calculations are based, are the war price and will change as soon as the war ends."[24] He immediately addressed this objection: "Let's now see whether, if peace were to bring back the old prices, the new manufactory could still compete with those of Europe." Other objections were raised by the narrator, in indirect speech, thus preempting remarks from the audience. For instance, before the fourth calculation, the narrator explained that there were "reasons to believe" that in this period, following the Hamiltonian report on manufactures, the U.S. government would "doubtless give all necessary encouragement to such an enterprise" against foreign (English) competitors, either by paying more for its gunpowder or by imposing taxes on imports. A dialogue was thus established.

The narrator then made sure to astonish his virtual witnesses by carefully examining their supposed objections, step by step, building suspense until the striking result of a

[22] Shapin, "Pump and Circumstance" (cit. n. 5).

[23] The technique of preemptively answering possible objections came from classical rhetoric and was extremely widespread in the eighteenth century: it was taught to every educated child. It was particularly prevalent in the world of natural philosophy, however, at least since Galileo's 1633 *Dialogue*, and among engineers; see, e.g., Frédéric Graber, "Obvious Decisions: Decision-Making among French Ponts-et-Chaussées Engineers around 1800," *Soc. Stud. Sci.* 37 (2007): 935–60.

[24] The end of the Quasi-War between the United States and France could push the demand for gunpowder down, and thus its price would increase; moreover, the price of saltpeter, imported from the main global production region, India, would decrease with increased freedom of trade.

pleasing bottom line number, reinforced every time by a concluding sentence that highlighted the unquestionable success of the operation. The repetition of the technique over ten pages created an effect of saturation that seemed to silence all other possible objections, and thus to achieve a fully demonstrative effect, beyond the simple illustration of likely profits: the future profitability of the manufactory was certain.[25]

Beyond this virtual witnessing it instated, the "Project" was also comparable to an experimental report in its focus on a specific object of experimentation: the gunpowder manufactory, understood as an economic entity. Of course, the manufactory was also a machine, made of a set of mills, and one where chemical processes would take place—but it was not experimented upon as such. Du Pont did present the technological features of the machine in detail, but he tellingly did so in a separate document, entitled "On the site and constructions necessary for the establishment of a powder manufactory" and attached as an appendix to the "Project."[26] No experiment was reported in this document. It described how the manufactory would be designed, not how it would behave under certain circumstances. It proceeded linearly, following the stream of the river along which the different buildings of the manufactory would be located, from the house of the director, located on a high point overlooking the entire site, to the storage room for the finished product, via two mills and other buildings housing specific processes. The reader was simply introduced to its specificities: those of an innovative technology of French origins. It was not just any gunpowder manufactory, but one that included two distinct types of mills, contrary to those of current American gunpowder manufacturers, and that incorporated architectural developments from the Manufacture des Poudres d'Essonnes, like widely separated buildings with thick walls and light roofs, so as to prevent potential explosions from causing excessive damage. These technological specificities were not the object of the "Project" itself.

Rather, the "Project" was concerned with the manufactory's economics. The available technologies were known, and so were the volumes of production and the related raw material inputs that went with them. What had to be demonstrated was the profits that they would yield, and that American market circumstances would allow. This was made clear in the very first paragraphs of the document, where the narrator explained that the high prices of labor and of raw materials in the United States had made industrial ventures unsuccessful in the country, but that "a manufactory in which nearly all the work is done by machinery, which would use foreign raw material, and which for those reasons could not feel the effects of the high price of national industries, would be sure of complete success." The narrator of the "Project" was showing his readers a series of four experiments conducted on, in today's jargon, a "business model." The object of experimentation was the cost and revenue structure of such a manufactory. Consistently, the experimental device used to study the manufactory was not a physical or chemical one, but an accounting one: a profit-and-loss account. Thanks to it, the narrator could manipulate the manufactory as an economic entity, performing various operations on it—increasing or decreasing certain costs and revenues—to see with what

[25] On the difference between illustration and demonstration in experimental practice, see Simon Schaffer, "Machine Philosophy: Demonstration Devices in Georgian Mechanics," *Osiris* 9 (1994): 157–82.

[26] E. I. Du Pont, "De l'emplacement et des constructions nécessaires pour l'établissement d'une fabrique de poudre," n.d. (1800), Group 3B (box 11), Longwood MSS, EMHL, HML.

profit levels it would respond. He could make the manufactory "speak," as if for itself, in the monetary language of business.[27]

The manufactory did not really speak for itself, however; it did not even exist yet! The land had not been purchased, the mills had not been built, the workforce had not been hired, and the raw materials had not been purchased. No gunpowder had been sold, let alone produced. The experiment that was being reported on had not taken place: it was being simulated, quantitatively, on the profit-and-loss account. Accounting is often accused of being excessively "creative," but this is especially the case when it is a proposed future, rather than an accomplished past, that is being "accounted" for. In some ways, this was an advantage for Du Pont—as it is for entrepreneurs in general.[28] He thus had some freedom in conducting his "experiment": the object of experimentation, virtual rather than actual, was not very recalcitrant.[29] And use this freedom he did, as was evident in the five successive drafts that he wrote for the "Project." The first simulation trial focused exclusively on a manufactory using the production process currently in place at competing gunpowder manufactories in the United States. Then as drafts went by, new hypotheses were simulated: the "old" (prerevolutionary) French production process was introduced, then the "new" one, then a number of hypotheses regarding the prices of gunpowder and saltpeter, until the low-profit simulation based on the US production process disappeared completely from the plan, and four main results were chosen—one high, one low, and two median ones, as we have seen. Tellingly, the numbers were also slightly massaged: the results of some of the profit calculations were at first rounded to the nearest tens of thousands, but then eventually brought back to the nearest thousand, by adjusting some of the cost and revenue figures that generated these results. Profit demonstration was a strategic enterprise, and accounting for the future made it possible to produce, select, and clean the displayed results quite freely.

Yet, although the practice of simulation gave him more freedom than the one enjoyed by natural philosophers conducting actual experiments, Irénée was not entirely different from them in this respect. Like them, he needed to be extremely careful in order for his manipulation of the experimental device to appear properly demonstrative.[30] He needed to be especially careful in erasing his personal interventions on the experiment, in fact, given that his readers would be well aware that the experiment had not actually taken place. He avoided rounding profit numbers excessively (for the first two simulations) and included nonrounded ones as well (for the last two) as a way of conveying an impression of accounting authenticity—that is, of truth to economic nature, if not of

[27] The ability of the "Project" to make the manufactory "speak" makes it comparable with Galileo's inclined plan as it has been analyzed in Isabelle Stengers, *L'invention des sciences modernes* (Paris, 2010), 84–94; see also Martin Giraudeau, "Inclined Plans: On the Mechanics of Modern Futures," in *Reset Modernity!*, ed. Bruno Latour (Cambridge, 2016), 286–92.

[28] Successful businesses are easier to promise on paper than to build in markets, which the critique of reckless "projectors" had repeatedly emphasized since the seventeenth century; see Yamamoto, *Taming Capitalism* (cit. n. 15); Simon Schaffer, "The Show That Never Ends: Perpetual Motion in the Early Eighteenth Century," *Brit. J. Hist. Sci.* 28 (1995): 157–89. The critique reemerged, formulated against the now standard "business plan," after the dot-com bubble burst; see, e.g., David E. Gumpert, *Burn Your Business Plan! What Investors Really Want from Entrepreneurs* (Needham, Mass., 2002).

[29] On the ability of scientific objects to resist in experiments, see Bruno Latour, "Des sujets récalcitrants: Comment les sciences humaines peuvent-elles devenir enfin 'dures'?," *La Recherche* 301 (1997): 88.

[30] Schaffer, "Machine Philosophy" (cit. n. 25).

objectivity—while still maintaining some clarity in the results.[31] The accounting tables he presented to the readers looked like ex post accounts of actual operations: should they invest in the venture, the readers of the project would one day receive such statements by mail, along with the related dividends on their invested capital.

Du Pont went further in ensuring the apparently impersonal nature of his conclusions. As the drafts of his plans evolved, they included more and more text: Du Pont intervened more in the narration so as to better remove himself from the scene. The textual parts of the project were indeed focused on justifying the accounting assumptions, by grounding them in empirical observation. The narrator did that by creating an "effect of reality" through the inclusion, within the report, of "little true facts," which readers either knew or could eventually verify should they want to, given that he authoritatively referred to them as undisputable "facts."[32] In the attached appendix, he described the mills by taking the reader for a stroll along a river, as if the manufactory had already been built. The narrator also brought in the existing gunpowder manufactories in the United States; he explained that he had visited some of them and commented on their inefficient production processes, the poor quality of their powder, and the decent profits they were still able to make in spite of this. The manufactory was finally inserted within an observable market context: the preliminaries of the Quasi-War were mentioned to justify the assumptions made on the future evolution of gunpowder and saltpeter prices, and so were the current importation of powder from England by the U.S. government, the major role of India in the international market for saltpeter, and so on.

Although the manufactory did not exist, it was nevertheless presented in the text as an established entity, inserted into actual contexts, rather than as the malleable figment of an entrepreneur's imagination. It could thus be experimented upon, and its response to the experimental operations performed on it would be reliable. If the narrator did intervene abundantly between accounting tables, it was simply to perform a translation from the readers' natural language to accounting, and back: he explained the costs and revenues put forward before the calculation took place and drew a conclusion from the resulting bottom line after it had taken place. He was thus an apparently neutral intermediary, rather than an active mediator, between the virtual witnesses and the proposed manufactory.

DEMONSTRATION IN BUSINESS AND SCIENCE

Why did Du Pont proceed in this way? In this section, I explore the social contexts in which he was operating, starting with the business circles for which he wrote the "Project," before moving to the scientific influences that may have shaped the demonstration techniques he used in it. First, I show that the demonstration device designed by Du Pont was intended to reassure investors with whom neither he nor the type of project he was proposing benefited from much prior credit. Du Pont was in a difficult credit situation and thus needed to come up with an original "instrument of credit" that would compensate for the specific resources he lacked.[33] However original, his "Project" nevertheless relied on existing demonstration techniques, which I suggest can be traced back to his apprenticeship with Lavoisier.

[31] Lorraine Daston and Peter Galison, *Objectivity* (New York, 2007), 55–113.

[32] Roland Barthes, "L'effet de réel," in *Littérature et réalité*, ed. Roland Barthes, Leo Bersani, Philippe Hamon, Michael Riffaterre, and Ian Watt (Paris, 1982), 81–90.

[33] Mario Biagioli, *Galileo's Instruments of Credit: Telescopes, Images, Secrecy* (Chicago, 2006).

Du Pont was a young man with little experience. He had been a student at the Régie des Poudres, which he mentioned in the plan. However, his three-year stint there ended before he was twenty years old when it was interrupted by revolutionary events.[34] He was therefore hardly qualified for designing and setting up an entirely new manufactory by himself. Aware that he lacked the necessary technical knowledge and experience, he returned to France between February and April 1801 to further his education, gather technological plans and drawings, and order machinery through his connections at the Essonnes powder works. He did have some "managerial" experience and a number of potentially useful business contacts, having run the family print shop with his father between 1791 and 1797. However, this experience did not necessarily seem relevant, as he did not mention it in the "Project." In a letter to his father's stepson, who was about to meet a potential investor in Paris on his behalf, Du Pont summarized the man's position on credit: "he [the potential investor] is always very keen on the prudence and the experience that are necessary to such commerce [as that of gunpowder]. In this respect, he and all of the other men of business that my brother will deal with in France are not predisposed in our favor."[35]

Du Pont had access to his father's networks and reputation and took full advantage of them. His father initially sent the "Project" to his friend Jacques Bidermann, a prominent cotton manufacturer and merchant, as well as a banker, originally from Geneva and later based in Paris.[36] Bidermann had already demonstrated his commitment to the Du Pont family. He invested in and was serving as the European correspondent for the company formed by Pierre Samuel Du Pont two years earlier, before the family immigrated to the United States, with the goal of founding an agricultural colony and conducting commercial business there.[37] The letter included a note asking that the project also be shared with Jean-Joseph Johannot and Jacques Roman, both Swiss citizens from Bidermann's personal circle who were involved in manufacturing. Johannot was also involved in public finance, and Roman happened to be Bidermann's brother-in-law.[38] Du Pont's letter mentioned other possible investors, including two other Swiss citizens close to Bidermann: Jacques Necker, the comptroller-general of finances of Louis XVI, who was also in touch with father Du Pont, directly and through their mutual friend, Madame de Staël; and Necker's brother, Louis Necker de Germany, a mathematician who had been involved in international trade. Adrien Duquesnoy, an associate in the large Catoire, Duquesnoy & Cie commercial house working primarily for the governmental salt farm, was also mentioned; Du Pont had known him at least since their election to the États-Généraux in 1789.[39] The elder Du Pont asked Bidermann to contact these investors regarding seven of his own proposals and only briefly mentioned Irénée's

[34] Irénée was also involved in gunpowder manufacturing in 1794 while he was serving at the Garde Nationale, but this was another brief experience.

[35] E. I. Du Pont to Jacques Bidermann, 10 May 1801, Group 3B (box 1), Longwood MSS, EMHL, HML. All translations are my own.

[36] P. S. Du Pont to Jacques Bidermann, 1 December 1800, Group 4A (box 1), Winterthur MSS, EMHL, HML. A translation can be found in B. G. Du Pont, *Life of Éleuthère Irénée du Pont from Contemporary Correspondence*, vol. 5, 1799–1802 (Newark, Del., 1925), 163–96.

[37] On Pierre Samuel's agricultural plan, see Martin Giraudeau, "Performing Physiocracy: Pierre Samuel Du Pont de Nemours and the Limits of Political Engineering," *J. Cult. Econ.* 3 (2010): 225–42, on 233–6.

[38] Louis Bergeron, *Banquiers, négociants et manufacturiers parisiens du Directoire à l'Empire* (Paris, 1999), 65–86.

[39] Du Pont to Bidermann, 10 May 1801 (cit. n. 35).

proposed gunpowder manufactory, as the family's eighth and last plan. In spite of this, none of the father's proposals caught the investors' attention. Thus, although the elder Du Pont's network and favorable reputation were useful, they were not enough to persuade the investors.

Another issue with Irénée's venture was the lack of creditworthiness of the type of business he was proposing. In his father's plans—the first, an agricultural and commercial one prepared between 1797 and 1799, and the seven commercial and financial ones he was proposing now—the elder Du Pont did not feel the need to carefully demonstrate the future profits he would be making: he simply described the projects and gave an indication of the overall increase in capital that the investors could expect at the end. The initial agricultural and commercial plan, for instance, promised that their capital would be multiplied by two, probably four, and perhaps ten or twenty over the company's twelve-year life span, and committed to the payment of standardized amounts of dividend in the meantime.[40] This was a common feature of investment proposals at the time.[41] Merchants, bankers, and wealthy individuals generally knew what a commercial, financial, or agricultural company could yield if managed by an able and reasonably trustworthy person. This was even the case for an establishment in the United States: Bidermann himself owned land in Kentucky and a commercial house in Philadelphia through his wife, the daughter of the Geneva banker Antoine Odier.[42] Some industrial ventures were also familiar: Johannot and Roman, the two men to whom, tellingly, Irénée Du Pont's "Project" was explicitly addressed, were both familiar with manufacturing—but in the textile industry. A gunpowder manufactory was an altogether different matter, especially one based in the United States. Gunpowder manufacturing in France had long been a state monopoly, conducted through the General Farms system until 1775 and then through the centralized administration of the Régie des Poudres. Investors in the country had no experience spending their capital on gunpowder production. Further, the shared understanding at the time, that the United States was not fertile ground for industrial ventures, was reinforced in various publications, starting with Benjamin Franklin's 1782 *Information to Those Who Would Remove to America*.[43] Du Pont was fighting an uphill battle with his proposal.

This specific, low-credit situation made Du Pont particularly cautious, and he relied on an arsenal of persuasion techniques. Well aware, as we have seen, of his lack of personal credit, he stressed the importance of face-to-face interactions with investors. This was the other reason why he visited France in February 1801 and made sure, when he returned to the United States, to direct his father's stepson to meet with some potential investors from his father's and Bidermann's circles. He even gave him instructions for these meetings. To compensate for his lack of personal credit, he stressed, "it is . . . very important while speaking to (potential investors) about the successes of our operations to inspire in them some confidence in our internal order."[44] However, this emphasis on

[40] P. S. Du Pont de Nemours, "Extrait d'un plan d'une opération rurale et commerciale à exécuter dans les États-Unis d'Amérique," n.d. (1797), Group 1 (469), Longwood MSS, EMHL, HML.

[41] Frédéric Graber and Martin Giraudeau, eds., *Les Projets: une histoire politique (17ème–21ème siècles)* (Paris, forthcoming).

[42] Bergeron, *Banquiers* (cit. n. 38).

[43] Benjamin Franklin, *Two Tracts: Information to Those Who Would Remove to America, and, Remarks Concerning the Savages of North America* (Dublin, 1784). This small pamphlet was initially printed in Paris in 1782, in French and in English, and was well known in France.

[44] Du Pont to Bidermann, 10 May 1801 (cit. n. 35).

the physical and vocal aspects of personal interactions in what we now refer to as the "pitching" of the project was a side note: the "Project" still played a central role. The demonstration device and the in-person demonstration were not exclusive of each other but complementary, something that is often overlooked in the literature emphasizing the importance of interpersonal exchanges in the establishment of entrepreneurial credit. They were intertwined. First, as Du Pont also explained to his father's stepson, "the memoir [i.e., the "Project"] I left with M. Biderman and the act of association will give my brother all of the information he may need to give a good idea of this operation and to get new subscribers."[45] The written plan was to shape the live demonstrations performed by Du Pont's collaborators, including his father's stepson and Bidermann, when he himself was unable to be there. Second, the "Project," left in the hands and under the scrutiny of potential investors, could by itself convey an impression of the "internal order" of its author: Du Pont, the young, inexperienced man, could be judged, and hopefully earn some personal credit, through the plan. Du Pont's notable efforts at demonstration in his project enhanced the credibility of his venture in three ways: directly, by "proving" the manufactory's future profitability, and indirectly, by helping his allies pitch the project, and by exhibiting his own personal seriousness as an entrepreneur.

Although the low-credit situation of Du Pont and of his venture explains his efforts to design a particularly convincing investment prospectus, it does not tell us where he found the specific demonstrative techniques that he deployed throughout the document, and particularly the use of the profit-and-loss account for simulation purposes. The answer to this question is necessarily tentative: there is no evidence of a model plan that Du Pont would have copied in writing his own plan. However, a number of leads can be followed so as to formulate hypotheses on the possible origins of Du Pont's demonstration device.

The approach Irénée developed in his "Project" does not seem to be modeled on typical investment proposals of the time; it appears to be quite original. No systematic survey of such proposals has been conducted to date for the period, but existing studies suggest that this kind of plan was not common, and no examples of similar cases can be found in the literature.[46] Patent or privilege applications, subscription proposals published in the press, or proposals circulated privately to investors would include accounting numbers, but they almost systematically put forward a single expected result, instead of, like Du Pont, experimentally exploring a number of possible results. They would also not focus much on profits, especially not through a profit-and-loss account. They would rather emphasize the benefits of the project for the public good (improvements), the subscribers' curiosity (wonders), or the investors' interests (dividends and capital gain).

This is consistent with the lessons from non-Whiggish accounting historians, who have shown that profit-and-loss accounts were almost nonexistent until the nineteenth century, be it in agriculture, industry, or trade, in Europe or the United States.[47] Balance

[45] Ibid. During his stay in Paris, Du Pont incorporated the new company formed in his name for gunpowder production there, so that he could receive investments. Blank spaces were left on the "act of association," where new names of subscribers could be added.

[46] For a review of the existing literature on project proposal practices since the seventeenth century, see Graber and Giraudeau, *Les Projets* (cit. n. 41).

[47] For broad-ranging reviews on the issue, see Yamey, "The 'Particular Gain or Loss'"; Levy, "Accounting for Profit" (both cit. n. 14). On the American case specifically, see also Naomi R. Lamoreaux, "Rethinking the Transition to Capitalism in the Early American Northeast," *J. Amer. Hist.* 90 (2003): 437–61.

sheets were certainly produced, where the net worth of an entity at the moment of reckoning was evaluated, and profits could technically be measured by calculating the difference in net worth between two consecutive balance sheets, but this was rarely done.[48] The shipping accounts of merchants would sometimes include "net profit" figures, used to split the returns between the investors for a given expedition, but such profit-and-loss accounts would always come ex post and were seldom produced at the scale of an entire company.[49] Even in a case like that of manufacturer Josiah Wedgwood and his 1772 cost accounting efforts, often put forward as an example, the cost and revenue measures were not made for the calculation, and even less so for the external reporting, of profit figures. The focus for Wedgwood, as for most other manufacturers, was on setting prices and assessing, for instance, whether they could be lowered.[50] It therefore seems extremely unlikely that Du Pont found inspiration in the common accounting practices of contemporary business people—or as a matter of fact in his own experience of business accounting while he ran the family print shop in the 1790s. If the accounts have unfortunately been lost, the correspondence with his father on the print shop's operations is abundant, and it never refers to actual profit calculations, even on specific jobs. Instead, the focus was on the necessity to control costs and set prices high enough, as well as on the need to find new customers, and be paid by them, so as to settle debts with suppliers.[51]

Another possible influence on Du Pont that requires a brief exploration is that of books promoting methods of wise administration in business. Although actual accounts did not allow that in practice, the accounting literature did recommend the calculation of "gain or loss" starting in the sixteenth century.[52] More importantly for my purpose here, authors writing on political economy had started looking into business accounting issues in the early eighteenth century. Richard Cantillon, theorizing the economic role of the entrepreneur in the 1730s, had started emphasizing that his activity consisted in "proportioning himself to his customer" in order to generate a profit. Around Francois Quesnay, the Physiocrats had focused on farmers and gone further by starting to formalize quantitatively the question of yield in agriculture, both in kind and in monetary terms. They developed accounts allowing them to measure the "net product" of agriculture, and to thus emphasize the advantages of certain expenses, or "advances," on agricultural profit levels.[53] One of the leading authors of the group happened to be Pierre Samuel Du Pont de Nemours, Irénée's father, who used these types of accounts at various times in his career.[54]

Interestingly, in 1771, the year Irénée was born, Pierre Samuel published two articles arguing against slavery. Using such accounts, along with figures gathered from actual

[48] Basil S. Yamey, "Scientific Bookkeeping and the Rise of Capitalism," *Econ. Hist. Rev.* 1 (1949): 99–113, on 108; Sidney Pollard, "Capital Accounting in the Industrial Revolution," *Bull. Econ. Res.* 15 (1963): 75–91.

[49] Gervais, "Why Profit and Loss Didn't Matter" (cit. n. 14), 34.

[50] Neil Mckendrick, "Josiah Wedgwood and Cost Accounting in the Industrial Revolution," *Econ. Hist. Rev.* 23 (1970): 45–67; Anthony Hopwood, "Accounting Calculation and the Shifting Sphere of the Economic," *Eur. Account. Rev.* 1 (1992): 125–43.

[51] B. G. Du Pont, *Life of Éleuthère Irénée du Pont from Contemporary Correspondence*, vol. 2, 1792–1794 (Newark, Del., 1925).

[52] Yamey, "The 'Particular Gain or Loss'" (cit. n. 14).

[53] Jean-Claude Perrot, "La comptabilité des entreprises agricoles dans l'économie physiocratique," *Annales* 33 (1978): 559–79.

[54] James J. McLain, *The Economic Writings of Du Pont de Nemours* (Newark, Del., 1977). See also Giraudeau, "Performing Physiocracy" (cit. n. 37), 229.

plantation accounts complemented by a number of assumptions, he calculated and compared the profits of two plantations: one relying on slave labor and one employing free laborers, to demonstrate the economic inefficiency of slavery, in addition to its moral unacceptability.[55] The similarity with Irénée's own use of a profit-and-loss account as the core experimental device in his demonstration is suggestive. Of course, one should note that Pierre Samuel did not rely on this same technique when writing his own plans. Also, there is no trace at all, in spite of the abundance of records, of Irénée's familiarity with Physiocratic thought or with his father's writings, or of Pierre Samuel's direct intervention in the preparation of Irénée's plan, the drafts of which were all written in his own hand. Yet, given the similarity between the two demonstrative uses of profit-and-loss accounts, it is impossible to reject the hypothesis that Irénée Du Pont found inspiration for his project in science, that is, in what his father had named the "new science of political economy."[56]

One last hypothesis, not incompatible with this previous one, must, however, be considered. There was one very specific sphere of activity where the use of a type of profit-and-loss accounting was well established in France in the last decades of the ancien régime: public finance. Concerned with controlling costs, and afraid of being taken advantage of by gunpowder manufacturers, the state required the Company of Tax Farmers to provide regular accounts of their revenues and expenses to the royal comptroller general.[57] The same form of accountability was required, from its formation in 1775, of the Régie des Poudres, the royal administration in charge of gunpowder production, where Du Pont was an apprentice from 1787 to 1791. In this so-called school of powders, he learned not only the chemistry and engineering behind gunpowder production but also accounting, and, like other apprentices, he had to complete an internship at the central accounting office at the Arsenal in Paris.[58] Du Pont was thus familiar with the revenue and expense statements of French gunpowder manufactories. Writing his "Project" for potential investors like former comptroller-general Jacques Necker, salt farm member Adrien Duquesnoy, and other individuals well versed in public finances, he may have felt compelled to use a form of calculation that was familiar to them.

This hypothesis is supported by the fact that it was not only the type of profit-and-loss account used in public finance that Irénée discovered at the Régie, but also a specific way of using it. At the General Farms, revenue and expense accounts were used for ex post reporting—and quite late in fact, with the accounting office taking about four years to deliver the accounts.[59] The situation was different at the Régie des Poudres, which had been formed in 1775 based on an "infinitely economical plan" designed and thus designated by Antoine-Laurent Lavoisier, the most active of the four directors.[60] Lavoisier, who had also been a general farmer since 1770, had established a large-scale ac-

[55] Caroline Oudin-Bastide and Philippe Steiner, *Calcul et morale: La pensée économique de l'anti-esclavagisme au XVIIIe siècle* (Paris, 2015).

[56] Pierre Samuel Du Pont De Nemours, *De l'origine et des progrès d'une science nouvelle* (Paris, 1768); Philippe Steiner, *La "science nouvelle" de l'économie politique* (Paris, 1998).

[57] George Tennyson Matthews, *The Royal General Farms in Eighteenth-Century France* (New York, 1958), 203–5, 217–9.

[58] On Du Pont's instruction at the Régie, see René Dujarric De La Rivière, *E.-I. du Pont de Nemours, élève de Lavoisier* (Paris, 1954); Patrice Bret, "Lavoisier à la Régie des poudres: Le savant, le financier, l'administrateur et le pédagogue," *Vie Sci.* 11 (1994): 297–317.

[59] Matthews, *The Royal General Farms* (cit. n. 57), 219.

[60] On the details of Lavoisier's biography, see Arthur Donovan, *Antoine Lavoisier: Science, Administration and Revolution* (Cambridge, 1996).

counting system at the Régie, designed not only for bottom-up reporting but also as an instrument of internal managerial control, for knowledge and control purposes. The twenty-five gunpowder manufactories throughout France had to send monthly and annual statements of revenues and expenses to the Paris headquarters. These statements, on printed templates, were prepared according to detailed guidelines that explained how to account for each revenue and expense item.[61] Compliance with these procedures was controlled, regularly, by regional commissioners and also occasionally by "inspectors." The manufactories were monitored through a centralized accounting system, which made it possible for the head office to compare their economic performances and to eventually favor some over others.[62] The brick-and-mortar manufactories existed in the provinces, but they were also replicated on paper on the desks of Lavoisier and his colleagues in Paris. There, they could manipulate them, consider possible changes in one or the other, and send managerial orders back to the provinces. This form of company-wide, profit-oriented manipulation of costs and revenues may well have shaped Du Pont's approach when he conducted profit simulations on his proposed gunpowder manufactory: he had already witnessed a comparable practice at the Régie's accounting office.

If this was indeed the case, then it was not late eighteenth-century French public finance that influenced Du Pont but rather Lavoisier's specific approach to public finance, which cannot be entirely separated from his scientific practices. Lavoisier seems to have run the accounting operations of the Régie like a large-scale experiment in the wild, the same way he did with gunpowder chemistry and engineering.[63] Lavoisier has often been described as an accountant in chemistry, but he was indeed also a chemist in accounting. He made this experimental approach to business explicit at a number of points in his career. As a general farmer, he explained in his accounting instructions for the tobacco farm that business "operations consist of an art where reasoning must always be guided by experiment," that is, by "the continuous comparisons (between) different manufactories," based on comparable metrics such as costs and expenses, and the resulting profit or loss.[64] Elsewhere, countering the idea that the financial effects of the interruption of saltpeter collection in private houses would be devastating for the gunpowder administration, he quipped: "the simplest calculation will destroy this assertion."[65] And he did not shy away from following through with such calculations, even if they were based on assumptions ("suppositions," in his words), as in the case of the *Richesse territoriale* where, continuing work that had been started by Pierre Samuel Du Pont, he extrapolated the net product of agricultural production for the entire king-

[61] Régie des Poudres et Salpêtres, *Principes de la Régie sur la Comptabilité* (Paris, 1785). Accounting records produced under this system can be found in the La Forte Archive, Carl A. Kroch Library, Cornell University, Ithaca, N.Y.

[62] The Régie, in this respect, appears as a surprisingly early example of the multidivisional firm as conceptualized in Alfred D. Chandler Jr., *Strategy and Structure: Chapters in the History of Industrial Enterprise* (Cambridge, 1962).

[63] To gather and compare the best processes, Lavoisier conducted his own experiments in his laboratory at the Arsenal, and he launched competitions and sent questionnaires to gunpowder manufactories and saltpeter refineries, before sending back instructions on new production methods. See Lucien Scheler, "Lavoisier et la Régie des poudres," *Rev. Hist. Sci.* 26 (1973): 193–222; Bret, "Lavoisier" (cit. n. 58); Seymour H. Mauskopf, "Lavoisier and the Improvement of Gunpowder Production," *Rev. Hist. Sci.* 48 (1995): 3–8.

[64] A.-L. Lavoisier, "Observations sur la Constitution et le Régime de la Ferme des Tabacs," n.d., Fonds Paulze et Lavoisier, Archives Nationales.

[65] Antoine-Laurent Lavoisier, *Œuvres de Lavoisier*, vol. 4 (Paris, 1892), 672.

dom based on assumptions regarding the costs and benefits of average farms. If business and economic experiments could seldom be controlled as well as physical and chemical ones, and often took the shape of simulations when the future was being discussed, they nevertheless had to be undertaken in the same way, and could prove just as useful to win an argument, or support decision-making. For Lavoisier, demonstrations in business and science were of a similar nature.[66] If Du Pont inherited part of his approach to demonstration from the Régie, then he was also inheriting it from Lavoisier's scientific practices.

In any case, thanks to his keen demonstrative efforts, Du Pont eventually managed to secure additional funds. He had tried to raise $36,000, and by the middle of 1801 Bidermann had committed to investing $6,000, as had both Necker de Germany and Duquesnoy, which altogether amounted to half of the capital that had been deemed necessary. In the following year, he worked hard to secure additional capital and alliances, including one with an American associate who would give him access to credit from Philadelphia bankers. In June 1802, he purchased a suitable plot of land near Wilmington, Delaware, on the Brandywine River. Construction of the mills started soon after that, and saltpeter was first refined there in July 1803. The first powder sale was recorded in May 1804, and the highly anticipated government purchases began in August 1805, generating revenues of $45,000.The economic experiment of the gunpowder manufactory was now a reality, and, of course, it was not as tightly controlled as the one that had been simulated on paper in the fall of 1800. Even as Du Pont was leaving France in the spring of 1801, he was aware that the diplomatic situation between France and the United States was shifting, which would lead to a decrease in gunpowder prices. For a decade, government purchases remained much lower than he had strategically assumed in the "Project," while competition proved harsher. Ultimately, though, the firm survived.

CONCLUSION

Irénée Du Pont de Nemours's 1800 "Project for the establishment of a war and hunting gunpowder manufactory in the United States" did not, in itself, permit the success of the DuPont Corporation—a very fragile "success" in its first few years, as we have seen. Nor does it explain, alone, the success of Du Pont's fund-raising efforts, which was also due to his family's networks and personal interactions with at least some of the investors. The claim of this essay is not that the new company was built "on" the plan, or even because of it. And yet, as we have seen, there was a plan. It was prepared, it was used, and it did seem to do something: it brought the project to the attention of

[66] It may even be remarked that Lavoisier relied in both business and chemistry on the exact same instrument. This instrument has been previously identified as the "balance" (Bernadette Bensaude-Vincent, "The Balance: Between Chemistry and Politics," *Eighteenth Cent.* 33 [1992]: 217–37; Norton Wise, "Mediations: Enlightenment Balancing Acts, or The Technology of Rationalism," in *World Changes: Thomas Kuhn and the Nature of Science*, ed. Paul Horwich [Cambridge, Mass., 1993], 207–56), but it is in fact a profit-and-loss account. We have seen it in his business practices, and we can also observe it in his chemistry. For instance, in the 1773 experiments on fluids, where his so-called balance sheet method in chemistry first emerged, the tables that he included in his laboratory notes and experimental reports were entitled "calculation," and they stated the "total weight of materials employed" and the "weight after combination," so as to produce by subtraction the result of what he called the "operation" (chemical reaction), this result being a "loss" (Frederic Lawrence Holmes, *Antoine Lavoisier: The Next Crucial Year: Or, The Sources of His Quantitative Method in Chemistry* [Princeton, N.J., 2014], 57).

distant readers, giving the proposed manufactory a tentative existence on paper; it helped prepare some of the entrepreneur's allies, and probably the entrepreneur himself, to pitch it; and it conveyed an impression of the entrepreneur's abilities. Thus, it both supplemented and supported other components of the entrepreneur's demonstrative arsenal. This essay is therefore primarily an invitation for scholars of demonstration practices to acknowledge the fact that entrepreneurs do prepare written proposals, and that their origins and roles in action must be accounted for.

Beyond this broad methodological lesson, which could certainly have been drawn from many other entrepreneurial plans, the Du Pont case is particularly instructive regarding the history of entrepreneurial calculations, and even business calculations in general. As I have shown, it appears to support the emergence around 1800, in the world of business, of the practice of profit-and-loss calculations, and more specifically ex ante profit-and-loss calculations. While businesses had long sought "profit," understood in a broad sense as the difference between low purchases of supplies and high sale prices, it was starting to be envisaged quantitatively, through a yearly estimation, at the scale of an entire company—rather than for temporally bound expeditions, for instance. As the study stressed, however, this did not indicate an increase in the "rationality" of the entrepreneur—perhaps at best an increase in the degree of control over costs and revenues, if the entrepreneur considered the plan an actual template for the project he was attempting to realize. Rather, the use of profit-and-loss calculations was an attempt by an entrepreneur who lacked personal credit and was proposing an unusual industrial venture in a foreign country to earn financial credit through new means. His use of an accounting technique not common in business practice, but certainly understandable by his audience, was thus deeply influenced by the context in which he was operating. He was caught in a preexisting web of expectations that he was trying to satisfy in an innovative way.

The innovation was far from radical, however. The Du Pont case is traditionally regarded as one of technology transfer between France and the United States—a transfer of production technology, like others that took place around the same time between England and the United States in the textile industry. Yet there was another technology transfer taking place in the Du Pont case—a transfer of demonstration technologies. Du Pont's device was indeed akin to an experimental report, made for the public demonstration of scientific results. In it, the narrator addressed an audience of virtual witnesses and preemptively countered their objections. At the same time, the narrator made sure to appear as a neutral intermediary between the audience and the manufactory, presented as responding by itself to the manipulations undertaken on it through the core experimental device, the profit-and-loss account. This reliance on quantification reinforced the impression of the narrator's neutrality. All of these demonstrative techniques appear to have had their origins not so much in the business practices of the time—or at best in those of the Régie, a state enterprise—but in scientific demonstration practices Du Pont was, or was likely to be, familiar with: those of his father, Physiocratic economist Pierre Samuel Du Pont de Nemours, and those of the mentor that his father had chosen for him, Antoine-Laurent Lavoisier.

The type of demonstration device produced by Du Pont with his "Project" would reappear over the following three decades, as part of the advice given by the French liberal economists around Jean-Baptiste Say to the employees of manufactories, who were encouraged to start their own independent "factories." The workers were advised to save money in order to acquire their own tools, but they were also instructed on the way to prepare a "factory project," in which they had to estimate their coming profits—or losses—under a number of assumptions, with the help of a profit-and-loss account.[67] Men without prior credit, like Du Pont, had to rely on thorough forms of economic demonstration to convince others, and themselves before that, that their projects were viable. Strikingly, Say's approach to economic matters was both an implicit continuation of the work started by the Physiocrats with farmers and an explicit extension of Lavoisier's chemistry. He explained, "We do not create objects: the mass of matter that the world is composed of cannot increase or diminish. All that we can do is reproduce the given matter under another form which makes it fit for whatever use that it did not previously have, or which increases the utility it previously had. Then there is creation, but of utility; and because this utility gives matter value, there is production of wealth. This is how the word production should be understood in political economy."[68] Say and his followers were standing at the same historical crossroads that Du Pont had encountered only a couple of years earlier, where the demonstration practices of entrepreneurs intersected with those of political economy and of Lavoisier's chemistry.

[67] Claude-Lucien Bergery, *Économie industrielle*, vol. 3, *Économie du fabricant* (Metz, 1830). On Bergery and his factory project guidelines, see François Vatin, *Morale industrielle et calcul économique dans le premier XIXe siècle: L'économie industrielle de Claude-Lucien Bergery (1787–1863)* (Paris, 2007).

[68] J.-B. Say, *Traité d'économie politique* (1803), bk. 1, chap. 1, cited in Alain Béraud, "Ricardo, Malthus, Say et les controverses de la 'seconde génération,'" in *Nouvelle histoire de la pensée économique*, vol. 1, *Des scolastiques aux classiques*, ed. Alain Béraud and Gilbert Faccarello (Paris, 2000), 365–508, on 400.

Lies, Damned Lies, and (Bourgeois) Statistics:
Ascertaining Social Fact in Midcentury China and the Soviet Union

*by Arunabh Ghosh**

ABSTRACT

Is there a correct way to ascertain social fact? As late as the 1950s, the scientific community remained divided over this question. Its resolution involved not just epistemological and theoretical debates on the unity or disunity of statistical science but also practical considerations surrounding state-capacity building. For scientists in places like the People's Republic of China (PRC) and the Soviet Union, at stake was the very ability to realize the kind of planned economic growth that socialist countries idealized. The solution they chose reformulated statistics explicitly as a social science, salvaging it from what they then dismissed as the tainted, bourgeois, and socially unproductive pursuit of mathematical statistics. This distinction—most tangibly understood as the rejection of all probabilistic methods—had implications for both the ways in which data was collected and the ways in which it was analyzed.

In this essay I investigate how "socialist" statistics came to be valorized as an explicitly anticapitalist antidote to the problem of accurate and correct knowledge production in the social world. Focusing on developments in the People's Republic of China (PRC) and the Soviet Union in the early 1950s, I explore the theoretical basis for this position. My goal is to demonstrate that socialist statistics posed a serious challenge to any notion of a universal statistical science and was as much an outcome of ideological confrontation as it was of theories of social and economic change. In so doing, I hope to disaggregate the history of statistics, which still tends to be coterminous with the rise of probabilistic thinking and the liberal-bourgeois concerns that, at least partly, drove them.

Two larger processes provide the backdrop to this story of midcentury statistics. The first is the overwhelming drive toward the modernization of statecraft, a trend

* History Department, CGIS South S135, 1730 Cambridge Street, Harvard University, Cambridge, MA 02138; aghosh@fas.harvard.edu.

I would like to acknowledge with thanks the valuable assistance of the volume editors, William Deringer, Eugenia Lean, and Lukas Rieppel, as well as the *Osiris* general editors and the anonymous reviewers. Matthew Jones, Fa-Ti Fan, and Henry Cowles read earlier versions of this essay and provided helpful comments. Work on this essay was supported by grants from the Social Science Research Council, the American Council of Learned Societies, and Columbia University.

 0369-7827/11/2018-0008$10.00

that manifested itself in the postwar years most noticeably through the development and deployment of a whole range of tools for calculation. Echoes of such moments in other historical contexts can be found in the essays by William Deringer and Martin Giraudeau in this volume.[1] As I explain below, the collection and collation of data about an ever-expanding range of activities were central to this particular moment in the history of modern statecraft and transcended ideological divides—real and manufactured—between capitalism and communism.[2] The second is the Cold War and the particular ideological and practical constraints it placed on scientific activity, in particular, by often elevating applied over basic research.[3] Together, these two processes gave shape to the intellectual and epistemological choices that are laid out in this essay. In other words, the organizational and institutional imperatives that lie at the heart of this story were universal—the desire of modern states for ever-increasing control—but they manifested themselves in particular and often dissenting forms. In unpacking one such adversarial manifestation, I focus here primarily on the theoretical justification, as put forward by certain Chinese and Soviet scientists, for the distinctiveness and correctness of socialist statistics. Not discussed in detail here are the implications this distinction had for the academic discipline of statistics and for statistical work more broadly, subjects that I cover in greater detail in my forthcoming book on how the PRC built statistical capacity during the 1950s.[4]

"A NEW TYPE OF STANDARDIZED STATISTICAL WORK"

With the establishment of the PRC in 1949, jubilant Chinese revolutionaries were confronted by the dual challenge of a nearly nonexistent statistical infrastructure and the pressing need to escape the universalist claims of capitalist statistics. Li Fuchun 李富春 (1900–1975), the deputy head of the Central Economic and Financial Commission of the PRC, captured these twin imperatives in a speech he delivered to statistical cadre at a national meeting in July 1951, in which he called for the wholesale repudiation of existing statistical thought and practice:

> In the past, China was a semicolonial, semifeudal country; strictly speaking, it did not possess any statistics [worth speaking of]. Statistics in Old China was learned from the Anglo-American bourgeoisie. This kind of statistics cannot serve as our weapon, it is unsuitable for [the tasks of] managing and supervising the country. . . . We need to build [a new] statistics for a new China.[5]

[1] William Deringer, "Compound Interest Corrected: The Imaginative Mathematics of the Financial Future in Early Modern England"; Martin Giraudeau, "Proving Future Profit: Business Plans as Demonstration Devices," both in this volume.

[2] On this connection, see also Julia Fein, "'Scientific Crude' for Currency: Prospecting for Specimens in Stalin's Siberia," in this volume.

[3] For a discussion of science and technology during the Cold War, see Naomi Oreskes and John Krige, eds., *Science and Technology in the Global Cold War* (Cambridge, Mass., 2014).

[4] Arunabh Ghosh, *Making It Count: Statistics and Statecraft in the Early People's Republic of China, 1949–1959* (Princeton, N.J., forthcoming); for the dissertation on which the book is based, see Ghosh, "Making It Count: Statistics and State-Society Relations in the Early People's Republic of China, 1949–1959" (PhD diss., Columbia Univ., 2014).

[5] Li Fuchun 李富春, "Zhongyang renmin zhengfu zhengwuyuan caizheng jingji weiyuanhui Li Fuchun fuzhuren zai quanguo caijing tongji huiyi shang de zhishi" [Directives delivered by Li Fuchun, deputy head of the Central Economic and Financial Commission of the National Administrative Council, at the first National Finance and Statistics Meeting], in *Tongji gongzuo zhongyao wenjian huibian—di yi juan* [Compilation of important documents relating to statistical work—volume 1] (Beijing, 1955), 1–5. Unless otherwise noted, all translations are mine.

By the time of Li's exhortation, the project to build a new statistics for a new China was already under way in the country's northeast region. In April 1950, the Northeast Statistical Bureau (NSB) had been established under the leadership of economist and statistician Wang Sihua 王思华 (1904–78). Later that summer, Wang addressed the first meeting of statistical workers in the northeast's history, proudly noting that it was also the largest such meeting in the nation's history.[6] This was clear evidence, according to him, that New China was already on the path toward achieving a planned economy, for no planned economy could be possible without the presence of scientific statistical work. In his address, Wang observed that since the promulgation of the "Decision Regarding Strengthening Statistical Work" on 10 April 1950, statistical work had entered a "new phase."[7] As a result, "statistical work in the northeast had begun to move from a state of decentralization to integration, from chaos to the correct path, from the nonscientific to the scientific." Wang explained that help offered by Soviet experts was critical to this change. The Soviet Union possessed over thirty years of experience, and it was their expertise and aid that allowed Wang to declare, "We have set up a statistical system, we possess method, [and] we possess confidence."[8]

A few years later, in December 1952, while addressing the third annual meeting of the NSB, Wang was even more emphatic. Over the previous three years, he would claim, the bureau had successfully established "a new type of standardized statistical work" [*xinxing zhengguihua de tongji gongzuo*].[9] Such work was the hallmark of socialist countries and was fundamentally different from the statistics of capitalist countries.[10]

Explicitly rejecting probabilistic methods such as large-scale random sampling, the system built under Wang consisted of two principal means of statistical data collection. The first, and by far the more significant, was the complete enumeration periodical report system [*quanmian dingqi tongji baobiao zhidu*].[11] The idea was that data would be collected primarily through a vast network of comprehensive and periodic statistical reports that would span all sectors of the economy. The periodicity of these reports could be as short as a ten-day cycle [*xun*], or as long as a year [*nian*], and included several possibilities in between: monthly [*yue*], seasonal [*ji*], semiannual [*bannian*], and so on. These reports were sent up through each successive level until they reached the headquarters of the NSB in Shenyang (and later Beijing). By 1954, this system had expanded to include data from industry, capital construction, agriculture,

[6] Wang Sihua, "Tongji gongzuo de renwu" [The tasks of statistical work], in *Tongji gongzuo: di yi xuanji* [Statistical work: the first anthology] (Shenyang, 1952), 7–14. A slightly edited version of this address was included in a collection of Wang's works published in 1986, where it states the speech was originally delivered in 1950. See Wang Sihua, *Wang Sihua tongji lunwenji* [Collected statistical papers of Wang Sihua] (Beijing, 1986), 1–14.

[7] Wang, "Tongji gongzuo de renwu" (cit. n. 6), 7. For a text of the decision, see "Guanyu jiaqiang tongji gongzuo de jueding" [Decision regarding strengthening statistical work], in *Tongji gongzuo: di yi xuanji* (cit. n. 6), 127.

[8] Wang, "Tongji gongzuo de renwu" (cit. n. 6), 8.

[9] Wang Sihua, "San nian lai dongbei tongjiju gongzuo zongjie" [Summary of statistical work performed by the Northeast Statistical Bureau over the past three years], in *Tongji gongzuo: di wu xuanji* [Statistical work: the fifth anthology] (Shenyang, 1953), 1–10. An edited version was published in Wang's collected works; see Wang, *Wang Sihua tongji lunwenji* (cit. n. 6), 31–49.

[10] Wang, "San nian lai dongbei tongjiju gongzuo zongjie" (cit. n. 9), 3.

[11] "Complete enumeration periodical report system" is how the Chinese themselves translated *quanmian dingqi tongji baobiao zhidu*. See File 105-00530-05: 11–12, Foreign Ministry Archives, Beijing (hereafter cited as "FMA").

trade, transport and communication, and private enterprise, among many others. When data was required with an irregular level of expediency, censuses [*pucha*] were also employed. For instance, a census of private industry was begun on 19 July 1950; it included six provinces and five cities and enumerated 39,539 private industrial units.[12]

Unlike in industry, statistical work in agriculture did not always permit the exhaustive enumeration of the periodic report system or the census method. Accordingly, surveys [*diaocha*] were designed to ascertain levels of agricultural production. These surveys were invariably based on some form of typical sampling [*dianxing diaocha*]. In designing them, the Chinese drew upon not so much Soviet expertise as their own experiences conducting rural surveys during the 1930s. In his memoirs, Xue Muqiao 薛暮桥 (1904–2005), the head of the State Statistics Bureau (SSB) for much of the 1950s, recalled that they operated by selecting typical cases [*xuan dian*] for closer analysis.[13] Using this system, 8,000 surveyors investigated 1,175 typical [*dianxing*] villages in the northeast. First, sown area was surveyed, and then areas were divided based on the nature of the harvest (good or bad); typical cases were selected from within these divisions and investigated, after which overall estimates were generated based on the relative weights of good and bad harvest areas. Such typical sampling formed the basis of several other surveys carried out in the northeast. These included surveys of rural grain surplus [*nongcun yuliang diaocha*], livestock [*xuchan diaocha*], and farms [*nongchang diaocha*].[14]

Taken together, these activities provide a catalog of the basic contours of statistical work in the northeast. Wang's "new type of standardized statistical work" made claims to three basic features. The first of these was "extensiveness" [*guangfanhua*]. For Wang this meant that socialist statistics, unlike its bourgeois version, had mass character [*dazhonghua*]. In other words, it ran through every sector of the economy and society:

> Not only did it include all branches of the national economy, but also culture, education, and health care; its content was extremely extensive, and the procurement of materials, the collation of numbers, depended on a comprehensive and unified reporting system.[15]

A unified reporting system also meant that each person and each element within the system was absolutely critical. Even if one person's work suffered from a fault [*maobing*], then the entirety of statistical work was at fault.

Accordingly, a second characteristic was the system's "completeness" [*zhengtihua*]. This completeness was a corollary to the extensiveness in that it sought to conceive of the statistical work of every unit as an integral part of a unified statistical apparatus. The key to achieving such completeness required standardization of forms and types of data so that comprehensive and comparable data were easily available. Equally important was the creation of norms for entering data on forms and for their accompanying textual explanations. Since the integrity of such an extensive and complete system could easily be threatened by a kink in any one level, it was imperative to create a uni-

[12] Data Center of the National Bureau of Statistics, "Huiwang: lishishang de dongbei tongjiju" [Looking back: the history of the Northeast Statistics Bureau], *China Stat.* (August 2010): 22–3.

[13] Xue Muqiao, *Xue Muqiao: huiyilu* [Xue Muqiao: memoirs] (Tianjin, 2006), 174.

[14] Data Center of the National Bureau of Statistics, "Huiwang" (cit. n. 12), 22–3.

[15] Wang, "Tongji gongzuo de renwu" (cit. n. 6), 13–4.

fied structure for reporting that permitted data (i.e., forms and tables) to be sent up and down different levels of the statistical and administrative system.[16]

Extensiveness and completeness were the basis for the third feature of socialist statistics, namely its "objectivity" [*keguanxing*]. Unlike bourgeois statistics, which was predicated on profit and meant that statistical data were frequently distorted [*waiqu*] to serve the interests of the few and protect their business secrets, the New Statistics was unafraid of exposing weakness. Quite to the contrary, it welcomed criticism and even took part in self-criticism in the pursuit of objectivity and truth [*zhenshixing*].[17]

These features—extensiveness, completeness, and objectivity—exemplified what was deemed scientific by Wang because only by counting everything could an objective sense of social reality be achieved.[18] Such exhaustive counting in turn was the basis of the socialist corrective to bourgeois statistics. Indeed, the implication was that a system that was socialist was by definition also scientific and therefore correct. The three terms—socialist, scientific, and correct—had become mutually interchangeable.[19]

ASCERTAINING SOCIAL FACT IN THE 1950S

Where did the inspiration for and confidence in exhaustive enumeration via a periodic report system come from? In the most concrete terms it came from the desire to emulate the success of the Soviet Union. To the Chinese (and many other socialist states after 1945), the Soviet Union "provided the only historical model for industrializing an economically backward country under socialist political auspices."[20] The Chinese Communist Party's attitude toward such "advanced experience" [*xianjin jingyan*] had been articulated by none other than Mao Zedong 毛泽东 (1893–1976) himself. In 1949, he had stated, "The Communist Party of the Soviet Union is our best teacher

[16] Ibid., 14.

[17] Ibid. It is worth noting that Wang's concerns about objectivity appear to center on the possible manipulation of data and not on the objectivity of measurement and representation themselves, which are the subject of Lorraine Daston and Peter Galison's seminal work; see Daston and Galison, *Objectivity* (Cambridge, Mass., 2007). In that sense, they resonate with Theodore Porter's formulation that "objectivity means knowledge that does not depend too much on the particular individuals who author it"; Porter, *Trust in Numbers* (Princeton, N.J., 1996), 229. In later articulations by Wang and other Chinese statisticians, a definition of objectivity emerges that valorizes exhaustive enumeration's ability to provide a like-for-like (i.e., one-to-one) representation of reality.

[18] For a brief but insightful discussion on the "reality" of statistics, esp. through an analysis of its practitioners, see Alain Desrosières, "How Real Are Statistics? Four Possible Attitudes," *Soc. Res.* 68 (2001): 339–55.

[19] Several years later, in 1957, the Indian statistician P. C. Mahalanobis would be impressed by these very qualities of the by then national statistical system, noting that it was "appreciably better than that in India in respect of coverage, availability, and accuracy of data required for purposes of planning and current policy decisions." See Mahalanobis, "Some Impressions of a Visit to China 19 June–11 July 1957," typed manuscript, 12–3, P. C. Mahalanobis Memorial Museum and Archives, Kolkata.

[20] Maurice Meisner, *Mao's China and After: A History of the People's Republic* (1977; repr., New York, 1999), 109. Mao had articulated much the same point in his essay "On Practice," where he noted: "The struggle of the proletariat and the revolutionary people to change the world comprises the fulfillment of the following tasks: to change the objective world and, at the same time, their own subjective world—to change their cognitive ability and change the relations between the subjective and the objective world. Such a change has already come about in one part of the globe, in the Soviet Union. There the people are pushing forward this process of change. The people of China and the rest of the world either are going through, or will go through, such a process." See Mao Tse-Tung, "On Practice," in *Selected Works of Mao Tse-Tung*, vol. 1 (Peking, 1952), https://www.marxists.org/reference/archive/mao/selected-works/volume-1/mswv1_16.htm (accessed 8 May 2018).

and we must learn from it."[21] The Soviet Union had responded with enthusiasm. A principal element of Soviet aid was sharing technical knowledge. By 1960, when the Sino-Soviet split became final, as many as 18,000 Soviet and Eastern European experts had spent time in China sharing their expertise.[22] A handful among them were statistical experts—consultants, administrators, and teachers—whose advice was instrumental in the organization of the NSB and, later, the SSB.

Even more significantly, however, the confidence in exhaustive enumeration came from a particular understanding of how to count and ascertain social reality, and the place of mathematics in that process. What the system of exhaustive enumeration did was to deny the usefulness of mathematics in general and mathematical statistics in particular. As Xue Muqiao, Wang's boss at the SSB after 1952, would recall in his memoirs, "Soviet statistical experts did not promote the use of surveys, [but rather] advocated making comprehensive arrangements for statistical reports."[23]

Embedded in this choice was a deep-seated criticism of statistics as it had evolved and was evolving at that time. Over the last two to three decades the history of quantification, probability, counting, and facticity in the early modern and pre-twentieth-century world has garnered increasing amounts of attention. The rich studies that have emerged have helped establish quantification and statistics as independent subjects of inquiry, situating them at the very heart of the modern and postmodern experience.[24] Among the many insights that have been offered by this rich literature is the significant role played by the state and by the study of the social world (in addition to the natural world), particularly through the expansion of bureaucracies and administrative duties, in the realization of what Theodore Porter has called the "statistical perspective."[25]

If these were essentially nineteenth-century developments, largely limited within the boundaries of nation-states or, in certain cases, imperial formations, the twentieth

[21] Mao Zedong (Mao Tse-tung), "'The People's Democratic Dictatorship,' Speech delivered 'In Commemoration of the 28th Anniversary of the Communist Party of China, June 30, 1949,'" in *Selected Works*, vol. 4 (New York, 1975), 411–24. This was also the speech in which Mao used the phrase "leaning to one side" to describe China's policy. On the speech and this policy, see Chen Jian and Yang Kuisong, "Chinese Politics and the Collapse of the Sino-Soviet Alliance," in *Brothers in Arms: The Rise and Fall of the Sino-Soviet Alliance, 1945–1963*, ed. Odd Arne Westad (Stanford, Calif., 1998), 246–94, esp. 247–50. For similar statements extolling the virtues of the Soviet Union, see also Thomas P. Bernstein and Hua-yu Li, eds., *China Learns from the Soviet Union, 1949–Present* (New York, 2010), 1.

[22] Shen Zhihua, *Sulian zhuanjia zai zhongguo (1948–1960)* [Soviet experts in China (1948–1960)] (Beijing, 2003).

[23] Xue Muqiao, *Xue Muqiao: huiyilu* (cit. n. 13), 174.

[24] See, e.g., Alain Desrosières, *The Politics of Large Numbers: A History of Statistical Reasoning*, trans. Camille Naish (Cambridge, Mass., 1998); Ian Hacking, *The Taming of Chance* (Cambridge, 1990); Lorenz Krüger, Lorraine J. Daston, and Michael Heidelberger, eds., *The Probabilistic Revolution*, vol. 1, *Ideas in History* (Cambridge, Mass., 1987); Silvana Patriarca, *Numbers and Nationhood: Writing Statistics in Nineteenth-Century Italy* (Cambridge, 1996); Mary Poovey, *A History of the Modern Fact: Problems of Knowledge in the Sciences of Wealth and Society* (Chicago, 1998); Theodore Porter, *The Rise of Statistical Thinking, 1820–1900* (Princeton, N.J., 1986).

[25] "What we might call the statistical perspective owed more to the work of census bureaus and statistical offices concerned with health, crime, trade, and education than to the pure theory of mathematical probability"; Theodore M. Porter, "Chance Subdued by Science," *Poetics Today* 15 (1994): 467–78, on 474. And in a review of two works on Russian and Soviet statistics, Michael D. Gordin has spoken of "the importance of statistics as a lens for examining the rise of Soviet governance," a claim that would apply more generally to the rise of governance as well. See Gordin, review of *Statistique et révolution en Russie: Un compromis impossible (1880–1930)*, by Martine Mespoulet, and *L'anarchie bureaucratique: Statistique et pouvoir sous Staline*, by Alain Blum and Martine Mespoulet, *Kritika*, n.s., 5 (2004): 803–10.

century presented challenges of a rather new nature. The experiences of World War I led to the desire for new international/global institutions of arbitration and peace. The most well-known example is the founding of the League of Nations. One of the activities in which the League was involved was the design and organization of internationally commensurable economic and financial data.[26] At the same time, the duties of the state also expanded far beyond their nineteenth-century norms. Fueled in part by the rise of Communism and by the experience of the Great Depression, which, in turn, facilitated the rise of Keynesian economics, the state began to take on ever-expanding responsibility in regulating the economy and providing services to its citizens.[27] From a technological standpoint, each of the World Wars also played their role, as they provided opportunities for the design, experimentation, and deployment of advanced techniques of data collection and analysis. Operations research, for instance, was a field that grew substantially on the basis of wartime exigencies.[28] After World War II, the earlier efforts of the League of Nations were renewed through the United Nations (UN), which established various institutions for global order and communication, including a Statistical Commission in 1947.[29]

Finally, the collapse of empires that gathered pace in the wake of World War II heralded the arrival of several new nation-states, each with its own modernizing mission. A principal way to be modern, and demonstrate one's modernity to others, was to know oneself and at the same time exhibit this self-knowledge to others. A statistical system that gathered and disseminated such knowledge using numbers, and that employed the latest methods, became the hallmark of a modern nation-state.[30] For a long time, a standard imperial argument had been that locals did not even possess an accurate and scientific means of ascertaining (social) fact. Empire's civilizing mission, thus, included this "gift" of fact and numerical representation. In the Chinese case, the discourse about the lack of facts was keenly felt. Almost on par with the nineteenth-century Western imperial assertion that China lacked a rational legal system, the lack of factual self-knowledge came to inform a discourse of national humiliation. As Tong Lam has shown, the response was the rise of a "social survey movement," the principal goal of which was to assess what China was—to make her legible, knowable, and thereby eventually changeable.[31]

[26] As early as August 1919, the league convened meetings to discuss international cooperation in statistics. It mooted the establishment of an International Statistical Commission, which met for the first time in Paris in October of the following year. For more, see J. W. Nixon, *A History of the International Statistical Institute, 1885–1960* (The Hague, 1960), 28–9, 42.

[27] For more on this, see, e.g., Desrosières, *Politics* (cit. n. 24). Mention can be made here of the rise of national income accounting, unemployment statistics, macroeconomic theory with its focus on employment, and economic planning.

[28] On the wartime histories of operations research in the United Kingdom and the United States, see Maurice W. Kirby, *Operations Research in War and Peace: The British Experience from the 1930s to 1970* (London, 2003); Randall Thomas Wakelam, *The Science of Bombing: Operational Research in RAF Bomber Command* (Toronto, 2009); and Charles S. Schrader, *History of Operations Research in the United States Army, Volume 1: 1942–1962* (Washington, D.C., 2006).

[29] For a history of the UN and its role in this process, see Michael Ward, *Quantifying the World: UN Ideas and Statistics* (Bloomington, Ind., 2004).

[30] In his discussion of planning in postcolonial India, for instance, Partha Chatterjee has shown how the act of planning for a new nation was formulated through the logic of numbers. Statistical indicators were precisely the metrics through which progress could be measured and contrasted with colonial underdevelopment. See Chatterjee, "Development Planning and the Indian State," in *The State, Development Planning and Liberalisation in India*, ed. T. J. Byres (New Delhi, 1998), 82–103.

[31] Tong Lam, *A Passion for Facts: Social Surveys and the Construction of the Chinese Nation-State, 1900–1949* (Berkeley and Los Angeles, 2011).

By the end of World War II, these developments, and the desire for a new world (order), contributed in part to what might be termed a more globalized epoch of statistics-driven numerical positivism.[32] Each of these developments depended upon and demanded new levels of data collection and analysis. This desire for data, for facts, was summed up by the UN Secretary General Trygve Lie in 1947. Addressing the inaugural session of the International Statistical Institute's twentieth meeting in Washington, D.C., Lie observed:

> The free exchange of information on economic and social affairs among all countries in the world is absolutely necessary to economic and social advancement. We cannot cure our troubles unless we know in the first place what those troubles are. Likewise we cannot achieve international understanding, which is the basis of advancement, unless the peoples of the world are given the facts about each other. . . . *There is no substitute for facts, for clear and systematically organized facts. They alone can be relied upon to measure resources and potentialities for progress and to direct policies and actions designed to achieve the objectives of all civilized people.*[33]

Speaking, as he was, at the preeminent gathering of international statisticians, the desire for "clear and systematically organized facts" was predicated on a form of numerical positivism—numbers alone could help understand realities on the ground, design solutions to problems, and plan for the future. The postwar years were, as a result, a time when several statistical innovations achieved tremendous legitimacy.[34] Among these innovations can be listed large-scale sample surveying, operations research, development theory (most notably W. W. Rostow's theory about the stages of economic growth), demographic transition theory, random sampling methods such as Monte Carlo, and the spread of economic planning.

A UNIVERSAL SCIENCE

Most of these innovations were made possible because of a specific shift in the understanding of statistics, quantification, and probability that occurred during the nineteenth century. Until the late nineteenth century, statistical activities broadly construed progressed along two nonintersecting tracks, one descriptive and the other analytic. The descriptive traced its origins to the seventeenth-century "political arithmetic" of William Petty among others, the German school of descriptive statistics (*Statistik*, from which the modern word originates), and through them to a much more fundamental and age-old global phenomenon—the desire of the state to collect information about its constituents, its land, and its produce. The analytic, which can be traced at

[32] On how some of these tendencies played out in the articulation of a particular "Cold War rationality" in the United States, see Paul Erickson, Judy L. Klein, Lorraine Daston, Rebecca Lemov, Thomas Sturm, and Michael Gordin, *How Reason Almost Lost Its Mind: The Strange Career of Cold War Rationality* (Chicago, 2013). See also Rebecca Lemov, *Database of Dreams: The Lost Quest to Catalog Humanity* (New Haven, Conn., 2015).

[33] Trygve Lie, "Opening Address to the International Statistical Institute's 1947 Meeting," in *Introduction: International Statistical Conferences, September 6–18, 1947, Washington, D.C.: Proceedings of the International Statistical Conference*, vol. 1 (Calcutta, 1951), 151; emphasis added.

[34] In 1986, Porter observed that statistical tools used in modern sciences had been worked out over the previous 100 years or so. The phase of invention was mostly in the nineteenth century and the phase of development during the late nineteenth and early twentieth centuries. Extending his periodization, we can label the rest of the twentieth century starting with World War II as the period of application and refinement. See Porter, *The Rise of Statistical Thinking* (cit. n. 24), 3.

least as far back as the sixteenth and seventeenth centuries, was principally concerned with concepts of chance, probability, and estimating and controlling error, primarily in an abstract, mathematical sense.[35] The shift that occurred in the nineteenth century involved a new understanding of error not as a reflection of chaos but rather as variations in a given natural order. As Porter notes, "the acceptance of indeterminism constitutes one of the most striking changes of modern scientific thought. With few exceptions, scientists and philosophers previous to the late nineteenth century would have agreed with Augustus De Morgan that to say an event occurs by chance is to say that it occurs for no reason at all."[36] Similarly, Ian Hacking has shown how the erosion of determinism made possible "autonomous laws of chance":

> The idea of human nature was displaced by a model of normal people with laws of dispersion. These two transformations were parallel and fed into each other. Chance made the world seem less capricious: it was legitimated because it brought order out of chaos. The greater the level of indeterminism in our conception of the world and of people, the higher the expected level of control.[37]

For Hacking and Porter, these developments owed as much to the study of natural phenomena as they did to social phenomena.[38] It was not error analysis per se, but rather the use of probability as a modeling tool to analyze variation in nature and society, a mutual cross-pollination across disciplines and intellectual pursuits, which contributed to the evolution of statistics as a distinctly mathematical field of inquiry. This, in brief, is the liberal story in which statistics emerges as a universal science, capable of studying and discerning patterns in both society and nature and employing a catholic approach to methodology—incorporating probabilistic and nonprobabilistic methods.[39] But statistics also has a parallel history, which, at least for a significant portion of the twentieth century, enjoyed great legitimacy and widespread use.

[35] This is no doubt a gross simplification of an era in which ideas about probability were in tremendous flux. As Lorraine Daston has shown, a key feature of this period was that probabilists did not differentiate between objective and subjective aspects of probability, the former a feature of the world and the latter based on a degree of judgment or certainty. For more, see Daston, *Classical Probability in the Enlightenment* (Princeton, N.J., 1988). See also Poovey, *A History of the Modern Fact* (cit. n. 24); and the discussion in Lukas Rieppel, Eugenia Lean, and William Deringer, "Introduction: The Entangled Histories of Science and Capitalism," in this volume.

[36] Porter, *The Rise of Statistical Thinking* (cit. n. 24), 149–50. Porter continues: "Still, this story is not simply one of a new appreciation of the empire of chance. Randomness first attained real standing in scientific thought not as a source of massive uncertainty, but as a small-scale component of an overarching order. The recognition of chance stemmed not from the weakness of science, but from its strength—or rather, its aggressive imperialism, the drive to extend scientific determinism into a domain that had previously been seen by most as the realm of inscrutable whimsy."

[37] Hacking, *The Taming of Chance* (cit. n. 24), vii.

[38] As Porter notes in pt. 2 of his book: "Much more central to the development of a mathematical statistics, however, was Quetelet's insight concerning the distribution of traits in human populations. His idea presents a specific connection of the greatest importance between social and mathematical statistics. It both illustrates and confirms the more general argument of this book concerning the role of social science in the development of statistical thinking." See Porter, *The Rise of Statistical Thinking* (cit. n. 24), 91. In a related way, Michel Foucault also analyzed this shift, though he attributed it more to the bourgeois drive for self-identification and the molding of society.

[39] See, e.g., the discussion in Alain Desrosières, ed., *Sampling Humans: The First Sample Surveys in Norway, Russia and the United States* (Berlin, 2001). It is perhaps instructive to recall that Porter's thesis, the basis for *The Rise of Statistical Thinking* (cit. n. 24), was entitled "The Calculus of Liberalism: The Development of Statistical Thinking in the Social and Natural Sciences of the Nineteenth Century" (PhD diss., Princeton Univ., 1981).

A CRITIQUE OF UNIVERSALITY

Lenin is famously said to have observed that socialism is first of all accounting.[40] Within the former USSR he enjoyed fame as a skilled statistician who used numbers to expose the oppression and injustices of imperialist czarist rule and, after 1918, as someone who vigorously promoted the development of state statistical institutions.[41] Stalin, too, is known to have remarked on the fundamental importance of statistical activities and data.[42] In accepting the centrality of statistics as a tool for understanding and shaping economic phenomena, Marxist and Soviet thinkers and practitioners shared much in common with thinkers in other parts of the world.[43] But in the selection of theory and methods to pursue these goals, they drew a sharp line between themselves and thinkers in a liberal tradition. Indeed, while Marx was likely cognizant and appreciative of developments in mathematics and statistics made possible by the works of such notable figures as Quetelet, Poisson, and Laplace, and by the numerical turn within German statistics, Soviet and Chinese statisticians were quick to point out that he restricted his own discussions on the efficacy of statistics solely to the realm of political economy.[44]

In the Soviet and Chinese statisticians' reading of Marx and Engels, and subsequently of Lenin and Stalin, there were no universal laws of economics and statistics.[45] Rather, different historical epochs and different social systems generated different statistical and economic laws. Accordingly, as A. Yugow noted in a 1947 review of Soviet statistics, it was felt that

[40] As Maya Haber notes, this was a popular paraphrasing of a sentiment Lenin had articulated in December 1917. See Haber, "Socialist Realist Science: Constructing Knowledge about Rural Life in the Soviet Union, 1943–1958" (PhD diss., Univ. of California, Los Angeles, 2013), 87; and Vladimir Il'ich Lenin, "How to Organize Competition?," in *Collected Works*, 3rd ed. (Moscow, 1977), 26:404–15, https://www.marxists.org/archive/lenin/works/cw/index.htm (accessed 8 May 2018).

[41] On Lenin as a statistician, see Samuel Kotz and Eugene Seneta, "Lenin as Statistician: A Non-Soviet View," *J. Roy. Stat. Soc. A* 153 (1990): 73–94.

[42] "Comrades, no constructive work, no state activity, no planning is conceivable without proper accounting. And accounting is inconceivable without statistics. Without statistics, accounting cannot advance one inch." Joseph Stalin "Organisational Report of the Central Committee," 24 May 1924, http://www.marxists.org/reference/archive/stalin/works/1924/05/24.htm (accessed 8 May 2018).

[43] "In the U.S.S.R., statistics cannot be separated from economics either in the field of theory, in the establishment of the methodology of study of new economic phenomena, or in practical work in the accounting and analysis of primary statistical data." See A. Yugow, "Economic Statistics in the U.S.S.R," *Rev. Econ. Stat.* 29 (1947): 242–46, on 243. Yugow is identified as a "Writer on Russian Economics, New York City," in the paratext to the journal issue.

[44] Xu Qian, "Tongji xueshuo fazhan bianhua de lunkuo" [An outline of the developments and changes in statistical theory], in *Tongjixue yuanli ziliao huibian* [Compilation of materials on statistical theory] (Beijing, 1983), 1–35, on 24.

[45] That Marx's own views on the matter were more complex perhaps requires little belaboring. Even so, his writings provided enough space for such a selective interpretation to proceed. Consider, e.g., his assertion in *Capital, Volume 1* that "every special historical mode of production has its own special laws of population, historically valid within its limits alone. An abstract law of population exists for plants and animals only, and only in so far as man has not interfered with them." But some twenty years earlier, he had also stated in the *Economic and Philosophic Manuscripts of 1844* that "History itself is a *real* part of *natural history*—of nature's coming to be man. Natural science will in time subsume under itself the science of man, just as the science of man will subsume under itself natural science: there will be *one* science"; emphasis in the original. See Karl Marx, *Capital, Volume 1: A Critique of Political Economy*, trans. Samuel Moore, Edward Bibbins Aveling, and Ernest Untermann, ed. Frederick Engels (London, 1906), 693; Marx, *Economic and Philosophic Manuscripts of 1844*, trans. Martin Milligan (New York, 2007), 111.

> the statistical methods applied in the accounting of the results of private economic activity, and the patterns established for economies in which the chief economic regulating forces are the laws of supply and demand, have proven inapplicable to statistical accounting in a planned state economy.[46]

The nature of the economy and social organization of a particular country, which, in turn, reflected a particular stage in its history, required its own sets of statistical and economic laws. The distinction between the laws that govern society, and thus appropriately belonged to social science, and those that govern nature, was one that could not be surmounted. Accordingly, they latched on to the numerous occasions when Marx identified statistics as a weapon that could be used in service to the proletariat classes, especially by providing trenchant critiques of existing political-economic conditions, and as a tool for the realization of workers' statistics. In such a formulation, statistics was fundamentally political arithmetic, and Marx's regard for William Petty, the originator of the phrase, as the founder of both statistics and political economy was seen as further evidence for the validity of the natural/social dyad.[47]

While Marx may not have enjoyed the political power necessary to put many of these ideas about the distinctions between the natural and social world into action, his self-appointed intellectual heirs found themselves in positions to do so. They were able to apply statistics to statecraft: to use it as a tool not only for analysis and critique but also for administering change. The manner in which they chose to do this had implications for the evolution of statistics. For instance, in picking up Marx's mantle, Lenin also gave primacy to statistics as a practical tool, going to the extent even of de-emphasizing academically oriented activities among statisticians.[48] He was especially keen on making the Soviet Central Statistical Board the primary source for all data relating to society and economy. In a letter to the board, Lenin noted that "the Central Statistical Board, which *lags behind* an unofficial group of writers, is a model bureaucratic institution. In about two years' time it may provide a heap of data for research, but that is not what we want." He proceeded to demand that "for our practical work we *must* have figures and the Central Statistical Board *must* have them *before anybody else*."[49]

The result, initially in the Soviet Union and subsequently in the places where its influence extended, was to create a distinction in statistical practice and, eventually, in statistics as a discipline. Fueled by such an understanding of human activity and

[46] Yugow, "Economic Statistics" (cit. n. 43), 243.

[47] "William Petty, the father of political economy, and to some extent the founder of Statistics"; see Marx, *Capital* (cit. n. 45), 299. For Marx, William Petty symbolized a leap in quality as far as statistical activities were concerned. See Xu Qian, "Tongji xueshuo fazhan bianhua de lunkuo" (cit. n. 44), 18–21. On Marx and his views on William Petty, see also Ted McCormick, *William Petty: And the Ambitions of Political Arithmetic* (Oxford, 2009). The epilogue, in particular, discusses what McCormick calls Marx's resurrection of Petty as the founder of modern political economy.

[48] In a letter dated 26 May 1921 to his friend, the economist, planner, and then chief of Gosplan, G. M. Krzhizhanovsky, Lenin observed: "The Central Statistical Board should be made into an organisation that does analysis for us, *current*, not 'scientific' analysis. . . . Statisticians must be our *practical assistants*, not engage in scholastics." See Lenin, *Collected Works* (cit. n. 40), 35:497–8, http://www.marxists.org/archive/lenin/works/1921/may/26gmk.htm (accessed 8 May 2018); emphasis in the original. The latter half of the passage above, about practical assistants, is also quoted in Kotz and Seneta, "Lenin as Statistician" (cit. n. 41), 87.

[49] V. I. Lenin, "To the Manager of the Central Statistical Board," 16 August 1921, in Lenin, *Collected Works* (cit. n. 40), 33:30–35, on 31, http://www.marxists.org/archive/lenin/works/1921/aug/16.htm (accessed 8 May 2018); emphasis in the original.

how to account for it, a definition of statistics that was specific, context based, and nonuniversal began to attain legitimacy by the 1930s:

> Statistics [thus] became an instrument for planning the national economy. Consequently, its basis is the Marx-Lenin political economy; it represents a social science or, in other words, a class science. The law of large numbers, the idea of random deviations, and everything else belonging to the mathematical theory of statistics were swept away as the constituent elements of the false universal theory of statistical science.[50]

Prior to the 1930s, Russian and Soviet statisticians had enjoyed a fair amount of freedom and saw themselves as participants in a common discipline of statistics, the bedrock of which was mathematics itself. Keenly aware of global developments in the field, they also actively sought engagement with methods devised outside the Soviet Union.[51] The application of theory to practical problems was a hallmark of this period.[52] But this was also the period during which numbers as measures of performance became absolutely critical. As Michael D. Gordin has observed in his review of two studies on Russian and Stalinist statistics, "For the Soviet state, given its self-professed character as a scientific state, these numbers become crucial for measures of how much socialism was an improvement over tsarist capitalism."[53] As a result, statisticians came under a tremendous amount of pressure, both in terms of how they understood their discipline and in terms of the actual numbers they produced and analyzed.[54]

By the late 1930s, *Vestnik Statistiki*, the leading statistics journal in the Soviet Union, had begun to reject submissions that used mathematical approaches in dealing with statistical problems.[55] As a result, over the course of the 1930s and 1940s, large

[50] S. S. Zarkovic, "Note on the History of Sampling Methods in Russia," *J. Roy. Stat. Soc. A* 119 (1956): 336–8, on 338.

[51] For an overview of how mathematical models of sampling were developed independently in the early Soviet Union, see Martine Mespoulet, "From Typical Sampling to Random Sampling: Sampling Methods in Russia from 1875 to 1930," *Sci. Context* 15 (2002): 411–25.

[52] For recent histories of statistics during czarist Russia and the early Soviet Union, see Martine Mespoulet, *Statistique et révolution en Russie: Un compromis impossible (1880–1930)* [Statistics and revolution in Russia: an impossible compromise (1880–1930)] (Rennes, 2001); Alain Blum and Martine Mespoulet, *L'anarchie bureaucratique: Statistique et pouvoir sous Staline* [Bureaucratic anarchy: statistics and power under Stalin] (Paris, 2003). A useful summary and review of both volumes is Gordin, review of *Statistique et révolution en Russie* and *L'anarchie bureaucratique* (cit. n. 25). For contemporary accounts of early Soviet statistical work and organization, see also A. Ezhov, *Soviet Statistics*, trans. V. Shneerson (Moscow, 1957); Ezhov, *Organisation of Statistics in the U.S.S.R.* (Moscow, 1967).

[53] Gordin, review of *Statistique et révolution en Russie* and *L'anarchie bureaucratique* (cit. n. 25), 807.

[54] For instance, Pavel Il'ich Popov, the first director of the Soviet Statistics Bureau, had argued in a letter to Stalin on 22 December 1925 that there was no difference between socialist and bourgeois statistics. He was subsequently removed from his post the following year. See ibid., 809, citing Blum and Mespoulet, *L'anarchie bureaucratique* (cit. n. 52).

[55] This claim is based on Zarkovic, "History of Sampling Methods in Russia" (cit. n. 50), 338; Eugene Seneta, however, observes that *Vestnik Statistiki* [Messenger of statistics], which began publication in 1919 as the organ of the Central Statistical Office (Tsentralnoe Statisticheskoe Upravlenie) of the USSR, endured a nearly twenty-year hiatus from 1930 until 1949. During that period, many of its functions, especially those pertaining to official statistics, were taken over by the journal *Planovoe Khozaistvo* [Planned economy]. It is likely this combined record that Zarkovic referenced in 1956. Indeed, a bibliographic record of the journal published in 1971 covered the fifty years from 1919 to 1968, clearly combining the two journals under the overall banner of *Vestnik Statistiki*. See Seneta, "A Sketch of the History of Survey Sampling in Russia," *J. Roy. Stat. Soc. A* 148 (1985): 118–25, on 122–3; and G. K. Onoprienko, A. N. Onoprienko, and V. S. Gelfand, *Bibliograficheskii Ukazatel*

numbers of statisticians withdrew from statistical work, finding a relative safe haven in universities or research institutes where they could continue their statistical research without identifying it as such.[56] The practitioners of statistics, in turn, grew increasingly ignorant of the latest statistical methods and their potential applications to statistical work.[57] Indeed, by the 1930s, the "most persistent difficulty for these statisticians stemmed from their dual task: they were supposed to provide accurate numbers for the five-year plans, but they also had to suppress real economic and demographic data that contradicted the official picture."[58] In Blum and Mespoulet's formulation, "in the hybrid that was Soviet Statistics, the Soviet triumphed over the Statistics."[59] This dichotomy extended into the 1950s and was clearly visible to outside observers. During his 1957 visit to the PRC, the Indian statistician Prasanta Chandra Mahalanobis informed his Chinese hosts that ever since 1947 the Soviet Union had not attended any sample survey activities organized by the International Statistical Institute, adding that they just did not attach importance to mathematical statistics.[60]

Confusion over the true definition and status of statistics, however, lingered until it was officially resolved in 1954. It was only through a national conference held in Moscow from 16 to 26 March that year that what had for the most part already occurred in practice was also given an official and final imprimatur.[61] K. V. Ostrovitianov, the vice president of the Soviet Academy of Sciences, subsequently published a summary of the conference.[62] Organized by the Soviet Academy of Sciences, the Central Statistical

Statei i Materialov po Statistike i Uchetu. Zhurnal Vestnik Statistiki za 50 let (1919–1968) (Moscow, 1971).

[56] Exemplary is the case of Andrey Kolmogorov (1903–87), recognized as one of the leading probabilists of his time, but who was rarely of interest to statisticians within the USSR. It is indeed possible that a general disciplinary demarcation may have also had the benefit of freeing up the harder sciences, most notably nuclear research, which could then use advanced mathematical statistics without falling prey to ideological battles. On Kolmogorov, see D. G. Kendall, "Andrei Nikolaevich Kolmogorov, 25 April 1903–20 October 1987," *Biogr. Mem. Fellows Royal Soc.* 37 (1991): 300–326; on Kolmogorov and Soviet statistics, see P. C. Mahalanobis, "Some Impressions of a Visit to China" (cit. n. 19), 5; on freeing up certain technical sciences, see J. M. and K. V. Ostrovitianov, "The Discussion on Statistics Summed Up," *Soviet Stud.* 6 (1955): 321–31, on 321. (This article is a translation of the Russian original, which was published in *Vestnik Akademii Nauk SSSR*, 8 (1954): 3–12. Ostrovitianov, the author of the report, is identified as "the chief organizer of Soviet economists.")

[57] Zarkovic, "History of Sampling Methods in Russia" (cit. n. 50).

[58] Gordin, review of *Statistique et révolution en Russie* and *L'anarchie bureaucratique* (cit. n. 25), 809.

[59] Ibid., 808.

[60] File 105-00530-06: 15, FMA.

[61] Ostrovitianov, "Discussion on Statistics" (cit. n. 56), 321.

[62] Ostrovitianov's report was published in *Vestnik Akademii Nauk SSSR*. All citations here are from a translation of the Russian original. In his commentary on the translation, J. M. noted that "Ostrovitianov's main point, broadly stated, is that, at the present time at any rate, the name *statistics* should be limited to the discipline and practice which handle the quantitative aspects of socio-economic phenomena; that this is the only statistical discipline deserving the status of science in its own right; and that statistical work on natural and technological phenomena is most conveniently regarded as a branch of mathematics, namely *mathematical statistics*"; Ostrovitianov, "Discussion on Statistics" (cit. n. 56), 321; emphasis in the original. *Vestnik Statistiki* had also published a detailed report of the conference: "Review of the Scientific Meeting on the Question of Statistics," *Vestnik Statistiki* 5 (1954): 39–95. Ostrovitianov would visit China as part of a Soviet Cultural Delegation later that year in October; see "Пребывание советско й делегации деятелей культуры в Китае" [Stay in China of Soviet delegation of cultural figures], *Izvestiia*, 9 October 1954, 3. "Банкет в честь советских специалистов в Китае" [Banquet in honor of Soviet specialists in China] and "Банкет у Го Мо-жо" [Banquet with Guo Moruo], *Pravda*, 6 October 1954, 4.

Board, and the Ministry of Higher Education, the conference's purpose was "to discuss matters in dispute and to examine the subject and method of statistical science."[63] In all, 760 invitations were issued across the various Soviet Republics.[64] The majority of those invited were academic and practicing statisticians, and they were joined by smaller numbers of economists, engineers, mathematicians, philosophers, and medical men. In all, sixty people spoke at the conference and twenty sent in written contributions. Three principal points of view emerged during the conference. The first held that statistics was a universal science that studies phenomena of society and nature. The second posited that statistics was a social science concerned with method, a system of principles for collecting quantitative information in order to characterize social phenomenon. The final view was that statistics was a social science that only studies productive relations in society, that is, the economy.[65]

Over the conference's ten days, the first two views were discussed and summarily rejected. Proponents of the first claim, which was no different from views in the "bourgeois liberal" West, were accused of a "gross error in attempting, with the help of a single science of statistics, to study by the same methods the phenomena both of nature and of social science."[66] This error was not just a scientific one; it also had moral dimensions, for, as Ostrovitianov declared in his report on the conference:

> The comrades who regard statistics as a universal science which studies nature and society make of it some kind of science over and above the [socioeconomic] classes, coldly indifferent to good and to evil, without any preference at all as between [socioeconomic] classes and between social structures.[67]

In support of this claim, Ostrovitianov offered a criticism of the false comparison made by other participants who drew upon similarities between the luminosity of stars and the grouping (categorization) of the peasantry:

> What can there be in common between the grouping of stars according to their luminosity and, let us say, Lenin's grouping of the peasantry according to their [socioeconomic] class characteristics? *Nothing, other than certain technical statistical devices used in the study of these totally different kinds of phenomena*. The grouping of stars according to luminosity requires no class analysis at all, whereas in investigating the differentiation of the peasantry we must make a [socioeconomic] class analysis of this phenomenon, proceeding from the postulates of historical materialism and political economy. Statistical devices and methods of investigation, the choice of characteristics for grouping, the combining of groups, were entirely without exception subordinated by Lenin to the task of [socioeconomic] class analysis of the rural population.[68]

Ostrovitianov conceded that quantitative relations in their pure form—equally applicable to "inorganic nature, to organic nature and to the domain of the social sciences"—could be studied by mathematics, more specifically by the subdomain of mathemat-

[63] Ostrovitianov, "Discussion on Statistics" (cit. n. 56), 323.

[64] Invitations went out to (name of modern republic in parenthesis): Leningrad (Russia), Kiev (Ukraine), Minsk (Belarus), Baku (Azerbaijan), Tashkent (Uzbekistan), Tbilisi (Georgia), Riga (Latvia), Yerevan (Armenia), Almaty (Kazakhstan), Kharkiv (Ukraine), Sverdlovsk (Ukraine), and Irkutsk (Russia).

[65] Ostrovitianov, "Discussion on Statistics" (cit. n. 56), 323–4.

[66] Ibid., 324.

[67] Ibid., 326.

[68] Ibid.; emphasis added.

ical statistics. But apart from this branch of mathematical statistics, and the socioeconomic science of statistics, there did not exist a third "non-mathematical but nevertheless universal 'general' statistics."[69] This distinction formed the basis of the rebuttal of the second claim, that statistics was a science of method. "Every social science," Ostrovitianov explained, "has not only its own subject, but also its own method of investigation." Furthermore, he noted that in political economy, the method employed was abstraction (historical materialism in the sphere of economic relations), and not, quoting Marx, a microscope or chemical reagents.[70] Thus, it was decided that only in the field of the study of social phenomena did statistics reach the level of an independent science, with its own subject and methods of investigation.[71]

And so, after almost two weeks of deliberation and debate, the conference was in a position to offer a definition of statistics:

> Statistics is an independent social science. It studies the quantitative aspects of mass social phenomena in full awareness of and therefore within their qualitative aspect; it investigates the quantitative expression of the laws of social development in the specific conditions of place and time.[72]

Statistics was thus to serve, in the words of Lenin, as the "rock of exact and indisputable facts," the very foundation on which rested not just political economy but all other social sciences as well. The effect, in the end, was to ratify and reify a distinction that was already fairly well entrenched: that between statistics and mathematical statistics.[73]

The 1954 resolution had a signal influence on the wider PRC statistical community. News of the resolution and the Soviet state-sanctioned definition of statistics, however, took time to filter through; a complete Chinese translation of the deliberations was not published until early 1955.[74] But its impact was evident in the more confident tone and tenor of the critiques that started emerging in 1955.[75] Most prominent among these was an essay by Xu Qian 徐前 and Liu Xin 刘新. The two had published an essay critical of bourgeois statistics in 1953 in the journal *New Construction* (*Xin jianshe*).[76] They published an updated version in the SSB's journal *Statistical Work*, explaining that their basic critique had been justified in light of the 1954 Moscow Con-

[69] Ibid. "Thus, there is no scientific grounds for a universal statistics as an independent discipline which studies both nature and society. It would be an extremely meagre science made up of some statistical devices which are used in all branches of knowledge, and would hang somewhere between socio-economic and mathematical statistics."

[70] It is indeed not clear how the act of abstraction can be compared to the use of a physical instrument such as a microscope or a substance such as a chemical reagent under the single category of method.

[71] Ostrovitianov, "Discussion on Statistics" (cit. n. 56), 327.

[72] Ibid., 328.

[73] On a related discussion concerning the status of other mathematical fields such as game theory (whose applications were also proscribed) and cybernetics (which flourished), see Erickson et al., *How Reason Almost Lost Its Mind* (cit. n. 32), 17–20.

[74] "Sulian tongji kexue huiyi jueyi" [Resolution of the Soviet Meeting on Statistical Science], *Tongji gongzuo tongxun* (February 1955): 1–4, 45.

[75] This is an impression based on the interviews I conducted with, among others, Renmin University Professor of Statistics Yuan Wei 袁卫 on 25 March 2011 in Beijing, People's Republic of China.

[76] Xu Qian 徐前 and Liu Xin 刘新, "Guanyu zichan jieji tongji lilun de pipan" [Criticisms of bourgeois statistical theory], *Xin jianshe* 1 (1953): 14–20; 2 (1953): 25–8.

ference.[77] In the ensuing months, additional essays were published demanding the eradication of the pernicious influence of bourgeois statistics and outlining the failings of bourgeois theories of index numbers.[78] The effect of these debates was to circumscribe statistics largely within the realm of application and divorce it entirely from mathematics.[79]

CHANCE TAMED; CHANCE REJECTED

Trygve Lie's desire for "clear and systematically organized facts" thus found an echo in Lenin's evocation of statistics as the "rock of exact and indisputable facts." But while such a desire united the Universalist claim and its Marxist (Soviet-Chinese) critique, the contrasting understandings of statistics within each positioned them at loggerheads when it came to questions of theory and method. The contradictory claims regarding statistics' universality outlined above were undergirded by two fundamentally different ontological positions on how social reality could be ascertained via quantification (table 1). The first, as we have seen, by embracing uncertainty, accepted its own limits. Chance and error were seen to be part of the natural order of things. Not everything could be counted, at least not accurately, and even more to the point, all attempts at counting generated errors. This did not mean that exhaustive enumeration or other nonprobabilistic methods were entirely abandoned.[80] Rather, the goalposts had shifted: intellectual energy was no longer channeled toward devising methods to count exhaustively but instead channeled toward devising methods for studying variation, and eventually toward how to accurately determine, control, and minimize the error that resulted during any count. Thus, probability, the law of large numbers, and general mathematical reasoning became crucial tools in devising new methods of counting and assessment.

This was in stark contrast to the second approach, which, fueled by the teleological inevitability of Marxist historical progression and by the confidence offered by dialectical materialism, was predicated on the absence of chance and uncertainty in the social world. Within such an ontological worldview there was no place for probability, the law of large numbers, or any kind of abstract mathematical thought. The social world had no chaos, only laws of production, population, and relations, all of which could be discovered via measurement. The result was an emphasis on complete enumeration—the counting of everything. In other words, just as the liberal tradition pivoted on the appreciation and eventual acceptance of uncertainty, chance, probability,

[77] Xu Qian and Liu Xin. "Guanyu zichan jieji tongji lilun de pipan" [Criticisms of bourgeois statistical theory], *Tongji gongzuo tongxun* 23 (August 1955): 30–6.

[78] Hu Daiguang 胡代光, "Ganqing zichan jieji tongji sixiang de liudu" [Eliminate the pernicious influence of bourgeois statistical ideology], *Tongji gongzuo* (October 1955): 25–9; and Wang Jianzhen 王健真, "Dui zichan jieji zhishu 'lilun' de pipan" [Criticisms of the bourgeois theory of index numbers], *Tongji gongzuo* (November 1955): 42–5.

[79] In her insightful discussion of science in China during the Cold War, Sigrid Schmalzer observes that the "overdetermined" dichotomy between applied and basic science has to be understood through the emphasis on self-reliance, which itself had at least two implications: the use of local methods and local independence form the center. Schmalzer, "Self-Reliant Science: The Impact of the Cold War on Science in Socialist China," in Oreskes and Krige, *Science and Technology* (cit. n. 3), 75–106, on 86.

[80] Through the end of the nineteenth century and beyond, many German and Italian statisticians continued to see statistics largely as social science, and data collection in U.S. agriculture continued to rely on exhaustive enumeration well into the 1930s; my thanks to Sarah Milov and one of the anonymous reviewers for pointing this out.

Table 1. *Two Approaches to Statistics*

Statistics	Socialist statistics
Statistics is a universal science	Statistics is a social science
Chance exists, randomness part of both natural and social order of things	No chance in the social world (only laws that can be ascertained)
All counting generates errors; can devise methods to ascertain and control error of any count	Can count everything accurately; the best count is the complete count
Key methods	
Law of large numbers	Exhaustive enumeration
Probability theory	Periodic reports
Variance, correlation, regression	Typical sampling
Random sampling	

and error in the world and in counting, so did the Socialist-Marxist critique base itself on the comfort of certainty, and the desire to enumerate and account for everything. Capitalism thus became a useful catch-all for the Communist critique of a catholic approach to counting that incorporated exhaustive and sampling-based (random and qualitative) methods as part of a complementary tool kit.

In taming and rejecting chance, both traditions were interested in increasing overall control. But they had very different expectations about what they could control. And each approach taken to its extreme could also lead to perverse social and economic outcomes. For instance, the political economists Peter Katzenstein and Stephen Nelson have argued that the emphasis on error, accuracy, and calculability in the West led to a fixation on risk, frequently at the cost of acknowledging the importance of uncertainty. Unlike the latter, assessment of the former was based on past experience and calculable and therefore could be used as the basis for decision making. For this very reason, uncertainty came to be discounted in the design of economic models, which over time contributed to an increasing disengagement with real-world actions.[81] To substantiate their claims for the importance of uncertainty, Katzenstein and Nelson analyzed the 2008 financial crisis, with its basis in increasingly sophisticated mathe-

[81] The distinction between uncertainty and risk in the sphere of economic activity was first articulated independently by the economists Frank H. Knight and John Maynard Keynes in the early 1920s. See Keynes, *Treatise on Probability* (1921; repr., New York, 1948); Knight, *Risk, Uncertainty, and Profit*, Hart, Schaffner, and Marx Prize Essays, no. 31 (Boston, 1921). For claims that among American economists, uncertainty has largely been written out in favor of risk, which is calculable, see Peter J. Katzenstein and Stephen C. Nelson, "Worlds in Collision: Uncertainty and Risk in Hard Times," in *Politics in the New Hard Times: The Great Recession in Comparative Perspective*, ed. Miles Kahler and David Lake (Ithaca, N.Y., 2013), 233–52; Nelson and Katzenstein, "Uncertainty, Risk, and the Financial Crisis of 2008," *Int. Org.* 68 (2014): 361–92. A more "orthodox" depiction of risk conquering uncertainty can be found in Peter Bernstein, *Against the Gods: The Remarkable Story of Risk* (New York, 1996). For a discussion of how risk came to be incorporated into economic thinking and capitalist practice in the United States in the nineteenth and twentieth centuries, see Arwen Mohun, *Risk: Negotiating Safety in American Society* (Baltimore, 2012); Jonathan Levy, *Freaks of Fortune: The Emerging World of Capitalism and Risk in America* (Cambridge, Mass., 2014); Dan Bouk, *How Our Days Became Numbered: Risk and the Rise of the Statistical Individual* (Chicago, 2015).

matical models that were supremely confident in their ability to ascertain and account for risk.[82]

Socialist statistics' fetishization of complete enumeration would also lead to certain systemic cul-de-sacs. But the illusion of control this method offered was extremely attractive. As A. Yugow noted in his review of Soviet statistics in 1947:

> Accounting of all economic phenomena on a national scale and the possibility of statistically following all economic phenomena in their complete economic cycle give Soviet statistics such great advantages in comparison with the statistics of other countries, that the forms, methods, and the results of the work of the statistical organs of the U.S.S.R. are highly instructive to all who work in the field of the theory and practice of economic statistics, even in cases when these data are not entirely exact and commensurable in particulars. The gathering of primary materials, the formulation of problems, and the effect of the plan upon economic factors in the U.S.S.R. provide an almost "laboratory" picture; in any case, they reflect the existing facts far more clearly and fully than in other countries, where the statistical study of national problems and those of individual branches is not only greatly complicated, in view of the technical incompleteness of the data, but is also deliberately distorted because of private interests, trade secrets, competition, etc.[83]

From a planning perspective, the promise of achieving data collection on a national scale, and thereby (in theory at least) ensuring that all manner of economic activity was accounted for, was hard to resist. It was also the kind of comprehensiveness that nonsocialist countries with their decentralized statistical apparatus found impossible to achieve.

SOCIALIST STATISTICS IN CHINA

In many parts of the world, and especially in much of the future Global South, these strengths of the Soviet system were seen as the basis for its economic performance. For instance, the Indian industrialists who drafted the Bombay Plan in 1944 wrote with admiration and aspiration:

> This [growth target for India] might appear to be too modest a goal for a planned economy to achieve, especially in view of the fact that in the U.S.S.R., within a period of 12 years since the beginning of the first Five-Year Plan, the national income is reported to have increased from 25 billion rubles to 125 billion rubles, i.e., five-fold.[84]

Even so, in India, such high regard for planned economic growth did not extend to an adoption of Soviet socialist statistics. In China, the story was more complicated, characterized both by energetic adoption and radical adaptation. In the paragraphs below, I suggest some key aspects of this adoption and adaptation. These and related themes are developed in greater detail in my forthcoming book on statistics and statistical work in 1950s China.

[82] On the role of risk and uncertainty management in precipitating two earlier financial crashes (the stock market crash of 1987 and the market turmoil that engulfed the hedge fund Long-Term Capital Management in 1998), see Donald A. MacKenzie, *An Engine, Not a Camera: How Financial Models Shape Markets* (Cambridge, Mass., 2006).

[83] Yugow, "Economic Statistics" (cit. n. 43), 244.

[84] *Memorandum Outlining a Plan of Economic Development for India* (Bombay, 1944), 29. This was given as the reason for hoping for a 200 percent increase in national income in fifteen years. Also quoted in Seymour E. Harris, "Appraisals of Russian Economic Statistics: Introduction," *Rev. Econ. Stat.* 29 (1947): 213–4, on 213.

The immediate years after 1949 in China were marked by an enthusiastic and comprehensive adoption of Soviet-style socialist statistics. Following the example set by Wang Sihua and the NSB, and through them of socialist statistics as espoused by the Soviet example, a centralized SSB was established in August 1952 under the leadership of Xue Muqiao with the express objective of overseeing all statistical activity in the country. By the summer of the following year, the central office in Beijing was joined by twenty-four regional, provincial, or city-level statistical offices, which employed a total of 1,275 statistical workers.[85] The numbers of statistical cadres rapidly ballooned, and within three years the national statistical apparatus had grown to include 200,000 cadres spread across all sectors of the nation. Responsible for the collection, analysis, and publication of all kinds of data, and aided by Soviet experts, the SSB worked closely with the National Planning Committee in formulating national policies such as the first five-year plan (1953–7). The key method for collecting data remained exhaustive enumeration through a complete periodic report system.[86] Even as late as 1957, by which time tensions between China and the Soviet Union had become increasingly evident, Xue Muqiao affirmed his faith in the Soviet model:

> [We must] seriously study the Soviet Union's advanced experience in constructing and managing a socialist economy, especially its experience in the planned management of the people's economy and in the establishment of planning and statistical work.[87]

But that is not to say that China's adoption of socialist statistics was entirely uncritical. In practice it incorporated significant Chinese innovation. For instance, unlike in the Soviet Union, where a similar unified chain of command was free from oversight by other party and government organs, in the People's Republic a statistical unit at any level was simultaneously also under the authority of the Party Committee at that level. In other words, a provincial bureau not only took its orders from the SSB in Beijing but was also a constituent part of the provincial People's Committee, which had a say in its organization. Thus, the provincial People's Committee could make its own requests for data to the provincial statistics bureau. A statistical worker at each level usually had more than one set of leaders to answer to. The hope was that such an arrangement would make the statistical system flexible enough to meet both local and national needs.[88]

An even starker example of autonomy is the extended series of exchanges begun with Indian statisticians in 1957. Reliance on exhaustive enumeration had generated

[85] These numbers and estimates are based on information culled from the National Bureau of Statistics, *Zhonghua renmin gongheguo tongji dashiji, 1949–2009* [Statistical chronicles of the People's Republic of China, 1949–2009] (Beijing, 2009), 7–35.

[86] "Among all the various methods of investigation, the complete enumeration periodical report system is the basic and [most] important method"; Wu Hui 吴辉, "Zhongguo de tongji diaocha fangshi" [China's statistical survey methods], in *Zhongguo shehui zhuyi tongji gongzuo de jianli yu fazhan* [The establishment and development of socialist statistical work in China], ed. China State Statistical Bureau's Institute of Statistical Science (Beijing, 1985), 31–42, on 32.

[87] Xue Muqiao, "Woguo de jihua gongzuo he tongji gonzuo bixu xiang sulian xuexi" [Our nation's planning and statistics work must learn from the Soviet Union], *People's Daily*, 12 November 1957, 7.

[88] Xue Muqiao, "Diyi ge wunian jihua qijian woguo tongji gongzuo de chubu jingyan he jinhou renwu" [The preliminary experience and recent tasks of statistical work during our country's first five-year plan period], *Tongji gongzuo* (November 1957): 1–21, on 8. A description of this system can also be found in State Statistics Bureau, *Tongji lilun yiban wenti jianghua* [Common problems in statistical theory] (Beijing, 1956), 18–9.

significant problems for the SSB by 1956, and they quickly sought a way out of the methodological cul-de-sac imposed by socialist statistics. Not surprisingly, therefore, the Sino-Indian exchanges were driven primarily by the Chinese desire to learn more about random sampling, something that the SSB had entirely eschewed up to that point.[89] These attempts were, in turn, overtaken by the tumult of the Great Leap Forward, when both the socialist approach and the approach to India were jettisoned in favor of an ethnographic model of social research credited to Mao Zedong.[90] The socialist method, however, made a return in the 1960s and remained dominant into the early post-Mao years.

CONCLUSION

Li Fuchun's dismissal in 1951 of statistics as it had been practiced up to that point indicated a crisis on two levels, one ontological and the other epistemological. What categories should the world be divided into? And what forms of knowledge can be deployed to ascertain these categories? In line with the discourse within the Soviet Union, as summarized by Ostrovitianov in 1954, the point of departure was the premise that there was no single universal reality for the known world; rather, different theories and methods were applicable to these different divisions—the social and the natural—of the world. These different theories and methods, in turn, promoted very different interests and led to very different social and economic outcomes in society. For Chinese thinkers, drawing upon Soviet and Marxist views, the natural and the social worlds were not the same and could not be understood using the same tools. They gained these views not only from reading Marx and Lenin but especially through Soviet statistical experts who were present in the PRC and through a vast array of translated material. The overall effect was to engender an overarching theoretical approach to all knowledge that prefigured what statistics could be. The practical impact of this theoretical distinction between the natural and social world was to idealize and attempt to construct a vast national network of statistical offices, from Beijing down to the county level, with the aim of capturing all necessary data. And while attempts at exploring alternate methods—with Indian statisticians, during the Great Leap Forward, and again during the Cultural Revolution (1966–76)—were repeatedly made, socialist statistics remained the dominant frame through which statistical activity was theorized and prosecuted until the early 1980s.

Under the shadow of the Cold War, this distinction became overdetermined as a contrast between capitalism and communism. Today, sixty years later, we can see that underlying all such projects, and cutting across the ideological divides, was the more deep-seated dream of establishing the kind of control that is characteristic of developmental thinking and modern state formation. That it could result in such radically different solutions not only helps us add nuance to any history of statistics but also adds caution to any celebration of big data as being fundamentally emancipatory, especially as the dream of total control is now shared by states and private corporations alike.

[89] For more on the Sino-Indian statistical exchanges, see Arunabh Ghosh, "Accepting Difference, Seeking Common Ground: Sino-Indian Statistical Exchanges 1951–1959," *BJHS Themes* 1 (2016): 61–82.

[90] This method echoed other similar valorizations of personal investigative authority that drew upon a researcher's personal experience conducting ethnographic research. See, e.g., Emmanuel Didier, "The First US Surveys: Representativeness between Sampling and Democracy," in Desrosières, *Sampling Humans* (cit. n. 39), 44–57.

ENTANGLED ONTOLOGIES

The Microbial Production of Expertise in Meiji Japan

*by Victoria Lee**

ABSTRACT

Microbes as an object of knowledge and the scientist as an institution of authority did not exist in Japan before the nineteenth century. This essay considers the formation of these two modern categories by looking at their boundaries in late Meiji Japan (1868–1912). Charting transformations in the landscape of brewing expertise, the processes that brewing technicians used to produce molds as commodities, and finally the critical reaction of the slime-mold naturalist Minakata Kumagusu who opposed the philosophical foundations of disciplinary science, it argues that the co-production of the microbe and the scientist as new categories reveals a convergence between imported European ideas and earlier Tokugawa-era (1603–1868) commercial developments. Their convergence in turn-of-the-century Japan is highly suggestive of the ways in which the modernity of scientific institutions is entangled with industrial capitalism.

INTRODUCTION

Microbes as a concept did not exist in Japan before the nineteenth century, but brewers of sake, soy sauce, miso, and other products had handled microbes—visible en masse as rice mold formations—with specialist skill, understanding them as essential raw materials in the brewing process, like water or rice. By the first decade of the twentieth century, brewers faced a new landscape of microbe species that were divided into "useful ones" [*yūeki naru mono*] and "harmful ones" [*yūgai naru mono*].[1] Within a species, microbe varieties were characterized by their physiological as well as morphological differences, which coincided with their role in different industries: the fungi used in the sake industry formed more sugars, whereas the fungi used in the soy-sauce

* Department of History, Ohio University, Bentley Annex 452, Athens, OH 45701-2979; leev@ohio.edu.

This essay includes material from my article, "Mold Cultures: Traditional Industry and Microbial Studies in Early Twentieth-Century Japan," in *New Perspectives on the History of Life Sciences and Agriculture*, ed. Denise Phillips and Sharon Kingsland (Cham, 2015). It is reprinted here with permission.

[1] Takahashi Teizō, *Jōzō bairon* [Theory of brewing molds] (Tokyo, 1903), chap. 1, pp. 3–4. On comparable trends in entomology, see Setoguchi Akihisa, *Gaichū no tanjō: Mushi kara mita Nihonshi* [Birth of the harmful insect: Japanese history through the insect] (Tokyo, 2009).

and tamari industries formed more amino acids, chemicals associated with protein breakdown and flavor.[2] These novel species were named by scientists, their existence conveyed to brewers in manuals and trade magazines, their strains to be distributed from national and prefectural experiment stations as well as made by brewing technicians.

Among the harmful were the "*hiochi* microbes" (*hiochi kin*; *hiochi* being the term for spoilage, when the sake "dropped" the fire put in during the heating process known as *hiire*).[3] Among the useful, specialists in the brewing industry had produced and sold the rice mold *kōji*, as well as its dried spores known as *moyashi* (or in the modern era, *tanekōji*) since the thirteenth century, which by the twentieth century were named *Aspergillus oryzae*. These were only a part of the diversity of "useful fermentation microbes" [*yūyō hakkōkin*] produced in Asia.[4]

This essay considers the formation of two modern categories in Japan by looking at their boundaries in the early twentieth century: the "microbe" as an object of knowledge, and the "scientist" as an institution of expertise. Both of these new categories relied on a notion of "nature" that had no Japanese-language equivalent in the Tokugawa period (1603–1868)—a word that referred to the whole of material reality as something universal, as well as something distinct from "society."[5] The microbe as an object of knowledge reflected a novel ontological division in the boundary between cellular life and the environment, which the new conception of nature allowed. The scientist as a socio-institutional category was a new kind of expert who specialized in nature. Yet both categories of "microbe" and "scientist" were not only linked to European categories by a self-consciously Western refashioning of Japan's political economy in the Meiji period (1868–1912); they were also shaped significantly by conceptions that had emerged amid the vibrant protocapitalism of the Tokugawa era.

The procedures and assumptions of commercial brewers that had been developing over several centuries and those of academic microbe scientists in turn-of-the-century Japan display a striking, suggestive convergence. In his study of early modern Japanese *honzōgaku* (materia medica), Federico Marcon points out the homology between species and commodities as abstractions concealing human labor. He argues that the production of species transformed the natural world into "a collection of objects to analyze, represent, manipulate, control, and produce . . . devoid of any metaphysical or sacred aura."[6] In the Meiji era, the parallel between the species of fermentation microbes produced by scientists and the commercial brands of spores produced by *tanekōji* makers became a literal one: both were microbes isolated, identified, and

[2] Teizo Takahashi, "A Preliminary Note on the Varieties of *Aspergillus oryzae*," *J. Coll. Agr. Imperial Univ. Tokyo* 1 (1909): 137–40; Teizo Takahashi and Takeharu Yamamoto, "On the Physiological Differences of the Varieties of *Aspergillus oryzae* Employed in the Three Main Industries in Japan, Namely Saké-, Shôyu-, and Tamari Manufacture," *J. Coll. Agr. Imperial Univ. Tokyo* 5 (1913): 151–61.

[3] Teizo Takahashi and Kin-ichiro Sakaguchi, *Summaries of Papers*, Committee of Commemorative Meeting of 35 Year's Anniversary of Professor Kin-ichiro Sakaguchi (Tokyo, 1958), 8, item 30.

[4] Saitō Kendō, "Higashi Ajia no yūyō hakkōkin" [Useful fermentation microbes of East Asia], *Tōyō gakugei zasshi* 23 (1906): 507–20.

[5] Federico Marcon, *The Knowledge of Nature and the Nature of Knowledge in Early Modern Japan* (Chicago, 2015), 276–7.

[6] Ibid., 296–7.

preserved by methods of pure culture, and characterized not only by their morphology as experienced by the senses or under a microscope, but by their physiological activity in the context of brewing processes. Thus microbes in Japan became living workers as much as pathogens.[7] Likewise, the role of the microbe scientist within the modern Japanese state was not like that of the early modern intellectual, but built instead on that of the technical specialist within the brewery: a manager of material production for capital accumulation, now on a national level.

Tracing the coproduction of the microbe and the scientist as new categories in light of the convergence between imported European ideas and Tokugawa-era commercial developments, I argue that this convergence in turn-of-the-century Japan is highly suggestive of the ways in which the modernity of scientific institutions is entangled with industrial capitalism. As Lee Vinsel and Sarah Milov discuss elsewhere in this volume, the entanglements of scientific investigation and capitalism created new entities for understanding: the accident-prone driver, the secondhand smoker, and, in this case, the microbe.[8] At the same time, the emergence of new entities came hand in hand with new modes of scientific expertise. For example, Vinsel traces the formation of a niche for applied psychologists in auto safety testing, while William Deringer and Paul Lucier examine the construction of mathematical and geological expertise respectively, in spaces opened up by financial and capitalistic enterprises.[9] What is especially revealing in the case of Japanese fermentation science is the continuity of both new categories—the microbe and scientific expertise—with earlier protocapitalist practices.

The first part of the essay charts the transformation in the landscape of brewing expertise following the rise of the microbe as a conceptual force and the formation of the scientist as a specialist authority. It then examines the significance of these new categories in the work of two figures who operated outside disciplinary institutions of science, the brewing technician Konno Seiji and the naturalist Minakata Kumagusu. Konno Seiji's brewing work illuminates the processes for producing molds as commodities and their continuity with the new scientific methods. Minakata Kumagusu's slime mold studies explicitly opposed the assumptions of disciplinary microbiology as well as science's established role in the state. Minakata's research highlights the emerging contours of academic science not because he exemplified them, but because he showily reacted against them, by developing alternative approaches to studying molds that directly contrasted with the work of academic scientists. The production of the conceptual and social boundaries of the new microbial entity cannot be separated from those of the new scientific expert; this essay offers a joint portrait of both.

[7] Victoria Lee, "Mold Cultures: Traditional Industry and Microbial Studies in Early Twentieth-Century Japan," in *New Perspectives on the History of Life Sciences and Agriculture*, ed. Denise Phillips and Sharon Kingsland (Cham, 2015), 231–52.

[8] Lee Vinsel, "'Safe Driving Depends on the Man at the Wheel': Psychologists and the Subject of Auto Safety, 1920–55"; Sarah Milov, "Smoke Ring: From American Tobacco to Japanese Data," both in this volume.

[9] William Deringer, "Compound Interest Corrected: The Imaginative Mathematics of the Financial Future in Early Modern England"; Paul Lucier, "Comstock Capitalism: The Law, the Lode, and the Science," both in this volume.

LANDSCAPES OF EXPERTISE

By the close of the Tokugawa period, soy-sauce, miso, and especially sake brewing accounted for by far the highest values of production in the entire nonagricultural manufacturing sector in Japan, easily surpassing textile weaving and raw silk production. At the beginning of the twentieth century, among those wealthiest people who came by their riches through industrial manufacturing, there were more brewers than any other occupation, and their numbers rivaled those in rising modern industries such as cotton spinning.[10] The largest brewers emerged during the second half of the Tokugawa period in Nada (near Kobe, west of Kyoto and Osaka) in the Kansai region of western Japan for sake and in Noda and Chōshi (east of Tokyo) in eastern Japan for soy sauce. Those breweries, which were located in rural areas, employed dozens of workers from the vicinity and were becoming increasingly mechanized.

In the late Tokugawa period, Japan was one of the most highly urbanized societies in the world, with numerous large cities and castle towns that formed part of a national network of consumption and distribution.[11] The leading breweries competed with each other on scale as well as quality to ship their goods to major urban markets, particularly the largest cities: Edo (renamed Tokyo after the Meiji Restoration), Kyoto, and Osaka. National markets grew first for sake, while the use of commercial rather than homemade soy sauce was initially more common in eastern than western Japan, and miso production was dominated by the home kitchen into the twentieth century.[12] However, brewing was a multilayered industry: outside of urban areas, village residents bought mainly from small or medium-scale local producers. In rural areas there was also widespread home brewing of unrefined sake [*nigorizake* or *doburoku*], which farm workers would drink early in the day as it was thought to give them energy for heavy labor on the farm.[13]

The brewing of rice into liquor, involving the *kōji* mold that grew on the rice, had been performed in Japan for perhaps two millennia. The specific origins of *kōji* brewing were not known, although there were clear connections with mold brewing of grains on the Asian continent. In the medieval period, specialist rice wine brewers emerged in Kyoto to supply the aristocracy for ceremonial or medicinal purposes, and by the fourteenth century, commercial sake was also produced in the countryside for public drinking on market days and special occasions.[14] Sake breweries developed especially in urban areas with access to rice, such as port cities in the Kansai region that saw large commercial rice transactions, or nearby temple towns that could sell their products in Kyoto. In the seventeenth century, the establishment of the Tokugawa shogunate based in Edo brought the growth of cities across the country, as sam-

[10] Masayuki Tanimoto, "Capital Accumulation and the Local Economy: Brewers and Local Notables," in *The Role of Tradition in Japan's Industrialization: Another Path to Industrialization*, ed. Masayuki Tanimoto (Oxford, 2006), 301–22, on 301–2.

[11] David L. Howell, "Urbanization, Trade, and Merchants," in *Japan Emerging: Premodern History to 1850*, ed. Karl F. Friday (Boulder, Colo., 2012), 356–65, on 356–7.

[12] Penelope Francks, "Inconspicuous Consumption: Sake, Beer, and the Birth of the Consumer in Japan," *J. Asian Stud.* 68 (2009): 135–64, on 149.

[13] Tanimoto, "Capital Accumulation" (cit. n. 10), 301–5.

[14] Katō Hyakuichi, "Nihon no sakazukuri no ayumi" [History of sake brewing in Japan], in *Nihon no sake no rekishi: Sakazukuri no ayumi to kenkyū* [History of sake in Japan: Sake brewing history and research], ed. Katō Benzaburō (Tokyo, 1977), 41–315, on 168; Francks, "Inconspicuous Consumption" (cit. n. 12), 153.

urai were required to live in castle towns to serve their domain lord. Though officially political authorities encouraged commoners merely to farm in order to produce taxes for their lords, over time the domains came to be chronically dependent on prominent merchants for loans and contributions, in return offering privileges such as the recognition of trade monopolies for certain commodities within regional markets. Thus the commerce that flourished in the Tokugawa period came to be dominated by merchants with ties to domain officials, initially wholesalers based in the cities.

By the mid-eighteenth century, the economy experienced a host of changes that historians have documented as a distinctively rural-centered, protoindustrial, protocapitalist transformation.[15] Around the country, there was a dramatic growth in the number of rural households that produced goods for sale in distant markets, typically engaging in side industries along with agriculture. Through such small-scale manufacturing activities, as well as moneylending and experimentation with farming techniques, a number of rural elites began to amass substantial wealth. Major urban centers declined or stagnated as "country places" rose on their outskirts, challenged the hold of city merchants, and became centers of vibrant consumption as well as production. Many domain authorities by this time had come to tolerate and even promote commercial growth, since they were increasingly dependent on commoner elites for funds and trading services. As interregional competition for national markets intensified, domain authorities could turn a blind eye when rural elites usurped urban monopolies, or encourage market-oriented production in the local region by aiding the introduction of new technologies, as well as inviting skilled experts from—or having observers travel to—more advanced regions.[16]

In this period Nada sake brewers, who usually began as wealthy landowners and rural merchants, competed with nearby urban establishments in Osaka, Nishinomiya, Ikeda, and Itami. Nada villages had several geographical advantages. They had access to water power that could drive rice-polishing machines, they were able to take advantage of a winter wage labor force when repeated restrictions by political authorities forced the concentration of sake brewing into the agricultural off-season (which incidentally produced a better-quality sake despite increasing the fermentation time), and they were located close enough to city markets as well as growing rural markets.[17] Across the country, refined sake was no longer a luxury reserved for the aristocratic and samurai classes; it came to accompany meals in restaurants and teahouses as well as in rural households when families entertained visitors. Izakaya bars flour-

[15] Thomas C. Smith, *Native Sources of Japanese Industrialization, 1750–1920* (Berkeley and Los Angeles, 1988); Akira Hayami, *Japan's Industrious Revolution: Economic and Social Transformations in the Early Modern Period* (London, 2015); Hayami Akira, "Keizai shakai no seiritsu to sono tokushitsu" [The emergence of the economic society and its characteristics], in *Atarashii Edo jidaishizō o motomete* [Searching for a new historical view of the Edo era], ed. Shakai keizaishi gakkai (Tokyo, 1977), 3–18; David L. Howell, *Capitalism from Within: Economy, Society, and the State in a Japanese Fishery* (Berkeley and Los Angeles, 1995); Kären Wigen, *The Making of a Japanese Periphery, 1750–1920* (Berkeley and Los Angeles, 1995); Edward E. Pratt, *Japan's Protoindustrial Elite: The Economic Foundations of the Gōnō* (Cambridge, Mass., 1999).

[16] Pratt, *Japan's Protoindustrial Elite* (cit. n. 15), chap. 1.

[17] Ibid., 71–2; Tessa Morris-Suzuki, *The Technological Transformation of Japan: From the Seventeenth to the Twenty-First Century* (Cambridge, 1994), 49–50; Andrew Gordon, *A Modern History of Japan: From Tokugawa Times to the Present*, 3rd ed. (New York, 2014), 33.

ished in Edo and along major roads, while many villages hosted small breweries or at least one or two drinking establishments.[18]

Morohaku sake, in which white rice instead of unrefined rice was used not only in the fermentation mash (where white rice had long been used) but also in *kōji* making, became widespread in the seventeenth century. *Morohaku* brewing encouraged broader changes in labor and technology within the sake industry.[19] The sharp increase in the amount of labor needed for rice polishing meant that rice polishers began to work separately from the brewing workers. Breweries began to rely on external technical specialists who were skilled in the brewing process, eventually leading to the rise of experts called *tōji* (head brewer). Whereas medieval *kōji* makers had been a separate specialist industry, in the Tokugawa period the *kōji* making process was integrated into the sake brewery. However, *moyashi* making—the making of dried *kōji* spores to be used as starters for *kōji* making—gradually emerged as an independent industry.

Brewing differed from other traditional side industries, such as textiles, in that it required relatively high levels of capital and labor. Sake and soy-sauce brewery owners needed a certain level of capital because they needed large tracts of land, buildings, and equipment such as tanks for the fermentation process, and because the brew took many months to mature. As for labor, the labor force was not family based, unlike in textiles. A contract-based, seasonal labor force came to the brewing house to brew in the winter months. The labor force was all male. Until the fifteenth century, sake brewing had been a female-gendered profession, and women had led a number of merchant guilds. By the middle of the Tokugawa period, the rise of religious ideas of pollution linked to women meant that women were banned from entering the brewery.[20] The brewery owner controlled but did not himself oversee the production process. In the case of both sake and soy-sauce brewing, he left the management of production to the *tōji* and the *kashira* (deputy), whom he hired, and interfered little. The *tōji* and *kashira* recruited the annually contracted workers as well as the day laborers and oversaw the day-to-day running of the brewing house. It was the *tōji*, rather than the brewery owner, who was master of the knowledge of the fermentation process.[21]

Reflecting the understanding of stages of material change in the sake brewing process, under the *tōji* and *kashira* the employees were divided into *kōji* specialists, *moto* specialists, rice-steaming specialists, cooks for all, and day laborers. The space of the brewing house mirrored these divisions, with its different rooms. The process began with the selection of the key raw materials, water and rice. Next was *kōji* making, which was done by adding purchased *moyashi* to steamed and dampened rice, letting it sit in the warm, humid *kōji* room to let it begin to grow, and then putting it into wooden boxes that would be lined up on shelves in the room and leaving it for a few days, resulting in a fine-smelling, green-yellow mold. The *moto* as a material was difficult to achieve well and guided the fermentation process. It was made by mixing *kōji*

[18] Francks, "Inconspicuous Consumption" (cit. n. 12), 155–6.
[19] Katō, "Nihon no sakazukuri no ayumi" (cit. n. 14), 212–5.
[20] Hitomi Tonomura, "Gender Relations in the Age of Violence," in Friday, *Japan Emerging* (cit. n. 11), 267–77, on 275.
[21] Tanimoto, "Capital Accumulation" (cit. n. 10); Mark Fruin, *Kikkoman: Company, Clan, and Community* (Cambridge, Mass., 1983), chap. 1.

with steamed rice and water in several stages and required regular stirring with an oar over many days. Then there was *moromi* making, in which *kōji* and *moto* were mixed with steamed water and rice. The ways in which this mash could be prepared were innumerable, and each brewery had its own style. Finally, the resulting liquid would be clarified by being squeezed out through a fine press into a large cask and would then be placed into storage. *Hiire* ("putting in fire," a heating process) would also be performed to discourage spoilage during storage.[22] With the first press of the season, the brewers would hang a large green ball of fresh cedar leaves outside the brewing house, which would gradually turn brown to signify the sake's maturation. Soy-sauce making was similarly based on *kōji* making followed by applying *kōji* to steamed soybean, wheat, and water in a variety of styles. Brewery workers believed that gods were responsible for the changes in the materials that made sake and aimed to preserve the cleanliness and sanctity of the brewing space. Inside the brewery there would be a shrine devoted to a sake god. The clapping of hands in worship in front of the shrine, the exclusion of women from the space, and the changing into and out of indoor sandals when entering or leaving the brewery were all everyday precautions to ensure a smooth brew amid many possible contingencies.[23]

The Meiji period was a difficult time for many brewers. In the 1850s, the United States and other Western powers coerced Japan into accepting unequal treaties by military force. Following the overthrow of the Tokugawa shogunate in 1868, the new Meiji government implemented a series of policies to promote the growth of industrial capitalism and national military modernization in order to avoid colonization by Western powers, adapting American and European institutional models under the slogan of "rich country, strong army" [*fukoku kyōhei*]. Having abolished the domains in favor of a centralized system of government, as well as the status distinctions between samurai and commoners and their accompanying occupational privileges and constraints, the government dismantled the early modern guild structure through which the domains and shogunate had controlled production and commerce, lifting all restrictions on who brewed and what amount they brewed. There was a proliferation of new sake brewers, especially landlords employing tenant labor, while brewers as a whole experienced unstable fortunes. The government taxed the alcohol industry heavily, frequently raising the tax in order to meet the demands of military preparation and infrastructure building. Cheap imported alcohol flooded the markets under the unequal treaties and put further pressure on brewers. The alcohol industry became the Meiji government's greatest source of tax revenue, exceeding the land tax by the end of the nineteenth century.[24]

[22] Katō, "Nihon no sakazukuri no ayumi" (cit. n. 14), 215–29.

[23] Ibid., 215–6.

[24] Fujiwara Takao, *Kindai nihonshuzōgyōshi* [History of the modern sake industry] (Kyoto, 1999), 185. From the late 1920s to the late 1930s, Marxist intellectuals looked back upon Meiji-era capitalism in an intense debate over how it should be situated analytically. The Kōza-ha (Lectures Faction), which followed the line given by the Comintern, emphasized the feudal elements remaining in the special or hybrid case that was Japanese capitalism and argued that state institutions were upheld by a base of semifeudal production relations in the countryside. The Rōnō-ha (Worker-Farmer Faction) took a different view that placed Japanese capitalism among the imperialist finance capitalisms of the world, and that saw the Meiji Restoration of 1868 as a bourgeois revolution with roots that reached back to the Tokugawa period. See Andrew E. Barshay, *The Social Sciences in Modern Japan: The Marxian and Modernist Traditions* (Berkeley and Los Angeles, 2004).

Private entrepreneurs, including rural elites, and the Meiji state alike invested in new capital-intensive projects during the state's push for what it saw as the late takeoff of the economy: modern banks and industries, railroads and other infrastructure. Moreover, at the local level, rural elites were nationalistic and enthusiastic as they spearheaded development efforts in the traditional sector, even though government policies favored export products over sake or soy sauce.[25] In the celebrated brewing regions of Kansai, wealthy brewing improvers published manuals and trade magazines to disseminate scientific principles and new European methods, such as the use of thermometers and salicylic acid, for scaling up production and making goods competitive for export. In other parts of the country, small and medium-sized brewers similarly formed discussion societies, trade and industry associations, and producers' cooperatives. They hoped to imitate the techniques of the Nada districts and standardize quality, in order to survive the competition in local or regional markets during the volatile economic conditions of the period.[26]

In the early Meiji decades, rural elites' hopes dovetailed with government initiatives, the latter of which included temporarily employing Western experts [*oyatoi gaikokujin*] as teachers and consultants, sending Japanese students abroad, establishing schools to teach Western-style disciplines and imperial universities to train a Japanese elite for government service, and building a series of state-owned model enterprises based on large-scale production. Rubbing shoulders with both Tokugawa political authorities and now Meiji officials, rural elites—who had sometimes been granted samurai privileges in the past—had high hopes for the greater role that they might play in the new Meiji regime. By the late 1880s, however, many of the state-owned enterprises had been sold off to the private sector (later to become the large industrial combines known as the *zaibatsu*) as a result of commercial failure, and prefectural agricultural schools, training centers, and experiment stations had been shut down. In addition, it was clear that a goal of state initiatives was to provide a source of employment for former samurai. A majority of the students at government schools or prefectural stations were samurai, not farmers, while high-ranking prefectural officials were generally samurai from other parts of the country.[27]

Led by the rural elite, the local trade and industry associations that supported improvement-of-industry movements soon turned critical of the government. In the later Meiji decades the state reversed its policies and, rather than solely prioritizing modern transplanted industries, began to encourage small- and medium-scale traditional industries in order to make them more competitive, both domestically and for export mainly to Asia. The government consolidated a network of middle-level institutions, including trade associations for disseminating information, technical colleges to train technicians for industry, and prefectural and national experiment stations that conducted research on behalf of small and medium-sized businesses.[28] As historians have argued more recently, rural-based, traditional industry continued to play a significant role in the Jap-

[25] Pratt, *Japan's Protoindustrial Elite* (cit. n. 15), 35.
[26] Fujiwara, *Kindai nihonshuzōgyōshi* (cit. n. 24), 185–99.
[27] Pratt, *Japan's Protoindustrial Elite* (cit. n. 15), 37–8.
[28] Morris-Suzuki, *Technological Transformation* (cit. n. 17), 98–103; Kaoru Sugihara, "The Development of an Informational Infrastructure in Meiji Japan," in *Information Acumen: The Understanding and Use of Knowledge in Modern Business*, ed. Lisa Bud-Frierman (London, 1994), 75–97.

anese economy well into the twentieth century alongside the modern factory system.[29] In this way, state institutions built on the existent structures and energy of local rural movements—even where they came under the supervision of bureaucrats in the Ministry of Agriculture and Commerce and were dominated by the scientifically trained, who were predominantly though not exclusively samurai—and consequently "made redundant many of the rural elites' traditional roles" by the first decade of the twentieth century.[30]

It was in this context of industrialization and the improvement of traditional industry that science institutions took shape. As James Bartholomew notes, early on Japan's first imperial university, Tokyo University, was "institutionally innovative" in incorporating not only a powerful faculty of medicine but also a strong college of engineering (1886) and agriculture (1890) from existing institutions. In government ministries, agricultural research was especially well supported.[31] At the turn of the century, the government worked with local elites to encourage the establishment of higher technical schools as well as agricultural and industrial experiment stations. The national Brewing Experiment Station was founded in Tokyo in 1904 under the Ministry of Finance, as part of a nationwide network of regional brewing experiment stations. Among other things, it surveyed breweries, ran training courses for *tōji*, undertook studies of raw materials, and promoted the use of pure-cultured yeasts in sake breweries in place of *moto* making, centering on yeasts distributed by the related Brewing Society.[32]

In manuals, trade magazines, scientific books, and scholarly journals, microbes [*kin*] began to appear alongside their common names and species names, with sketches of their appearance under a microscope. They became the living forces of the brewing process—the *kōji* mold that turned starch into sugar, or the wild *kōbo* (yeasts) cultivated in the *moto* that transformed sugar into alcohol. Scientists argued that the quality of the sake was connected to the purity of the *kōji*, and brewers' ability to keep the *kōji* free from contamination.[33] Some argued that the yeasts of the *moto* created specific flavors of sake.[34] (Brewers themselves did not accept this simplistic narrative, and to scientists it quickly became clear that there were also symbiotic lactic acid bacteria involved in the ecology of the *moto*.[35]) A few of the scientists who named the earliest brewing microbes included German botanist Hermann Ahlburg (a foreign consultant teaching at the Tokyo Medical School, the predecessor of Tokyo University's Faculty of Medicine) and German agricultural chemist Oskar Kellner (a foreign consultant teaching

[29] Tanimoto, *The Role of Tradition* (cit. n. 10); Masayuki Tanimoto, "From Peasant Economy to Urban Agglomeration: The Transformation of 'Labour-Intensive Industrialization' in Modern Japan," in *Labour-Intensive Industrialization in Global History*, ed. Gareth Austin and Kaoru Sugihara (London, 2013), 144–75.

[30] Pratt, *Japan's Protoindustrial Elite* (cit. n. 15), 6. On the class background of scientists in the early twentieth century, see James R. Bartholomew, *The Formation of Science in Japan: Building a Research Tradition* (New Haven, Conn., 1989), 52–63.

[31] Bartholomew, *Formation of Science* (cit. n. 30), 93.

[32] Lee, "Mold Cultures" (cit. n. 7).

[33] Shimoyama Jun'ichirō, "Seishu no jōzō ni tsuite" [On the brewing of sake], in *Jōzō taikashū* [Anthology of great brewing experts], vol. 1, ed. Hirayama Kōnosuke (Tokyo, 1902), 1–10; Kozai Yoshinao, "Nihonshu jōzō no kairyō ni tsuite" [On the improvement of sake brewing], *Jōzō zasshi* 305 (1901): 23–7.

[34] Shimoyama Jun'ichirō, "Seishu no jōzō" (cit. n. 33).

[35] Furukawa Sōichi, "Nyūsankin to kōbo no kyōson to kyōsei" [The coexistence and symbiosis of lactic acid bacteria and yeast], *Seibutsu kōgaku kaishi* 90 (2012): 188–91. Thanks to Furukawa Yasu for the reference.

at the Komaba Agricultural School, the predecessor of Tokyo University's Faculty of Agriculture), who named *Aspergillus oryzae* as the *kōji* mold of sake.[36] Saitō Kendō (a doctor in botany from the Faculty of Science of Tokyo Imperial University) named *Saccharomyces soja* as a soy-sauce yeast.[37] Yabe Kikuji (official appraiser at the Ministry of Finance, and a doctor in agricultural chemistry from the Faculty of Agriculture of Tokyo Imperial University) named *Saccharomyces sake* as a sake yeast.[38]

All these scholars [*gakusha*] with doctoral degrees [*hakushi*] in specialist disciplines were considered to be scientists; they worked in government ministries, experiment stations, technical schools, or universities, often in a number of these, during their lifetimes.[39] A microbe scientist could also have been trained in a university faculty of engineering as a doctor in applied chemistry (Tsuboi Sentarō, discussed below), a doctor in pharmacy (Shimoyama Jun'ichirō, cited above), as well as a doctor from a university faculty of medicine. When in 1902 Saitō Kendō was tasked with investigating the microbes around a possible site for the national brewing laboratory, nobody in Japan was an expert on "fermentation microbes" [*hakkōkin*]. Saitō copied methods from German, Danish, and Japanese books on brewing science. His research was primarily taxonomic in aim, with the goal of elucidating and classifying new microorganisms. Much of it was pioneering work because "at the time the kinds of wild fungi produced in Japan were completely unknown."[40] Yet like the microbe species named above, the organisms that he found mapped onto the spaces and stages of the established brewing process. The next two sections discuss academic science's striking convergence with Konno Seiji's manufacturing research, as well as its divergence from Minakata Kumagusu's epistemologically driven questions. These trends were a result of the reorganization of investigation in Japan that came hand in hand with the introduction of the microbial entity in modern disciplinary science.

PURE CULTURE PRODUCTION

By the turn of the twentieth century, to improve the brewing process using the study of microbes meant applying the techniques of pure culture. Konno Seiji, the original founder of today's *moyashi* companies Akita Konno Shōten and Kobe-based Konno Shōten in the Meiji period, was a pivotal figure in bringing pure culture of *moyashi* into the brewing industry.[41] Born in 1882, Seiji was one of the sons of a brewing family in Kariwano in Akita Prefecture, the oldest after his elder brother died when Seiji was five. In the snowy town on the Sea of Japan side of Honshū, Seiji's father was the kimono-clad *tōji* of the family's soy-sauce factory. However, a fire completely destroyed the factory when Konno Seiji was young.[42] After Seiji completed his studies

[36] Murakami Hideya, "Sake to kōji" [Sake and kōji], in Katō, *Nihon no sake no rekishi* (cit. n. 14), 319–92, on 321.

[37] Saitō Kendō, *Tōkyō zeimusho kantokkyoku Saitō Kendō chōsa* [Surveys by Saitō Kendō for the Tokyo Tax Office and Inspectorate] (Tokyo, 1905).

[38] Murakami Hideya, "Kinkabu" [Microbial strains], in *Kōjigaku* [Kōji science], ed. Murakami Hideya (Tokyo, 1986), 48–81, on 57–8.

[39] On the Japanese doctoral degree, see Bartholomew, *Formation of Science* (cit. n. 30), 50–2.

[40] Saitō Kendō, *Hakkō biseibukki* [An account of fermentation microorganisms] (Osaka, 1949), 224.

[41] *Moyashi* is the traditional term for what is now often called *tanekōji*.

[42] Konno Eiichi (nephew of Konno Seiji), Konno Hiroshi (president of Akita Konno Shōten and grandnephew of Konno Seiji), and Konno Kenji (former president of Akita Konno Shōten and nephew of Konno Seiji), interview by Victoria Lee, 20 February 2012, Kariwano, Daisen-shi, Akita-ken, Japan.

at Akita Middle School, he left the cold northern prefecture to study the scientific principles of brewing. At the time the only college in Japan that had a Brewing Department was Osaka Higher Technical School, located in the heart of the metropolitan merchant capital.[43]

Surrounded by the traditional brewing districts of Kansai, Osaka Higher Technical School's Brewing Department trained technicians from breweries all over the country. The department had been established in 1897 in response to calls from brewers to create an independent subject for brewing, unlike at Tokyo Higher Technical School, for example, where training in the use of microscopes was part of the Applied Chemistry Department. It was rumored that the manager of Osaka Beer (the predecessor of Asahi Beer) prodded the prefectural government's decision by buttonholing a high-ranking official after a nationwide meeting of the Association of Sake Brewing.[44] At the school, Konno Seiji studied under Brewing Department head Tsuboi Sentarō. Tsuboi, a graduate of the Applied Chemistry Department from the Imperial College of Engineering (later the Faculty of Engineering at Tokyo Imperial University), saw the department's research as bringing scientific ideals and actual practice [*jicchi*] together.[45] At the time, Tsuboi's laboratory was working on the pure culture of *tanekōji* as well as yeast for application in industry. Tsuboi's advertisement in the back pages of *Jōkai* in 1902 asking brewers to buy pure-cultured *moyashi* made by his college laboratory appeared alongside those of established, commercial *moyashi* makers licensed by the Ministry of Agriculture and Commerce. These commercial firms claimed in their various advertisements that the pure-cultured *tanekōji* of their respective *moyashi* makers, the fruit of laborious research efforts and enthusiastically tested technology, drew high praise in the "twentieth-century brewing world."[46] In the picture that both Tsuboi and commercial *moyashi* makers painted in their advertisements, the application of science placed their products at the cutting edge of the industry.

Konno Seiji graduated in the spring of 1905 and entered Kawamata Shōyu, one of the largest soy-sauce companies in western Japan, whose factory was part of the chimneyed cityscape of Sakai, just south of Osaka. As chief technician of Kawamata, Konno was busily occupied with the mechanization of the factory.[47] He was a man so obsessed with the precision of watches that he would make charts of how late each one ran to record its reliability, checking its performance in horizontal and vertical directions.[48] Apprentices remember that he kept the factory very clean.[49] That autumn,

[43] Akita Konno shōten kabushiki kaisha, ed., *Konno moyashi 101nen no ayumi* [History of 101 years of Konno Moyashi] (Daisen, Akita, 2011), 26–7.

[44] Hyakushūnen kinen jigyōkai, ed., *Hyakunenshi: Ōsaka daigaku kōgakubu jōzō, hakkō, seibutsu kōgakuka* [Hundred-year magazine: Department of Brewing Science/Fermentation Science/Biological Engineering, Faculty of Engineering, Osaka University] (Suita, 1996), 10–3.

[45] Tsuboi Sentarō, "Jōzōkai no kakumei jigyō" [Revolutionary projects in the brewing world], *Jōkai* 23 (1903): 45–8.

[46] Advertisements, *Jōkai* 10 (1902).

[47] Akita Konno shōten, *Konno moyashi* (cit. n. 43), 27–8.

[48] Ibid., 28–9. On clocks, industry, and work discipline, see E. P. Thompson, "Time, Work-Discipline, and Industrial Capitalism," *Past & Present* 38 (1967): 56–97. On the history of clocks and time measurement in Japan, see Yulia Frumer, *Making Time: Astronomical Time Measurement in Tokugawa Japan* (Chicago, 2018); Takehiko Hashimoto, "Japanese Clocks and the History of Punctuality in Modern Japan," *East Asian STS* 2 (2008): 123–33.

[49] Kawamata kabushiki kaisha, ed., *Murasaki: Sakai no shōyuya Kawamata, Daishō 200nen no ayumi* [Murasaki: 200-year history of Kawamata/Daishō, soy-sauce brewer of Sakai] (Sakai, 2000), 90.

while directing the newly opened Kawamata Shōyu Brewing Experiment Station, Konno Seiji isolated an excellent *kōji* microbe ("*Kawamata kin*"), which the company began to use for their soy sauce. In 1909 Seiji isolated a microbe suitable for sake, and the following year while keeping his position at Kawamata he and two brothers, Shigezō and Kenkichi, opened a shop (Konno Shōten) in Kyoto and began selling "*sake moyashi Konno kin*" and other microbes as pure-cultured *tanekōji* to *kōji* makers and sake, miso, and soy-sauce companies. The shop soon moved back to Osaka and opened another department for selling tools and machinery, many of which Seiji played a leading role in developing and patenting at Kawamata. Konno Shōten also published a trade journal, *Jōzōkai* (Brewing world), and later opened a soy-sauce *moyashi* branch in Sakai and a sake *moyashi* branch in Nada.[50]

These developments were under way well before the national Brewing Experiment Station in Tokyo developed a method for the pure culture of *kōji* microbes, on which the related Brewing Society published their first report in 1911.[51] By then, other companies were already rapidly adopting the use of pure culture. The largest soy-sauce companies in Kantō, such as Noda Shōyu (later Kikkōman), Yamasa Shōyu, and Higeta Shōyu, also began to make *tanekōji* in-house by the 1910s.[52] In fact, the head technician at Higeta Shōyu had interned under Konno Seiji at Kawamata before he first began isolating and pure-culturing *kōji* microbes for soy-sauce *tanekōji* at Higeta in 1912.[53] Subsequently, new specialist *tanekōji* companies appeared in the late 1910s and 1920s to supply smaller soy-sauce makers.[54]

Unlike the yeasts that the Brewing Society worked to maintain, the distribution of *kōji* microbes was already under the private oligopoly of *tanekōji* companies, who specialized in preparing what were dried microbial spores that would seed *kōji* making elsewhere. The *tanekōji* sector had distant roots in the medieval *kōjiza* (*kōji* groups), who held lucrative monopolies over *kōji* making and thereby controlled the source of the entire medieval brewing economy. The shogunate had frequently banned *kōji* making by unlicensed houses, partly to regulate tax collection but also to minimize brewing activity in order to suppress wastage of valuable rice. Where the monopoly system weakened, specialist *kōji* makers continued to supply brewing houses who preferred not to make *kōji* in-house.[55]

[50] Akita Konno shōten, *Konno moyashi* (cit. n. 43), 26; Kawamata, *Murasaki* (cit. n. 49), 88–9.

[51] Narahara Hideki, "Moyashi" [Moyashi], in Murakami, *Kōjigaku* (cit. n. 38), 32–47, on 35; Chikudō Shō, "Tanekōji ni tsuite (1)" [On tanekōji (1)], *Jōzō kyōkaishi* 6, no. 7 (1911): 47–52; Chikudō Shō, "Tanekōji ni tsuite (2)" [On tanekōji (2)], *Jōzō kyōkaishi* 6, no. 8 (1911): 32–41.

[52] Murai Toyozō, "Tanekōji konjaku monogatari" [Story of tanekōji past and present], *Shushi kenkyū* 7 (1989): 39–44, on 40; Fukuoka-ken shōyu kumiai, ed., *Fukuoka-ken shōyu kumiai nanajūnenshi* [Seventy-year history of the Fukuoka Prefecture Soy-Sauce Association] (Fukuoka, 1979), 158. Kikkōman also claims that they pioneered pure-cultured *tanekōji* in 1904 and that the practice spread from there. Nakadai Tadanobu, "Kikkōman ni okeru shōyu jōzō gijutsu kaihatsu no rekishi" [History of the development of soy-sauce brewing technology at Kikkōman], *Chiba-ken kōgyō rekishi shiryō chōsa hōkokusho* 4 (1995): 1–11, on 4.

[53] Yamazaki Yoshikazu, "Higeta shōyu gijutsu enkaku no kaisō" [Reflections on the history of Higeta Shōyu technology], *Chōmi kagaku* 21 (1974): 2–6, on 2.

[54] Murai, "Tanekōji konjaku monogatari" (cit. n. 52), 40.

[55] One such incident occurred in Kyoto in the mid-fifteenth century, when the shogunate attempted to revoke the monopoly law in response to wider discontent, and the struggle with Kitano Tenmangu shrine, which dominated *kōji* making in and around the capital, resulted in most of the shrine burning down. Koizumi Takao, *Kōji kabi to kōji no hanashi* [Kōji mold and the story of kōji] (Tokyo, 1984), 105.

In the Meiji period, as the sake improvement movement grew, more and more brewing companies requested *tanekōji* from specialist makers rather than making the starter in-house, and by the end of the period almost no sake brewers in Nada were making *tanekōji* themselves, though in-house manufacture was still prevalent in regions on the periphery of the sake economy.[56] It had become common practice to shake off the spores of the *kōji* from a good brew, dry them and use them as seeds in the next round of *kōji* making; the spores in this case were called *tomokōji*. However, the original spore starter—at this point sold by *tanekōji* houses—was tricky to generate.[57] In theory, if a sake or soy-sauce company built a *kōji* room, it was possible to make original starter spontaneously, by putting steamed soybean and ground wheat (to grow suitable microbes for soy sauce) or steamed rice (for sake) in an open *kōji* box and then leaving it on the shelf of the *kōji* room to wait for mold to enter from the air. In a long-standing *kōji* room, plenty of good *kōji* microbes should have settled there and be floating in the air. But how could one get to that point, and where did *kōji* microbes come from? Moreover, how could one maintain good *kōji* after finding it? With successive culturing, any good *kōji* would become old, produce fewer spores, and become contaminated by other molds like *kekabi* or *kumonosukabi*. The color of the spores would darken and turn black, the mold would have a strange smell, and the taste of the sake or soy sauce would worsen.[58]

It is likely that *tanekōji* companies who pure cultured *kōji* microbes were in the minority in the early twentieth century. For example, Kōjiya Sanzaemon in Kyoto, who sold *moyashi* under the label Biokku and claimed lineage from a *kōjiza* licensed by the Ashikaga shogunate (1336–1573), adopted pure culture technology much later, in 1951.[59] In the post–World War II period, however, the number of *moyashi* companies shrank. Compared with the hundred or so that existed at the beginning of the twentieth century, by the last decades of the twentieth century there were fifteen *tanekōji* companies, of which six distributed nationally. Though the companies were all small-scale, with fewer than fifty employees each, the concentration of the industry implies the level of technology needed to stay competitive.[60]

Konno Seiji's nephew Konno Kenji became head of Akita Konno Shōten much later, by which time the main branch of the company had moved back to Kariwano as a result of rice shortages in Kansai during World War II. Kenji has a childhood memory of watching an apprentice of Seiji's, Ueno Shiejirō, "making *genkin*," or generating the original starter microbes.[61] Ueno had not attended a technical school and had learned these methods under Konno Seiji during the war. Starting with a mass of spores floating in water, Ueno used a syringe to deposit a droplet of the spore mixture into a container of pure water, repeating until he had a very dilute mixture. Then Ueno would draw a mark on the cover glass and look at the sterile colored liquid through the microscope, to see whether there was only a single spore on the dish. If there was, he sterilized a piece of filter paper by splashing alcohol on it and used it to suck up the spore. He expanded the single spore into a colony by culturing it on rice grains, in other words making *kōji*

[56] Narahara, "Moyashi" (cit. n. 51), 35.
[57] Ibid., 34.
[58] Kawamata, *Murasaki* (cit. n. 49), 86–8.
[59] Murai, "Tanekōji konjaku monogatari" (cit. n. 52), 42.
[60] These figures are for 1985. Narahara, "Moyashi" (cit. n. 51), 36.
[61] Konno Eiichi, Konno Hiroshi, and Konno Kenji, interview by Lee (cit. n. 42).

within a flask, using wide-bottomed flasks that Konno Seiji had specially designed to increase the area for culturing. Then he subjected the pure colony to testing, investigating properties such as the formation of proteases, amylases, acid-resistant amylases, and so on. If the colony was strong, Ueno would select it for preserving and, taking spores from that colony, repeat the entire process perhaps hundreds of times. In this manner, by thoroughly investigating weak and strong microbes using the single-spore method, only the microbes with the very best qualities would be propagated and made into *genkin*. Droplet by droplet, taking spores wrapped in single droplets, one could cultivate them and bring up their descendants.

The single-spore method was also crucial to preserving the selected microbe. Otherwise, when successively cultured, the strain would quickly degrade.[62] What protected the products of *tanekōji* makers who employed pure culture methods was partly their reputation, but also the fact that other makers did not have the technology to maintain the strains even if they physically possessed them. The expense of maintaining high-quality strains also meant that brewing factories increasingly preferred to purchase *tanekōji* from specialist makers.

The reason for the swift adoption of pure culture technologies in the *tanekōji* industry was that they were upgraded versions of technologies that *tanekōji* makers had already been using. Since the products that they sold were dried microbial spores, they had long followed practices for identifying and isolating "good" cultures, mainly relying on sensory means. By inspecting the color of the spores, one could tell what kind of mold the *kōji* consisted of as well as how old the *kōji* was, as the yellow or yellow-green spores tended to darken to brown with successive culturing and with time. From the smell, one could tell how dry the spores were and the method of production.[63] This varied widely between makers, for example, in the geographical source of the ash used, how they stacked the *kōji* boxes during culturing, and the way and degree to which they dried the spores.[64] If the *moyashi* maker put the spores in his mouth he could make similar distinctions through their taste and hardness. He could also check them for bacterial colonies. Finally, the maker could actually make *kōji* and see how smoothly it went and then ask breweries to try out the *tanekōji* and see how the sake, soy sauce, or miso tasted. By these means, the *moyashi* maker could select the best spores. Makers had also, since the late thirteenth century, attempted to store, maintain, and propagate the mold cultures as purely as possible by adding special ash.[65] In a report in 1903, Tsuboi Sentarō noted that if one went to the places where *kōji* was made, sometimes the workers would first sprinkle camellia ash onto the rice, and then after mixing in the *kōji* on which were stuck all kinds of microbes and bacteria and bringing the whole thing into the *kōji* room, somehow only the mold microbes would multiply.[66] New scientific methods allowed *tanekōji* makers to fulfill the same aims with a much higher degree of control.

Most important, *tanekōji* makers and academic scientists not only shared common tools and techniques; they shared similar intellectual concerns. University laboratories, government-run experimentation stations, and the thousands of brewing houses

[62] Ibid.
[63] Narahara, "Moyashi" (cit. n. 51), 42–3.
[64] Ibid., 35.
[65] Ibid., 32–3.
[66] Tsuboi, "Jōzōkai no kakumei jigyō" (cit. n. 45), 47.

that specialized in *tanekōji*, *kōji*, sake, or soy sauce shared a concern with isolating, identifying, and preserving individual microbial strains and investigating their properties. Academic scientists depended upon *tanekōji* makers and other brewers to provide them with their objects of study, the microbes, which they then studied and preserved in a similar fashion. It was the shared intellectual concerns between academia and industry that helped to drive the adoption of new technologies in a dynamic private sector, a sector that in turn shaped the way academic researchers thought about problems and the research objects they used.

THE BOUNDARIES OF THE CELL

During the same period, the naturalist and folklorist Minakata Kumagusu (1867–1941) sought to use microbes' variety and life cycles to illuminate the ways in which the identity of the living organism could be understood. Minataka's studies staked out a position that directly opposed the emerging purposes and methods of microbe scientists. In reaction to academic mold studies, Minakata's work stood against the taxonomic fixation on species, against the assumption that microbes should be understood as bounded entities, and against the notion that scholars should be functionary experts instead of tackling more wide-ranging philosophical questions. Minakata was famed internationally for his studies of cryptogams (or spore-producing lower plants that do not bear flowers or seeds), particularly slime molds, which he liked to examine under an old-fashioned single-lens microscope.[67] He began collecting specimens during his four-year visit to the United States, beginning at the age of nineteen, during which he traveled from Michigan to Florida and even ventured to Cuba. He continued collecting over his eight-year sojourn in London from 1892 to 1900, and during the subsequent decades of long walks in the forests of his native Wakayama Prefecture in Japan.

He published fifty-one contributions in the scientific journal *Nature*, mostly during his time in London. By the 1910s, *Nature* had begun to reject his letters with the explanation that they were too diffuse. He went on to publish hundreds of essays in *Notes and Queries*, spanning subjects from the natural sciences to history to ethnography.[68] Among other things, Minakata was an eccentric in refusing to conform to academic science's sharply emerging disciplinary boundaries. His alternative studies on slime molds cast light on those boundaries, which defined science as a modern institution in Japan, far more than its amenability to exchange with the international scientific community. Together with his followers he amassed a substantial collection of slime molds and continued to send specimens to the naturalists Arthur and Gulielma Lister at the British Museum over the years.

Minakata held no degrees and worked entirely outside the Japanese academy. He was a vehement activist against the Meiji state's aggressive reach in Wakayama, specifically, the government's destruction of thousands of older forest shrines in order to merge them with large modern shrines, in the campaign to foster Shintō as the

[67] Nakazawa Shin'ichi, *Mori no barokku* [Baroque of the woods] (Tokyo, 1992), 227.

[68] Yoshiya Tamura, "The English Essays of Minakata Kumagusu: Centering on his Contributions to *Nature*," *Japan Foreign Policy Forum* 16 (2013), http://www.japanpolicyforum.jp/archives/society/pt20131007043828.html (accessed 23 May 2016).

official state ideology.[69] As numerous letters to Japanese acquaintances testify, behind Minakata's fascination with slime molds lay his interest in epistemological and metaphysical questions about life. Is there such a thing as a species? What is the nature of life and death?

Minakata came to believe that there were no true species, and that new species and new varieties of slime molds were nothing more than states of variation. In 1928, he wrote in a letter to a disciple that there were few clear true species, and instead there were many intermediates that did not belong under anything. Of those, which were real species, which were varieties, and which were different states? In between similar varieties, there were still more countless intermediate varieties. The more one looked, the more it became clear that in heaven and earth, there was not one thing that could be certified as a species; in the natural world, there was absolutely no such thing as a genus or species. To reach this realization was the point to their painstaking studies. Moreover, Arthur Lister's catalog of slime molds, Minakata pointed out, contained few new species, and rather many new varieties, and these supposedly advanced slime mold morphology. The achievements of slime mold research had become the discovery of new varieties, rather than new species. He encouraged his correspondent not to send away research materials to others but to settle matters at his own leisure—giving up hope of discovery or invention, for to discover a new species was nothing more than child's play.[70]

The ambiguous status of slime molds between the plant and animal kingdoms particularly fascinated Minakata. Despite the slime mold's visually appealing form in the plantlike part of its life cycle, which attracted naturalists, Minakata insisted on the significance of its animal-like characteristics during its other, amorphous phase. He described the slime mold's ability in its amoeba-like form to extend leg structures and go and swallow up solid bodies.[71] In a 1911 letter, he wrote that he had studied phenomena of life and death and matters of the soul as they concerned this organism for fifteen years. In a 1931 letter, he related insights from a conversation twenty-two years earlier regarding his slime mold work with an expert Buddhist acquaintance, Tsumaki Jikiryō. In the letter, Minakata described the slime mold's life cycle in light of the Mahāyāna Mahāparinirvāṇa Sūtra, an eschatological text with origins in first-century south India.[72] It was like when the shade was extinguished, light was born and consumed it, and when light was extinguished, darkness was born. It was like when a guilty person faced death, in hell the masses were expectant of one new life; when the guilty person recovered in vitality, in hell people clamored that the unborn child was on the verge of being miscarried; and when the guilty person finally died, the people in hell celebrated the smooth delivery of the child with relief.

In the slime mold's protoplasmic form, it ate rotten wood and dead leaves. After a long time, depending on various fates of light, heat, humidity, or wind, it could not stop at its protoplasmic form and welled up, offshoots becoming stalks, other offshoots

[69] On Minakata Kumagusu's protests against the state, see Julia Adeney Thomas, *Reconfiguring Modernity: Concepts of Nature in Japanese Political Ideology* (Berkeley and Los Angeles, 2001), 188–93; Mark Driscoll, *Absolute Erotic, Absolute Grotesque: The Living, Dead, and Undead in Japan's Imperialism, 1895–1945* (Durham, N.C., 2010), 1–21.

[70] Nakazawa, *Mori no barokku* (cit. n. 67), 253–5.

[71] Ibid., 246–8.

[72] On Minakata Kumagusu's thought and Buddhism, see G. Clinton Godart, *Darwin, Dharma, and the Divine: Evolutionary Theory and Religion in Modern Japan* (Honolulu, 2017), 92–103.

clambering up it and some becoming spores, some becoming walls surrounding the spores, some becoming filaments tying stalks, spores, and walls together. In the wind they dried, and in an instant the walls broke apart and the spores dispersed and flew, to someday transform into their protoplasmic form and prepare to propagate in another place. But while not yet dry, if there was a great wind or rain, the offshoots would in an instant cover their tracks and retreat to the original protoplasmic form, avoiding catastrophe by hiding under a tree or leaf. If the weather reverted to fine, the protoplasmic form would well up once again to make spore sacs. Minakata reflected that people scorned the slime mold in its protoplasmic form as a shapeless, phlegm-like semifluid. When under the microscope lens people watched the form bearing beautiful spores, walls, and stalks, they celebrated. If those offshoots once again became fluid and the slime mold reverted to its protoplasmic form, people would think that the slime mold was dying. The protoplasmic form seemed to have no function at all, like it was dead. Yet, Minakata pointed out, it was the protoplasmic form that was active, eating. When the form became stalks, spores, sacs, and filaments, it did not act at all; it was dying—solidifying and changing into another form.[73]

For Minakata, what the slime mold's life cycle challenged was the independent identity of the organism's life, which biologists in Europe—and later, Japan—had come to characterize as the autonomous interior maintenance of a bounded, organized individual, taking animals as their primary objects.[74] The notion of life as a self-directing ensemble of functions to resist death went beyond medical or animal studies, since cell theory drew on the same ideas.[75] Minakata's account offered a different view. A person observing the slime mold under the microscope could see that the life of the slime mold's protoplasmic form, which appeared like the metabolic animals that physiologists dissected, was different from the life of the slime mold's reproductive form, which appeared like the shapely plants that naturalists appreciated. Yet the two lives shared an interconnection that could not be seen from the vantage point of either kind of observer, and which went deeper than both. To Minakata, the slime mold's life cycle illustrated that individual life itself was a superficial phenomenon, and that the interconnectedness of things governed the organism at a more fundamental level.

In contrast, the assumptions about microbes that academic science adopted in the Meiji period allowed scientists to buttress directly the state's vision of *fukoku kyōhei* (rich country, strong army). For the scientists who imported the foreign concept of the microbe and adapted it for the purposes of rapid national industrialization, the existence of species was not a question in itself. As Christina Matta has shown, in Europe the philosophical foundations for bacteriology had been laid as early as the 1860s and emerged from cryptogamic botany rather than medicine. These developments, she

[73] Nakazawa, *Mori no barokku* (cit. n. 67), 260–2.

[74] On this point I follow the interpretation of Minakata's thought in ibid., 263–9. The literature on comparative anatomy and physiology is extensive; see, e.g., Michel Foucault, *The Order of Things: An Archaeology of the Human Sciences* (New York, 1970); Foucault, *The Birth of the Clinic: An Archaeology of Medical Perception*, trans. A. M. Sheridan Smith (New York, 1975); John E. Lesch, *Science and Medicine in France: The Emergence of Experimental Physiology, 1790–1855* (Cambridge, Mass., 1984).

[75] William Coleman, *Biology in the Nineteenth Century: Problems of Form, Function, and Transformation* (1971; repr., Cambridge, 1977), chap. 2; J. Andrew Mendelsohn, "Lives of the Cell," *J. Hist. Biol.* 36 (2003): 1–37; Laura Otis, *Müller's Lab: The Story of Jakob Henle, Theodor Schwann, Emil du Bois-Reymond, Hermann von Helmholtz, Rudolf Virchow, Robert Remak, Ernst Haeckel, and Their Brilliant, Tormented Advisor* (Oxford, 2007).

argues, were driven by experimental study, or the aim to apply the methods of the physical sciences in botany instead of relying on morphology alone. It was in this period that Ferdinand Cohn developed a taxonomic system for classifying bacteria into distinct, stable species, based on physiological identifiers where morphological properties were not sufficient.[76] Subsequently, this intellectual framework became important to agricultural as well as medical science.[77]

For medical bacteriologists, as Olga Amsterdamska has argued, the stability of species was a "grounded assumption" that allowed efficient research strategies for engaging in laboratory investigations on disease, even when scientists knew that there were reasons to question the assumption.[78] In Japan, the same was true for studies on industrial processes. Laura Otis has observed that the cellular boundary rose as a prominent image in nineteenth-century European medical thought.[79] Rudolf Virchow pushed a vision of cells as free, self-bounded, self-responsible units in a body, like people in a liberal society. Louis Pasteur's experiments helped drive people to see microbes as a living force, by making germs easily visible to his society and encouraging people to erect barriers against them. Later, Robert Koch's goal of "aggressive intervention in the interest of saving lives" led him to assume a one-to-one specificity between a species and a disease, as well as to ensure pure culture by growing colonies only from single microbes so that the specific identification could be guaranteed.[80] While Tokugawa-era Japanese, for example, had lacked the language to convey directly the notion of bodily nature as autonomous mechanical necessity, in Meiji Japan scientists adopted European conceptions of cellular identity in their microbial work, enabling their role to serve state intervention in the lives of people and industries.[81]

In the Meiji era, the parallel between species and commodities—between scientists' methodologies for making species of fermentation microbes, and brewers' methodologies for making commercial brands of spores—became a literal one. Scientists, with each new agent they identified as a cause of chemical change, sought to bring the processes of brewing further under control. Their procedures of investigation were like those of *tanekōji* makers, and unlike those of unsupported naturalists such as Minakata Kumagusu. Federico Marcon argues that a vision of "the subsumption of knowledge under production and the subsumption of scholars under a state apparatus" had already begun to develop in the closing years of the Tokugawa period, with the emergence of the notion of a centralized political economy that preceded and influenced the Meiji vision. But as he also observes, the conception of "nature" as a category that referred

[76] Christina Matta, "The Science of Small Things: The Botanical Context of German Bacteriology, 1830–1910" (PhD diss., Univ. of Wisconsin–Madison, 2009).

[77] Similarly, Christoph Gradmann argues that concepts of purity and the stability of species were not prioritized in medical bacteriology until the late 1870s; see Gradmann, "Isolation, Contamination, and Pure Culture: Monomorphism and Polymorphism of Pathogenic Micro-Organisms as Research Problem 1860–1880," *Perspect. Sci.* 9 (2001): 147–72. On species in immunology, see Pauline M. H. Mazumdar, *Species and Specificity: An Interpretation of the History of Immunology* (Cambridge, 1995).

[78] Olga Amsterdamska, "Medical and Biological Constraints: Early Research on Variation in Bacteriology," *Soc. Stud. Sci.* 17 (1987): 657–87, on 667.

[79] Laura Otis, *Membranes: Metaphors of Invasion in Nineteenth-Century Literature, Science, and Politics* (Baltimore, 1999).

[80] Ibid., 29.

[81] Frederik Cryns, "The Influence of Hermann Boerhaave's Mechanical Concept of the Human Body in Nineteenth-Century Japan," in *Dodonæus in Japan: Translation and the Scientific Mind in the Tokugawa Period*, ed. W. F. Vande Walle and Kazuhiko Kasaya (Leuven, 2001), 343–63.

to the whole of material and phenomenal reality—as something universal, and distinct from "society"—was new to Meiji Japan, even if analogous conceptions had begun to develop in the late Tokugawa period.[82] Only in the Meiji period did "nature" come to be a coherent domain, one that industrial and state actors perceived could be exploited toward the goal of capital accumulation.

The modern scientist in Japan specialized in nature as an occupation. In the new ontological separation between "nature" and "society" in Japan described by historian Robert Stolz, the scientist had no particular authority to speak on philosophical and moral questions.[83] This made him different from the intellectuals of the early modern era, and from thinkers like Minakata Kumagusu, who in the modern period rejected such separations as subjective. Though the microbe came with a set of philosophical assumptions, microbe scientists separated the material from epistemological and metaphysical problems. The Meiji-era social and political order set rigid educational and career paths for these new specialists.[84] Within the political-economic framework of the state, they were managers of production, like *tōji* in a brewery. In their studies on fermentation microbes, scientists mimicked the activities of the experts who worked for brewing companies, as they themselves became technical experts who worked for the state.

CONCLUSION

The surprising convergence of assumptions or procedures from European microbiology, which were imported during the Meiji period, with commercial practices in Japanese traditional industry that had been developing over several centuries is highly suggestive of the ways in which the modernity of scientific institutions is bound up with industrial capitalism. The philosophical foundations that underlay the routes for producing microbes as species found a striking parallel in the goals that drove existent, specialist methods of producing microbes as commodities. Thus, for *moyashi* makers such as Konno Seiji, the introduction of the microbe as a concept and pure culture as a technique did not fundamentally change the aims of production or its labor organization, though it enabled makers to identify and refine the capacity of molds with greater precision. In turn, protoindustrial and protocapitalist developments from the Tokugawa period found direct expression in Meiji scientific institutions, as the paths for producing microbial commodities and microbial species became equivalent through close exchange—both intellectual and material—between brewers and scientists. Because of this, "microbe" [*kin*] referred as much to fungi as bacteria.

Similarly, state policies to promote a capitalist economy co-opted the local initiatives of long-standing experts in industry to improve manufacturing. Newer scientific

[82] Marcon, *Knowledge of Nature* (cit. n. 5), 276–7.

[83] The point regarding the implications for scientific authority within the framework built by the Meiji state is my own. Stolz instead emphasizes the gradual breakdown of this ontological separation and the emergence of an environmental consciousness after the Ashio Mine pollution disaster. Robert Stolz, *Bad Water: Nature, Pollution, and Politics in Japan, 1870–1950* (Durham, N.C., 2014), chap. 1.

[84] On the contrast with Tokugawa-era intellectual paths, see Bartholomew, *Formation of Science* (cit. n. 30), chap. 2; Ellen Gardner Nakamura, *Practical Pursuits: Takano Chōei, Takahashi Keisaku, and Western Medicine in Nineteenth-Century Japan* (Cambridge, Mass., 2005), 166–70; Janine Tasca Sawada, *Practical Pursuits: Religion, Politics, and Personal Cultivation in Nineteenth-Century Japan* (Honolulu, 2004), chap. 4.

experts had an even clearer role to play in the political economy of the Japanese state. The government took an interventionist approach to what it saw as Japan's capital-poor, late-developer economy. Modern science was institutionalized at the same time as a network of imperial universities, technical colleges, national and prefectural experiment stations and research programs, scholarly and trade journals, and industrial associations took shape around government policies. This resulted in a social constellation of scientific expertise in Japan that did not make a strong distinction in definition or hierarchy between science and medicine, engineering, or agriculture. As highlighted by the contrast between scientific questions and the goals of inquiry of the nonacademic naturalist Minakata Kumagusu, the new conception of "nature" enabled the cellular division between the independent living force and the mastered environment and made scientists managers of production. Thus the scientist did not have special authority to speak on metaphysical and epistemological issues, unlike the early modern intellectual. While microbes became living workers, the modern scientist became a technical expert for the state.

"Safe Driving Depends on the Man at the Wheel": Psychologists and the Subject of Auto Safety, 1920–55

*by Lee Vinsel**

ABSTRACT

In the first decades of the twentieth century, deaths from automobile accidents quickly mounted, and influential figures, like Herbert Hoover, sought ways to control this icon of industrial capitalism and its users. These early regulatory efforts opened up the new field of automotive safety, a crowded market for ideas full of both buyers and sellers of potential solutions. This essay examines the 1920s and 1930s, as one profession, psychology, entered and sought to influence this emerging field, which thoroughly entangled science and capitalism. It describes how psychologists used a committee in the National Research Council to find positions of power. It argues that the psychologists' successes and failures were largely determined through a dialectical process between the psychologist's skills, other powerful professions, like engineering, and available patronage and funding. The psychologists' greatest success came through positing a novel entity—the accident-prone driver. Yet by the late 1930s, the most influential psychologists had turned against this idea, criticizing less prestigious colleagues who promoted it to industry and government. Established psychologists worried mostly that self-interested, junior colleagues were overselling their ideas and aligning too closely with corporate capitalism, thereby undermining the young profession's already tenuous credibility.

In October 1922, 10,000 children walked through the streets of Manhattan for a "safety week." Such events, which drew thousands, were common in American cities during this period.[1] Of the children marching in New York, 1,054 were part of a group known as the "Memorial Division." Each of them represented a child who had died in an accident in the city in 1921. The first American died from an automobile accident in 1899.

* Science and Technology in Society, Virginia Tech, 233 Lane Hall, 280 Alumni Mall, Postal Code: 0247, Blacksburg, VA 24061; leevinsel@gmail.com.

Much of the research that undergirds this essay was done while I was a fellow at the Lemelson Center for the Study of Invention and Innovation. Thanks to all of the staff for their support, and especially to Eric Hintz. This essay benefited from comments from James McClellan III, Andrew Russell, the participants in Matt Stanley's New York City History of Science Working Group, and, most of all, from Jeremy Blatter.

[1] Peter D. Norton, *Fighting Traffic: The Dawn of the Motor Age in the American City* (Cambridge, Mass., 2011), 42–3.

Thereafter, deaths grew at a staggering rate. By 1913, 4,200 people a year were being killed by cars, and by 1922, that number had nearly quadrupled to 15,300.[2] Much of this resulted from the ever-increasing number of cars on the road, an effect of mass production and falling prices associated with Fordism. In 1900, there were 8,000 cars registered in the United States. By 1913, the number had increased to 1.26 million, and by 1922, there were 12.27 million vehicles crisscrossing American roads.[3] Deaths mounted. Public frustration built. But relatively few solutions were on offer beyond public shaming and raising awareness. At the Manhattan safety week in 1922, the New York City health commissioner, Royal S. Copeland, stepped forward and offered a prayer: "May the memory of these children cause us to devise methods to save their companions and others who will bless us in the days to come."[4]

In 1924, the U.S. Secretary of Commerce, Herbert Hoover, held the first National Conference on Street and Highway Safety to find the kinds of methods Copeland sought. The automobile was a dominant symbol of industrial and consumer capitalism, and auto accidents raised key questions about how to deal with technological risks and hazards in a capitalist and liberal democratic society. Hoover was a firm believer in laissez-faire government and opposed federal regulation. Instead, he envisioned a form of governance that scholars would later call "associationalism": cooperative voluntary associations would form a kind of private government that would aggregate expertise and solve dilemmas without state intervention.[5] At most, the role of the federal government was to help form and guide these associations, rather than apply force. In practice, this meant that Hoover held conferences to draw together groups and interests and find the best solutions.[6] These gatherings created markets for ideas that could reward problem solvers with wealth and status. The emerging field of auto safety fit Hoover's model perfectly. It was a kind of interstitial space between government and industry. No single level of government—local, state, or federal—controlled it, nor did any industry. Indeed, the automobile industry itself was largely absent. Which professional groups would control the field was also an open question.

When the history of regulation emerged as a subfield within the history of business and policy from the 1960s through the 1980s, it typically focused either on financial

[2] Susan B. Carter, Scott Sigmund Gartner, Michael R. Haines, Alan L. Olmstead, Richard Sutch, and Gavin Wright, eds., *Historical Statistics of the United States*, millennial ed. (Cambridge, 2006), "TABLE Df448–456 Motor vehicle accidents, death rates, and deaths, by type of accident: 1913–1996."

[3] Ibid., "TABLE Df339–342 Motor vehicle registrations, by vehicle type: 1900–1995."

[4] "10,000 Children Join Safety March," *New York Times*, 10 October 1922, as quoted in Norton, *Fighting Traffic* (cit. n. 1), 43.

[5] Ellis W. Hawley, "Herbert Hoover, the Commerce Secretariat, and the Vision of an 'Associative State,' 1921–1928," *J. Amer. Hist.* 61 (1974): 116–40; Brian Balogh, *The Associational State: American Governance in the Twentieth Century* (Philadelphia, 2015). I describe the associational role of Hoover's Department of Commerce with respect to auto safety in greater detail in Lee Vinsel, "Virtue via Association: The National Bureau of Standards, Automobiles, and Political Economy, 1919–1940," *Enterprise & Soc.* 17 (2016): 809–38.

[6] Throughout this essay, I describe auto safety as a new arena, domain, ecology, or field that emerged at the national level in the early 1920s. Here, I have been guided and influenced by recent discussions around field theory, which explores how fields emerge and become structured and how actors, both established and new—or "incumbents" and "challengers"—vie for power within them. For a recent synthetic treatment of field theory, see Neil Fligstein and Doug McAdam, *A Theory of Fields* (Oxford, 2012). See also Edward O. Laumann and David Knoke, *The Organizational State: Social Choice in National Policy Domains* (Madison, Wis., 1987).

regulation or on post-1960s regulation of safety and pollution.[7] Since that time, scholars have built upon this foundation, filling gaps and offering an overall richer account. The multidisciplinary literature on experts in regulation is robust, including a large body of work on how experts and economic interests shaped thinking and action around risk, accidents, and hazards.[8] Moreover, particularly in recent years, historians have turned more and more to how city and state governments shaped technologies and markets.[9] Yet studies of associationalism, a dominant form of American governance in the 1920s and 1930s, have yet to learn fully from this more recent work. Its early statements remained overly focused on industrial trade associations and did not recognize the sheer diversity of associations that Hoover formed and drew together, and the literature did not take into account how individuals and groups attempted to shape the areas that Hoover opened up.[10]

This essay examines one group, professional psychologists, who attempted to answer Copeland's prayer "to devise methods" by applying scientific thought to the problem of auto accidents.[11] Psychologists were just one group of professionals acting as "research entrepreneurs" who sought to influence this emerging regulatory field.[12] Because money and status are at stake, such professional jockeying is common in regulation, especially in new policy domains, but we are rarely able to reconstruct the thoughts and actions of strategizing individuals within professional groups with much, if any, detail. From the mid-1920s through the late-1930s, psychologists used a committee within the National Research Council's Division of Anthropology and Psychology to find a place within auto safety. The records of this committee—the Committee on the Psychology of the Highway—give us a clear view into how psychologists planned, strategized, and attempted to become influential.

This essay seeks to explain a paradox. On the one hand, psychologists achieved some influence in auto safety. Their ideas were widely reported in the popular press and enshrined in publications of the U.S. federal government that were meant to inform policy. On the other hand, the psychologists themselves believed their efforts

[7] Thomas K. McCraw, *Prophets of Regulation* (Cambridge, Mass., 2009); David Vogel, "The 'New' Social Regulation in Historical and Comparative Perspective," in *Regulation in Historical Perspective*, ed. Thomas McCraw (Cambridge, Mass., 1981). See also Steven W. Usselman, *Regulating Railroad Innovation: Business, Technology, and Politics in America, 1840–1920* (Cambridge, 2002).

[8] Importantly, "risk" was not an actor's category used by the psychologists in this study, but the topics here clearly connect to the wider literature on risk and safety. Mark Aldrich, *Safety First: Technology, Labor, and Business in the Building of American Work Safety, 1870–1939* (Baltimore, 1997); Jonathan Levy, *Freaks of Fortune: The Emerging World of Capitalism and Risk in America* (Cambridge, Mass., 2012); Arwen Mohun, *Risk: Negotiating Safety in American Society* (Baltimore, 2012); Dan Bouk, *How Our Days Became Numbered: Risk and the Rise of the Statistical Individual* (Chicago, 2015).

[9] William J. Novak, "The Myth of the 'Weak' American State," *Amer. Hist. Rev.* 113 (2008): 752–72; Richard R. John, *Network Nation* (Cambridge, Mass., 2010). Norton's *Fighting Traffic* (cit. n. 1) expertly describes how various interests shaped automotive safety at the level of the city.

[10] Hawley's "Herbert Hoover" (cit. n. 5) was particularly focused on trade associations and missed much of Hoover's other associational activities. Balogh's *The Associational State* (cit. n. 5) attempts to synthesize more recent trends in history and acts as an invitation to enrich our account of associational forms of governance.

[11] John C. Burnham, *Accident Prone: A History of Technology, Psychology, and Misfits of the Machine Age* (Chicago, 2009); Jeremy Todd Blatter, "The Psychotechnics of Everyday Life: Hugo Münsterberg and the Politics of Applied Psychology, 1887–1917" (PhD diss., Harvard Univ., 2014).

[12] On the research entrepreneurs, Charles E. Rosenberg, *No Other Gods: On Science and American Social Thought*, rev. ed. (Baltimore, 1997), 160–7.

largely ended in failure. They believed this for two reasons. First, their influence never achieved the breadth they desired. Second, they came to believe the idea that received the widest circulation—accident proneness—was intellectually bankrupt. I argue that this paradox arose through a kind of dialectic between the capacities of professional psychologists—and how psychologists and others understood these capacities—and the contours and power structures within the emerging field of auto safety. Repeatedly, the opportunities that were presented to psychologists pushed them down a narrow road toward the question of how to weed out problematic drivers. The interests of capital provided and shaped these opportunities. Potential rewards encouraged psychologists, especially young, ambitious ones, to overpromise solutions, and in the end, this overpromising and the professional threat it posed led psychologists to turn on one another.

This essay proceeds through three chronological sections, which mirror its argument. First, in the mid-1920s, the National Research Council (NRC) psychologists responded to the call of Hoover's auto safety conference and sought positions of influence in the emerging field of auto safety. They encountered difficulties, however. The economic and political ecology of auto safety strongly shaped their opportunities. More powerful professions, like engineering, inhabited some niches; other niches lacked patrons and the necessary sustenance of research funding. Through a Goldilocks-like process, the psychologists found that the only niche in which they had both jurisdiction and research support was in psychological testing—both written tests and tests of activity, like driving. At the same time, eminent psychologists constantly worried that young colleagues would oversell the promises of psychology—often to make profit—thereby undermining the profession's credibility.

Second, from the late 1920s through the mid-1930s, virtually the only research funding NRC psychologists could find was aimed at finding better ways to weed out problematic drivers. The firms and professional groups that funded this research almost always stood to gain financially from such methods. During this time, the NRC psychologists coalesced around a novel object of science—the "accident-prone" driver, a small percentage of individuals who caused the preponderance of crashes. Accident proneness assuaged capitalism. It suggested that the problem of auto accidents lay not with companies filling roads with machines beyond human control, or companies poorly designing dangerous cars, but rather with a small population of incompetents. Psychologists thus played a part in redefining, or constructing, the subject of auto safety. At the same time, psychologists were playing a dangerous game. They cast themselves as experts in ways that fit with the interests of corporate capitalists as well as with—in the case of some psychologists—their own desire to make a buck. Overselling, fraudulence, hucksterism, and quackery constantly threatened psychology.

Finally, in the mid-1930s, the NRC work on accident-prone drivers reached its apotheosis. It was featured both in the public press and in significant government publications. Yet the tension within the profession around younger and improperly trained individuals overselling solutions and threatening the profession's credibility led to breakdown: leading psychologists at the NRC turned against the notion of accident proneness itself, and by the late 1930s the NRC's efforts had fallen to pieces. But there is an irony here. Beginning with Ralph Nader's 1965 book, *Unsafe at Any Speed*, and then carried through the historiography of auto regulation, writers who focused on automobile safety often claimed that the auto industry was responsible for the pre-1960s emphasis on drivers as the cause of accidents.[13] The story of the NRC psychologists and their work on auto safety leads to a more complex view: it suggests that many groups unconnected to the auto industry were arguing that governance should focus on drivers, often because doing so was in the interests of those groups.

Like the other essays in this volume, the argument here attempts to better elucidate the relationship between—or the entanglement of—science and capitalism. Often, experts succeed at boundaries between scientific disciplines and other capitalist venues, especially businesses, by positing novel ontological entities of understanding. In this essay, that entity is the accident-prone driver. In her contribution to this volume, Sarah Milov tells a similar tale of what Ian Hacking called "making up people" in her story of the "secondhand smoker," and she too finds that the category traveled far beyond its authors' original intentions. Similarly, Victoria Lee traces the emergence of the "microbe" as an important concept at the boundary between fermentation science and the brewing industry. Moreover, several essays in this volume examine forms of expertise that emerged around capitalist markets and attempts to control them. For instance, again, Lee shows that notions of the microbe and the idea of fermentation science—and the fermentation scientist—arose basically hand in hand. William Deringer examines how mathematicians tried to find a place for their calculative expertise within the contested realm of finance. Paul Lucier describes how scientists offered expertise for a fee in American mines, which served as enlarged trading zones, situated competitions among miners, engineers, men of science, lawyers, and capitalists. In this way, the essays in this volume show not only that science and capitalism are thoroughly entangled, but that scientists and their concepts, instruments, and claims to authority often play important, constitutive roles in emerging or hotly contested capitalist markets.[14]

AUTOMOBILE SAFETY AS A POTENTIAL VENUE FOR PSYCHOLOGY

In the beginning of the twentieth century, psychology was a young, uncertain discipline that was enthusiastic yet often unsure of itself and frequently of questionable authority.[15] During those years, the subfield of applied or industrial psychology, or "psychotechnics," was an ambitious emerging discipline looking to increase its status and territory. Often this search involved shopping for new areas or venues in which psychology could be applied. Importantly, these venues included both private and public institutions. At Harvard University, a founder of the field, Hugo Münsterberg, and his students worked for corporations, including streetcar, electric power, motion picture,

[13] Ralph Nader, *Unsafe at Any Speed: The Designed-In Dangers of the American Automobile* (New York, 1965). This book was in part a historical explanation of why the United States featured so many deadly and debilitating automobile accidents, with the automakers playing the part of villains. Unfortunately, Nader's too-simple story was picked up uncritically in the auto regulation historiography. The best example is Joel W. Eastman, *Styling vs. Safety: The American Automobile Industry and the Development of Automotive Safety, 1900–1966* (Lanham, Md., 1984). The historian Peter Norton has been calling the Naderite interpretation into question for some time, both in his dissertation and in Norton, "Four Paradigms: Traffic Safety in the Twentieth-Century United States," *Tech. & Cult.* 56 (2015): 319–34.

[14] Sarah Milov, "Smoke Ring: From American Tobacco to Japanese Data"; Victoria Lee, "The Microbial Production of Expertise in Meiji Japan"; William Deringer, "Compound Interest Corrected: The Imaginative Mathematics of the Financial Future in Early Modern England"; Paul Lucier, "Comstock Capitalism: The Law, the Lode, and the Science," all in this volume.

[15] C. James Goodwin, *A History of Modern Psychology* (New York, 2015).

and advertising companies, as well as for governmental organizations like courts and public schools.[16] Even working for nonbusiness organizations, like governments and nonprofit groups, Münsterberg and his disciples often focused on the problems of industrial capitalism. This generation of applied psychologists met with some success. Most importantly, the First World War was a transformative moment. As Daniel Kevles, John Carson, and others have described, psychologists such as Walter Dill Scott, Walter Van Dyke Bingham, and Robert M. Yerkes achieved, as Carson puts it, "the most important early success of American mental testers" by using intelligence testing for the U.S. military during the war.[17]

The U.S. war effort also gave birth to the NRC, which was organized as part of the National Academies of Science and which included a Psychology Committee. The NRC was reorganized after the war in 1919, and the Psychology Committee became the Division of Anthropology and Psychology. For years afterward, this division—which was typically chaired by eminent psychologists—became a powerful tool that psychologists used to find new opportunities for their profession. And one opportunity they spied in those first postwar years was the emerging field of auto safety. Indeed, the topic was hard to avoid at a time when auto accidents, injuries, and deaths covered the front pages of newspapers across the nation. In 1922, the psychologist Raymond Dodge, the chairman of the NRC division, gave a talk titled "The Human Factor in Highway Regulation and Safety" before the Highway Research Board, an engineer-dominated organization formed as a partnership between the NRC and the U.S. Bureau of Public Roads.[18] Dodge used the talk to outline potential safety topics that psychologists could explore: the nature and placement of signs, the nature and time of electric signals, the visual limitations of drivers and pedestrians, minimum reaction times and their relationship to speed laws, the development of standardized tests for driver licensure, and education for safety. But no support was forthcoming for the ideas Dodge outlined; his proposal seemed to go nowhere.

The NRC psychologists did not address auto safety in a serious way until two years later. That year, 1924, Secretary of Commerce Herbert Hoover held the first National Conference on Street and Highway Safety, which brought together experts and concerned parties from all over the nation. Hoover's goal was to connect people with problems with people with solutions without the federal government stepping in and disrupting the ways of the free market through legislation and regulation. The psychologist R. S. Woodworth, the new chairman of the Division of Anthropology and Psychology, heard rumors about the upcoming conference and thought it would be worthwhile to get representation for psychology within it. He contacted Ernest Greenwood, one of Herbert Hoover's assistants at the Department of Commerce, and offered help with the "study of automobile accidents from the psychological angle, and accordingly we should like to be informed regarding any movement on foot looking to-

[16] Blatter, "Psychotechnics" (cit. n. 11).

[17] John Carson, "The Science of Merit and the Merit of Science: Mental Order and Social Order in Early Twentieth-Century France and America," in *States of Knowledge: The Co-Production of Science and the Social Order*, ed. Sheila Jasanoff (New York, 2004), 181–205, on 182. See also Daniel J. Kevles, "Testing the Army's Intelligence: Psychologists and the Military in World War I," *J. Amer. Hist.* 55 (1968): 565–81.

[18] Raymond Dodge, "The Human Factor in Highway Regulation and Safety," in *Highway Research Board Proceedings*, ed. William Kendrick Hatt (1923), 73–8.

ward the prevention of such accidents in which we might properly have a share."[19] Greenwood responded enthusiastically and invited the psychologists to attend the conference and join some of the committees. Woodworth and the psychologist Knight Dunlap, a professor at Johns Hopkins, attended the first Hoover conference, which was held in late December 1924. That same month, at the annual meeting of the American Psychological Association, thirty-five psychologists put forward and adopted a resolution to form the Committee on the Psychology of the Highway within the NRC.[20] Psychologists had walked onto the emerging national field of auto safety, though their position within it was uncertain.

Woodworth soon began planning the committee's shape. Although applied psychologists worked in a wide variety of fields, no matter what they examined—whether it was firing a cannon, driving an electric streetcar, or working on the factory floor—they divided problems into the same sets of subtopics, including the psychology of vision, signals, attention, reaction times, and education, just as Dodge had done in his 1922 paper. When Woodworth planned the new auto safety project, he envisioned a series of subcommittees, each dedicated to a different topic. His proposed subcommittees covered a broad spectrum, from designing tests for taxi drivers to creating the most effective "publicity devices," like posters and warning signs.[21] He also pondered including a subcommittee that would consider what psychology had to say about redesigning automobiles in the name of safety. While his process of selection is unclear, Woodworth narrowed the eight possible subcommittees to three: the Subcommittee on Tests, the Subcommittee on Signals, and the Subcommittee on the Causes of Accidents.

Each of the three subcommittees is interesting in its own way, and each imparts essential lessons about the fraught place of professional psychology in the emerging field of auto safety during the 1920s. As the field of auto safety emerged at the national level, a number of subfields arose within it, most of which were orchestrated by some preexisting authority.[22] The fate of each of the NRC subcommittees largely rested on the shape of these subfields, especially whether other professions claimed jurisdiction over them and/or whether the subfields had adequate resources and sustainable funding.[23]

Woodworth asked Dunlap to head the Committee on the Psychology of the Highway.[24] Dunlap accepted and also took charge of the Subcommittee on Signals. He had

[19] R. S. Woodworth to Ernest Greenwood, 17 October 1924, Folder "Anthropology & Psychology: Committee on Psychology of Highway: Beginning of Program—1922–1925," National Academy of Sciences, Washington, D.C. (hereafter cited as "NAS").

[20] R. S. Woodworth to Vernon Kellogg, "Memorandum: Conference on the Possibility of Psychological Research Directed Towards Problems of Highway and Traffic Safety and Efficiency," 14 January 1925, Folder "Anthropology & Psychology: Committee on Psychology of Highway: Beginning of Program—1922–1925," NAS.

[21] R. S. Woodworth, "Tentative Outline of Projects," 26 January 1925, Folder "Anthropology & Psychology: Committee on Psychology of Highway: General—1925–1926," NAS.

[22] One of the key elements of Fligstein's and McAdam's theory of fields is that fields are nested, like Russian dolls; see Fligstein and McAdam, *A Theory of Fields* (cit. n. 6). What we see here is that several subfields—often with separate professional jurisdictions—began to open up within the emerging field of auto safety.

[23] In this way, the experiences of the NRC subcommittees fit the model that Andrew Abbott laid out in *The System of Professions: An Essay on the Division of Expert Labor* (Chicago, 1988).

[24] For a summary of Dunlap's dissertation work, see Knight Dunlap, "Tactual Time Estimation," in *Harvard Psychological Studies*, vol. 1, ed. Hugo Münsterberg (New York, 1903), 101–21.

a long-held interest in that topic. Almost immediately after beginning work on that subcommittee, however, Dunlap found resistance. Engineers from a number of different organizations, including the Bureau of Public Roads, had been standardizing signs for years and had recently begun working on the standardization of signals. The engineers found little reason to listen to the psychologists, unless the psychologists supported what they were already doing. As one engineer told Dodge, "I would be much interested, however, in having your suggestions if for no other reason than to determine whether our practical developments have been in line with the results of your investigations."[25] The psychologists encountered this phenomenon repeatedly: their input was seen as worthwhile if it legitimated the ongoing actions of those in authority. Otherwise, they were ignored. The subcommittee held one meeting, in June 1925, at which its members discussed possible research projects, but they were unable to generate interest in or get funding for their work, and by late 1926, the subcommittee had evaporated.[26]

The Subcommittee on the Causes of Accidents fared a little better than the one on signs and signals. This particular subcommittee had an odd structure, in that it was technically a part of two organizations, the NRC Committee on the Psychology of the Highway and Hoover's National Conference on Street and Highway Safety. Woodworth invited the famed industrial psychologist, W. V. Bingham, to head the subcommittee, and Bingham accepted. The NRC was proud of the Subcommittee on the Causes of Accidents. Not only was an NRC psychologist heading this important effort on the national stage, but he was able to get two other psychologists, George M. Stratton and J. McKeen Cattell, onto it as well. After meeting and communicating frequently in 1925, the subcommittee published its final report in March 1926, which argued that more research needed to be done. The subcommittee soon led to frustration and misery for Bingham, however. The leadership of the Committee on the Psychology of the Highway resolved to make every effort to get a research budget from the National Academy of Sciences and also to approach Hoover for funding.[27] But the money never materialized either from the NRC or from Hoover, and this committee also died.

On a surface level, the NRC Subcommittee on Tests for Drivers did little better than the other two, but its influence was longer lasting. The subcommittee was headed by the eminent psychologist and editor of *Science* J. McKeen Cattell. Over half a decade earlier, Cattell had been fired from Columbia University for opposing the draft in World War I. In 1921, using the money he had won from a wrongful termination suit, he founded the Psychological Corporation, a private, contract research endeavor dedicated to the "advancement of psychology and the promotion of the useful applications of psychology."[28] In founding the Psychological Corporation, Cattell believed that two of its core functions would be supplying standardized psychological tests

[25] E. W. James to Knight Dunlap, 31 December 1924, "Committee on Psychology of Highway: Subcoms: Signals," NAS.

[26] "Meeting of Sub-Committee on Signs and Signals of the Committtee on the Psychology of the Highway," 20 June 1925, "Committee on Psychology of Highway: Subcoms: Signals," NAS. The latest correspondence on record for the subcommittee is a letter that Dunlap sent to its members on 14 May 1926, "Committee on Psychology of Highway: Subcoms: Signals," NAS.

[27] Bingham to Vernon Kellogg, 16 October 1926, "Committee on Psychology of Highway: National Street and Highway Safety Conference, 1924–1926," NAS.

[28] J. McKeen Cattell, "The Psychological Corporation," *Ann. Amer. Acad. Polit. Soc. Sci.* 110 (1923): 165–71.

throughout the country and conducting contract research within businesses.[29] The corporation would also create jobs for the many graduates coming out of psychology programs. The other members of the committee, whom Woodworth had approached in March 1925, were the psychologists Fred A. Moss of George Washington University, David Weschler of Columbia University, Morris Viteles of the University of Pennsylvania, and Adolph Judah Snow, a "consulting industrial psychologist" working in Chicago.[30] The members met at Cattell's office in New York to plan the committee's research on 14 May 1925.[31] The priority, they decided, was to settle on a set of standardized tests, including both written tests and apparatus tests, which could be used to sort good drivers from bad.

That Cattell was both head of the subcommittee and head of the Psychological Corporation was no coincidence. Cattell, probably in conjunction with Woodworth, laid out an ambitious vision for the subcommittee that relied on the corporation's structure and organizational capabilities. By 1925, the corporation had branches in over a dozen U.S. cities, and Cattell used these offices to reach out to local taxi companies. The ultimate aim—once the subcommittee had agreed on a standardized set of tests—was to have trained psychologists administer the tests for taxi companies at the various branches, generating revenue for the corporation and creating a means of filtering out bad drivers for the taxi firms.[32]

The end result of Cattell's efforts via the Psychological Corporation was that all of the committee members, except Cattell himself, were placed in taxi companies. From the start, the committee members' efforts were poorly coordinated, likely because they had greater incentives to go their own way and develop their own tests than to collaborate. Yet their efforts did culminate in real products. Each of the committee members (again, except Cattell) published at least one paper on the topic, describing the tests they developed and reporting positive results in decreasing accidents at the taxi companies.[33]

Here, it is worth briefly noting Adolph Judah Snow's role in the subcommittee. In many ways, actors during the mid-1920s found Snow's work to be the most exciting on the topic of driving, but tensions that surrounded Snow foreshadowed future problems in automotive psychological research. Snow had already been working at the Yellow Cab Company in Chicago when the Sub-Committee on Tests for Drivers formed.[34] That the Chicago Yellow Cab Company was an early mover on this research is not altogether surprising. In 1923, the firm had bought and installed the city's first traffic light system in an effort to reduce accidents and pedestrian deaths.[35] Hiring Snow was just another step along this path. Unlike the other members, Snow commu-

[29] Ibid., 168–70.

[30] R. S. Woodworth to Morris Viteles, 30 March 1925, "Committee on Psychology of Highway: Subcoms: Tests," NAS.

[31] "Memorandum for Dr. Woodworth from the meeting of the Sub-Committee on Tests for Drivers," "Committee on Psychology of Highway: Subcoms: Tests," NAS.

[32] Ibid.

[33] Fred A. Moss, "Standardized Tests for Automobile Drivers," *Public Personnel Stud.* 3 (1925): 147–65; Adolph Judah Snow, "Tests for Chauffeurs," *Indust. Psychol.* 2 (1926): 30–45; Morris S. Viteles, "Transportation Safety by Selection and Training," *Indust. Psychol. Mon.* 2 (1927): 119–28; David Weschler, "Tests for Taxicab Drivers," *Personnel J.* 5 (1926): 24–30.

[34] See Snow's letter accepting membership in the committee, written on Yellow Cab Company letterhead; Snow to R. S. Woodworth, 1 April 1925, "Committee on Psychology of Highway: Subcoms: Tests," NAS.

[35] Norton, *Fighting Traffic* (cit. n. 1), 135.

nicated with the NRC on Chicago Yellow Cab Company stationery. He was the only committee member who was not employed by an elite institution. Moreover, Snow did not yet have his doctorate, and the thesis he was writing was on Newton's place in the history of science, not on psychology.[36] In the eyes of some other psychologists, Snow was a marginal character who lacked proper training and, furthermore, might have been a threat to psychology's credibility.

In an attempt to figure out how Snow's work was proceeding, Woodworth contacted L. L. Thurstone, a fellow professor of psychology at the University of Chicago. After meeting Snow and inspecting his work, Thurstone repeatedly warned Woodworth about the threat Snow posed: "I am frankly a little suspicious about Snow's competence to handle his job with the Yellow Cab Company. He seems to have made a hit with the officials, but I am not at all sure that he is able to handle his data adequately."[37] A later psychologist disparagingly summarized the tests Snow developed this way: "The driver was given an intelligence test, [and] placed in a room to rearrange gasoline, TNT, lighted candles, and rickety furniture. [Thereby,] his judgment of what constituted safety in general was assumed to be measured." Knight Dunlap later wrote, "A considerable number of energetic young psychologists who were not entirely familiar with practical traffic conditions have devised ingenious practical tests of driving ability, and have urged . . . their adoption. . . . Even the hoary reaction-time has been trotted out of its desuetude, and touted as a useful test."[38] The NRC so far had examined such tests, found them wanting, and warned authorities away from them; as Dunlap wrote, "it would have been a cause for humiliation on the part of psychologists if we had not." Professional psychologists were excited by the opportunities opening in the field of auto safety, but they also knew that such opportunities came with risks, including young and improperly trained practitioners overpromising potential outcomes. The professionals policed their boundaries to preserve their credibility.

The testing subcommittee fell into inactivity in 1926 and was officially discontinued in 1927, but the work of the NRC Sub-Committee of Tests for Drivers set the shape of subsequent psychological research into auto safety, despite the protests of some psychologists, like Dunlap, who never tired of railing against the dominance of testing research.[39] This subcommittee succeeded, in whatever limited sense, largely because of the structure of the emerging field of auto safety. In the domain of signs and signals, psychologists had bumped into the jurisdiction of engineers, and in the broader arena of the causes of accidents, psychologists had failed to find long-term support. Only in their traditional domain of testing did the psychologists find promise. The many possible research paths open to psychologists—including both the reconstruction of the world and construction of the subject—narrowed to a focus on drivers and their behaviors. Importantly, it was the capitalist market for ideas that drove them down this narrow path. While the psychologists would continue to work with and seek to influence government organizations and law, repeatedly their only available patronage came

[36] A. J. Snow, "Matter and Gravity in Newton's Physical Philosophy: A Study of Natural Philosophy in Newton's Time" (PhD diss., Columbia Univ., 1926).

[37] L. L. Thurstone to R. S. Woodworth, 22 June 1925, "Committee on Psychology of Highway: Subcoms: Tests," NAS.

[38] Dunlap, "Report of the Committee on the Psychology of the Highway, 1931," 5, "Committee on Psychology of Highway—Annual Reports," NAS.

[39] J. McKeen Cattell, "Sub-Committee on Tests for Drivers," 15 April 1926, "Committee on Psychology of Highway: Subcoms: Tests," NAS.

from private interests seeking public influence and corporations looking to control their workers.

PSYCHOLOGISTS AS EXPERT WEEDERS

By 1927, the original three subcommittees had all dissolved, and it was unclear whether the committee itself would continue. In late 1927, however, the committee was approached by the chairman of the motor vision commission of the American Optometric Association. The association offered $1,000 to fund psychological research on "the vision limits of automobilists."[40] As one of the committee's later annual reports cynically put it, the optometrists "believed that good vision is a very desirable personal trait; and that those who had none too much of it should seek for artificial means of betterment, such as well-fitted glasses . . . either voluntarily or under coercion from administrative authorities."[41] In other words, the optometrists were hoping for research that would support their view that laws should require glasses for drivers with bad eyesight. From this moment until the mid-1930s, the psychologists only received money from groups that were interested in weeding out problematic drivers, and the psychologists' goal and role became to put this weeding out on scientific footing. Typically, as in the case of the optometrists, the funders stood to gain financially from the psychologists' findings.

For more than six years, much of the NRC's research money would fund the work of the psychologist Alvhh Lauer. Lauer began this work while a doctoral student of the leading behaviorist psychologist Albert Paul Weiss, who had a lab at Ohio State University. Knight Dunlap directed the optometrists' funds there at the end of 1927.[42] The first step Weiss and Lauer took was the construction of a new instrument for psychological experiments. Weiss's assistants bought a secondhand "steering wheel, clutch, and brake" pedal to build a simple driving simulator that could be used for observing driver reactions. With the simulator, Weiss and Lauer began working under the research paradigm of accident proneness, an idea then emerging on both sides of the Atlantic. The notion of accident proneness was that there was a portion of the population—often associated with poverty and mental feebleness as well as notions of gender, race, and ethnicity—who caused the vast majority of accidents.[43] As Weiss put it in his 1929–30 annual report to the NRC, "There is a class of drivers who contribute more than their proportion of accidents."[44] If these drivers could be filtered out, or at least removed from potentially hazardous activities, industrial and public accidents would fade, and capitalism, including the mechanization of transportation and work, would function more smoothly. In this way, driving simulators first emerged as scientific instruments meant to create representations of normal and abnormal drivers rather than as simulation technologies for training drivers, which they later became. With the simulators, Weiss and Lauer hoped to categorize individuals who "resemble their crime

[40] Paul Brockett to Vernon Kellogg, 5 October 1927, and Knight Dunlap to A. B. Fletcher, 7 October 1927, "Committee on Psychology of Highway: General, 1927–1939," NAS.

[41] H. M. Johnson, "Report of the Committee on the Psychology of the Highway, 1939," 1, "Committee on the Psychology of the Highway—Activities Summary," NAS.

[42] Knight Dunlap, Memorandum for Doctor Kellogg, 2 December 1927, "Committee on the Psychology of the Highway—Work at Ohio State University," NAS.

[43] Burnham, *Accident Prone* (cit. n. 11).

[44] A. P. Weiss, "Report of the Committee on the Psychology of the Highway, 1929–1930," "Committee on the Psychology of the Highway—Annual Reports, 1926–1939," NAS.

before they commit it," as Michel Foucault memorably described such efforts.[45] The simulator and other instruments attempted to put the image of dangerous drivers on scientific, empirical grounding.

Weiss and Lauer's research at Ohio State culminated in a 1930 monograph, *Psychological Principles in Automotive Driving*. The book consisted of a series of studies on groups of drivers who had been involved in automobile accidents and groups of drivers who had not. The goal, the authors explained, was to develop a standardized series of tests for use throughout the country. They argued that "state laws and city ordinances almost universally require a high standard of mechanical perfection" in automobiles but that drivers rarely had to meet any standard, certainly not a high one.[46] They went on: "It is in developing these necessary scientific standards for examining and training drivers . . . that research is so important, just as research in strength of materials and soundness of design was important in developing the highly refined testing and inspecting techniques," which "eliminated mechanical failures" that caused automobile accidents.[47] Standardization was the mode of governance of the day, but whereas engineering societies were standardizing objects, the psychologists attempted to standardize subjects—and the instruments for holding subjects up to that standard.

The Ohio State studies would have continued for some time, but Weiss fell ill and retired from his work. Lauer took a job as a professor of psychology in the education school of Iowa State University. Dunlap requested that the NRC funding follow Lauer to this new position so that he could continue the work, and the request was granted.[48] Dunlap prayed that Lauer's research would move beyond the narrow confines of testing and filtering to include work on the psychology of perception and signs, particularly the design of license plates. To encourage this broadening, Dunlap convinced the U.S. Bureau of Roads to transfer its collection of license plates from around the United States to Lauer's new lab in Ames, Iowa.[49]

Yet Lauer never really took Dunlap up on this cause. His energy increasingly came to focus on the driving simulator, and the potentials it had for creating a normalized picture of drivers. Convinced that "subjects around a psychological laboratory are not a fair representation of the driving population," he took his driveometer, the driving simulator and psychological testing apparatus, and set it up at the Iowa State Fair.[50] There, among the smells wafting from food booths, the whistles and chimes emanating from games and rides, Lauer and his colleagues fulfilled their wish "to try the equipment on a non-academic population." Shortly thereafter, the Chrysler Corporation gave Lauer a truck to convert the driveometer into a mobile laboratory. He drove it to Detroit where he experimented on, among others, Chrysler's company stunt drivers, who were to perform tricks in the firm's new line of cars at the 1933 Chicago World's Fair, "A Century of Progress."

[45] Michel Foucault, *Abnormal: Lectures at the Collège de France, 1974–1975* (New York, 2007).

[46] Albert P. Weiss and Alvhh Lauer, *Psychological Principles in Automotive Driving* (Columbus, Ohio, 1930), 2.

[47] Ibid., 4.

[48] Dunlap to Madison Bentley, Chairman of the Division of Anthropology and Psychology, 17 February 1931, "Committee on the Psychology of the Highway—Work at Ohio State University"; Madison Bentley to Vernon Kellogg, Permanent Secretary of the National Research Council, n.d., "Committee on the Psychology of the Highway—Work at Iowa State College," NAS.

[49] Dunlap to Madison Bentley, 17 February 1931 (cit. n. 48).

[50] Alvhh R. Lauer, *The Psychology of Driving: Factors in Traffic Enforcement* (Springfield, Ill., 1960), vii, xiii, 8–9.

The Committee on the Psychology of the Highway was proud of Weiss's and Lauer's work, and in his annual reports, Dunlap boasted of Lauer's research. Nonetheless, the Committee on the Psychology of the Highway had its conflicts. The optometrists were not happy with the direction Weiss's and Lauer's research was taking. They pressured Dunlap to use their money in some other way that better reflected their professional stake in automotive safety. Meanwhile, Dunlap, who had headed the committee since 1925, folded under his many pressures and resigned from his chairmanship, handing it to Lauer.[51]

When Lauer took control of the committee, tensions only increased. Lauer was prone to grand visions, and, like Adolph Judah Snow before him, he provoked professional psychologists' worries about losing credibility and also made leaders at the NRC uncomfortable. His first problematic proposal was a cross-country "promotional tour" through the American West and South with his mobile driving laboratory, at the behest of the American Optometric Association. Lauer saw the trip as an opportunity to further standardize tests based on the driving simulator by experimenting on a wide variety of people from around the country, but the trip furthered the interests of optometrists in ways others found worrisome. One optometrist tried to interest another in supporting Lauer. Lauer "has a daring, original, and very marvelous demonstration of the visual, mental, and physical re-actions necessary to drive a motor vehicle. . . . Crowds gather and newspaper cameras flash," he wrote.[52] The optometrist believed the trip would draw attention to the necessity of good vision and, thus, the need for glasses. Lauer also offered free advertising in *Popular Mechanics*, *Motor News*, and other publications to a maker of truck trailers and campers if the company loaned him two trailers for the trip. "By virtue of the fact that we as a safety group using these trailers would be giving them a silent but powerful endorsement," he wrote the company.[53]

Although the cross-country tour never came to fruition, it was the first in a series of proposals that raised hackles within the circle of professional psychologists and leaders at the NRC.[54] By 1934, Lauer envisioned creating "an institute [of] psychotechnology" at "some large eastern university," which would cost at least $1.3 million and would conduct research "for a period of 50 years."[55] Dunlap, however, was very skeptical of Lauer's proposal.[56] The Committee on the Psychology of the Highway had barely been able to raise a few thousand dollars a year for research. How was it going to find more than a million? Dunlap himself was exhausted with fund-raising and had lost a good deal of hope.

Lauer's plans traveled further and further from those of the NRC's leadership. In March 1934, Lauer had written the NRC suggesting, among other things, the publica-

[51] Dunlap to Edwin Silver, 1 May 1931, "Committee on the Psychology of the Highway—Work at Iowa State College," NAS.

[52] Jaques to H. J. Godin, 2 January 1933, "Committee on the Psychology of the Highway—Promotional Tours," NAS.

[53] Lauer to Wolfe Bodies, Inc., 14 April 1934, "Committee on the Psychology of the Highway—Promotional Tours," NAS.

[54] A. T. Poffenberger to Lauer, 27 February 1934, "Committee on the Psychology of the Highway—Promotional Tours," NAS.

[55] Lauer to Poffenberger, 18 June 1934, "Committee on the Psychology of the Highway—Psychotechnologic Institute Proposed," NAS.

[56] Dunlap to Lauer, 24 May 1934, "Committee on the Psychology of the Highway—Psychotechnologic Institute Proposed," NAS.

tion of three sets of tests for drivers, ranging from a "simple combination" for governments to a "complex set" for companies with fleets of automobiles. In the same letter, Lauer proposed patenting psychological testing instruments, which would be sold to governments and firms, with the goal of turning "royalties toward further research." Lauer warned the NRC, "If we do not, some commercial companies will do it as soon as the work gets under way and capitalize on our efforts."[57] Lauer understood that regulations often create markets, and he envisioned demand for his instruments if states, corporations, and other centers of control adopted his tests. The secretary of the Division of Anthropology and Psychology responded to Lauer curtly. It would not authorize the publication of standard driving tests under the NRC's name, she wrote, and "the general attitude is much against the Council holding patents."[58] Lauer's priorities were not the NRC's.

Lauer stepped down as the chairman of the Committee on the Psychology of the Highway in 1934, although organizational papers do not record his reasons. Still, by that time, the NRC had its own reservations about Lauer's actions. As a report of the Committee on the Psychology of the Highway later noted,

> Dr. Lauer's activities had passed the initial problem of devising and validating tests, and were including an effort to have certain tests adopted by administrators of traffic laws and regulations, by employers, and by others. Since it is not the primary business of the Council under its present organization to urge action upon governmental organizations and elsewhere, but rather to find facts, it seemed to some members of the committee that the Council should not sponsor this phase of Dr. Lauer's work, although it should seek support for his efforts and those of others to establish facts. This position seemed to be the more strongly justified because the tests he had developed had not been validated.[59]

Lauer went on to be involved sporadically with the NRC through the 1950s. In the state of Iowa and in the world of driving education and driving psychology, Lauer remained influential for many years, but he exerted that influence separately from the NRC, whose leadership distanced themselves from his work. Like Adolph Judah Snow, Lauer was a more or less marginal actor—based in the cornfields of Iowa, instead of at elite East Coast institutions—and he formed relationships with businesses and governments that other NRC psychologists avoided. This tension only increased in the next, final stage of the Committee on the Psychology of the Highway.

THE FALL OF ACCIDENT PRONENESS AMONG PROFESSIONAL PSYCHOLOGISTS

This third and final major stage of the NRC Committee on the Psychology of the Highway began in the 1933–4 academic year, when Knight Dunlap and W. C. Shriver, an engineer with Chrysler Corporation, began a subcommittee examining the biographical records of commercial drivers.[60] Lauer had forged the committee's relationship with Chrysler a few years earlier. One part of Dunlap's and Shriver's initial investiga-

[57] A. R. Lauer to Marian Hale Britten, 25 March 1934, "Committee on the Psychology of the Highway—General, 1927–1939," NAS.

[58] Marian Hale Britten to A. R. Lauer, 4 April 1934, "Committee on the Psychology of the Highway—General, 1927–1939," NAS.

[59] Johnson, "Report" (cit. n. 41), 5.

[60] Committee on the Psychology of the Highway, Annual Report 1938–1939, "Committee on the Psychology of the Highway—Subcoms: Study of Biographical Records," NAS.

tions resulted in the report titled "Reducing Accidents in Commercial Driving," a study of driving records and accidents at an unnamed public utility in the Midwest.[61] Shocked by rising company accident tolls in the mid-1920s, executives in the firm had set up an informal "traffic court," which was "made up of employees representing each department" of the company. The court reviewed each accident at the company and "attempted to analyze and present to the offending driver the underlying conditions or infractions contributing to the accident in question."[62] The traffic court was not very successful, however. The company's accidents increased.

The firm's next remedy was to identify those drivers—twenty-two in all—who were repeatedly involved in auto accidents, after it was found that "more than 50% of the accidents involved men having two, three, or four accidents within a year."[63] The company tried to educate these men in safety-mindedness, but this tack also failed. In an effort to help, psychologists examined five of the frequent accident offenders and found that four of them were "definitely subnormal in intelligence, ranking on about the level of an eight year old child."[64] The fifth man had high intelligence but bad vision. The psychologists then did a wider study of those involved in accidents at the company and found that "as was to be expected, in a larger number of instances . . . the level of intelligence of the men was such that it would be extremely difficult to impart to them sufficient instructions to make them safe in the driver's seat of an automobile." Education would not work, so the firm began removing those who were involved in multiple accidents and placing them in other lines of work. Following this path, the company reduced its accidents from 173 in 1929 to forty-five in 1933.[65]

In 1936, Johnson published an article based on Dunlap's and Shriver's research on commercial drivers titled "Born to Crash" in *Collier's Weekly*.[66] Johnson reported on the findings from "Reducing Accidents in Commercial Driving" as well as studies that the NRC conducted at three other large companies. The notion of accident proneness took center stage. As Johnson stated boldly in the opening of the article, "I am going to present evidence that there exists a class of drivers who are accident prone—that is, drivers who habitually have accidents whatever the condition of the highway, the state of the weather, or the speed at which they are traveling." Johnson held out little hope that the kinds of tests Lauer and others were developing would be much help in foreseeing which drivers would be repeat accident offenders because the methods could not predict a "driver's future performance." Instead, he believed the best method was retroactive. "By taking off the road the worst 1 per cent, 2 per cent, 5 per cent of the drivers who are now on it," policy makers could save as many as 12,000 of the 36,000 lives that would likely be lost that year.

In 1936, the same year that Johnson's article was published in *Collier's*, the U.S. Congress passed a directive ordering the Bureau of Public Roads to spend $75,000 researching highway safety. Through the Highway Research Board, some of this money

[61] "Reducing Accidents in Commercial Driving," "Committee on the Psychology of the Highway—Subcoms: Study of Biographical Records," NAS.
[62] Ibid., 1.
[63] Ibid., 2.
[64] Ibid., 4.
[65] Ibid., 3.
[66] H. M. Johnson, "Born to Crash," *Collier's*, 25 July 1936. I have relied on an offprint in "Committee on the Psychology of the Highway—Subcoms: Study of Biographical Records," NAS.

went to the Committee on the Psychology of the Highway. Representatives from the committee visited Connecticut, found that the state had the best accident records in the nation, and decided to use those records as the basis of a study.[67] The report for Congress needed to be written quickly, however, so Johnson penned a study based solely on biographical records. This report became "The Accident Prone Driver," the final part of a six-part Congressional document titled *Motor-Vehicle Traffic Conditions in the United States*.[68] "The Accident Prone Driver" was influential: it was likely the highest-level government document based on the notion of accident proneness. Engineers and policy makers took it up, and the Bureau of Public Roads promoted its vision as a way to ameliorate accidents.[69] It seemed that the NRC psychologists and their ideas had arrived. Accident proneness had found an eminent audience in Congress, and their work was being distributed throughout the nation.

The publication of "The Accident Prone Driver" was met by a great irony, however. As soon as Johnson sent off the report, things began to fall apart for the idea of accident proneness, at least for the NRC researchers. Johnson may have suspected that problems were on the way. He certainly did not shy away when they arrived. Many professional psychologists, including Johnson, worried greatly about how close young psychologists, like Lauer, were to the corporations who sponsored their work. A frequent subject of these anxieties was Harry R. DeSilva. Like Lauer, DeSilva apparently did some training with Albert Paul Weiss, although the exact relationship between DeSilva and Weiss is not clear.[70] After Weiss died, DeSilva received some of his equipment, including at least one of the primitive driving simulators. In 1936, DeSilva moved from a position at MIT to the Albert Russel Erskine Bureau for Street Traffic Research at Harvard University. Founded in 1925, the bureau was named after and funded by the president of the Studebaker Corporation.[71]

DeSilva's work focused on creating various driving tests, particularly tests that were based on driving simulators. In 1936 or 1937, DeSilva and his colleagues began holding a series of "driver clinics" in eighteen states.[72] In the clinics, DeSilva and his assistants would run subjects through a battery of four tests, based on two psychological instruments, a driving simulator (the Vigilance Test) and a device for testing vision (the Universal Visual Test). As DeSilva wrote, "Assembling many tests to form a clinic permits us to find a weakness in everyone. Once a clinic supervisor pierces the armor of conceit of Mr. Average Driver, he can often get him to break down and talk about himself," thereby opening "the way to self-study and self-discovery."[73] DeSilva liked to give anecdotes about this kind of jolting and armor piercing. He believed that this kind of humiliation at the clinics deflated unhealthy hubris and that a driver "went

[67] Memorandum for Mr. Crum, 7 August 1936 (unsigned but likely H. M. Johnson), "Committee on the Psychology of the Highway—Cooperation with the Highway Research Board," NAS.

[68] U.S. Bureau of Public Roads, *Motor-Vehicle Traffic Conditions in the United States* (Washington, D.C., 1938).

[69] See Bureau of Public Roads, *Highway Accidents, Their Causes, and Recommendations for Their Prevention* (Washington, D.C., 1938).

[70] DeSilva and a coauthor, Willis D. Ellis, acknowledged their debt to Weiss in "Changing Conceptions in Physiological Psychology," *J. Gen. Psychol.* 1 (1934): 145–59.

[71] Norton, *Fighting Traffic* (cit. n. 1), 165–9.

[72] Harry R. DeSilva and Ralph Channell, "Driver Clinics in the Field," *J. Appl. Psychol.* 22 (1938): 59–69.

[73] As quoted in Harry R. DeSilva, "Mechanical Tests for Drivers: Are They of Value in Promoting Safety?," *Tech. Rev.* 40 (1938): 309–11, 326–8, on 326.

away vowing never again to brag about or get into trouble as a result of his supernormal ability."[74]

Some of DeSilva's activities brought him close—some thought uncomfortably so—to industry interests. DeSilva held the driving clinics at the "courtesy of the Automotive Safety Foundation."[75] The Automotive Safety Foundation was founded in 1937 by the Automobile Manufacturers Association, the primary auto industry trade organization. Paul Hoffman, the president of the Studebaker Corporation, headed the foundation. Hoffman would go on to write *Seven Roads to Safety* (1939), a work much reviled in auto safety circles for putting almost all blame for road injuries and deaths on drivers and none of it on automotive design.[76] Much later, in *Unsafe at Any Speed* (1965), Ralph Nader would attack the Automotive Safety Foundation for pushing the auto industry's agenda and retarding the prospects of real safety reform in the United States.[77] Some of DeSilva's other events were sponsored by the Aetna Casualty and Surety Company.[78] Aetna would go on to create its own driving simulators "as a means of helping make automobile drivers more safety conscious."[79]

All of this grew to be too much for Johnson. He wrote a review of *Psychology and the Motorist* (1938), in which he criticized the auto safety field.[80] Johnson attacked solution peddlers like DeSilva, finding "that the safety movement is infested by racketeers, each having some special gadget or service to sell at high profit"; "that among the propagandists for safety are many evangelists, who know little about human causes of accidents, who disregard the fact that" available information does not "yield a valid conclusion about many important questions for which they offer guaranteed answers"; "that certain 'driver-clinicians' have misled the public as to the significance and importance of certain personal traits, especially reaction time"; and "that many newspaper campaigns against traffic offenders and dangerous driving practices, sponsored by safety organizations, lunch-clubs, and the like, are ineffective, being planned for news value and prestige rather than for any important social effects."[81] Later, after he visited one of DeSilva's driving clinics, Johnson wrote that "at the exit-gate there stood a very well-dressed, and otherwise prosperous-looking, individual who showed me my 'skill-profile,' pointed out my 'weaknesses,' remarked that each was a sign of danger (notwithstanding my long accident-free record), and handed me a little leaflet of his insurance company as he politely bowed me out of a public gate."[82]

For Johnson and others at the NRC, the critique of the tests developed by DeSilva, Lauer, and others went deeper than the level of attacking conflict of interest to the level of scientific method and credibility. Growing increasingly uncomfortable with the idea

[74] DeSilva and Channell, "Driver Clinics" (cit. n. 72), 69.

[75] See ibid., n. 1.

[76] Paul Gray Hoffman, with the assistance of Neil McCullough Clark, *Seven Roads to Safety: A Program to Reduce Automobile Accidents* (New York, 1939). See Eastman, *Styling vs. Safety* (cit. n. 13), esp. 141–3, for a heavily Nader-influenced take on Hoffman's book.

[77] Nader, *Unsafe at Any Speed* (cit. n. 13), esp. 183–6.

[78] "Drivers 'Not So Good,' 9,000 Auto Tests Show," *New York Times*, 24 May 1936.

[79] See "News," *Insurance Field* 66 (1937): cxxxix.

[80] H. A. Toops and S. E. Haven, *Psychology and the Motorist* (Columbus, Ohio, 1938); H. M. Johnson, review of *Psychology and the Motorist*, by H. A. Troops and S. E. Haven, *Psychol. Bull.* 35 (1938): 561–4.

[81] Johnson, review of *Psychology and the Motorist* (cit. n. 80), 562.

[82] H. M. Johnson, review of *Why We Have Automobile Accidents*, by Harry DeSilva, *Amer. J. Psychol.* 57 (1944): 436–43, on 440.

of accident proneness, Johnson hired Percy W. Cobb, a psychologist and statistician who had a long history of working on the psychology of the automobile.[83] Cobb created a sophisticated, multipart argument against the notion of accident proneness and tests for it, parts of which never found their way into print. First, Cobb did a statistical analysis, sampling from 29,531 Connecticut drivers, and found that because of the many complex factors that led to auto accidents, even the perfect driving test could only account for **a** small number of accidents.[84] (Importantly, Cobb used the same body of records that Johnson would have used if he had not been in a rush to put together the Congressional report. In other words, if Johnson had had more time, he may have realized the problem with the notion of accident proneness and never published "The Accident Prone Driver.") Second, Cobb conducted two sets of driver's tests on 3,663 Connecticut residents: one set was Lauer's, the other, DeSilva's, adding up to seventy-three tests in all.[85] Cobb compared the test scores with the subjects' driving records and found a correlation of only 0.35. Johnson concluded, "Thus, the test scores are useless as a means of diagnosing the accident liability of individual drivers." The tests could not be relied on to weed out potentially problematic drivers at licensing facilities, as Lauer and DeSilva hoped they would be. For this purpose, accident proneness was a bust.[86]

When Johnson wrote the 1939 annual report of the Committee of the Psychology of the Highway, he styled it as a narrative history of the committee, from its inception to the present. He struck a depressing tone. The committee had limped along for years. Its members, Johnson wrote, "are unanimously in favor of disbanding unless it is possible to proceed, for they see no use, at present, in continuing an inactive committee."[87] It had left many failures in its wake, from Dunlap's inability to influence sign and signal standards to the coalescence of Lauer and DeSilva around the bankrupt notion of accident proneness. Johnson recounted each of these defeats, blow by blow. At the end, Johnson wrote, "The work of the committee has now reached a stage at which it can be discontinued gracefully, or carried on. I make no categorical recommendation." In the 1940s, Johnson and Cobb entered into a period of heated polemics against DeSilva. After DeSilva's magnum opus, *Why We Have Automobile Accidents*, was published in 1942, Johnson excoriated it in a review in the *American Journal of Psychology*, claiming that the work could only be understood against the "background of fact, faith, and pressure-group propaganda against which it was written."[88] Regardless of whether he was fair, Johnson was on the winning side, and as John Burnham has examined, the notion of accident proneness fell into disfavor among experts. For the Committee of the Psychology of the Highway, 1939 came with the sense of an ending, and its members never again reformed under that name.

[83] See Woodworth to Percy W. Cobb, 4 June 1925, "Committee on the Psychology of the Highway: Subcoms: Tests"; Report of the National Research Council (Washington, D.C., 1937), 64.

[84] Percy W. Cobb, "The Limit of the Usefulness of Accident Rate as a Measure of Accident Proneness," *J. Appl. Psychol.* 24 (1940): 154–9.

[85] This research of Cobb's was never published, but Johnson summarized it in the Committee on the Psychology of the Highway Final Report of 1939 and referred to an unpublished manuscript of Cobb's in his review of *Why We Have Automobile Accidents* (cit. n. 82), n. 10.

[86] Burnham, *Accident Prone* (cit. n. 11), 168.

[87] Johnson, "Report" (cit. n. 41), 1.

[88] Johnson, review of *Why We Have Automobile Accidents* (cit. n. 82), 436.

CONCLUSION: PSYCHOLOGISTS AND THE LONG POLITICAL HISTORY OF AUTO REGULATION

Even if the main body of professional psychologists abandoned auto safety and the idea of accident proneness, their ideas were still out there, and others took up the cause. Increasingly, the auto industry–funded Automotive Safety Foundation put out propaganda pinning safety on the backs of drivers. By the 1950s, the lines were definitively drawn: auto safety advocates pushed for safety standards and safety devices like seatbelts, while industry representatives argued that the only hope was better driver training. When activists, like Ralph Nader, and historians looked back at the history of auto safety—without the benefit of archival sources, like those of the NRC committee—they came to believe that the auto industry itself had given birth to this driver-centric view. What they missed was that many groups—psychologists, insurance executives, optometrists, safety movement leaders, and others—had put forward this view because it fit both their understanding and their interests. Psychologists offered a wealth of ideas about who drivers were and how they could be sorted and controlled. Although the mainstream of psychology walked away from these ideas, they lived on, still influencing the world in a ghostlike state that only later political upheaval would overthrow.

From the moment that psychologists entered the field of auto safety by responding to Herbert Hoover's first conference in 1924 to the day the U.S. Congress published "The Accident Prone Driver" in 1938, psychologists were drawn down an increasingly narrow road of research to the question of how to weed out problematic drivers. As they had done since the first years of the twentieth century, psychologists tried to apply their methods to the problems of capitalism, including the increasing mechanization of human activity. Yet, they were never able to achieve the expansive vision they initially planned around auto safety. If psychologists in the 1920s and 1930s had their druthers, historians looking back today would see their influence in a wide variety of subfields around auto safety. Psychologists would have published prominent reports on the causes of accidents; they would have shaped the design of road signs and traffic signals; perhaps they would have taken up influential positions within the Highway Research Board or even the Society of Automotive Engineers. It was economic interests—the aspirations of other professional groups like optometrists, the worries of business leaders who employed professional drivers, and the potentials of markets around new regulations—that determined the psychologists' opportunities. In the end, these opportunities also undermined psychologists, or at least caused them to go to war with one another, as the promise of rewards seduced some individuals into overselling their wares. Among other things, this story draws our attention to the rich connections between the history of regulation and what the editors of this volume have called the entanglement of science and capitalism: if regulation seeks to influence capitalism and remove its sharper edges, it is also true that markets and the interests of capital ultimately influence how regulatory domains are structured, including both the opportunities and risks for individual scientific disciplines, like psychology.

Comstock Capitalism:
The Law, the Lode, and the Science

*by Paul Lucier**

ABSTRACT

The term "Comstock capitalism" describes new commercial, legal, and scientific conditions emergent in the silver mining industry of the early 1860s in western Nevada. On the Comstock, the first joint-stock mining companies in the American West were incorporated, and stockholders, "the speculative interest," underwrote exceptional investments in large labor forces, powerful machines, cutting-edge engineering, and, most importantly, incessant litigation. In high-stakes court cases over mining rights, men of science, as expert witnesses, played central roles in the takeover of silver mining by big, well-financed companies. The development of Comstock capitalism essentially rested on the consolidation of scientific theory. Where once there had been numerous silver veins, geologists found a new object of nature—an immense single deposit called the Comstock Lode.

> Imagine a corps of German engineers in control at the time the Comstock lode was discovered. What would have been the result? . . . In the long run it is possible that the supposed system of the German engineers would have proved more economical. . . ; but it must be remembered that the mainstay of development, the money. . . , depended upon the speculative interest, which looked to brilliant and speedy discoveries of great bonanzas, and that this interest could hardly have supported the delay which would have resulted from a system more economical but less suited to the time and the locality.
>
> —Clarence King (1883)[1]

THE PECULIAR SYSTEM OF AMERICAN SILVER MINING

It takes "a gold mine to work a silver mine" forewarns a timeless Spanish adage. For unlike gold, silver is not stranded as pure, unalloyed nuggets in the soft sands and gravel beds of rivers where it can to be panned out and pocketed; rather, silver occurs admixed in mineral veins and embedded in the hard rocks of mountains. In a word, silver must be mined, and mines are very expensive. That truism would not be lost on

* 35 Bethany Road, Wakefield, RI 02879; paullucier2@gmail.com.

I appreciate the insightful comments of the participants and organizers of the workshop on "Science and Capitalism" as well as the helpful suggestions of two anonymous reviewers and of the *Osiris* general editors. As always, I am especially thankful to Andrea Rusnock. Funding for this research was provided by a fellowship from the National Endowment for the Humanities.

[1] Clarence King, "Introductory Remarks," in S. F. Emmons and G. F. Becker, *Statistics and Technology of the Precious Metals*, 10th U.S. Census (Washington, D.C., 1885), vii–xiv, on ix.

anyone—over the millennia—who has ever owned, invested, or worked a silver mine, but it is a real financial fact often forgotten when discussing the American West, where the romance of the self-reliant prospector—the mythical gold miner—can still be confounded with the hard rock labor of the silver miner.[2] American silver mines were costly industrial organizations, in terms of money, miners, and machines, and the first of these big capitalist enterprises appeared in western Nevada on the Comstock Lode during the early 1860s.

The Comstock Lode, famous in its own time and today, was the first major silver mining district in the United States, and for over twenty years, from its "discovery" in 1859 through the mid-1880s, it ranked as the largest and richest silver mining district in the world. The lode yielded upwards of $325 million of silver (over $8 billion today) extracted from tunnels, shafts, stopes, and galleries exceeding 250 miles, an enormous underground city where thousands of miners sweltered at depths over 3,000 feet in temperatures reaching an unimaginable 160°F at the height of its production.[3] In his now-classic *Mining Frontiers of the American West* (1963), Rodman Paul characterized the Comstock as the "advanced school" of American mining, not so much for its immense size and richness, but for its sophisticated business model. Capitalized in the millions of dollars and issued to investors in thousands of shares, Comstock mining companies emerged as the first western ventures to organize and manage themselves as joint-stock firms.[4] Of the innumerable stock companies, many, if not most, never turned a profit or even opened a mine; still, the going operations required substantial and continuous cash reserves to pay for the large labor forces (at premium wages of $4 per shift), powerful machinery (steam engines, hoists, and cables), and cutting-edge engineering (square-set timbering, pumps, and ventilators) needed to deal with the immense depth, searing water, and stifling air of the mines. "Technologically, economically, and sociologically," Paul concluded, "the Comstock Lode represented a big and abrupt stride beyond the farthest limits reached in California during the 1850s."[5]

Contemporaries would have agreed. Clarence King, probably the best-known geologist in Gilded Age America, regarded the Comstock as "incomparably the most valuable metal deposit known to modern times."[6] King's prestige and celebrity rested on a series of brilliant exploits in the West: mountaineering in California (1864–6), leading the Geological Exploration of the Fortieth Parallel (1867–72), and making a dramatic public exposé of the Great Diamond Hoax in 1872. In consequence, he

[2] The mythical prospector is persistent enough to appear on the cover of Elliott West's revised, expanded edition of Rodman Wilson Paul, *Mining Frontiers of the Far West, 1848–1880*, rev. ed. (1963; repr., Albuquerque, 2001).

[3] George F. Becker, *Geology of the Comstock Lode and Washoe District*, USGS Monograph vol. 3 (Washington, D.C., 1882). On heat and mining conditions, see John A. Church, *The Comstock Lode and Its Formation and History* (New York, 1879), 176–220.

[4] Western historians can often overstate the financial novelty of Comstock mining companies. On the East Coast, hundreds of large joint-stock companies were likewise organized to "mine" the newly discovered petroleum fields of Pennsylvania in the early 1860s. See Paul Lucier, *Scientists and Swindlers: Consulting on Coal and Oil in America, 1820–1890* (Baltimore, 2008).

[5] Paul, *Mining Frontiers* (cit. n. 2), 57. In his revised, expanded edition, Elliott West supplemented Paul's business history with the stories of diverse peoples displaced and disrupted by those businesses—Native Americans, immigrants, women, children, and miners. West represented the new western historians through his concern for and focus on social history. The economic and technical aspects of western mining, those parts of the history literally focusing on "working capital and scientific knowledge," West left untouched. That phrase comes from Martin Ridge's "Foreword" to the 2001 edition.

[6] Clarence King, *First Annual Report of the USGS* (Washington, D.C., 1880), 37–47, on 39. George F. Becker's first report on the Comstock appeared herein.

became, in 1879, the first director of the newly created United States Geological Survey (USGS), the permanent federal bureau charged with investigating geological structure and mineral resources and production.[7] King's expertise on American mining derived in no small measure from his experience on the Comstock, which he first visited in 1863 on his way west to California; he then returned to it in the winter of 1867–8 on his way back east with the geological corps of the Fortieth Parallel Exploration. King wrote a detailed geological report on the mining operations of the Comstock Lode (1870), and in 1880, as part of the tenth U.S. Census, he supervised a further reexamination of the mines and their productions.[8] King believed the Comstock exemplified "the best characteristics of American mining practice," especially its "engineering enterprise," an artful amalgam of technical ingenuity informed by comprehensive scientific studies like his own. Capital made that combination work, and King, as quoted in the epigraph, celebrated the speculative interest—"stockholders, clamorous for speedy profits," who sunk large sums of cash in mining shares so that Comstock companies could drive their tunnels faster and dig their shafts deeper in search of silver bonanzas. King, like many other geologists interested in advancing the industry, saw virtually no downsides (say, to the miners or to the environment) to free market mining. For him, shareholders' "desire for immediate returns" determined the "peculiar conditions" of American silver mining—its "originality," "boldness," and "speed."[9]

In stark contrast, silver mines found in "foreign countries," by which King meant the German states, operated under "government ownership and supervision." Those mines epitomized "stability and steadiness," King's euphemisms for "plodding" and "safe." King's harsh comments came in direct response to foreign engineers who criticized the peculiarities of American silver mining as crude, hasty, and wasteful. King wanted to laud the advantages of the speculative interest, without which, he concluded, mining became dull and bureaucratic, thus lacking the alleged enterprise but also the undoubted excitement and excesses of American practice.[10] Putting aside King's defensiveness, government-owned systems of mining, with all their concomitant consistency and regularity, have served as more reliable subjects for scholarly studies of the interconnections among science, technology, and mining, as the historians John A. Norris and Hjalmar Fors have shown in their work on early modern Germany and Sweden, respectively. For Pamela O. Long, the Erzgebirge mines on the border of Saxony and Bohemia represent ideal "trading zones," arenas in which artisans and learned men clustered and melded together skilled and scholarly cultures. Likewise, Ursula Klein has identified German mining with the commingling of art and science, particularly the sciences of chemistry and metallurgy.[11] Arguably, then, the administrative interest—gov-

[7] Mary C. Rabbitt, *Minerals, Lands, and Geology for the Common Defense and General Welfare*, vol. 2, *1879–1904* (Washington, D.C., 1980).

[8] Thurman Wilkins and Caroline Hinkley Lawson, *Clarence King: A Biography* (Albuquerque, 1988); James Gregory Moore, *King of the 40th Parallel* (Stanford, Calif., 2006); Robert Wilson, *The Explorer King* (New York, 2006); Martha Sandweiss, *Passing Strange: A Gilded Age Tale of Love and Deception across the Color Line* (New York, 2009).

[9] King, "Introductory Remarks" (cit. n. 1), viii.

[10] Ibid.

[11] John A. Norris, "*Aus Quecksilber und Schwefel Rein*: Johann Mathesius (1504–65) and *Sulfur-Mercurius* in the Silver Mines of Joachimstal"; Hjalmar Fors, "Elements in the Melting Pot: Merging Chemistry, Assaying, and Natural History, Ca. 1730–60"; Ursula Klein, "Chemical Expertise: Chemistry in the Royal Prussian Porcelain Manufactory," all in *Osiris* 29 (2014): 35–48, 230–44, 262–82; Pamela O. Long, "Trading Zones in Early Modern Europe," *Isis* 106 (2015): 840–7.

ernment officials looking for steady development—has proved to be a most accommodating adit for science to enter mining.

In the American system of silver mining, science found its way underground by a different route. The heady pursuit of profits, as King boasted, spurred joint-stock mining companies to bold engineering achievements, but capitalism also drove those Comstock companies toward another, somewhat less celebrated, although equally characteristic, condition of American mining—litigation. Indeed, it might be a truth universally acknowledged among Americans that wherever and whenever locators made mining claims, lawsuits inevitably followed. From this perspective, American mines can be treated as a different kind of trading zone, one typified by competition and contestation more often than by coordination and commingling. For historians of science and technology, studies of court cases typically involve intangible assets such as patents and intellectual property,[12] whereas legal contests over mining claims deal with real property like land, mines, and minerals. On the Comstock, rival joint-stock companies competed for the right to mine silver, and money paid by the speculative interest funded expensive and seemingly interminable legal actions. So, if the Comstock represented a big step, technologically, economically, and sociologically, beyond earlier mining districts, it also constituted a new condition legally.

Comstock capitalism, to coin a term, describes these new conditions. While it refers specifically to the American system of silver mining in the West during the early 1860s, historians of capitalism will recognize some of its features as characteristic of a broader narrative about the transition from market competition toward hierarchical corporate control.[13] What is significant and different about Comstock capitalism is the role of science in that transition. Unlike the foreign systems of government-owned mines, science became entangled in Comstock capitalism through private engagements of professional experts in high-stakes court cases over mining rights.[14] Litigation proved to be the preferred passageway for men of science to enter the mines and study the geology of silver.[15] In providing geological knowledge to help the largest joint-stock companies win their legal cases against smaller competitors, men of science accelerated, if not enabled, the corporate takeover of the Comstock. Moreover, corporate control needed the consolidation of scientific theory. In the climax to the Comstock cases, geologists combined the once numerous and distinct veins of silver into one big entity called the Comstock Lode.

To analyze the development of Comstock capitalism, this article focuses on two of the most sensational court cases: *Ophir Silver Mining Company v. Burning Moscow Mining Company* in 1863, and *Gould & Curry Silver Mining Company v. North Po-*

[12] Stathis Arapostathis and Graeme Gooday, *Patently Contestable: Electrical Technologies and Inventor Identities on Trial in Britain* (Cambridge, Mass., 2013); Tal Golan, *Laws of Men and Laws of Nature: The History of Scientific Expert Testimony in England and America* (Cambridge, Mass., 2007).

[13] Naomi R. Lamoreaux, Daniel M. G. Raff, and Peter Temin, "Beyond Markets and Hierarchies: Toward a New Synthesis of American Business History," *Amer. Hist. Rev.* 108 (2003): 404–33.

[14] Since the early nineteenth century, well-funded mining companies had routinely consulted chemists and geologists, who, for a fee, provided expert advice on the underground and largely unseen nature of minerals, such as coal, oil, copper, and precious metals. See Lucier, *Scientists and Swindlers* (cit. n. 4).

[15] A somewhat different interpretation of the relations between science and mining can be found in Karen Clay and Gavin Wright, "Gold Rush Legacy: American Minerals and the Knowledge Economy," in *The Evolution of Property Rights Related to Land and Natural Resources*, ed. Daniel Cole and Elinor Ostrom (Pittsburgh, 2012), 67–95.

tosi Gold and Silver Mining Company in 1864. Both lawsuits concerned the key geological question of the nature of the silver veins constituting the Comstock. Only the latter case, however, relied extensively on expert scientific witnesses, in large part because of the dramatic decision handed down in the former case. In addition to men of science, the *Gould & Curry v. North Potosi* case also relied on a court-appointed referee. While not unprecedented, the judgment of a referee reflected both the financial importance of the latter case and the resolution by big joint-stock companies that they could no longer afford the uncertainties inherent in a legal system designed, initially, by miners to protect miners against monopolies. The outcome of the latter case would not only undercut the miners and the smaller companies; it would also confirm the corporate control of the Comstock and the existence of the newly christened Comstock Lode. More broadly, Comstock capitalism would determine the content and enactment of the first federal law concerning mining rights in the American West.

THE "DISCOVERY" OF THE COMSTOCK

According to Dan De Quille, the author of a *History of the Big Bonanza* (1877), perhaps the most entertaining account of the Comstock, two Irish prospectors, Peter O'Riley and Patrick McLaughlin, discovered silver on the northeastern slope of Sun Peak (soon to be renamed Mount Davidson) in May 1859.[16] Like many prospectors since the early 1850s, O'Riley and McLaughlin had come over the Sierra Nevada from California in the early spring months, when there was enough rainfall and melting snow, in order to pan for gold in the gulches that ran down the mountains of the Virginia range of western Utah territory.[17] They were shoveling sand and gravel into their rocker—a long wooden box topped with a sieve and lined with rifles for sifting out gold—when a tall, thin Canadian named Henry T. P. Comstock accosted them. "Old Pancake" was known as a fast-talking fraud, yet Comstock managed to trick O'Riley and McLaughlin into giving him and his partner, Emanuel "Manny" Penrod, shares in the discovery. Together the four prospectors staked a claim and called it the Ophir.[18] Gold was what they were looking for, but some heavy blue-black dirt kept clogging their rocker. In June 1859, the prospectors sent a bag of the "blue stuff" back to California, where the mining expert Melville Atwood assayed it and found it to be rich in silver ore.[19] The rush to Washoe (as the mining district would be called) began the minute Atwood's secret report became public, which, according to De Quille, was that very evening.

The "discovery" of the Comstock is a colorful and compelling story, and a version of it has been told for virtually every mining district in the West. According to T. A. Rickard, the doyen of American mining engineering in the nineteenth century, "The fortuitous character of these events is nothing new in the history of mining; blind Fortune has often stubbed her toe against rich ore since the miner first invoked her aid in

[16] Dan De Quille was the pen name of William Wright, a journalist for the *Territorial Enterprise*, the Virginia City newspaper. Mark Twain also worked for the paper and wrote the "Introductory" to De Quille, *History of the Big Bonanza* (Hartford, Conn., 1877).

[17] Nevada Territory was organized by an act of Congress on 2 March 1861; prior to then, the region embracing the Comstock, known as Washoe, was the westernmost part of Utah Territory.

[18] Ophir was a common name for mines. In the Old Testament, Ophir was a land rich in gold; 1 Kings 9:28, 10:11, 22:48.

[19] Alamarin B. Paul, a developer of the Washoe process of pan amalgamation, recounted that the early assays showed the blue-black stuff to be "sulphuret of silver" (silver sulfide) worth $3,000 per ton with an additional $876 in gold. *Mining and Scientific Press* 45 (1882): 392.

his search for wealth."[20] But if the discovery was commonplace and the subsequent rush inevitable, the relation between the two—the cause and the effect—are analytically suspect and dangerously misleading. As Kent Curtis has shown, a gold rush is not caused by a serendipitous strike. Gold was discovered in countless places, innumerable times. Other factors were needed.[21] In the discovery of the Comstock, a critical factor was the assay by Melville Atwood, a well-known English chemist, metallurgist, and geologist working in Nevada City, California. Quite plausibly, Atwood could have claimed that he had found the abundant riches of the Comstock, as well as supplied a scientific basis for the subsequent rush to Washoe.[22]

The four original claimholders, like most prospectors, had neither the knowledge nor the money, nor much inclination, to develop their Ophir claim, so they sold out to a group of California investors led by George Hearst.[23] The investors formed a partnership, Ophir Mining Company, and started digging, thereby initiating mining operations on the Comstock. By November 1859, the partners had hired pack animals to carry ore over the Sierra Nevada and back to San Francisco, where company assayers estimated its worth at more than $6,000 in bullion per ton. On 28 April 1860, Hearst and his partners incorporated their mine according to California law as the Ophir Silver Mining Company, the first joint-stock company on the Comstock. They issued 16,800 shares (at a par value of $300 per share) amounting to a nominal capital of $5,040,000. The company elected a board of directors, appointed a superintendent of the mine, and hired a small workforce of miners.[24]

The example of the Ophir Silver Mining Company was quickly imitated, and by the end of 1860, another thirty-seven Comstock mines had been incorporated in California. According to Eliot Lord, the official USGS historian of the Comstock: "In quick succession the most promising claims in the Washoe district were owned and controlled by stock companies."[25] From the very start, then, Comstock mining was led by substantial corporations, not by intrepid miners. "It is hardly to be doubted that silver mines of great value have really been discovered," the *New York Times* reported in March 1860:

[20] T. A. Rickard, *A History of American Mining* (New York, 1932), 96.

[21] Kent Curtis identified three factors: government force (military control over an area), infrastructure (roads), and communications. Curtis, "Producing a Gold Rush: National Ambitions and the Northern Rocky Mountains, 1853–1863," *West. Hist. Quart.* 40 (2009): 275–97; Curtis, *Gambling on Ore: The Nature of Metal Mining in the United States, 1860–1910* (Boulder, Colo., 2013).

[22] The role of assayers in western mining is important but hitherto largely unexamined, and, of course, not all assayers were scientific or honest. "I am confident some of these learned gentlemen in the assay business," observed J. Ross Browne, "could have detected the precious metals in an Irish potato or a round of cheese for a reasonable consideration." See Browne, "A Peep at Washoe," *Harper's New Monthly Magazine* 22 (December 1860): 1–17; 22 (January 1861): 145–62, on 162; 22 (February 1861): 289–305.

[23] George Hearst, future U.S. senator and father of the newspaperman William Randolph Hearst, had rushed over the Sierra Nevada to Washoe in the summer of 1859 and purchased a one-sixth interest in the Ophir claim, the beginnings of his mining fortune. Grant H. Smith, *The History of the Comstock Lode, 1850–1997*, rev. ed. with new material by Joseph V. Tingley (Reno, Nev., 1998).

[24] Investments in Comstock companies were more often quoted in terms of dollars per foot (or fraction thereof), corresponding to the original length of the mining claim. The Ophir, e.g., issued 12 shares per foot (12 × 1,400 feet = 16,800 shares); Gould & Curry issued 4 shares per foot (4 × 1,200 feet = 4,800 shares). The *Mining and Scientific Press* printed weekly share and per foot prices along with recent assessments (per foot) and capital valuations for all the large Comstock companies.

[25] Eliot Lord, *Comstock Mining and Miners*, U.S. Geological Survey (Washington, D.C., 1883), 97. In the newly found petroleum regions of western Pennsylvania, hundreds of mining claims were also controlled by joint-stock companies; see Lucier, *Scientists and Swindlers* (cit. n. 4).

> It is more probable, however, that these mines will prove to be mines for the investor and not for the adventurer; for the capitalist who commands machinery, and not for the gleaner upon the surface. . . . It is to be hoped that this may be so; for mining [can then be] a steady organized industry taking its place among regular occupations of the people and prosecuted with all the aids of science.[26]

The kind of mining industry devoutly to be wished involved science and capital working together, presumably in opposition to unpredictable prospecting, and toward dependable development (as in Germany?). Yet the newspaper reports themselves (the *Times* entitled one "The Wonders of Washoe") only fed "a certain flush of speculation" and excited "speculative nerves."[27]

Incorporation also led to legal conflicts over mining claims. "Before the close of the year 1860," Lord explained, "work upon the principal claims had reached a point where collision was inevitable, and then the geological character of the district and the distribution of the ore deposits became a problem of absorbing interest."[28] The collision referred to litigation, and this singular feature of the American capitalist system of mining highlights the contrast with the German system, where an official bureaucratic administration obviated such costly legal contests. Competition between Comstock mining companies effectively drove claimants into court and compelled them to open up more of the underground in order to prove their claims. More importantly for historians of science, the legal competition between mining companies indexed and influenced competing scientific theories about the subterranean nature of Comstock silver.

In practice, the inevitability of the Comstock lawsuits had everything to do with the distinctive circumstances surrounding the establishment of American mining districts. In the early days of the California Gold Rush, miners had written and enforced their own rules or codes governing mining districts. At that time, the federal government had enacted no laws for the management of public mineral lands. In 1847, in fact, Congress had abandoned a forty-year experiment of leasing mineral lands because it had proved both unmanageable and unprofitable. The federal government and subsequently the territorial authorities in Utah and later Nevada followed a policy of "non-interference."[29] The early mining codes in California and Nevada were based on the principle of equity and on a determined effort among the miners to prevent the establishment of monopolies. Significantly, these efforts were transcribed into restrictions on the size and content of mining claims. Initially, mining codes applied specifically to placers, the sands and gravel beds along rivers and streams that were panned for gold. Placer mining claims were measured in only one dimension, the distance along the water, because determining length was simple compared to surveying an-

[26] "The Mines of Washoe," *New York Times*, 13 March 1860. Another *Times* editorial predicted that "California capital is inadequate to development of the silver mines; [but] Washoe offers a splendid field for Eastern capitalists." See "The Wonders of Washoe," *New York Times*, 19 May 1860.

[27] "The Mines of Washoe"; "The Wonders of Washoe" (both cit. n. 26). The rush to incorporate spread far beyond Washoe. Altogether more than a thousand mining companies were incorporated in California in 1860, an unprecedented number that set the stage for many speculative fevers in mining stocks. Maureen A. Jung, "Capitalism Comes to the Diggings: From Gold-Rush Adventure to Corporate Enterprise," *Calif. Hist.* 77 (1998/1999): 52–77.

[28] Lord, *Comstock Mining* (cit. n. 25), 97.

[29] Joseph Ellison, "The Mineral Land Question in California," *Southwest. Hist. Quart.* 30 (1926): 34–55; Gary D. Libecap, "Government Support of Private Claims to Public Minerals: Western Mineral Rights," *Bus. Hist. Rev.* 53 (1979): 364–85; Libecap, *The Evolution of Private Mineral Rights: Nevada's Comstock Lode* (New York, 1978).

gles. As a result, placer claims extended an undetermined distance to either side of a river or stream and to an unspecified depth.[30]

In the course of shoveling, prospectors also discovered gold in quartz veins embedded in hard rock. Quartz mining became another type of claim, but it too was measured in only one dimension—along the length of the vein, or, as it was sometimes called, the ledge. As with placer mining claims, there were no sidelines to a quartz-mining claim; the width was undetermined, as was the depth. A quartz-mining claim extended as far down as the gold-bearing vein went, which, theoretically, meant to the center of the earth. In this respect, miners followed geologists in describing gold-bearing quartz as fissure veins—narrow mineral deposits bounded by two rock walls. Geologists thought fissure veins originated deep within the earth; however, their source was not as important to the miners as their occurrence and the ability to identify them on the surface (so-called indications) and to track their continuity underground. By custom and later by code, a miner could follow a fissure vein along all "its dips, spurs, and angles" (the technical and legal description of the underground nature of a vein) unless the vein crossed the end lines of an adjacent quartz-mining claim. Allowing a claimant to follow a vein along its dips, spurs, and angles thus conformed to common practice and scientific theory.[31]

In the spring of 1859, the pioneer prospectors, including the Ophir discoverers, set up the Washoe mining district; they wrote the rules and requirements for locating claims, identifying and classifying them, fixing their size and place, recording them, and holding and working them.[32] In Washoe, the four original locators of the Ophir had initially staked out a placer-mining claim. When the assay showed rich silver ore, they changed their plans and filed a quartz-mining claim. Per the Washoe code, each partner received 300 feet along the vein, and, as discoverers of the Comstock, they received an extra claim, for a total equaling 1,500 feet. (The partners subsequently deeded 100 feet of their claim to the Mexican Gold and Silver Mining Company.) By the end of 1860, nearly twenty mining companies had laid claim to over 12,000 feet along the main Comstock vein in the Washoe district (fig. 1; table 1).

On the main vein, the Washoe miners' code was clear-cut; the lengths of the quartz-mining claims were fixed and officially recorded, and the mines continuously worked. In effect, these claims were uncontestable, and these Comstock companies tended not to sue one another. On land adjacent to these main companies and on ledges running near to the main Comstock vein, there was a great deal of legal and scientific confusion, hence controversy. The surface outcrop of the Comstock vein varied in width from 2 to 200 feet. Nearby, prospectors located other quartz veins (most narrower and containing less silver ore) separated from the main one by intervals of different rock.

[30] Rossiter W. Raymond called the miners' codes "the law of the lariat" for their rough justice and rudimentary measuring; see Raymond, "The Law of the Apex," *Eng. Mining J.*, 2 and 9 August 1884, 74–5, 89–91; and Raymond, "Comparison of Mining Conditions To-day with Those of 1872, in Their Relation to Federal Mineral-Lands Laws," *Bull. Amer. Inst. Mining Engineers* 88 (1914): 577–84, on 577.

[31] For an explanation of the consensual (American) science of vein formation, see James Dwight Dana, *Manual of Geology: Treating of the Principles of the Science with Special Reference to American Geological History* (Philadelphia, 1863), 711–6.

[32] The specificity of "discovery" and "development" found in American mining codes derived from earlier Spanish mining legal codes; see Charles Howard Shinn, *Land Laws of Mining Districts* (Baltimore, 1884); Andrea G. McDowell, "From Commons to Claims: Property Rights in the California Gold Rush," *Yale J. Law & Hum.* 14 (2002): 1–72. The Washoe mining district was organized at a miners meeting on 11 June 1859 at Gold Hill, on the southern portion of the Comstock Lode. The first sections of the code appear in De Quille, *Big Bonanza* (cit. n. 16), 40; the entire code was published in Lord, *Comstock Mining* (cit. n. 25), 42–4.

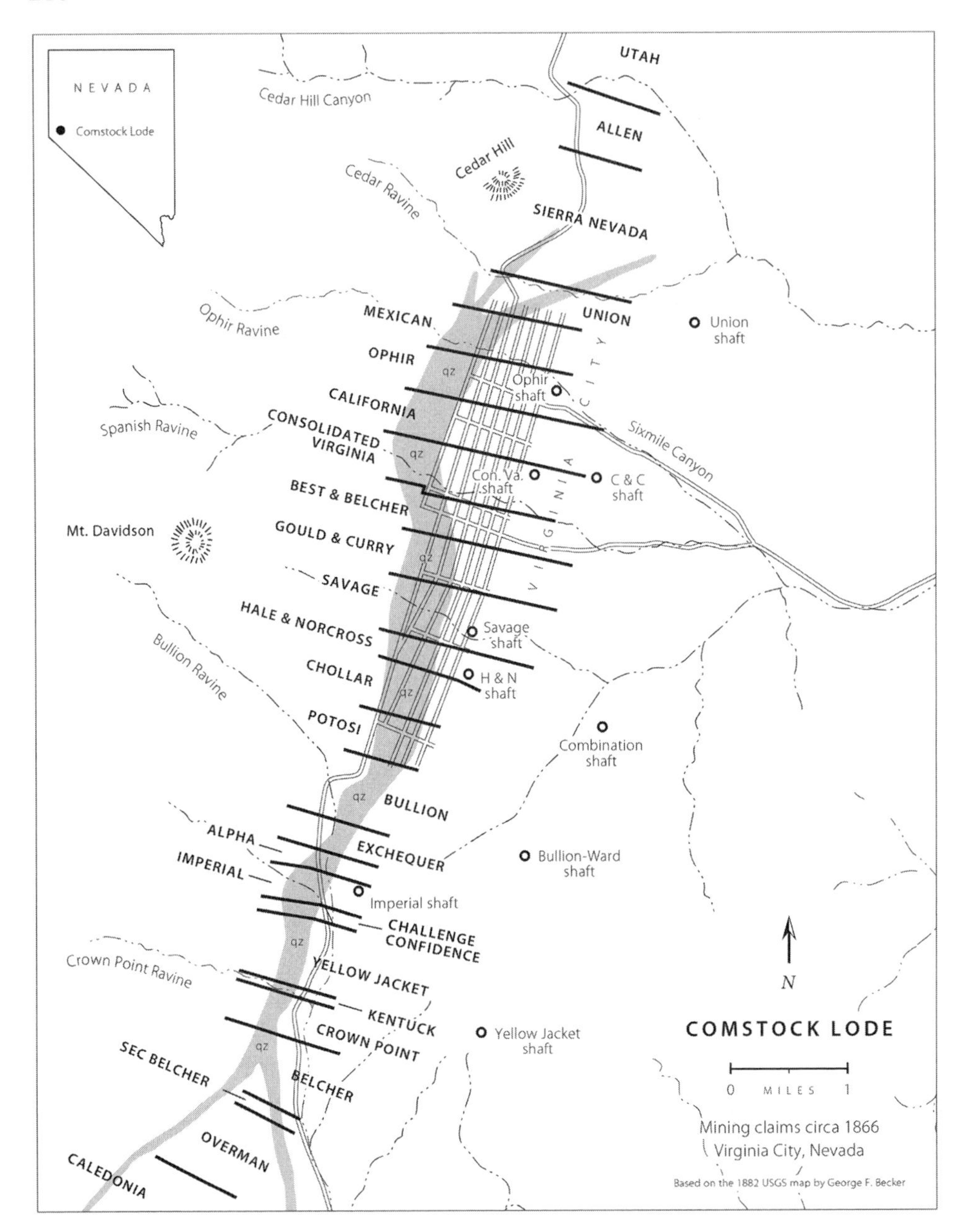

Figure 1. *Trace of the Comstock Lode along the surface based on Becker*, Geology of the Comstock Lode (*cit. n. 3*).

Other companies were subsequently organized on these nearby quartz-mining claims. Per the Washoe code, a locator could claim only 300 feet along one vein. The miners included these restrictions in order to prevent one big company from controlling the entire Comstock vein and Washoe district. Many ledges and limited lengths thus meant multiple companies on the main Comstock vein as well as around it.

Many ledges also corresponded to the theory of independent fissures. Midcentury geologists thought each fissure continued in depth to some unknown source, and, accordingly, miners regarded the Comstock as a mineralized zone comprising many

Table 1. *Mining Claims (N to S) on Northern Part of Comstock Lode*

Company	Length of claim (feet)
Utah	1,000
Allen	925
Sierra Nevada	1,959
Union	500
Mexican	100
Ophir	1,400
California	600
Consolidated Virginia	710
Best and Belcher	250
Gould & Curry	1,200
Savage	771
Hale and Norcross	400
Chollar-Potosi	1,434
Bullion	940

Note. The Comstock mining companies of 1866 were nearly identical to those originally established in 1859 and 1860, the principal exceptions being the California and Consolidated Virginia, which were formed later from the mergers of smaller claims. A full list, including Gold Hill district companies, appeared in James D. Hague's report on the Fortieth Parallel Survey (cit. n. 74).

ledges, not as a single lode. By the end of 1860, upwards of 17,000 quartz-mining claims were recorded in the 3-mile area around the main Comstock vein.

"The interminable litigation," as Eliot Lord described the situation, did not refer to the aboveground features of these quartz-mining claims (although plenty of cases did contest the locating, fixing, and recording of claims), but rather to the underground geology and the bedeviling question of dips, spurs, and angles (fig. 2). "The Comstock Ledge was in a mess of confusion," explained J. Ross Browne, a Treasury Department agent visiting Washoe in early 1860:

> The shareholders had the most enlarged views of its "dips, spurs, and angles;" but those who struck croppings above or below [the main vein] were equally liberal in their notions; so that, in fine, every body's spurs were running into every body else's angles. . . . It was a free fight all around, in which the dips, spurs, and angles might be represented thus—after the pattern of a bunch of snakes.[33]

THE *OPHIR V. BURNING MOSCOW* CASE

The Ophir Silver Mining Company owned 1,400 feet along the north-south trend of the main vein. The cropping or surface exposure of the Comstock vein in the Ophir

[33] Browne, "A Peep at Washoe" (cit. n. 22), 161. Many in Washoe complained of Browne's "burlesque map." "The results of an actual survey," Browne responded years later, "are precisely the same as those produced by a bundle of straw well inked and pressed upon a sheet of paper." J. Ross Browne, "Washoe Revisited," *Harper's New Monthly Magazine* 30 (May 1865): 681–96; 31 (June 1865): 1–12, on 8; 31 (July 1865): 151–61.

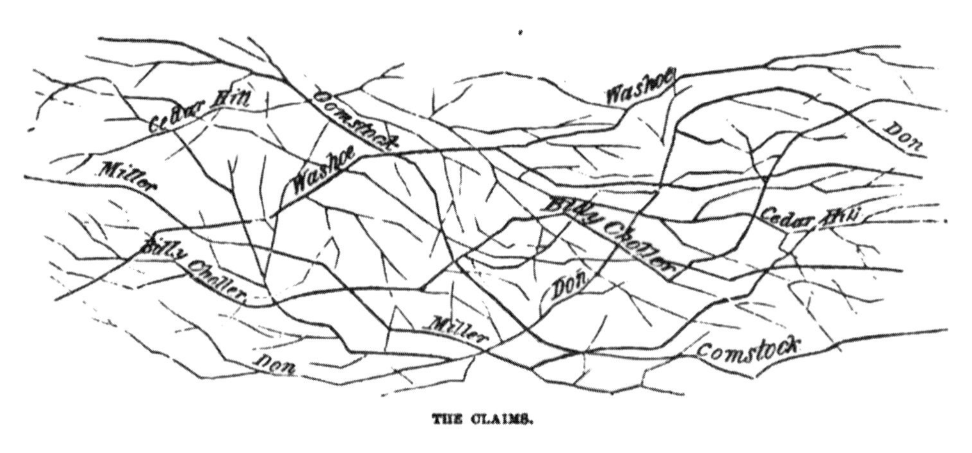

Figure 2. *The claims. Reprinted from Browne, "A Peep at Washoe" (cit. n. 22), 161.*

mining claim was about 2 feet wide. The quartz vein itself was soft and crumbling, and miners referred to it as sugar quartz. Sugar quartz did not require blasting with black powder; it could be taken out using a miner's pick and shovel. The Ophir mine thus began as an open pit. Miners referred to such pits as "gopher" holes, and Thomas Fitch, the Washoe County district attorney, described the early years of mining on the Comstock as a land of "Ophir holes, gopher holes, and loafer holes."[34] As the Ophir miners dug deeper, their vein began to widen. At 50 feet in depth, their hole began to flood, which required pumping, so the first steam-powered Cornish water pumps had to be hauled over the Sierra Nevada from San Francisco to be installed at the Ophir. By the end of 1860, the mine had reached a depth of approximately 180 feet, thereby necessitating the first steam-powered hoisting equipment, also from San Francisco. At that level, the Comstock vein had widened to nearly 50 feet, much bigger than any gold-bearing quartz vein found in California. It also began to dip at a 45° angle to the west, that is, toward Mount Davidson. To the west of the Ophir lay a group of smaller veins or ledges owned by a number of smaller mining companies. These nearby ledges were separated from one another by several feet of yellowish rock called porphyry. As the Ophir miners followed the dip of their main vein, it became apparent that they were headed toward the ledges of these other companies to the west, and one of these companies had a dazzling name, Burning Moscow Mining Company.

Burning Moscow claimed a ledge, 23 feet wide, lying to the west of and distinct from the main Comstock vein. In the summer of 1862, Burning Moscow began sinking a shaft and removing ore, which proved to be rich in silver. As the Burning Moscow mine grew deeper, the company's stock began to rise. The Ophir Company watched these developments closely, especially the fall in its own stock price, and in March 1863, Ophir filed suit in the First District Court in Virginia City against Burning Moscow to recover what it claimed to be its silver ore from a spur off its main vein.

The legal contest between Ophir and Burning Moscow became one of the most famous and costly cases precisely because it went to court. Before 1863, the bigger joint-stock companies located along the main Comstock vein had been able to buy

[34] Thomas Fitch, "Nevada," *Harper's New Monthly Magazine* 31 (August 1865): 317–23, on 321.

Table 2. *Leading Mining Companies Involved in Lawsuits Prior to 1867*

Company	Suits as plaintiff
Ophir	28
Yellow Jacket*	24
Savage	22
Gould & Curry	20
Overman*	18
Chollar-Potosi	14
Crown Point*	12
Bullion	11
Hale and Norcross	2

Note. The Yellow Jacket, Overman, and Crown Point companies were located on the southern part, or Gold Hill section, of the Comstock Lode. Lord, *Comstock Mining* (cit. n. 25), 177.

up or force a compromise upon the smaller companies through the threat of costly litigation. As table 2 shows, the largest Comstock companies filed the most lawsuits.

In the *Ophir v. Burning Moscow* case, it was the size of the rival companies that forced the issue. Each had a capital stock in the millions ($3 million for Burning Moscow), and a large number of shareholders, which meant even more money to spend on litigation. According to California law, a joint-stock firm could levy an assessment on its stockholders whenever and for whatever amount the directors thought necessary for the purposes of conducting business.[35] As litigation became a cost of doing business on the Comstock, assessments became routine features of joint-stock companies. In fact, all but six Comstock companies levied assessments on their stockholders in excess of the dividends they paid out, including the Ophir, which made thirty-five assessments totaling $2,689,400, while making only twenty-four dividend payments equaling $1,595,800.[36]

Both Ophir and Burning Moscow thus had the requisite "sinews of war." According to Lord, "wealthy Californians who had invested fortunes in these mining claims saw the imminent peril of their savings." Such desperation, Lord continued, "obliterated considerations of abstract morality."

> Some [stockholders] contented their consciences with taking no active part in the struggle except furnishing the sinews of war, and closed their eyes and ears to the unpleasant

[35] The San Francisco Stock and Exchange Board was organized on 11 September 1862. William W. Cook, *A Treatise on Stock and Stockholders and General Corporation Law*, 2nd ed. (Chicago, 1889).

[36] Lord calculated the dividends and assessments for 103 Comstock mining companies that were still in business as of 30 June 1880. See Table III in the appendix of Lord, *Comstock Mining* (cit. n. 25), 419–20. J. Ross Browne, U.S. Commissioner of Mines in 1866, made one of the first tabulations of dividends and assessments and calculated total dividends equal to $1,794,400 and assessments of $1,232,380. In that year, Ophir levied assessments of $55,500 and paid out $0 in dividends, while Gould & Curry paid out $252,000 in dividends and levied $0 in assessments. Browne, *A Report upon the Mineral Resources of the States and Territories West of the Rocky Mountains*, U.S. Department of the Treasury (Washington, D.C., 1867), 367–71.

> sights and sounds beyond the Sierras. Every well-meaning man washed his hands of the stain as far as this was in his power, and eagerly shook off the burden of responsibility upon less fastidious shoulders, for agents could always be found to carry out a plan.[37]

The plan for both companies was to use their cash to exploit the worst aspect of the legal practices in Virginia City.[38]

Virginia City, the boomtown that spread on top of and around the Comstock, proved to be a rich culture for lawsuits. "The whole [Washoe] district is racked with litigation," Browne observed.[39] According to one mining engineer, "Our principal industry is litigation."[40] Browne estimated "about every tenth man in Washoe is a lawyer,"[41] and another visitor thought the Comstock was "the most fruitful field for law yet developed in the United States."[42] "Two evils therefore beset Washoeites," Browne concluded, "many ledges and many lawyers."[43]

Like most early mining cases, *Ophir v. Burning Moscow* was brought on the side of the law, rather than in equity, meaning the litigants chose to try their case before a jury of miners instead of a judge or a court-appointed referee. The reason seemed obvious to everyone in Virginia City: it was far easier and more effective to corrupt juries than it was to persuade legal authorities by legal means. "Chicanery won more suits than eloquence and learning, and bribery and corruption more than solid merit," explained R. M. Clarke, the attorney general of Nevada. "The practice of law had to some extent degenerated into the practice of villainy."[44] In the practice of bribery, there were two principal targets—the witnesses and the jurors.

> Witnesses were manufactured by wholesale, and testimony to suit the requirements of a case was bought and sold with scarcely a pretense of secrecy. No facts were so clear and well established that they could not be controverted by a troop of hired liars, and the trials became conflicts, in which witnesses were pitted against each other on the ground of numbers rather than of competence or character, for a hundred asservations of ignorant, prejudiced, and corrupt men were relied upon to outweigh the careful reports of trained observers.[45]

Lord continued:

> The same corrupt influences were brought to bear more secretly . . . upon juries impaneled. In a district where everybody speculated in mining claims, it was practically impossible to obtain an unprejudiced jury, but the bias was intentionally increased in many instances by direct bribing.[46]

Eliot Lord may have protested too much. Other contemporary reporters, such as those from *The Mining and Scientific Press* and the *New York Times*, as well as later historians, regarded the Comstock cases as less corrupt largely because these observers

[37] Cited in Lord, *Comstock Mining* (cit. n. 25), 134.
[38] On corruption and swindling in western mining, see Dan Plazak, *A Hole in the Ground with a Liar at the Top: Fraud and Deceit in the Golden Age of American Mining* (Salt Lake City, 2006).
[39] Browne, "Washoe Revisited" (cit. n. 33), 156.
[40] Cited in Smith, *History of the Comstock Lode* (cit. n. 23), 66.
[41] Browne, "A Peep at Washoe" (cit. n. 22), 295.
[42] Merlin Stonehouse, *John Wesley North and the Reform Frontier* (Minneapolis, 1965), 169.
[43] Browne, "Washoe Revisited" (cit. n. 33), 156.
[44] R. M. Clarke was attorney general of Nevada from 1867 to 1870; cited in Lord, *Comstock Mining* (cit. n. 25), 136n.
[45] Ibid., 134.
[46] Ibid., 135.

focused their attention on the character of the other protagonists—the judges and the lawyers.

The richest lawyer in Virginia City was the fastidious William Morris Stewart. Born in New York, Stewart grew up in Ohio, where he taught mathematics as a young man before enrolling at Yale College in 1849. The following year Stewart left Yale and moved to California in search of gold. His career as a miner was brief; and in 1852, he took up the law in Nevada City, California. When he learned of the Comstock in early 1860, he set out for Virginia City. Stewart was smart and hardworking, and he was a master of mining law, particularly the rules and regulations of the Washoe district. Between 1861 and 1864, Stewart earned nearly $200,000 a year, over half a million in total, by representing the big Comstock companies, including Ophir in its suit against Burning Moscow.[47]

The presiding judge of the First District Court of Nevada Territory was John Wesley North. A New York lawyer and ardent Republican, North had been a delegate to the Republican National Convention in Chicago that had nominated Abraham Lincoln, and upon Lincoln's election, the president had appointed North to be the surveyor of the newly organized territory of Nevada. As the government official responsible for fixing and recording the lengths of mining claims, North's position required judicious skill. In August 1863, President Lincoln appointed him to the court upon the unanimous recommendation of the Virginia City Bar Association.[48] That fall, the convention to draft a state constitution for Nevada also elected him president. Both positions put North on a direct collision course with Stewart.

The political battles between North and Stewart make for a fascinating story, but one that cannot be detailed here.[49] Suffice it to say that Stewart regarded Judge North as stubborn and corruptible; however, most observers described North as conscientious and concerned for the miners and the small mining companies. North often went to the mines to see for himself the points in dispute, and, although he had no training in geology or engineering, he did have the experience of a surveyor, which he employed when rejecting Ophir's petition for an injunction against Burning Moscow in December 1863:

> It is difficult to see how these two bodies of quartz, separated at one point by 50 or 55 feet of porphyry, as appears both from the weight of evidence and from my personal examination, and at another point by 90 feet of the same material [i.e., porphyry], can be one and the same ledge. In view of the facts, at least, I cannot hold that they are proven to be one, and without the fact being proven the plaintiff [Ophir] falls far short of proving title

[47] The Belcher Mining Company, e.g., paid him stock equal to a 100-foot claim as a fee, which Stewart sold for $100,000. The Yellow Jacket Mining Company paid him $30,000 in fees. For Stewart's biography, see George Rothwell Brown, ed., *Reminiscences of Senator William M. Stewart of Nevada* (New York, 1908).

[48] After the Nevada Territory was established in 1861, President Lincoln appointed three justices to the Territorial Supreme Court: Chief Justice George Turner, and Associate Justices Horatio N. Jones and Gordon N. Mott. John W. North replaced Mott. North was at the center of most of the litigation and of Nevada politics, for President Lincoln counted on him to keep the territory loyal to the Union. North served as a trial judge in three districts, hence "riding the circuits," until, beleaguered by volatile politics and mining disputes, he resigned in August 1864. Stonehouse, *John Wesley North* (cit. n. 42).

[49] Abraham Lincoln favored a tax on mining (including the newly founded industry of petroleum mining) to raise revenues for the Union Army. Stewart opposed the tax, as did the big silver mining companies. North supported a federal tax. The issue of taxes is probably why the first state constitution, drafted by North, was rejected in the spring of 1864. Neither North nor Stewart nor the proposed tax were part of the second state constitution, which passed, and Nevada became a state on 31 October 1864, three days before the General Election. As a result, Abraham Lincoln got three more Electoral College votes from pro-Union Nevada.

> to the [ledge] on which defendant's [Burning Moscow's] works are situated. At the depth where this controversy arises, the evidence on both sides shows that there are several and distinct ledges.[50]

According to Lord, this judgment delivered "a staggering blow" to Stewart, the big Comstock companies, and the one-ledge theory.

In reaching this solid decision, North revealed another distinctive feature of western mining law and the Washoe code. Burning Moscow's mine was located on its own ledge, but that ledge was a mere 50 feet from the Ophir's main vein. As noted above, quartz-mining claims had no sidelines, but they also provided no ownership of the land; a claimant held only the right to mine the mineral vein. By law, the federal government owned the land; thus every prospector, claimant, and mining company was, technically, guilty of trespass. By legal precedent in California and in Washoe, the federal government allowed a claimant an easement to the land, that is, a prior right to use the land, which meant permission to enter the land, open and operate a mine, and erect mills and other buildings on the property. Rossiter W. Raymond, a leading expert on mining law, argued that such easements encouraged the rapid development of mining, without the time-consuming burden of leasing titles or purchasing properties.[51] Nonetheless, as a result, companies like Ophir did not have legal means to prevent the entry of another company, like Burning Moscow, or the locating of another claim, or even the digging of another mine very nearby. Big joint-stock companies thus had only two courses of action against aggressive neighbors: (1) threaten them with expensive lawsuits (hence the numerous cases in which big Comstock companies were the plaintiff) or (2) drive their own underground works as fast and as deep as possible to prove an identity—the one-ledge theory. In effect, corporate capitalism and its attendant geological science pursued the same objective—a single Comstock lode.

Judge North had reasoned that 50 feet of porphyry was sufficient proof to distinguish the main Comstock vein from Burning Moscow's ledge, but there was even more science to his decision. Geologists, including Josiah Dwight Whitney, the director of the California Geological Survey, who had visited Washoe in September 1861 and again in July 1862, had predicted that the main Comstock vein would follow an eastward dip down the slope and away from Mount Davidson, not a "false" westward dip toward the mountain. As early as September 1859, Melville Atwood, the assayer who first determined the richness of the Comstock silver, had published a similar opinion in a letter to the *Sacramento Union*:

> The course of the Ophir vein appears to be a few degrees west of magnetic north, dipping westward at an angle of about forty-eight degrees, but the dip being a false one, you may not sink far before it will change.[52]

In fact, by the time Judge North decided in favor of Burning Moscow, the Ophir mine had gone deep enough to discover that the main vein had changed directions. Oddly

[50] Cited in Lord, *Comstock Mining* (cit. n. 25), 144.

[51] Rossiter W. Raymond served as U.S. Commissioner of Mines (1868–75), after J. Ross Browne, and as editor of the *Engineering and Mining Journal* (1867–90). Raymond became one of the foremost scholars on mining law. For a comprehensive history of the American system, see Raymond, *Mineral Resources of the States and Territories West of the Rocky Mountains* (Washington, D.C., 1869), Part II: The Relations of Government to Mining, 175–223.

[52] Cited in Lord, *Comstock Mining* (cit. n. 25), 144.

enough, this change meant that Ophir's operations could continue, now in an easterly direction, without threat from Burning Moscow's diggings. But the other big Comstock companies soon found numerous new potential competitors to their east. For Gould & Curry Mining Company, the evident threat came from North Potosi Gold and Silver Mining Company. Accordingly, in May 1864, Gould & Curry brought a suit before Judge North against North Potosi.

THE *GOULD & CURRY V. NORTH POTOSI* CASE

The one-ledge theory was a geological interpretation of the nature of the main Comstock vein in depth. According to Lord, it was the scientific debate over "dips, spurs, and angles" of the main vein that made "the extraordinary litigation [on the Comstock] . . . one of the most curious and instructive chapters in the history of the mining industry in America."[53] Eliot Lord, it will be recalled, was the official USGS historian of the Comstock. Prior to joining the USGS, he had been a newspaper reporter for the *New York Herald* and the *New York World*; he then got a job with the U.S. Census Office preparing the 1880 Census, where he caught the attention of Clarence King, who was supervising the volume on *Statistics and Technology of the Precious Metals*. At the time (1879), King had just been appointed director of the USGS, and he hired Lord to write an official history—*Comstock Mining and Mines* (1883)—the only time the USGS has commissioned a nonscientific report on a mining district. King took Lord to visit the Comstock, but Lord wrote most of his book in San Francisco, the home of wealthy bankers and stockholders. Perhaps not unexpectedly, in his USGS monograph, Lord favored the large mining companies, William Stewart, and the geologists who consulted for the Comstock capitalists. Equally predictably, modern scholars of the Comstock have been critical of Lord's biases. Grant Smith, perhaps the most influential historian, thought Lord "overdoes it in devoting 60 of 414 pages of his valuable book to [mining litigation]."[54] Yet an unintended consequence of such dismissals of Lord's work has been a downplaying of the science involved in the Comstock cases.[55]

The *Gould & Curry v. North Potosi* case, like *Ophir v. Burning Moscow*, had its origins in geology. By the end of 1863, the big companies were working at depths between 500 and 700 feet, levels at which the main vein did indeed dip eastward, as predicted, down the slope of Mount Davidson and directly underneath Virginia City. Initially, Gould & Curry had not paid much attention to the mines on the city's eastern flats, but now it seemed their vein was headed toward North Potosi's ledge.[56] In their suit, Gould & Curry claimed "twelve hundred [1,200] feet, north and south, on what is known as the Comstock vein . . . with all its dips, spurs and angles." North Potosi, a company half the size of Gould & Curry, responded that their ledge was "seventy-five [75] feet to the east of, and a separate and distinct ledge from, that

[53] Ibid., 129.

[54] Smith, *History of the Comstock Lode* (cit. n. 23), 70.

[55] Stonehouse likewise dismissed Lord's "alleged history of the Comstock." See Stonehouse, *John Wesley North* (cit. n. 42), 67. Charles Howard Shinn, however, very much liked Eliot Lord's graphic book and included a chapter on mining litigation; see Shinn, *The Story of the Mine, as Illustrated by the Great Comstock Lode of Nevada* (New York, 1897).

[56] The Gould & Curry mine was a combination of two claims originally located by Alva Gould and "Old Abe" Curry. San Francisco capitalists, including George Hearst, organized the joint-stock company on 25 June 1860.

set forth and described in Plaintiff's complaint as the Comstock vein."[57] According to Judge North, the case was straightforward: "Is, then, the ledge claimed by Defendant [North Potosi], a portion of the vein indicated by the Gould & Curry croppings, or is it a distinct and separate vein?"[58]

That question was no different from so many other Comstock cases. What made *Gould & Curry v. North Potosi* significant was the manner in which it was answered. Unlike earlier Comstock cases, *Gould & Curry v. North Potosi* was to be decided in equity—before a court-appointed referee. While not unprecedented, the reliance on a referee reflected a dramatic change in the financial climate. In the spring of 1864, Comstock share prices crashed. "Startling," the *Mining and Scientific Press* declared.

> We may well use the expression . . . when we consider the manner in which Gould & Curry, and in fact almost all the leading stocks "tumbled." . . . Stocks were low enough before this crash, but when we witness such a line of stock as Gould & Curry "knocked" . . . it will, as a matter of course, bring all the other stocks down with it, and what little confidence is left in operators will soon be entirely lost, and the market will consequently be dead for some time.[59]

The *New York Times* was even more alarmist. "Mining stocks have gone completely out of sight, deeper than the depths of the deepest shafts and tunnels of Washoe."[60] The cause of the fall was unknown, at least to the *Mining and Scientific Press*, but the consequences were clear; continuous litigation was too costly.[61]

Gould & Curry and North Potosi thus needed a timely and trustworthy resolution, so they turned to a referee, "one of those honest gentlemen in whom everybody has confidence," Browne explained. Judge North appointed John Nugent, a highly regarded San Francisco lawyer. "As a referee," Browne continued, "he is bound to decide according to law and evidence."[62] Both sides believed the best evidence was scientific.[63] According to Nugent,

> For the solution of this question both parties deemed it necessary to call into the investigation men of science and those of practical experience by means of whose combined knowledge and observation, as well as by reference to scientific authorities, light should be caused to dawn upon a subject hitherto surrounded with considerable darkness and doubt.[64]

For North Potosi, John Allen Veatch, a physician and naturalist, who had moved to California in the mid-1850s, served as the principal expert witness. Between 1858 and 1861, Veatch had been the curator of conchology at the California Academy of Sciences, but since 1862, he had lived in Virginia City, where he practiced medicine and

[57] [John Nugent], *Opinion of Referee, August 22, 1864. Gould & Curry Silver Mining Co. vs. North Potosi Gold and Silver Co.*, District Court, First Judicial District, Storey County, Nevada Territory (Virginia City, Nevada Territory, 1864), 3.

[58] Ibid., 5.

[59] *Mining and Scientific Press*, 14 May 1864, 388.

[60] "California Gossip: The Panic in Mining Stocks," *New York Times*, 3 July 1864.

[61] At the same time, North Potosi was contending an equally costly case, in law, against another big Comstock company, Savage. While that case also involved an underground investigation by "scientific gentlemen," it resulted in a hung jury, perhaps another inducement to North Potosi to use a referee in the Gould & Curry case. *Mining and Scientific Press*, 23 April 1864, 265.

[62] Browne, "Washoe Revisited" (cit. n. 33), 159.

[63] On expert witnesses in mining cases, see Clark C. Spence, *Mining Engineers and the American West: The Lace-Boot Brigade, 1849–1933* (New Haven, Conn., 1970), chap. 6.

[64] Nugent, *Opinion of Referee* (cit. n. 57), 5.

geology. North Potosi also called upon men of practical knowledge, superintendents and foremen of several Comstock mines in competition with Gould & Curry, who testified to the localized habits of ledges and the surrounding rock that separated them.[65]

For its expert witnesses, Gould & Curry turned to three distinguished men of science: William Ashburner, William Phipps Blake, and Benjamin Silliman Jr. Ashburner was a consulting geologist working in California and Nevada. He had graduated from the Lawrence Scientific School at Harvard and then attended the École des Mines in Paris. In 1860, he had been hired as an assistant to the California Geological Survey under Whitney, but by 1862, after only eighteen months, Ashburner had left the survey to pursue a career in consulting. William Blake was also a well-respected consulting geologist. He had studied at the Yale Scientific School under Silliman and graduated in 1852. The following year Blake had joined the Pacific Railroad survey under Lt. R. W. Williamson and traveled to California. In 1863 he was appointed the first professor of mineralogy, geology, and mining at the newly founded College of California (in Berkeley). Benjamin Silliman Jr. was professor of general and applied chemistry in the Yale Scientific School and a highly regarded consulting chemist, who had headed to California in early spring 1864.[66] His trip had been arranged and paid for by a group of California bankers and stock speculators headed by William Ralston, who would found the Bank of California in July 1864 as part of a grand financial scheme to take control of the Comstock.[67] Silliman arrived in San Francisco in April but wasted no time in the city before taking a steamboat to Sacramento and then a stagecoach over the Sierra Nevada and on to Virginia City. The details of Silliman's Comstock consulting—the mines he visited, the people he met, and the companies he consulted for—are unknown. But in a letter to President Abraham Lincoln, Judge North praised the work Silliman did, and Nugent relied heavily on Silliman's geological expertise.[68] In short, Silliman was the star of the case.

Nugent began his referee report with a summary of the scientific points on which all or most of the expert witnesses agreed.

First, the Comstock vein was a true fissure vein. A fissure vein meant that it was filled with a mineral substance (in this case, sugar quartz), which had its origin in some deep-seated source, and it was presumed the fissure vein extended indefinitely downward.

Second, as a true fissure vein, it "must have two walls."[69] The Comstock vein was usually found between two very different rocks. To the west of the main vein, mean-

[65] C. C. Thomas, later superintendent of the famous Sutro Tunnel, was a practical expert for North Potosi. Gould & Curry called upon its mine superintendent, Charles Bonner, to testify.

[66] The story of Silliman's western adventure has focused on his engagements in southern California, around Los Angeles and Santa Barbara, where he consulted for various petroleum companies. Silliman's oil engagements would eventually lead to the heated showdown with Josiah Dwight Whitney in the National Academy of Sciences over the proper role of science and men of science in capitalist enterprises. See Lucier, *Scientists and Swindlers* (cit. n. 4).

[67] The Bank of California was a joint-stock corporation capitalized at $5 million (50,000 shares at $100 per share). The bank opened a branch in Gold Hill, near Virginia City, in September 1864, and William Sharon became the Nevada agent.

[68] Silliman accepted at least one other Comstock commission. See Benjamin Silliman Jr., *Report on the Empire Mill and Mining Co. of Gold Hill, on the Comstock Lode, in Nevada* (San Francisco, 1864).

[69] The best explanation and diagrams of headwalls, footwalls, and hanging walls is Otis E. Young Jr., *Western Mining* (Norman, Okla., 1970). Strangely enough, Lord provided a paltry abstract of this "far-reaching" case because, he assumed, Nugent's referee report was "buried from sight in the vaults of a San Francisco bank." Lord, *Comstock Mining* (cit. n. 25), 167.

ing up the slope of Mount Davidson, lay syenite, a hard rock often mixed with hornblende. To the east, meaning down the slope toward Virginia City, lay porphyry, a soft, friable rock often mixed with feldspar. In addition to the sugar quartz, the syenite, and the porphyry, there was a fourth rock, a very curious deposit often found in the mines. Between the syenite and the sugar quartz and between the porphyry and the sugar quartz lay sheets of dark bluish clay. These clay sheets could be several feet thick in places.[70]

Third, the syenite marked the western or footwall of the Comstock vein. The quartz vein never crossed into the syenite.

The question on which the expert witnesses disagreed, was where was the eastern or hanging wall of the Comstock vein?

For the defendants, North Potosi, there were two possible answers. The first, and weaker, of the two arguments was that a sheet of thick clay found between the main Comstock vein and North Potosi's ledge represented the eastern wall. Blake, under cross-examination, conceded that the clay sheet may have marked the eastern wall, but Ashburner and Silliman dismissed this theory and argued instead that the clay had formed after the quartz vein, and therefore it could not be the wall that contained the quartz vein. Nugent, in his referee report, agreed with Ashburner and Silliman and decided that the clay deposit did not represent the eastern wall.[71] The stronger argument made by the North Potosi experts was that the 75 feet of porphyry marked the eastern wall of the Comstock vein and clearly separated the main vein from North Potosi's ledge. By this reasoning, the geological situation was nearly identical to that in the *Ophir v. Burning Moscow* case.

Silliman, however, disproved this argument by taking the referee, the other expert witnesses, and, in effect, the entire case underground. While scientific examinations in mines were not uncommon, they could be dangerous. J. Ross Browne vividly recounted his harrowing experience of climbing down steep ladders and creeping through dark holes into narrow dismal passages of mud and wet walls, where "ore came tumbling down more or less all the time." "Miners, like sailors," Browne concluded, "grow indifferent to danger."[72] Silliman's underground examination, while apparently safe, would nonetheless have been difficult, not least because of the darkness. By April 1864, the Gould & Curry mine was the deepest on the Comstock. Nevertheless, peering through gloom at the 500-foot level, Silliman decided that crosscuts and drifts from the main Gould & Curry shaft eastward toward the North Potosi mine showed that the intermittent porphyry was not 75 feet thick, but rather only 6–7 feet.[73] In court, Silliman argued that the Gould & Curry vein and the North Potosi ledge were dipping toward one another. They were not two separate and distinct fissures, but rather one large V-shaped vein, whose two branches were coming closer and closer in depth. Silliman predicted the two branches would eventually merge, and in 1868, after the geological corps of the Fortieth Parallel Survey reexam-

[70] Nugent, *Opinion of Referee* (cit. n. 57), 5.

[71] The expert witnesses proposed three theories to explain the deposition of the thick sheets of clay: (1) from the surface by springs or floods, (2) from below by steam and vapor, and (3) from mechanical motion and attrition of the walls. The third theory was favored by Silliman and subsequently by most geologists, including Ferdinand von Richthofen, Clarence King, and George F. Becker.

[72] Browne, "Washoe Revisited" (cit. n. 33), 152–3.

[73] Nugent, *Opinion of Referee* (cit. n. 57), 8.

ined the Gould & Curry mine, Clarence King confirmed that the merger had indeed occurred between the 600-foot and 800-foot levels.[74]

What apparently separated the Gould & Curry vein from the North Potosi ledge was a triangular block of porphyry, and Silliman identified this porphyry wedge as a "horse." A common feature in many mines, a horse was a barren spot of nonpaying rock, but Silliman had never found one so large. "The 'Horse' in the mine," the *New York Times* reported, "seems to prove that there is no more ore to be had," and, in fact, its presence had helped to pull down the price of Gould & Curry stock and to precipitate the panic. "Professor Silliman has examined into the matter," the *Times* continued, "and says the mine is good, the ore abundant."[75] Nugent was likewise reassured by Silliman's explanation of how the Comstock vein had originally come from a deep source, filling the chasm and branching around the porphyry horse. "The ledges then are not separated," Nugent decided, "by the masses of country rock [porphyry] relied upon by the Defendant [North Potosi] as a division."[76]

In August 1864, Nugent made his decision public: there was only one vein, and all its dips, spurs, and angles belonged to Gould & Curry. Judge North granted an injunction preventing North Potosi from mining its ledge. Gould & Curry had won the case.

THE COMSTOCK LODE

> From October, 1862, until March, 1864, speculation ran riot, and the Territory of Nevada was converted into one vast swindling stock exchange. . . . To sell out, to speculate, to gamble was the object of all. What wonder that when the bubble burst, as it did in the spring of 1864, the distrust was as wide-spread as the disaster brought. . . . Silver mining, like every other business, requires to be managed as a business not as a speculation. . . . The character of the mining in Nevada, requiring a large outlay of capital before return, has produced the present system of incorporated companies.
>
> —J. Ross Browne (1865)

The conclusion of the *Gould & Curry v. North Potosi* case marked the culmination of the interminable litigation. It also marked the consolidation of a new system of mining—Comstock capitalism—along with a new interpretation of the Comstock vein and, in a relatively short time, a new foundation for U.S. mining law.

"The report of the Referee, Nugent, is worthy of perusal of all who are interested in claims upon the Comstock," the *Mining and Scientific Press* advised. "It is a learned and able argument in support of the one ledge theory."[77] "If the one-ledge theory is recognized," another Virginia City observer warned, "mines and mining in this region would receive a check from which the country could never recover."[78] Many ledges had meant many companies; one ledge dealt a fatal blow to all the small companies located near the main vein. Whereas before 1864, the lack of surface sidelines

[74] Clarence King's geological report on the Comstock Lode is found in James D. Hague's *Mining Industry* (Washington, D.C., 1870), Volume III of the Fortieth Parallel Survey Reports, 11–96, on 45.
[75] "California Gossip" (cit. n. 60).
[76] Nugent, *Opinion of Referee* (cit. n. 57), 10.
[77] *Mining and Scientific Press*, 27 August 1864, 132.
[78] "More of the 'One Ledge' Theory," *Mining and Scientific Press*, 25 April 1864.

to quartz claims had allowed, and almost induced, rival companies to locate ledges and begin mining, now those missing property boundaries encouraged the big companies to grab the little mines in the underground by their dips, spurs, and angles. Litigation, obviously, reflected competition; the one-ledge theory indicated consolidation. The system of large incorporated companies that Browne identified continues to the present. Comstock capitalism became American mining.

Still, as with any regular business requiring a large outlay of capital, mining relied on expertise; and on the Comstock, much of that came with the one-ledge theory. Essentially, scientific consensus led to corporate consolidation. Silliman drove home this point, again, in the fall of 1864 in a consulting report for another big joint-stock Comstock company, wherein he introduced the term "Lode"—with a capital "L"—to designate the main vein. The Comstock Lode encompassed all the dips, spurs, and angles; the Lode defined the profitable mineralized zone.[79] Although not everyone immediately embraced Silliman's emphatic capital "L," other men of science, capitalists, newspaper editors, and the general public soon referred to the silver mines in the Washoe district as the Comstock *lode*.

Among the early adopters of the big Lode was the German geologist Ferdinand von Richtofen, who in 1866 produced the first investigation of the Washoe district: *The Comstock Lode: Its Character and the Probable Mode of Its Continuance in Depth*.[80] Richtofen had graduated from the University of Berlin in 1856 and arrived in San Francisco in 1862 with a little English and a lot of curiosity about gold and silver.[81] He befriended Whitney, who helped to polish his English, and he undoubtedly met with Silliman, who probably introduced him to Ralston of the Bank of California. Ralston and a group of capitalists commissioned Richtofen to study the proposed route of a 4-mile-long tunnel to be driven 1,600 feet under Mount Davidson in order to provide drainage and ventilation to the big Comstock mines. The Sutro Tunnel was a bold engineering project, precisely the kind that Clarence King would later praise as characteristic of American mining.[82] In scientific terms, *The Comstock Lode* proved equally influential. Richtofen identified and classified Washoe's volcanic rocks and tried to find all the horses that separated the main vein into so many ledges. He also plumbed, like geologists before him, the "indefinite continuity" of the fissure vein in depth and, consequently, reckoned on the "constantly increasing" capital that would be required to mine it. Richtofen's report would become the guide to further mining operations, and, just as important, the scientific outline for future U.S. government investigations, including King's Fortieth Parallel Survey, of the Comstock Lode.[83]

If the Comstock Lode exemplified the raveling of science and capital, the legal strands still needed tying together. In 1865 the dean of Comstock legal counselors,

[79] Silliman, *Report on the Empire Mill and Mining Co.* (cit. n. 68).

[80] Ferdinand Baron Richthofen, *The Comstock Lode: Its Character, and the Probable Mode of Its Continuance in Depth* (San Francisco, 1866).

[81] After six years, Richtofen left California for China; see Shellen Wu, "The Search for Coal in the Age of Empires: Ferdinand von Richtofen's Odyssey in China, 1860–1920," *Amer. Hist. Rev.* 119 (2014): 339–63.

[82] Adolph Sutro raised $3 million, but he did not complete his tunnel until 1878, by which time the big Comstock mines had gone deeper.

[83] Richtofen later published his classification in Richtofen, *The Natural System of Volcanic Rocks* (San Francisco, 1868). Clarence King described the Comstock "Lode" as the entire mineralized zone, 200–800 feet in width; see King's report in Hague, *Mining Industry* (cit. n. 74), 11–96, on 38.

William Stewart, arrived in Washington as the first Senator from the new state of Nevada. He immediately acted to establish the Committee on Mines and Mining and set about drafting a federal law on mining that incorporated all his experience and expertise of the Comstock.[84] In March 1866, Congress passed the first federal mining law. The Mining Act of 1866 was a significant piece of legislation and a major accomplishment for Stewart. It codified the rules that would govern western mining, many of which remain in place today, including, most importantly, the free and open exploration and occupation of federal lands—without fear of trespass. Most of its sections copied the old miners' codes. Thus, for instance, it specified that no claim could exceed "two hundred feet along the vein," but "the discoverer of the lode" could claim an additional two hundred feet. And, perhaps not surprisingly, the new federal law confirmed the claimant's right to follow "a vein or lode with its dips, angles, and variations, to any depth" and into "the land adjoining." The federal law, like the Washoe code, did not specify the lateral extent of a mining claim. Claimants had a right to mine and an easement to the land, just as the big Comstock mining companies had exercised. But if the Mining Act of 1866 was supposed to sanction the old mining codes into a new federal law, it did so by betraying the miners themselves. Despite Senator Stewart's rather disingenuous denials, the details of the federal law worked to assist the big mining companies at the expense of the lone prospectors and small miners. To retain a claim, for instance, the claimant had to expend $1,000 a year in labor or improvements, an amount beyond the means of all but the well-capitalized companies. Furthermore, there was no restriction on the number of claims a corporation could hold.[85] As Rossiter Raymond, the foremost expert on American mining law, later summarized the situation, the Act "was simply intended to legalize existing mining conditions, especially at Virginia City, Nevada, which was in 1866 the most productive locality of "quartz-mining" in the West." The conditions in Virginia City were decidedly capitalist—a system of large incorporated mining companies—and as Raymond reminded his readers, Virginia City "is situated on and along the outcrop of what came to be known afterward as the great Comstock lode."[86]

[84] The House, likewise, created a Committee on Mines and Mining in December 1865, and both Senate and House committees became defunct in 1946.

[85] Because of Senate Committee rules, the actual bill had a nonobvious title: "An Act Granting the Right of Way to Ditch and Canal Owners over the Public Lands, and for other Purposes." For a complete history of the political maneuvering, see Gregory Yale, *Legal Titles to Mining Claims and Water Rights, in California, under the Mining Laws of Congress, of July, 1866* (San Francisco, 1867); Carl J. Mayer, "The 1872 Mining Law: Historical Origins of the Discovery Rule," *Univ. Chicago Law Rev.* 53 (1986): 624–53.

[86] Raymond, "Comparison of Mining Conditions" (cit. n. 30).

Organizing the Marketplace

*by Lukas Rieppel**

ABSTRACT

This essay engages a classic debate about the way nineteenth-century biology was informed by contemporaneous developments in political economy, and vice versa. However, rather than argue for a convergence between classical liberalism and Darwinian evolution, this essay traces the way that the concept of organization moved between both fields of discourse. Compared to the theory of evolution by natural selection, the logic of organization was far more teleological and, often, authoritarian. Instead of asserting that competition for access to scarce resources among autonomous agents yields adaptive outcomes in the population at large, it held that organic entities inexorably tend to develop from a state of simplicity to one of complexity. Moreover, it stressed the production of hierarchical structures wherein the whole was privileged over its constitutive parts, in biology as well as society. It is my thesis that the logic of organization proved especially attractive in debates about the transition from free-market to corporate capitalism, providing a powerful means to describe, discuss, and dispute the centralization, rationalization, and bureaucratization that contemporary observers often took to be characteristic of a distinctly modern political economy.

Around the turn of the twentieth century, the United States underwent a period of rapid business consolidation during which a large number of private enterprises were absorbed into a few corporate firms. At the time, this transformation was often described as a transition from free markets ruled by fierce competition to a more orderly system of rational planning, centralized oversight, and bureaucratic control. To many observers, a similar trajectory could be seen nearly everywhere one chose to look. Much like the solar system had formed out of an inchoate, nebulous mass over astronomical time, free-living cells adopted a communal life history to reap the benefits of multicellularity over biological time. So too, human societies seemed to exhibit an inexorable tendency toward increasingly organized complexity. Why should the political economy be any different? Thus, when J. P. Morgan's right-hand man George W. Perkins reflected on his own efforts to merge the Tennessee Coal, Iron, and Railroad Company with U.S. Steel after the Panic of 1907, he mused that "all that man has done in society [will never be] so complete a form of organization, so vast a trust, so centralized a form

* Department of History, Brown University, 79 Brown Street, Providence, RI 02912; lukas_rieppel@brown.edu.

In writing this essay, I received invaluable advice, provocation, and criticism from numerous friends and colleagues. In addition to the anonymous reviewers, my coeditors William Deringer and Eugenia Lean, and all the other contributors to this volume, I would especially like to thank members of Princeton's Modern America Workshop, particularly Angela Craeger, Andrew Edwards, Erika Milam, and Daniel Rodgers. Lynn Nyhart and Andrew Reynolds also offered invaluable advice at an earlier stage in the essay's development.

of control . . . as . . . that all-including system of perfect organization called the Universe."[1] As this and countless other examples demonstrate, the corporate reconstruction of the U.S. economy was often described, debated, and discussed as part of a much larger tendency for things to become increasingly organized over time.

This essay draws on the concept of organization to engage a classic debate at the intersection of science and capitalism: in what ways and to what extent has the history of biology been informed by contemporaneous developments in political economy, as well as vice versa? But rather than rehearse well-worn claims about the remarkable convergence between classical liberalism and Darwinian evolution, I hope to offer a new way to frame the debate. In particular, I want to resist the temptation to assume that both fields of discourse converged on a shared set of explanatory models wherein the self-interested behavior of more or less autonomous agents competing for access to scarce resources yields adaptive outcomes in the population at large, as if by the workings of some invisible hand. Instead, I want to ask what happens if we trace the history of another, very different and in some ways even conflicting concept—namely, organization—across both intellectual endeavors. Rather than tracking the way competition moved between biology and political economy during the long nineteenth century, I will attend to the concept of organization instead.

One thing that made the concept of organization so powerful was its capaciousness and its malleability. This allowed it to be interpreted in a number of different and at times even conflicting ways. But that did not mean it was fundamentally incoherent. Rather, I want to show that a particular logic of organization took shape over the course of the long nineteenth century. This logic was deeply teleological, and it held that organic entities inexorably tend to develop from a state of simplicity to one of complexity. Moreover, that developmental process was usually understood to eventuate in the production of hierarchical structures wherein the whole was privileged over its constitutive parts. Rather than emphasizing the notion of a population composed of autonomous individuals competing for access to scarce resources, the logic of organization therefore primarily stressed the importance of functionally integrated part/whole relationships.

Previous historians have not failed to notice the deep and pervasive conceptual overlaps between evolutionary biology and political economy. Indeed, attempts to trace the connection between the two have been a mainstay in the history of science.[2] Nor has the claim that evolutionary theory was primarily leveraged to naturalize free markets by celebrating the power of unfettered competition gone unchallenged.[3] In-

[1] George W. Perkins, "The Modern Corporation," in *The Currency Problem and the Present Financial Situation*, ed. Edwin R. A. Seligman (New York, 1908), 153–70, on 155.

[2] See, e.g., Adrian J. Desmond and James R. Moore, *Darwin* (New York, 1991); Silvan S. Schweber, "Darwin and the Political Economists: Divergence of Character," *J. Hist. Biol.* 13 (1980): 195–289; Robert M. Young, *Darwin's Metaphor: Nature's Place in Victorian Culture* (Cambridge, 1985). For a discussion of evolutionary metaphors in economic thought, see Philip Mirowski, ed., *Natural Images in Economic Thought: "Markets Red in Tooth and Claw"* (Cambridge, 1994); Mary S. Morgan, "Evolutionary Metaphors in Explanations of American Industrial Competition," in *Biology as Society, Society as Biology: Metaphors*, ed. Sabine Maasen, Everett Mendelsohn, and Peter Weingart (Boston, 1995), 311–37.

[3] The classic account of "social Darwinism" is Richard Hofstadter, *Social Darwinism in American Thought: 1860–1915* (Philadephia, 1945). For critical responses to Hofstadter's thesis, see Robert C. Bannister, *Social Darwinism: Science and Myth in Anglo-American Social Thought* (Philadelphia, 1979); Carl Degler, *In Search of Human Nature: The Decline and Revival of Darwinism in American Social Thought* (New York, 1991); Richard J. Evans, "In Search of German Social Darwinism: The History and Historiography of a Concept," in *Medicine and Modernity: Public Health and Medical Care in*

stead, the historiography now recognizes a whole spectrum of social and political implications that could be drawn from biology, including a staunch defense of mutualism, socialism, and cooperation.[4] What has been less widely appreciated, however, is the extent to which evolution was invoked to explain the transition to corporate capitalism that many industrial economies underwent during the late nineteenth and early twentieth centuries. It is my thesis that the logic of organization proved especially attractive in the specific context of these debates, providing a powerful means to describe, discuss, and dispute the centralization, rationalization, and bureaucratization that contemporary observers often took to be characteristic of a distinctly modern political economy.[5]

Compared to historians of science, business historians and historians of technology have had more to say about the corporate reconstruction of industrial economies during the nineteenth and early twentieth centuries.[6] Much of this literature is heavily indebted to Alfred D. Chandler, whose work had a transformative impact on a whole generation of scholarship.[7] But whereas Chandler primarily chalked the proliferation of large corporate firms up to the gains in efficiency that could be reaped by tapping new economies of scale and of scope, others have offered a competing set of explanations.[8] Recent historians of capitalism have been especially skeptical of the functionalist assumptions that undergird Chandler's narrative, often insisting on the role of the

---

Nineteenth- and Twentieth-Century Germany, ed. Manfred Berg and Geoffrey Cocks (Cambridge, 1997), 55–80; Mike Hawkins, *Social Darwinism in European and American Thought, 1860–1945* (New York, 1997).

[4] See Adrian J. Desmond, *The Politics of Evolution: Morphology, Medicine, and Reform in Radical London* (Chicago, 1989); Piers J. Hale, *Political Descent: Malthus, Mutualism, and the Politics of Evolution in Victorian England* (Chicago, 2014); Daniel Philip Todes, *Darwin without Malthus: The Struggle for Existence in Russian Evolutionary Thought* (New York, 1989). For a discussion of organicist theories in particular, see James Elwick, *Styles of Reasoning in the British Life Sciences: Shared Assumptions, 1820–1858* (London, 2007); Gregg Mitman, *The State of Nature: Ecology, Community, and American Social Thought, 1900–1950* (Chicago, 1992); Mitman, "Defining the Organism in the Welfare State: The Politics of Individuality in American Culture, 1890–1950," in Maasen, Mendelsohn, and Weingart, *Biology as Society* (cit. n. 2), 249–78; Paul Weindling, *Health, Race, and German Politics between National Unification and Nazism, 1870–1945* (Cambridge, 1989).

[5] See Max Weber, *Die protestantische Ethik, und der Geist des Kapitalismus* (Tübingen, 1905); Weber, *Wirtschaft und Gesellschaft* (Tübingen, 1922). See also Werner Sombart, *Der moderne Kapitalismus* (Leipzig, 1902).

[6] See Louis Galambos and Joseph A. Pratt, *The Rise of the Corporate Commonwealth: U.S. Business and Public Policy in the Twentieth Century* (New York, 1988); David A. Hounshell, *From the American System to Mass Production, 1800–1932* (Baltimore, 1984); Hounshell and John K. Smith, *Science and Corporate Strategy: Du Pont R&D, 1902–1980* (Cambridge, 1988); Kenneth Lipartito and David B. Sicilia, eds., *Constructing Corporate America: History, Politics, Culture* (Oxford, 2004); David F. Noble, *America by Design: Science, Technology, and the Rise of Corporate Capitalism* (New York, 1977); Steven Shapin, *The Scientific Life: A Moral History of a Late Modern Vocation* (Chicago, 2008).

[7] Alfred D. Chandler, *The Visible Hand: The Managerial Revolution in American Business* (Cambridge, Mass., 1977); Chandler, *Scale and Scope: The Dynamics of Industrial Capitalism* (Cambridge, Mass., 1990).

[8] Economic historians tend to stress the importance of transaction costs. See Ronald Coase, "The Nature of the Firm," *Economica* 16 (1937): 386–405; Naomi Lamoreaux, Daniel Raff, and Peter Temin, "Beyond Markets and Hierarchies: Toward a New Synthesis of American Business History," *Amer. Hist. Rev.* 108 (2003): 404–33; Oliver E. Williamson, "The Modern Corporation: Origins, Evolution, Attributes," *Journal of Economic Literature* 19 (1981): 1537–68.

state in creating a regulatory framework that helped promote corporate consolidation.[9] An especially striking example is Richard White's history of the transcontinental railroads, which argues that far from establishing effective command and control structures, these large and unwieldy business ventures only managed to avoid going under by drawing on generous public bailouts and other forms of government intervention.[10] For many recent historians of capitalism, then, the proliferation of large, capital-intensive, and vertically integrated industrial firms was anything but the inevitable outcome of an impersonal market rewarding gains in efficiency of various kinds.

But if large corporate firms were so unwieldy and cumbersome, how can we account for the period's widespread belief that an increasingly centralized business landscape represented a hallmark of modernity? My suggestion is that we frame this development as part of a larger enthusiasm for the logic of organization. As such, we need not postulate any inexorable law of progress or efficient market hypothesis to understand why so many late nineteenth- and early twentieth-century observers genuinely believed themselves to inhabit a world that seemed to be getting more organized over time. Rather, the enormous prestige enjoyed by the logic of organization, as well as the continuity between biological organisms and higher-level social assemblages it explicitly sought to establish, helped to ensure that a vision of society as an increasingly complex hierarchy of part/whole relationships would become a kind of self-fulfilling prophecy. In other words, the logic of organization had a performative dimension, offering a compelling conceptual resource to those who sought to create exactly the kind of reality that it described. The fact that this served to put those with a claim to specialized expertise in a position of power only made it all the more effective, insofar as it allowed them to claim a privileged role for themselves in the creation and management of a more organized world.

A SCIENCE OF ORGANIZATION

Like many other contributions to this volume, my essay begins by tracing the emergence of a new kind of thing. In much the same way that a social process involving the work of scientists brought the accident-prone driver, "scientific crude," the secondhand smoker, and the Comstock Lode into being, the organization as a distinct object or entity has a history all of its own.[11] Moreover, I want to suggest that the organization, understood as a definite thing in the world, came into being at precisely the time that a new science of life was created as well. Not only did the new science of life take as its primary task the empirical study and conceptual elucidation of organization, but, in so doing, its development into a distinct branch of natural philosophy effectively

[9] See Naomi R. Lamoreaux and William J. Novak, eds., *Corporations and American Democracy* (Cambridge, Mass., 2017); Peter B. Evans, Dietrich Rueschemeyer, and Theda Skocpol, eds., *Bringing the State Back In* (Cambridge, 1985); William Roy, *Socializing Capital: The Rise of the Large Industrial Corporation in America* (Princeton, N.J., 1997); Martin J. Sklar, *The Corporate Reconstruction of American Capitalism, 1890–1916: The Market, the Law, and Politics* (Cambridge, 1988).

[10] Richard White, *Railroaded: The Transcontinentals and the Making of Modern America* (New York, 2011).

[11] Lee Vinsel, "'Safe Driving Depends on the Man at the Wheel': Psychologists and the Subject of Auto Safety, 1920–55"; Julia Fein, "'Scientific Crude' for Currency: Prospecting for Specimens in Stalin's Siberia"; Sarah Milov, "Smoke Ring: From American Tobacco to Japanese Data"; Paul Lucier, "Comstock Capitalism: The Law, the Lode, and the Science," all in this volume.

refashioned organization from merely being a property or feature of living things to becoming a thing in itself: the organism.

Following Michel Foucault, historians often trace the origins of the modern life sciences back to the turn of the nineteenth century, when naturalists began using the word "biology" to designate a philosophical inquiry into "living nature."[12] As much as anything else, the new science of life was a science of organization. Of course, living matter had been identified as organized matter as far back as classical antiquity. But the precise meaning and significance of that claim changed a great deal during the intervening years. In his writings on natural history, for example, Aristotle used the word for a tool or an instrument—ὄργανον or *organon*—to indicate that different parts of an animal's body had a purpose, such as the heart pumping blood. But it did not escape notice that, in order to function properly, the heterogeneous parts of an animal's body had to be coordinated and adapted to one another as well. Indeed, over time it was this latter feature that would increasingly come to be stressed. As a result, the word "organ" eventually came to be associated not just with any and all instruments, but the pipe organ in particular, celebrated for the way that its stunning array of diverse elements could be made to yield harmonious sounds. In a related vein, the description of living things as "organized beings" came to be understood as a way of emphasizing their functional integration over the purpose of individual elements. Thus, by 1701, Nehemiah Grew could insist that "every Body should have its Organism," meaning that all "the Parts of the Organ, be fitly Cized, Shaped, and set together," whereas Diderot and d'Alembert's *Encyclopédie* defined the concept of *organisation* as "the arrangement of parts that make up living bodies."[13]

The notion that living matter was organized matter was therefore increasingly understood to mean that living things constituted a functionally integrated whole. An especially clear articulation of this view was offered by Immanuel Kant, who described the plant and animal body as an "organized and self-organizing being" composed of many diverse elements, each of which must be "conceived as if it exists only through all the others, thus as if existing for the sake of the others and on account of the whole."[14] This way of thinking became even more widespread during the nineteenth century. Georges Cuvier, for example, articulated a conception of the "organized being" as "a unique and closed system, in which all the parts correspond mutually, and contribute to the same definitive action by a reciprocal reaction."[15] Around the same

[12] Michel Foucault, *The Order of Things: An Archaeology of the Human Sciences* (New York, 2002); Gottfried Reinhold Treviranus, *Biologie: Oder Philosophie der Lebenden Natur für Naturforscher und Aerzte* (Göttingen, 1802).

[13] Nehemiah Grew, *Cosmologia Sacra: Or a Discourse of the Universe as It Is the Creature and Kingdom of God* (London, 1701), bk. II, chap. 1, 34, 32–3; Jean d'Alembert and Denis Diderot, eds., *Encyclopédie, ou, Dictionnaire Raisonné des Sciences, des Arts et des Métiers* (Paris, 1751), 9:629b. For more on this history, see Tobias Cheung, *Die Organisation des Lebendigen: Die Entstehung des biologischen Organismusbegriffs bei Cuvier, Leibniz und Kant* (Frankfurt, 2000); Cheung, "From the Organism of a Body to the Body of an Organism: Occurrence and Meaning of the Word from the Seventeenth to the Nineteenth Centuries," *Brit. J. Hist. Sci.* 39 (2006): 319–39; Karl M. Figlio, "The Metaphor of Organization: An Historiographical Perspective on the Bio-Medical Sciences of the Early Nineteenth Century," *Hist. Sci.* 14 (1976): 17–53.

[14] Immanuel Kant, *Critique of the Power of Judgment*, ed. Paul Guyer (Cambridge, 2000), 245–6. See also John Zammito, "Organism: Objective Purposiveness," in *Immanuel Kant: Key Concepts*, ed. Will Dudley and Kristina Engelhard (New York, 2014), 170–83.

[15] Martin J. Rudwick, *Georges Cuvier, Fossil Bones, and Geological Catastrophes: New Translations and Interpretations of the Primary Texts* (Chicago, 1997), 217.

time, the emergence of Romantic *Naturphilosophie* saw a number of German intellectuals embrace Kant's biological holism with particular fervor, transforming it from a regulative principle into an empirical object of study. As this happened, the word "organism" began to be used as a noun rather than an adjective, designating a kind of thing rather than a feature or property. In his *Lehrbuch der Naturphilosphie*, the first edition of which was published in 1809, for example, Lorenz Oken explained that by "organism" he meant any "individual, complete, self-contained, body that moves and excites itself," speculating that even the earth in its entirety constituted an organic unity properly so-called.[16] Similarly, the influential physiologist Johannes Müller wrote that "the organism constitutes a composition of unequal elements dominated by the unity of the whole."[17] As Tobias Cheung has convincingly argued, the turn of the nineteenth century thus witnessed a transformation of the organism from a general "principle of order" into a "generic name for individuals as natural entities or living beings."[18] At precisely the time the new science of biology truly began to take shape, the organism came into being as its primary unit of analysis.

The organism emerged as a discrete and meaningful entity in its own right once the new science of biology elevated organization from an attribute of living things into a pressing problem of knowledge in its own right. But that did not answer the question of how organization was generated and reproduced. If the living body was fundamentally an organized body, what might account for the origin and maintenance of organization over time, and by what means was it reliably re-created from one generation to the next? One possibility was to conclude that organization was a primitive feature of life, always already present inside of the germ, perhaps in some kind of latent or invisible form. Another was to reject the idea that embryogenesis involved no more than growth and unfolding, insisting that it constituted a process of genuine development. The latter conviction led naturalists such as Buffon, Needham, and Wolff to speculate that a subtle, occult, attractive, or penetrating force might explain how the diverse parts of an embryo gradually emerged and arranged themselves during gestation. Citing Newton's success in postulating the force of gravity to explain the organization of our solar system, wherein each planet traced a precise, elliptical orbit around a centrally located sun, they asked why an analogous principle could not be inferred from the regular appearance of organization in the plant or animal body? By the time Kant wrote his third critique at the very end of the century, his contemporary Johann Friedrich Blumenbach even proposed that all living matter was endowed with a *Bildungstrieb*, an innate drive that "bestows on creatures their form, then preserves it, and, if they become injured, where possible restores their form."[19] By invoking an analogue to Newtonian gravity such as the *Bildungstrieb*, it was possible to explain the organization of living things without simply taking its preexistence for granted, or having to invoke some sort of Aristotelian soul. Just as important, however, was that it also allowed naturalists such as Blumenbach to avoid the implausible claim that complex and apparently purposive organization could be produced by mere matter in motion. Only thus could the new field of biology claim to be a gen-

[16] See Lorenz Oken, *Lehrbuch der Naturphilosophie* (Jena, 1831), 144.
[17] See Johannes Müller, *Handbuch der Physiologie des Menschen für Vorlesungen*, 3rd ed. (Koblenz, 1837), 1:20.
[18] Cheung, "From the Organism of a Body" (cit. n. 13), 319.
[19] Johann Friedrich Blumenbach, *Über den Bildungstrieb* (Göttingen, 1781), 12–3.

uine source of knowledge, one that was able to account for the teleological features of living things without running afoul of the period's accepted standards of scientific explanation.[20]

In addition to embryogenesis, Blumenbach speculated his *Bildungstrieb* might also explain the development of life over geological time. Judging from the strange-looking creatures being unearthed in the fossil record, he reasoned that some kind of revolution or catastrophe must have destroyed all of "pre-Adamite creation" at some point in the deep past, only to see the *Bildungstrieb* repopulate the globe from inorganic materials again.[21] In broad strokes, this resembles a theory promulgated around the same time by the Parisian zoologist Jean-Baptiste de Lamarck. Drawing on the age-old notion that all of creation could be arranged into a single, great chain of being, Lamarck argued that more "perfect" organisms had developed out of their much simpler predecessors via the actions of "subtle" or "imponderable fluids," concrete examples of which included electricity as well as caloric (a substance thought to be responsible for the production of heat). Much like Blumenbach did with the *Bildungstrieb*, Lamarck argued these subtle fluids were agents of organization, altering the solid parts of an organism when the two came into contact and inducing the latter to grow in complexity. Indeed, the power that inhered in these fluids was so great they could even breathe life into inert matter, thereby replenishing the lowest rungs of creation as simple animalcules gradually evolved into more complex organisms.[22]

The thesis that a close parallelism obtained between the stages of an individual's life cycle and the arrangement of species into a natural hierarchy eventually became a canon of nineteenth-century biology. Initially, it found particular favor among practitioners of Romantic *Naturphilosophie* like Lorenz Oken, but it was soon taken up by a far more diverse cast of characters, too, including the German anatomist Johann Friedrich Meckel and the French physiologist Etienne Serres, who confidently proclaimed that the "entire animal kingdom can, in some measure, be considered ideally, as a single animal, which, in the course of formation and metamorphosis of its diverse organisms, stops in its development, here earlier and there later."[23] Of course, there were detractors as well, most notably the Baltic embryologist Karl Ernst von Baer, who argued that close observational scrutiny showed the developing embryo to proceed from homogeneity of structure to heterogeneity of structure rather than ascending a hierarchical chain of being.[24] But that did not cause the recapitulation doctrine, as it came to be called, to disappear. In fact, exactly the opposite happened, and later naturalists increasingly drew parallels between the embryonic stages of modern animals and the adult stages of their prehistoric ancestors, not currently existing but less highly orga-

[20] See Timothy Lenoir, *The Strategy of Life: Teleology and Mechanics in Nineteenth Century German Biology* (Dordrecht, 1982). For a conflicting account, see Robert J. Richards, *The Romantic Conception of Life: Science and Philosophy in the Age of Goethe* (Chicago, 2002). See also Shirley A. Roe, *Matter, Life, and Generation: Eighteenth-Century Embryology and the Haller-Wolff Debate* (Cambridge, 1981).

[21] Johann Friedrich Blumenbach, *Beyträge zur Naturgeschichte*, 2nd ed. (Göttingen, 1806), 13, 19.

[22] Jean-Baptiste de Lamarck, *Philosophie Zoologique* (Paris, 1809). For more on Lamarck and his theory of evolution, see Pietro Corsi, *The Age of Lamarck: Evolutionary Theories in France, 1790–1830* (Berkeley and Los Angeles, 1988); Snait Gissis and Eva Jablonka, eds., *Transformations of Lamarckism: From Subtle Fluids to Molecular Biology* (Cambridge, Mass., 2011).

[23] Antoine Étienne Serres, "Principes d'Embryogenie, de Zoogenie, et de Tetratogenie," *Mém. Acad. Sci.* 25 (1860): 1–943, on 834.

[24] Karl Ernst von Baer, *Über Entwickelungsgeschichte der Thiere: Beobachtung und Reflexion*, 3 vols. (Königsberg, 1828).

nized creatures. Thus, by the second half of the nineteenth century, recapitulation had become a keystone of progressivist biological theory, espoused in various guises by everyone from Ernst Haeckel and August Weismann in Germany, to Louis Agassiz and Edward Drinker Cope in the United States, as well as Sir Richard Owen and the noted popularizer Robert Chambers in England.[25]

The enormous success of recapitulation fed directly into the nineteenth century's increasingly widespread enthusiasm for evolutionary progress. Just as the vast majority of naturalists agreed that embryogenesis generated an increasingly "perfect" organism, so too did it become common to hold that whole species inevitably grew more and more organized over geological time. Among the most influential advocates of the notion that all of creation tended toward higher complexity was Herbert Spencer, whose "development hypothesis" could arguably be characterized as the period's grand unified theory of organization. Explicitly modeled on von Baer's claim that a simple, fertilized egg advanced "from homogeneity of structure to heterogeneity of structure" as it developed into an adult, Spencer projected this basic trajectory from simplicity to complexity onto all aspects of life, including the evolution of language, society, culture, commerce, and art, as well as the universe in its entirety: "From the earliest traceable cosmical changes down to the latest results of civilization," Spencer argued, "progress essentially consists" of a "transformation of the homogeneous into the heterogeneous."[26] For Spencer, an entity's degree of complexity and the interdependence among its constituent parts thus served as an indicator of its normative status, rendering nearly everything either more primitive or more highly evolved.[27]

So far, I have stressed the degree to which nineteenth-century biologists agreed that a basic, even defining feature of life was its capacity to grow increasingly organized over time. But the logic of organization did not just pertain to the development of individual plants and animals, nor even the evolution of an entire species. Often, it was applied to much larger living assemblages, too. For example, it became common to compare the emergence of modern society to the evolution of multicellular organisms from their single-celled ancestors. To understand how the logic of organization came to inform broader debates about social life, I want to take a step backward in time and examine the history of the cell, especially the notion that metazoan organisms could be usefully likened to a cell state. In so doing, I want to stress that not everyone drew the same political implications from this analogy. In particular, whereas the multicellular organism was sometimes described in strikingly liberal terms, the cell-state metaphor also gave rise to a more corporatist vision according to which living things naturally grouped themselves into progressively larger assemblages whose functional integration required each part to give up a measure of its autonomy for the sake of the whole.[28]

[25] See Stephen Jay Gould, *Ontogeny and Phylogeny* (Cambridge, Mass., 1977), 1–209.

[26] Herbert Spencer, "Progress: Its Law and Cause," *Westminster Rev.* 67 (1857): 445–85, on 446–7.

[27] For a historical overview of progressivist theories of evolution, see Michael Ruse, *From Monad to Man* (Cambridge, Mass., 1996).

[28] See Scott Lidgard and Lynn Nyhart, eds., *Biological Individuality* (Chicago, 2017); Nyhart and Lidgard, "Individuals at the Center of Biology: Rudolf Leuckart's *Polymorphismus der Individuen* and the Ongoing Narrative of Parts and Wholes," *J. Hist. Biol.* 44 (2011): 373–443; Andrew Reynolds, "Ernst Haeckel and the Theory of the Cell State: Remarks on the History of a Bio-Political Metaphor," *Hist. Sci.* 46 (2008), 123–52; Paul Weindling, "Theories of the Cell State in Imperial Germany," in *Biology, Medicine and Society 1840–1940*, ed. Charles Webster (Cambridge, 1981), 99–155.

When Kant described living things as organized beings during the 1790s, he had anatomical arrangements in mind, explicitly citing "the structure of a bird, the hollowness of its bones, the placement of its wings for movement and of its tail for steering," in making the argument that "teleological judging is rightly drawn into our research into nature."[29] Before long, however, organization was read even deeper down into the plant and animal body. During the first decades of the nineteenth century, a number of physiologists and anatomists began to suspect that (i) all organic matter is composed of cells, which (ii) represent the most basic units of life and (iii) always stemmed from another, preexisting cell.[30] The articulation of the cell theory had a huge impact on how biologists understood the structure of metazoan organisms. It not only implied that all plants and animals were fundamentally composed of the same basic units, but that each of those building blocks were themselves living things, meaning that complex biological individuals could be decomposed into smaller and simpler ones.

With the success of the cell theory, questions about biological organization came to be bound up with questions about social organization, and biologists almost immediately began to enlist social metaphors to help them make sense of the relationship between parts of the metazoan body. Expounding on what Matthias Schleiden had termed the "double life" of cells—"one independent" and the "other intermediary, since it has become an integrated part of the [whole]"—Theodor Schwann likened the metazoan body to a kind of "state" in which "each cell is a citizen."[31] At the outset, the task of elaborating upon this metaphor fell most profoundly to the pathologist and pioneering advocate of public health, Rudolf Virchow. Writing in the *Archiv für pathologische Anatomie*, Virchow compared the organism to "a free state of individuals with equal rights though not with equal endowments, which keeps together because the individuals are dependent upon one another."[32] It is no coincidence that Virchow also agitated for liberal reform during the German revolutions of 1848, helping to found the progressive Deutsche Fortschrittspartei, which he represented in the Reichstag during the 1880s and early 1890s. Given to voicing his political views without hesitation, Virchow frequently clashed with the authoritarian government of Otto von Bismarck. As is well known, this antipathy to centralized control manifested itself in Virchow's ideas about multicellularity as well. At one point, for example, he insisted "the life of a people is nothing more than the sum of the lives of the individual citizens. So it is also in the little country, which the body of every plant and every animal represents."[33] According to Virchow, the metazoan organism thus constituted a cell state in which power clearly devolved to the constituent parts.

[29] Immanuel Kant, *Critique of Judgment* (Oxford, 2007), 233–4.

[30] See Henry Harris, *The Birth of the Cell* (New Haven, Conn., 1999); Jan Sapp, *Genesis: The Evolution of Biology* (Oxford, 2003); Daniel J. Nicholson, "Biological Atomism and Cell Theory," *Stud. Hist. Phil. Biol. Biomed. Sci.* 41 (2010): 202–11; Andrew Reynolds, "The Redoubtable Cell," *Stud. Hist. Phil. Biol. Biomed. Sci.* 41 (2010): 194–201.

[31] Theodor Schwann and M. J. Schleiden, *Microscopical Researches into the Accordance in the Structure and Growth of Animals and Plants*, trans. Henry Smith (London, 1847), 165, 192, 231–2; François Jacob, *The Logic of Life: A History of Heredity* (New York, 1973), 119.

[32] Rudolf Virchow, "Cellular-Pathologie," *Arch. Pathol. Anat. Physiol. Klin. Med.* 8 (1855): 3–39, on 25.

[33] Rudolf Virchow, "Alter und Neuer Vitalismus," *Arch. Pathol. Anat. Physiol. Klin. Med.* 9 (1856): 1–55, on 16.

Virchow's tendency to read liberal political theory into the metazoan organism induced him to dismiss what he called a "monarchial principle in the body."[34] But he did not therefore downplay the importance of organization in the multicellular organism. "Multitudinous as the parts may be," Virchow argued in an address before the Berlin Academy of Natural Sciences in 1859, the body is "a real commonwealth in which each part . . . requires the other, and does not win its full significance outside of the whole."[35] Indeed, Virchow even argued that disease ought to be understood as a breakdown of the cellular commonwealth: "As in the lives of nations, so in the lives of individuals the state of health of the whole is determined by the well-being and close interrelation of the individual parts," he argued adding that "disease appears when individual members begin to sink into a state of inactivity disadvantageous to the commonwealth, or to lead a parasitic existence at the expense of the whole."[36] Liberal it may well have been, then, but Virchow's biology nonetheless emphasized the importance of maintaining the integrity of the body politic as a self-organized entity.

Virchow was a pathologist and cancer biologist, so it is no surprise that he would have focused on what happens to the metazoan body when the social organization of its cells falls apart. But to a morphologist and embryologist such as Ernst Haeckel—who studied under Virchow during the 1850s—the origins and maintenance of biological organization were far more pressing concerns. In search of a way to address this issue, Haeckel developed a powerful synthesis of the cell-state metaphor and the theory of evolution. In his two-volume magnum opus, *Generelle Morphologie der Organismen*, Haeckel argued that just as a single cell developed into a fully formed adult during ontogeny, phylogeny described the process in which single-celled organisms evolved into complex, multicellular wholes. Hence, individual cells were not just elementary organisms in the sense that they exhibit all the basic properties of life. In Haeckel's synthetic vision, the metazoan body was literally an organized collection of single-celled animals that evolved a communal life history.[37] Moreover, in a series of popular lectures, Haeckel explicitly departed from Virchow's liberal conception of the self-organized cell state to endorse a more authoritarian vision. "Although the cells in every multicellular organism exercise a certain degree of autonomy, they are nonetheless subordinated to the whole," Haeckel maintained, arguing that the more highly evolved an organism was, the more centralized and autocratic its cell state would be. Indeed, he even went so far as to argue that animals resembled a cellular monarchy, whereas plants could be compared to a republic. Haeckel contrasted both forms of political arrangement to ephemeral agglomerations of otherwise free-living cells such as slime molds, which he likened to unruly barbarians. "If the plant and animal organism can be compared to a well-organized *Kulturstaat*," he mused, "the loose heap of cells formed by protists

[34] Rudolf Virchow, "Über die Reform der Pathologischen und Therapeutischen Anschauungen durch die Mikroskopischen Untersuchungen," *Arch. Pathol. Anat. Physiol. Klin. Med.* 1 (1847): 207–55, on 216. See also Laura Otis, *Membranes: Metaphors of Invasion in Nineteenth-Century Literature, Science, and Politics* (Baltimore, 1999), esp. 22.

[35] Rudolf Virchow, "Atoms and Individuals," in *Disease, Life, and Man: Selected Essays*, trans. Lelland J. Rather (Stanford, Calif., 1958), 120–42, on 141; Virchow, "Atome und Individuen," in *Vier Reden über Leben und Kranksein* (Berlin, 1862), 35–76, on 72.

[36] Virchow, "Atoms and Individuals" (cit. n. 35), 153.

[37] Ernst Haeckel, *Generelle Morphologie der Organismen*, vol. 1, *Allgemeine Anatomie der Organismen* (Berlin, 1866).

can, at best, be compared with the crude and uncultivated hoards formed by members of primitive tribes."[38] Unlike Virchow's liberal emphasis on personal autonomy, then, Haeckel envisioned a highly evolved cell state as a collection of "well-raised" or "well-trained citizens" ruled by a "mighty centralized government, the nerve center or brain." "The more highly developed the animal," he concluded, "the stronger and more centralized the cellular monarchy is, and the more powerful the commanding brain."[39] According to Haeckel, evolution not only produced increasingly complex levels of organization but also brought with it a reduction of individual liberty.

Ernst Haeckel and Rudolf Virchow certainly ranked among the most influential biologists writing about the cell state during the nineteenth century, but they were hardly alone. During the second half of the nineteenth century, the problem of biological organization gave shape to an entire discursive field, one that drew in a wide range of people working across disciplinary lines and extending beyond Germany's still amorphous borders. As this happened, a vigorous debate about multicellularity began to emerge, and the question of exactly what kind of politics obtained in the cell state became a major point of contention. These difference became especially clear when the analogy between cells in the body and citizens in a state was pushed to suggest that human societies resembled a multicellular organism as well as the other way around.

One particularly noteworthy example was Haeckel's student Oscar Hertwig, who sought to develop a subtler conception of the way individuals related to one another in a collective while embracing his teacher's corporatist vision of the multicellular organism. According to Hertwig, the most important feature of biological development—regardless of whether it was applied to the creation of a single organism or a larger social assemblage—was the fact that each part exerted a causal influence on every other part in producing the whole. This led him to reject all manner of reductionist theories that privileged the autonomy of individual agents over the good of the whole. For example, in a polemic directed at Weismann's theory that a preexisting organized substance within the germ cells controlled the developmental process, he insisted that biological organization only emerged epigenetically. Just as in human society, Hertwig claimed, the development of organized complexity required the active cooperation of each individual, for it was only by coming together that each part of a larger assemblage could specialize and diversify. For Hertwig, the multicellular organism and modern society therefore both constituted an irreducibly complex social arrangement, leading him to critique not only preformationist theories of biological development such as Weismannism but also laissez-faire economic policies that he felt gave undue weight to the interests of individual sovereign selves. Unlike Haeckel, however, Hertwig refused to fully embrace an authoritarian alternative to classical liberalism, always attempting to steer a middle path that stressed the willing cooperation among parts in a whole.[40]

[38] Ernst Haeckel, *Das Protistenreich: Eine Populäre Uebersicht über das Formengebiet der Niedersten Lebewesen* (Leipzig, 1878), 16–7, 18.

[39] Ernst Haeckel, *Gesammelte Populäre Vorträge aus dem Gebiete der Entwickelungslehre* (Bonn, 1878), 156–7.

[40] For Hertwig's critique of Weismann, see Oscar Hertwig, *Zeit- und Streitfragen der Biologie*, 2 vols. (Jena, 1894). For his organicist social theory more broadly, see Hertwig, *Der Staat als Organismus: Gedanken zur Entwicklung der Menscheit* (Jena, 1922). See also Paul Weindling, *Darwinism and Social Darwinism in Imperial Germany: The Contribution of the Cell Biologist, Oscar Hertwig, 1849–1922* (Stuttgart, 1991).

In England, Herbert Spencer emerged as an even more conflicted proponent of organicist social theory. Spencer argued that if "an ordinary organism may be regarded as a nation," then "a nation of human beings may be regarded as an organism."[41] Like many nineteenth-century social theorists (including Saint-Simon, Comte, and Durkheim), Spencer conceived of society as much more than a simple analogy to the biological organism; it *was* one in every sense of that word.[42] For Spencer, this was simply a consequence of his universal "development hypothesis," which could be expected to give rise to a nested hierarchy of biological individuals. "Life in general," he speculated in the *Principles of Biology*, "commenced with minute and simple forms" that transitioned "to organisms made up of groups of such units, and to higher organisms made up of groups of such groups."[43] As the process continued, more and more complex entities would gradually emerge, from the family to the primitive tribe, only to culminate, eventually, in the modern state.[44] At the same time, Spencer was at pains to square his organicist social theory with his commitment to individual liberty, which led him to emphasize that notwithstanding all their similarities, one "cardinal difference" obtained between human societies and lower levels of biological organization. Whereas "consciousness is concentrated" among only a small subset of the cells that make up the metazoan body, he noted, "it is diffused throughout" all of society. Because every man, woman, and child "possess[es] the capacities for happiness and misery," it followed that "society exists for the benefit of its members; not its members for the benefit of society."[45] And, elsewhere, he elaborated that whereas "it is well that the lives of all parts of an animal should be merged in the life of the whole, the welfare of citizens cannot rightly be sacrificed to some supposed benefit of the State."[46] Thus, while Spencer embraced an organicist social theory, he clearly recoiled from its more authoritarian implications.

THE INDUSTRIAL ORGANISM

Spencer's awkward and somewhat tortured attempt to reconcile organicist social theory with a commitment to personal liberty points to a profound tension between holism and individualism that resonated across much of the late nineteenth century's intellectual landscape. In the remainder of this essay, I want to examine the way this tension played itself out in debates about the political economy of modern, corporate capitalism. In so doing, I will suggest that there, too, the logic of organization did essential explanatory work. This was particularly so in the United States, which im-

[41] Herbert Spencer, *The Principles of Sociology* (London, 1874), 473.
[42] See also Virchow, "Atoms and Individuals" (cit. n. 35), 144.
[43] Herbert Spencer, *The Principles of Biology* (London, 1864), 204.
[44] See also Albert Schäffle, *Bau und Leben des Socialen Körpers*, 4 vols. (Tübingen, 1881); Émile Durkheim, *De la division du travail social: Etude sur l'organisation des sociétés supérieures* (Paris, 1893).
[45] Spencer, *Principles of Sociology* (cit. n. 41), 479–80.
[46] Herbert Spencer, "The Social Organism," in *Essays: Scientific, Political, and Speculative*, vol. 1 (London, 1868), 384–428, on 396–7. For more on Spencer's organicism, see James Elwick, "Herbert Spencer and the Disunity of the Social Organism," *Hist. Sci.* 41 (2003): 35–72; Elwick, "Containing Multitudes: Herbert Spencer, Organisms Social and Orders of Individuality," in *Herbert Spencer: Legacies*, ed. Mark Francis and Michael Taylor (New York, 2015), 89–110; Tim S. Gray, *The Political Philosophy of Herbert Spencer: Individualism and Organicism* (Aldershot, 1996); Gray, "Herbert Spencer: Individualist or Organicist?," *Polit. Stud.* 33 (1985): 236–53.

pressed many observers as having evolved from an agrarian backwater into an industrial powerhouse with almost unimaginable speed during precisely this period. Not only that, but it was often remarked that in the United States, unfettered competition seemed to be giving way to rational planning and managerial oversight as small private enterprises were increasingly replaced by large, complex, and functionally integrated corporate firms.[47]

In some ways, the logic of organization already informed the work of classical political economists at the end of the eighteenth century. It is well known, for example, that Adam Smith characterized "civilized society" by its remarkable "division of labor," the result of which was that each of us "stands at all time in need of the cooperation and assistance of great multitudes."[48] Others emphasized the complex interdependence of society's different productive elements, too, as Rousseau's invocation of the age-old analogy between the "body politic" and the "organized, living body" in his entry on "Economie" for Diderot and D'Alambert's *Encyclopédie* made abundantly clear.[49] But Smith and his eighteenth-century contemporaries tended to agree that social organization was an aggregate outcome of individual choices, insisting that it "is not from the benevolence of the butcher, the brewer, or the baker, that we expect our dinner, but from their regard to their own interests."[50] Thus, although classical political economists such as Smith clearly thought of society as organized, they did not claim that it inexorably developed from a state of simplicity to one of complexity. Nor did they primarily understand it as a functionally integrated entity whose parts were compelled to give up a measure of their autonomy for the sake of the whole.[51]

Contrast this with an influential textbook written by Alfred Marshall just over a century later, which described the development of modern society from simple and primitive tribes, no better than "a pack of wolves or a horde of banditti." Over time, Marshall observed, human beings were gradually endowed with a more "noble patriotism" while their "religious ideals are raised and purified," until, eventually, the entire collective was best adapted to its environment. According to Marshall, a truly advanced stage of civilization could therefore be characterized by the willingness of each individual "to sacrifice himself for the benefit of those around him." Whereas Smith was primarily concerned with the division of labor, then, Marshall was far more interested in its integration, emphasizing the "growing intimacy and firmness of connections between the separate parts of the industrial organism." In a similar vein, he argued that just as "a cathedral is something more than the stones of which it is made" and "a person is something more than a series of thoughts and feelings, so the life of society is something more than the lives of its individual members."[52] What makes these claims especially revealing is that despite often being credited as one of the founders of neo-

[47] Besides Chandler, see also Louis Galambos, "The Emerging Organizational Synthesis in Modern American History," *Bus. Hist. Rev.* 44 (1970): 279–90; Jonathan Levy, "The Trust Question," in *Freaks of Fortune* (Cambridge, Mass., 2012), 264–307; Daniel T. Rodgers, "In Search of Progressivism," *Rev. Amer. Hist.* 10 (1982): 113–32; Robert Wiebe, *The Search for Order, 1877–1920* (New York, 1967).

[48] Adam Smith, *An Inquiry into the Nature and Causes of the Wealth of Nations* (London, 1776), 17.

[49] Jean-Jacques Rousseau, "Economie," in d'Alembert and Diderot, *Encyclopédie* (cit. n. 13), 5:338.

[50] Smith, *Wealth of Nations* (cit. n. 48), 17.

[51] See Jonathan Sheehan and Dror Wahrman, *Invisible Hands: Self-Organization and the Eighteenth Century* (Chicago, 2015).

[52] Alfred Marshall, *Principles of Economics* (London, 1890), 322, 324, 87, 330.

classical economics, Marshall explicitly cited Ernst Haeckel, Herbert Spencer, and the organicist sociologist Albert Schäffle (in addition to Darwin) as inspirations for his own work, arguing that a thorough grounding in modern biology was indispensable for the study of political economy.[53]

These sentiments were hardly unique to Marshall, and the second half of the nineteenth century saw numerous political economists and practicing capitalists alike look to the new science of biology for inspiration. A particularly notable example was Andrew Carnegie, who ranked not only as one of the period's wealthiest capitalists but also as one of its most enthusiastic proponents of industrial organization. Carnegie is a particularly good case in point, because his zeal to consolidate a vast empire of steel led him to construct precisely the kind of centralized business landscape that political economists such as Marshall would argue could only be understood in thoroughly biological terms.

Andrew Carnegie's wholehearted embrace of organization was informed by his well-known obsession with Herbert Spencer. In fact, Carnegie was so fanatical in his adoration of the Victorian philosopher that he even contrived to meet Spencer in person by booking passage aboard the same ocean liner from England to the United States. The ploy proved successful, and the two struck up a friendship that Carnegie treasured for the rest of his life, showering Spencer with lavish praise, expensive gifts, and, eventually, an anonymous pension. But the relationship was hardly one-sided. As Carnegie recalled in his autobiography, Spencer repaid the wealthy industrialist many times over by supplying a productive new philosophical outlook to live by. "'All is well since all grows better' became my motto, my true source of comfort," he recalled, having grown certain there was no "conceivable end" to humanity's "march to perfection."[54] Where did this relentless march lead? Carnegie was unequivocal on this point: "Everywhere we look we see the inexorable law ever producing bigger and bigger things." In particular, he was keen to celebrate the "concentration of capital" as "an evolution from the homogeneous to the heterogeneous" and therefore "another step in the upward path of development." For Carnegie, it was not only inevitable but laudable that small, independent producers would inevitably give way to much larger, often vertically integrated conglomerates. Or, to echo the way he described the process of industrial evolution himself, "The day of small concerns within the means of many able men seems to be over, never to return."[55]

In no small part, the new business landscape whose development Carnegie ascribed to evolutionary dynamics was of his own making. For example, because the Bessemer process of steel manufacturing employed in many of Carnegie's plants required an uninterrupted supply of high-quality ore that contained almost no phosphorus, a fluctuation in the supply chain or noticeable drop in quality had the potential to bring his whole operation to a standstill. But rather than stand by and watch costs mount as expensive machinery sat idle, Carnegie began to integrate backward and operate his own mines. As his onetime assistant James H. Bridge described it, Carnegie thereby transformed his manufacturing enterprise into a "mammoth body" that "owned its own

[53] See also Camille Limoges and Claude Menard, "Organization and the Division of Labor: Biological Metaphors at Work in Alfred Marshall's *Principles of Economics*," in *Natural Images in Economic Thought: "Markets Read in Tooth and Claw,"* ed. Philip Mirowski (Cambridge, 1994), 336–59.

[54] Andrew Carnegie, *Autobiography of Andrew Carnegie*, ed. John Charles Van Dyke (Boston, 1920), 339. See also David Nasaw, *Andrew Carnegie* (New York, 2006), 221–32.

[55] Andrew Carnegie, "Popular Illusions about Trusts," *Century Magazine* 60 (1900): 144–5.

mines, dug its ore with machines of amazing power, loaded it into its own steamers, landed it at its own ports, transported it on its own railroads, distributed it among its many blast-furnaces, and smelted it with coke similarly bought from its own coal-mines and ovens, and with limestone from its own quarries."[56] Not only that, but by the end of the century, Carnegie also began to integrate forward into the production of finished goods, threatening to encroach on some of his most powerful rivals and initiate a fierce struggle for survival. In hopes of averting a costly price war, the investment banker from Manhattan, J. P. Morgan, decided to buy Carnegie out and create U.S. Steel, consolidating two-thirds of the whole industry into one massive enterprise.[57]

Vertical integration allowed late nineteenth-century capitalists such as Carnegie to conquer vast areas of geographic space. But the process also involved the creation of almost equally vast new areas of institutional space. In search of a way to maintain oversight and control of their sprawling business empires, industrialists such as Carnegie implemented an ambitious bureaucratic machinery to centralize and to rationalize administrative decision making. Among other things, this involved the adoption of a military-style line-and-staff organizational structure designed to ensure that orders would flow smoothly down the chain of command. It also involved the adoption of formalized reporting mechanisms in hopes of ensuring that information would reliably flow back in the other direction. Particularly noteworthy was the development of new bookkeeping techniques, of which cost management accounting was arguably the most important. Whereas traditional double-entry bookkeeping had been developed to assess a firm's relationship to the outside world by tracking everything coming into and out of the shop, cost accounting was designed to put detailed and up-to-date information about an organization's internal workings at management's fingertips. It was hoped that doing so would allow management to make informed decisions without personally having to set foot on a factory floor and engage in face-to-face conversations with numerous employees.[58]

Here, too, Carnegie made for an especially striking example. Having served as a superintendent on the Pennsylvania Railroad during his twenties, he learned the value of careful bookkeeping early in life. As one of his biographers describes them, the account books he encountered during his time at the railroad were a veritable "statistician's paradise," whereas another explains that "manically detailed data were compiled, month by month" and "station by station" on "every category of expenditure."[59] Years later, when he first entered the iron manufacturing business, Carnegie was shocked by the lackadaisical record keeping of rival firms. "It was a lump business," he explained, "and until stock was taken and the books were balanced at the end of the year, the manufacturers were in total ignorance of results." Feeling "as if we were

[56] James Howard Bridge, *The "Carnegie Millions and the Men Who Made Them": Being the Inside History of the Carnegie Steel Company* (London, 1903), 169.

[57] Chandler, *Scale and Scope* (cit. n. 7), 127–40; Glenn Porter, *The Rise of Big Business, 1860–1920* (Wheeling, Ill., 2006).

[58] See H. Thomas Johnson and Robert Kaplan, *Relevance Lost: The Rise and Fall of Management Accounting* (Boston, 1991); Margaret Levenstein, *Accounting for Growth: Information Systems and the Creation of the Large Corporation* (Stanford, Calif., 1998); Johnathan Levy, "Accounting for Profit and the History of Capital," *Crit. Hist. Stud.* 1 (2014): 171–214; JoAnne Yates, *Control through Communication: The Rise of System in American Management* (Baltimore, 1989).

[59] Harold C. Livesay, *Andrew Carnegie and the Rise of Big Business* (New York, 2000), 45; Nasaw, *Andrew Carnegie* (cit. n. 54), 64.

moles burrowing in the dark," Carnegie implemented immediate reforms that "would enable us to know what our cost was for each process," with particular attention paid to "what each man was doing, who saved material, who wasted it, and who produced the best results."[60] Before long, Carnegie had become so fixated on costs that he made a point of personally scrutinizing his company's books on a regular basis, insisting that cost sheets be sent to him at least once a month (sometimes every week), even when he was traveling overseas. In this way, Carnegie found that he could ensure the profitability of his firm from afar, hardly ever having to leave New York City and travel to the actual site of his steel mills in Pennsylvania.

As the preceding example helps demonstrate, the logic of organization was brought into an industrial context by business leaders like Carnegie who were eager to integrate the operation of their factories more fully in order to gain additional control over all aspects of their production process. The logic of organization thus did not simply travel from one discourse into another. Much more important was its deliberate implementation by people who hoped that its material instantiation would pay dividends in the form of increased revenues. However, the corporate reconstruction of North America's political economy was far from uncontroversial. Populists from the western United States who rode the period's antimonopoly sentiment to national prominence during the 1880s and 1890s were particularly outspoken critics of the trend. Named after their support of the "people's party," populists tended to be rural farmers of limited means who strongly objected to the increasing consolidation of social, political, and financial power in the hands of moneyed elites from the East.[61] Indeed, populists even formed cooperative societies so that farmers and small manufacturers could pool their resources and stand up to the large railroad cartels they often accused of price-fixing schemes.[62] Thus, much like the debates over the cell state I outlined in the previous section largely turned on the question of whether individuals had to sacrifice their autonomy for the good of the whole, the populist campaign against corporate consolidation largely centered on whether ordinary citizens would retain the power of self-determination in the face of powerful corporate firms.

At the same time, the sentiment that North America's increasingly centralized business landscape undermined the promise of social mobility became increasingly widespread. As a result, social reformers largely shifted from advocating laissez-faire policies to demanding direct government intervention, hoping that stricter regulations might safeguard the integrity and accessibility of the marketplace. Calls to help level the economic playing field only grew more intense as some 1,800 industrial firms consolidated into just 157 corporations during the "Great Merger Movement" from 1895 to 1905.[63] As the so-called trust question was propelled onto the national stage, the U.S. Congress even began to hold hearings in which numerous business leaders, including J. D. Rockefeller, Andrew Carnegie, and J. P. Morgan, were publicly called to account for unfair and collusive behavior. By 1914, public distrust of corporate monop-

[60] Carnegie, *Autobiography* (cit. n. 54), 135.

[61] See Noam Maggor, *Brahmin Capitalism: Frontiers of Wealth and Populism in America's First Gilded Age* (Cambridge, Mass., 2017).

[62] Charles Postel, "A Farmer's Trust," in *The Populist Vision* (Oxford, 2007), 103–33. See also Sidney A. Rothstein, "Macune's Monopoly: Economic Law and the Legacy of Populism," *Stud. Amer. Polit. Develop.* 28 (2014): 80–106.

[63] Naomi Lamoreaux, *The Great Merger Movement in American Business, 1895–1904* (Cambridge, 1985).

olies had grown so fierce that the future Supreme Court Justice Louis D. Brandeis could write a best-selling screed taking direct aim at the "Curse of Bigness," excoriating an exclusive cadre of elite bankers and capitalists for making a fortune by cheating the common man out of the just rewards for his hard work and efforts.[64]

The period's widespread enthusiasm for trust-busting notwithstanding, it is noteworthy just how few of its most ardent proponents really took issue with the basic logic of organization that stood at the core of the issue. Although the suspicion that large conglomerates could reap ill-gotten gains by stifling competition fueled much of the period's legislative agenda, the Supreme Court's affirmation of the "rule of reason" to the Sherman Anti-Trust Act at the turn of the century showed that the latter was not primarily understood as a bulwark against corporate consolidation per se. Instead of combating the fact of monopoly power, the focus of antitrust legislation always remained more on checking its potential to erect barriers of entry into a market and other supposedly artificial restraints upon trade. Indeed, there was widespread agreement that "natural monopolies" would inevitably develop in industries with particularly high fixed costs. Even the period's most conspicuous trustbusters agreed that, under certain circumstances, monopoly power could actually work to the benefit of society. Theodore Roosevelt, for example, famously gave his explicit blessing to J. P. Morgan's scheme of merging the Tennessee Coal, Iron and Railroad Company with its main rival, U.S. Steel, on the grounds that it would help dampen the calamitous effects of the 1907 financial panic.[65] Not by coincidence, it was precisely this move that George W. Perkins would later defend by insisting that the merger only served to bring North America's political economy further in line with a universal tendency for everything to progress from disorder to organization.

At least in part, the logic of organization became so entrenched in American law and policy because a wide range of intellectuals embraced it as well, agreeing that competitive free market capitalism was no more than a transitory stage in the evolution of modern civilization. Hence, the consolidation of numerous small businesses into large, multidivisional corporate firms such as U.S. Steel was often described as no more than a step in the progressive march toward an industrial future wherein competition had become a thing of the past. The American engineer Charles W. Baker, for example, reasoned that whereas a "system of competition" may have been well "adapted" to "the formative period of civilization," the time had come to abandon "the cruelly terse 'survival of the fittest,'" which "was never meant to control the wondrously intricate relations of the men of the coming centuries."[66] According to rosy optimists such as Baker, the dawn of corporate capitalism represented the development of "a vast organism in which each individual, each community, each State, each nation has its prosperity and destiny indissolubly interwoven with the prosperities and destiny of every other one." For that reason, the period's most powerful "monopolies" could be seen as a harbinger for good things to come, having "grown out of the chaos with competition, or, as the student of the natural sciences expresses it, the survival of the fittest, as its mainspring."[67] Similarly, the social reformer Frank Parsons insisted that "monopoly means cooperation

[64] Louis D. Brandeis, *Other People's Money: And How the Bankers Use It* (New York, 1914).
[65] See Sklar, *Corporate Reconstruction* (cit. n. 9).
[66] Charles Whiting Baker, *Monopolies and the People* (New York, 1889), 187.
[67] Baker, "Public Control," quoted in William Dwight Porter Bliss, "Monopolies," in *The Encyclopedia of Social Reform* (New York, 1897), 888–94, on 889.

instead of conflict, wise management instead of planless labor, economy instead of waste."[68]

Nor were these sentiments confined to the United States. In the United Kingdom, for example, the widely read political economist John Hobson described how, under the "pressure" of "economic forces," industrial firms inevitably underwent "an expansion in size, a growing complexity of structure and functional activity, and an increased cohesion of highly differentiated parts." This evolutionary dynamic was reinforced by the tendency among captains of industry to realize they could increase profits by forming cartels, "raising the cooperative action so as to cover the whole" and thereby "reducing the competition to zero." As a consequence, it was not long before entire sectors of industry were consolidated into a few immensely large syndicates, forming a kind of "company of companies" in which each firm was "incorporated as single cells in the larger organism." This posed a pressing problem for all of society, as "the economic object of a Trust is to substitute monopoly for competitive prices." But while Hobson did regard it as absolutely incumbent upon those engaged in the "scientific study of modern industry" to "discover" how the "forces of conscious reform" could tame the industrial beast, he also cautioned against turning back time. Instead of trying to undo the process of industrial integration, he judged it far more expedient that it be pushed to completion by having the state socialize the nation's productive capacity entirely. An argument to the contrary was an argument against nature, he insisted, for "the progressive socialization of certain classes of industry" was just as "natural and necessary" as the process by which "machine-industry superseded handicraft and crystallized in ever larger masses," so that the "indictment against social control over industry is an indictment against a natural order of events."[69]

An especially striking example of the degree to which faith in the evolution of organizational efficiency could make strange bedfellows between utopian socialists and wealthy capitalists was King C. Gillette. Gillette would eventually make his fortune in the mass production of a disposable safety razor, but before he even filed his first patent application or built a multinational company, Gillette began writing a series of idiosyncratic books that might be described as nineteenth-century science fiction if their author was not so earnest and sincere in bringing his dreams to fruition. Part screed against the wastefulness of unregulated competition and part blueprint for an industrial new world order, Gillette's first book advocated the incorporation of a "united stock company" that would organize the production and distribution of all the world's goods and services both more efficiently and also more equitably than free markets possibly could. By rationalizing the planet's productive capacities and replacing mindless competition with long-range planning, Gillette had no doubt that, given a devoted army of skilled and benevolent managers such as himself, a powerful corporation such as the one he envisioned could make poverty and corruption a thing of the past. This company, which he would actually go on to charter in Arizona Territory under the name "World Corporation," was initially tasked with purchasing shares in other, existing concerns. By reinvesting the dividends from its expanding stock holdings to

[68] Frank Parsons, *The Public Ownership of Monopolies* (Philadelphia, 1894), quoted in Bliss, "Monopolies" (cit. n. 67), 891.

[69] J. A. Hobson, *The Evolution of Modern Capitalism: A Study of Machine Production* (London, 1895), 115–6, 126, 130, 137, 352, 361–2.

purchase more of the same, Gillette thought it would eventually acquire an ownership stake in the globe's entire productive capacities. At that point, it would have achieved its end goal of becoming a total governance structure, with sufficient power and scope to regulate all human affairs to everyone's mutual benefit from the top down.[70]

It is easy to poke fun at Gillette's plans for a "World Corporation" as the naive pipe dream of an unruly capitalist, but its basic premise that large corporate firms served an enlightened governance function could not be dismissed out of hand. After all, not only did other industrialists such as Carnegie (who liked to describe himself as an evolutionary instead of a revolutionary socialist) share a similar optimism about the power of organization, but so too did some of the period's most respected political economists. Whereas mid-nineteenth-century American views on matters of commerce were piously liberal, the next generation of social theorists was more likely to have been steeped in the teachings of the German Historical School. Many received their graduate training in that nation's renowned universities, and they often criticized the methodological principles of their predecessors as sterile, overly mathematical, and highly deductive, making them incapable of capturing the dynamics of an evolving system that was constantly changing. By 1885, this group had become sufficiently cohesive that they banded together to form the American Economic Association, whose official platform denounced "the doctrine of *laissez-faire* . . . [as] unsafe in politics and unsound in morals."[71] Among the group's most interesting and influential members was Richard T. Ely, who had studied with Karl Knies in Heidelberg. In a fascinating book that was first published in 1903—*Studies in the Evolution of Industrial Society*—Ely narrated the history of modern capitalism as a process of "integration, a binding together." On this view, the competitive theories of classical liberals like Smith and Ricardo only applied to a relatively primitive, agrarian society such as the one they had inhabited themselves. During more recent times, however, the "abuses that appeared with the factory system" had rightly caused a "reaction against the *laissez-faire* policy," leading, eventually, to the "trust movement," which resulted in "the integration of allied industries." The power of Darwinian natural selection notwithstanding, Ely therefore concluded that "Men learn to act together in increasingly large number and for increasingly numerous purposes." By creating greater and more inclusive social organizations, the evolutionary process ensured that competition would be relegated to higher and higher levels of individuality, from the cell to the organism to the family group until, eventually, it would disappear almost entirely.[72]

[70] See King C. Gillette, *The Human Drift* (Boston, 1894); Gillette, *World Corporation* (Boston, 1910). For a best-selling work of real science fiction that bore a striking resemblance to Gillette's vision, see Edward Bellamy, *Looking Backward, 2000–1887* (Boston, 1888).

[71] "Platform of the American Economic Association," in *Report of the Organization of the American Economic Association* (Baltimore, 1896), 5–6. See also Howard Brick, *Transcending Capitalism: Visions of a New Society in Modern American Thought* (Ithaca, N.Y., 2006); Mary S. Morgan, "Competing Notions of 'Competition' in Late Nineteenth-Century American Economics," *Hist. Polit. Econ.* 25 (1993): 563–604; Daniel T. Rodgers, *Atlantic Crossings: Social Politics in a Progressive Age* (Cambridge, Mass., 1998), 76–111.

[72] Richard Theodore Ely, *Studies in the Evolution of Industrial Society* (New York, 1903), 9, 15–7, 60, 64, 89, 134, 154.

CONCLUSION

By the very end of the long nineteenth century, then, the logic of organization had come to inform nearly all aspects of life in an industrial society such as the United States. It structured not only the way people understood the evolution and development of biological species, but also the running of large corporate firms and the reforms being implemented on a municipal, state, and federal level. With remarkable literalness, nearly everything was turned into an object of organization, subject to managerial oversight, rational planning, and top-down control. But in the wake of World War II, increasingly vocal attacks on the logic of organization began to emerge from all sides of the political spectrum, particularly in the United States. Around the same time that Joseph Schumpeter's valorization of entrepreneurial independence and creativity really began to take hold, for example, William Whyte published his withering takedown of the conformism, lack of imagination, and opposition to spontaneity in the large corporate firm. Writing in the vein of a naturalist, he described the poor species that inhabited this unenviable ecological niche—aptly calling it the "organization man"—as one in the thralls of a "bureaucratic ethic," slavishly beholden to "a belief in the group as the source of creativity; a belief in 'belongingness' as the ultimate need of the individual; and a belief in the application of science to achieve the belongingness."[73] Of course, it was precisely the same period that also saw a resurgence of interest in classical liberalism among academic economists. The rise of the so-called Chicago School of neoclassical economists is well known, but it is nonetheless worth quoting from Milton Friedman's full-throated defense of individual liberty, if only to emphasize the degree to which some of his deepest antipathies were directly targeted at the logic of organization: "The organismic, 'what you can do for your county' implies that government is the master or the deity, the citizen, the servant or the votary," he observed, adding that "to the free man, the country is the collection of individuals who compose it, not something over and above them."[74] From the perspective of Cold War America, few things, it seems, had come to look so outdated and dangerously naive as the nineteenth-century obsession with the progressive development of ever more inclusive part/whole relationships.

Tellingly, this was no less the case in biology than in political economy. There, too, top-down explanations in which the whole determines the behavior of its parts were increasingly replaced by bottom-up explanations wherein even the most exquisitely complex organizations could be derived from the (often self-interested) behavior of its constituent entities. For example, postwar biologists inspired by the mathematical models of population geneticists such as R. A. Fischer mounted a vigorous and remarkably successful campaign against group selection, the idea that evolution favored individuals with traits that were good for the collective or species at the expense of themselves.[75] In its place, a new generation of theoretically inclined biologists such as John Maynard Smith, William D. Hamilton, and George C. Williams, among others, articulated a new conception of life and its history, one in which altruistic acts

[73] William H. Whyte, *The Organization Man* (New York, 1956), 7.

[74] Milton Friedman, *Capitalism and Freedom* (Chicago, 1962), 1. See also Angus Burgin, *The Great Persuasion: Reinventing Free Markets since the Depression* (Cambridge, Mass., 2012).

[75] See George C. Williams, *Adaptation and Natural Selection: A Critique of Some Current Evolutionary Thought* (Princeton, N.J., 1966).

of self-sacrifice were reinterpreted as selectively advantageous from a genic perspective, effectively displacing the organism from its central place in biology. As this happened, the process of evolution was fundamentally altered, now taking place on the level of genes rather than individuals, let alone higher-order social assemblages. Indeed, in the hands of its most skilled and most controversial proselytizer, the organism came to be seen as no more than a puppet in life's splendid drama, with self-replicating molecules cast as the ones who were truly pulling the strings. "We are survival machines," Richard Dawkins insisted, "robot vehicles programmed to preserve the selfish molecules known as genes."[76] Given this remarkable transformation taking place before their very eyes, is it any surprise that so many mid- to late twentieth-century historians would stress the degree to which biologists and political economists made common cause in advancing the claim that autonomous agents competing for access to scarce resources drove large-scale, world-historical change? But, as I have tried to suggest here, from the vantage point of the late nineteenth century, the logic of organization would have been seen as a far more powerful tool with which to understand all aspects of life, including the economic behavior of our multicellular selves.

[76] Richard Dawkins, *The Selfish Gene* (New York, 1976), xxi.

"Scientific Crude" for Currency:
Prospecting for Specimens in Stalin's Siberia

*by Julia Fein**

ABSTRACT

Responding to the Soviet state's call to expand export for currency during the First Five-Year Plan (1928–32), Siberian scientific personnel—pointing to Siberia's importance in the history of global science and exploration—created and promoted a new category of commodity, calling it "scientific" or "museum crude." These uniform sets of objects of the natural and human sciences of Siberia represented a departure from existing international specimen trade in that the expeditions to extract scientific/museum crude relied on institutions and techniques of state socialism. In this vision, scientific goods became a nationalized resource subject to state planning and capitalization. However, once collected, these commodities came up against characteristically Soviet/Stalinist barriers: after being trumpeted as a contribution to Soviet trade, most of these collections were stopped at the border because of a shift in Soviet political culture. Instead of contributing profits to Soviet international trade, therefore, scientific crude's accumulation in Moscow and Leningrad functioned as a way of requisitioning value from the periphery to the center, not unlike state treatment of other resources. Suggesting that scientific/museum crude represents the ultimate logic of capitalism as applied to the scientific spheres, this essay argues that state socialism itself created the institutional spaces that fostered and then shut down this extreme iteration of capitalism-science entanglement.

During Joseph Stalin's First Five-Year Plan (1928–32), the centralized Soviet state called upon its subjects to look at their regions through foreign eyes. Specifically, citizens and institutions were mobilized to view the country's potential products from the point of view of foreign individuals and agencies possessing internationally viable currencies. Foreign currency was vital for Soviet purchase abroad of tractors, factory equipment, and other accoutrements of modernization that were unavailable domestically. This extractive, self-prospecting drive to convert national wealth into foreign money began with signals from the state but was carried out by socialism's enthusiastic participants. The campaign emphasized not only maximizing quantities of existing types of export—authorities were already administering the sale of gold and other metals, religious heritage objects, and works of art—but also exhorting Soviet people to create new understandings of what constituted an export commodity to begin with.[1]

* Independent scholar; jfein@uchicago.edu.

[1] On the art, religious object, and gold trades, see Anne Odom and Wendy Salmond, eds., *Treasures into Tractors: The Selling of Russia's Cultural Heritage, 1918–1938* (Washington, D.C., 2009); E. A. Osokina, *Zoloto dlia industrializatsii: "Torgsin"* (Moscow, 2009).

Even at the peak of its anticapitalist ideological zeal, therefore, the early Soviet Union relied on international markets to supply itself with capital goods, and on Soviet citizens to think creatively about what constituted commodifiable national wealth.

Responding to the state's call to expand export, Siberian scientific personnel created and promoted a new category of commodity, calling it "scientific" or "museum crude" (*nauchnoe/muzeinoe syr'ë*).[2] They legitimated this move by invoking Siberia's longstanding contributions to the sciences since the eighteenth century. World war, revolution, and civil war had interrupted the flow of Siberian goods into international scientific markets for over a decade, and the region's rich resources in the natural and human sciences could therefore be expected to command high prices. To profit from this imagined demand, Siberian scientific agencies and museums proposed a five-year plan to organize expeditions specifically to acquire maximally uniform "sets" of specimens for export to foreign museums and scientific institutions.[3] These sets of ethnographic, archaeological, zoological, mineralogical, and botanical objects consisted of one to two hundred well-labeled things, photographs, and maps that, together, were supposed to provide a representative snapshot of scholarly knowledge about Northern Asia. Locally contracted collectors also kept careful documentation of their acquisitions to maximize the collections' scientific and thus monetary value. The prices of these sets were to be established by state trade agencies when they arrived in Moscow for transport abroad.

When the project began in 1929, subjecting scientific collecting expeditions to planning and capitalization by the state aligned perfectly with Soviet political and economic morality. However, the expeditions to "produce" scientific crude were shut down after only two field seasons and politically denounced five years after they began. With completed collections mostly stopped in the capitals before crossing the border, many of the objects were reinvented as national patrimony and removed both from international market circulation and from regional museums. Since Moscow and Leningrad institutions refused to return the collections to Siberia, scientific crude functioned—against its inventors' intentions—as a mode of cultural requisitioning from regions to centers.

This essay is about the origins, conduct, and demise of this ambitious regional attempt to realize the commodity category "scientific/museum crude" by planning for the creation of an international market in these specimen "sets." The episode of scientific entrepreneurship described in this essay occurred in a rural country lacking a robust legacy of capitalist development. Moreover, the USSR regarded itself at the time in question as being in direct political-ideological and economic opposition to capitalism. Regardless of being part of a purportedly Marxist state, however, the historical actors in this essay applied extractive, profit-oriented principles to regional resources based on numerous factors, including an understanding of capitalist investment, risk, and projections of future profits as applied to objects of science; notions of "salvage

[2] The Russian word being translated here as "crude" could also be rendered as "raw, unprocessed material." I chose the English "crude," although almost exclusively applied to oil, to evoke the deliberately extreme conceptualization of these things along the lines of a natural resource subject to planned extraction to meet a calculable yet market-responsive demand, in contrast to the more culturally sensitive market for art and icons.

[3] This scheme is alluded to in passing in a small number of Russian-language works, and nowhere in English. Its most sustained treatment of this campaign occupies one page in V. S. Sobolev, *Dlia budushchego Rossii: deiatel' nost' Akademii nauk po sokhraneniiu natsional'nogo kul'turnogo i nauchnogo naslediia 1890–1930 gg.* (St. Petersburg, 1999), 128–9.

anthropology" and knowledge of international markets in particular kinds of museum collections;[4] an understanding of how advertising and promotion create demand for novel goods; and a savvy, strategic command of professional rewards available to scientists willing to facilitate the exchange of specimens for money. The fact that the Soviet state and its scientists fall squarely within these regimes of international capitalist profit seeking indicates that Soviet actors—even during the USSR's most extreme self-articulation under early Stalinism—operated fully within the long-unfolding, global history of accumulation and exchange of objectified and commodified scientific knowledge for money.

In this essay, I argue that it was the Soviet inventors of the "scientific crude" commodity category who suggested the ultimate logic of capitalism in the scientific/cultural spheres. While paintings, indigenous masks, or animal skins had long been bought and sold, it was a new idea to apply an industrial projection model of production to artifacts of the natural or human sciences, subjecting them to planned extraction—capitalized by the state—for an international market as if they constituted a bulk natural resource. Soviet regional scientists took "capitalist" ways of seeing the world to lengths not observed in the West because of—not despite—the peculiar imperatives of state socialism.[5] They did so even as contemporary Western museums were seeking to remove collections from market circulation, and even as Western scientists increasingly asserted that they served truth rather than profits. I argue in this essay that state socialism itself created the institutional spaces that fostered the extreme, explicit iteration of capitalism-science entanglement represented by scientific/museum crude. This essay thus joins Arunabh Ghosh's contribution to this volume in maintaining that we should take seriously socialist actors' self-understandings regarding the differences between their scientific practices and those of the West—as these can reveal underlying characteristics about science and capitalism that remain invisible when we only examine Western cases—while simultaneously examining the substance of those trumpeted differences with a critical eye.[6] In his foundational work on commodities, Igor Kopytoff alluded to the fact that it was precisely in socialist (state) economies that the commodification we associate with capitalism was observed to "expand into novel areas."[7] A socialist case study may therefore be more sharply illustrative than a capitalist one of the complexity of market logics and values regimes inherent to the scientific marketplace.

This essay begins by briefly situating the Soviet invention of scientific/museum crude within broader histories. I then document the planning and execution of this endeavor to join state-led economic planning to the specimen trade, illustrating the larger issues at stake in these collecting projects and their fates. I explain how creation and valuation

[4] James J. Hester, "Pioneer Methods in Salvage Anthropology," *Anthropol. Quart.* 41 (1968): 132–46. For an exploration of competitive global markets in ethnographic goods in the formative years of the inventors of scientific crude, see H. Glenn Penny, *Objects of Culture: Ethnology and Ethnographic Museums in Imperial Germany* (Chapel Hill, N.C., 2003).

[5] On Western scientists' assertions during the period under discussion, see Harold J. Cook, "Sciences and Economies in the Scientific Revolution: Concepts, Materials, and Commensurable Fragments," in this volume.

[6] Arunabh Ghosh, "Lies, Damned Lies, and (Bourgeois) Statistics: Ascertaining Social Fact in Mid-century China and the Soviet Union," in this volume.

[7] Igor Kopytoff, "The Cultural Biography of Things: Commoditization as Process," in *The Social Life of Things: Commodities in Cultural Perspective*, ed. Arjun Appadurai (Cambridge, 1988), 64–91, on 72–3.

of Soviet scientific objects as international commodities and national resources crystallized new political frictions, in turn causing these collections to linger on the uneasy border of commodification before their preemptive removal from commercial circulation.[8] This turnaround occurred against a backdrop of changes in both the international market for museum goods and Soviet state rhetoric and policy between 1929 and 1934. The essay's conclusion posits the specificity of issues raised by Soviet scientific crude within larger histories of scientific entrepreneurship and big data.

CREATING SCIENTIFIC CRUDE

Russian museum development had broadly followed a Western norm, beginning with Peter the Great's Kunstkamera (1714), for the contents of which he was particularly indebted to the state-directed exploration of Siberia by German scholar-explorers. The eighteenth century saw the intensive development of the Russian Academy of Sciences museums, followed by regionally oriented scientific museums. These regional museums began in Siberian cities associated with high wealth potential (mining or trade), subsequently appearing throughout the Russian Empire following the Great Reforms (1861–74), and especially blossoming in Siberia during the 1870s to 1890s thanks to the influx of educated but impoverished political exiles provided by the early revolutionary movement. Many of these exiles participated in the museum sciences as a way of applying their educations and supplementing their incomes.

By the mid- to late nineteenth century, participation in the museum sciences could be profitable for Russians as for Americans and other Westerners (though British scientific culture retained a preference for independently wealthy, disinterested practitioners).[9] Income was to be made either through buying and selling specimens—"bird skins are capital," wrote one American in 1874[10]—or through offering one's expert services in valuation, interpretation, or commissioned collecting in a kind of "scientific consulting" practice.[11] Despite the gulf between late nineteenth-century Russian and American capitalist development in terms of railroad infrastructures, futures markets, entrepreneurship, and commercial enterprise, Russian scientific collectors and museum workers were full members of a transnational culture of commercially inflected science. As scientists traded, bought, and sold specimens—with or without individuals or specialized businesses as middlemen—the market was a key vehicle by which knowledge traveled from the local to the global level.

By the twentieth century, states and institutions were much more involved than hitherto in issues of scientific knowledge and value. World War I had ushered in new kinds

[8] On valuation as a lens onto social frictions, see Viviana Zelizer, "Human Values and the Market: The Case of Life Insurance and Death in 19th-Century America," *Amer. J. Sociol.* 84 (1978): 591–610; Jens Beckert and Patrik Aspers, eds., *The Worth of Goods: Valuation and Pricing in the Economy* (New York, 2011).

[9] Paul Lucier, "Commercial Interests and Scientific Disinterestedness: Consulting Geologists in Antebellum America," *Isis* 86 (1995): 245–67.

[10] Mark Barrow, "The Specimen Dealer: Entrepreneurial Natural History in America's Gilded Age," *J. Hist. Biol.* 33 (2000): 493–534, on 523.

[11] On scientific consulting, specimen dealing, and other kinds of profitable scientific activities, see Susan Sheets-Pyenson, *Cathedrals of Science: The Development of Colonial Natural History Museums during the Late Nineteenth Century* (Kingston, Ontario, 1988); Lucier, "Commercial Interests" (cit. n. 9); Penny, *Objects of Culture* (cit. n. 4); Paul Lucier, "The Professional and the Scientist in Nineteenth-Century America," *Isis* 100 (2009): 699–732.

of cooperation among state agencies, businesses, and scientific-technical expertise in all belligerent countries. With the imperatives of intensified economic management required to remain in a long, industrial war came increased state need for information about natural and human resources, and this self-knowledge remained a central feature of modern sovereignty after the war ended.[12] In Russia, specialists in the Russian Academy of Sciences had voluntarily formed the Commission for the Study of Natural Productive Forces (KEPS) to help with the war effort.[13] With language taken loosely from Karl Marx and Friedrich Engels, the conceptualization of wealth potential as "natural-productive forces"—including human bodies—depended on an understanding of regions as primary economic units to know and mobilize these "forces."[14] Invocations of natural-productive forces and their organized maximization were used to argue for regional as well as centralized state control during the revolutionary and civil war years. The later creation of scientific crude—a category eliding differences among shamans' drums, seeds of exotic northern plants, and ancient arrowheads—as a raw resource with hard currency value came out of the epistemological and institutional framework of natural-productive forces in the context of shifting science-state-capital relations common to modern states following World War I.

The case of scientific crude demonstrates how Siberian research institutions applied planning and projection techniques for state socialist industrial production to scientific collecting expeditions. This required adapting the accounting and accountability norms for investment and output (returns) in a command economy to the "production" of "raw" museum goods from "nature." The process of pricing fundamentally defines the emergence of new objects into a commodity form, even as prices are critical communication tools telling buyers that the commodity is desirable.[15] In the case of nineteenth-century trade in dinosaur fossils described elsewhere by Lukas Rieppel, price negotiations played a critical role in conveying the emergence of a new value regime—a commercial market in a particular kind of substance dug up from the ground (fossils) based on their desirability to scientists located far away.[16] In the case of Siberian scientific crude for export, the state-generated communications about prices—based on a mix of real and imagined knowledge about international market values in museum goods—likewise marked a shift to a new value regime, in this case toward the expansion of a state command economy beyond industry, into every conceivable area of Soviet life.

Two weeks after the stock market crash of 1929 that devastated the economies of the North Atlantic and continental Europe, Joseph Stalin celebrated the twelfth anniversary of the Bolshevik Revolution of 1917 with the declaration that the Soviet Union was making its "Great Break" into socialism. The Great Break discontinued the legacy of Lenin's compromise between capitalism and communism in the form of the New

[12] See Ghosh, "Lies" (cit. n. 6).

[13] On KEPS, see Alexander Vucinich, *Empire of Knowledge: The Academy of Sciences of the USSR (1917–1970)* (Berkeley and Los Angeles, 1984), 101–6.

[14] The fundamental work on productive forces was Karl Marx and Friedrich Engels, *The German Ideology*, ed. Christopher John Arthur (1846; repr., New York, 1970).

[15] See Marion Fourcade, "Price and Prejudice: On Economics and the Enchantment (and Disenchantment) of Nature," in Beckert and Aspers, *The Worth of Goods* (cit. n. 8), 41–62, on 44–5.

[16] Lukas Rieppel, "Prospecting for Dinosaurs on the Mining Frontier: The Value of Information in America's Gilded Age," *Soc. Stud. Sci.* 45 (2015): 1–26, on 7. On prices as communication, see also Patrik Aspers and Jens Beckert, "Value in Markets," in Beckert and Aspers, *The Worth of Goods* (cit. n. 8), 3–38, on 27.

Economic Policy—declared in 1921 and called "state capitalism" by Lenin himself—and ended the Leninist practice of protecting scientific experts and encouraging regional scientific institutions to proliferate and flourish.[17] As the United States, Great Britain, Germany, and France tumbled successively into depression, the Soviet Union's economy grew under Stalin's leadership.[18] During the First Five-Year Plan, Soviet power battled to modernize the country's economic base and cultural level through industrialization, collectivization, and the purging of subjects perceived as disloyal. These subjects included scientific and technical specialists as well as kulaks and other real and imagined political enemies. Indeed, throughout the rest of Stalin's lifetime, scientists—like the general population—would be alternately mobilized and destroyed for the alleged sake of securing socialism's national security.

Soviet citizens on a day out at the regional studies museum in Novosibirsk in 1930 learned of another in a long series of enemies to be eliminated: "export-illiteracy."[19] The pedagogy of the centrally mandated *Liquidation of Export Illiteracy!* exhibition was aimed as much at those designing it as at the visitors. In 1929, the State Import-Export Trade Bureau, Gostorg, had begun to lean on the Soviet Union's scientific-technical apparatus to broaden existing conceptions of "raw material" [*syr'ë*] for export and to extract those new resources.[20]

Early in 1930, the journal *Soviet Regional Studies* [*Sovetskoe kraevedenie*] had featured a piece by ichthyologist Aleksei Berezovsky publicizing the idea of conducting archaeological and ethnographic collecting expeditions for export. He insisted that the expeditions be conducted by specialists and people experienced in museum work in order to maintain a strict standard of scientific value. Berezovsky noted this scheme's Siberian origins during 1929 and pointed out the particular suitability of Russian Asia for the mining of these "scientific exports."[21] The category of "museum *collections*" was already included in the central planning of new exports and consisted of now well-known sales of arts, antiquities, and other items already in Soviet museums and zoos. Siberian museum or scientific crude, by contrast, would be funneled into international exchange almost immediately after its extraction from "nature."

The invention of scientific/museum crude and the plan to tap regional museum workers' expertise to create this export good originated in the Society for the Study of Siberia and Its Natural Productive Forces based in Novosibirsk, a city re-creating itself as the new central node of long-established Siberian scientific networks.[22] The

[17] On Soviet science and state policy in the 1920s, see Kendall Bailes, *Science and Russian Culture in an Age of Revolutions: V. I. Vernadsky and His Scientific School, 1863–1945* (Bloomington, Ind., 1990); Vera Tolz, *Russian Academicians and the Revolution: Combining Professionalism and Politics* (Houndmills, Basingstoke, 1997); James Andrews, *Science for the Masses: The Bolshevik State, Public Science, and the Popular Imagination in Soviet Russia, 1917–1934* (College Station, Tex., 2003).

[18] On the importance of the Great Depression to Soviet understandings of their own economic growth, see Stephen Kotkin, *Magnetic Mountain: Stalinism as a Civilization* (Berkeley and Los Angeles, 1995); Steven Marks, "The Word 'Capitalism': The Soviet Union's Gift to America," *Society* 49 (2012): 155–63, on 156.

[19] Perepiska po razvitiiu nauchno-issledovatel'skoi raboty po ekportu, State Archive of Novosibirsk Region (hereafter cited as "GANO"), f. R-217, op. 1, d. 123.

[20] Gostorg to Nauchno-tekhnicheskoe upravlenie of Vesenkha (NTU VSNKh) SSSR, 1929, Russian State Archive of the Economy (RGAE), f. 3429, op. 7, d. 3645, ll. 18–18ob.

[21] A. Berezovskii, "Ob uvelichenii eksporta," *Sovetskoe kraevedenie* 1–2 (1930): 27–30.

[22] The stated goal of the organization, established in 1925, was to study the natural and human riches of the region and give rational assistance in economic development of those resources. "Obshchestvo po izuchenii proizvoditel'nykh sil Sibiri," *Sovetskaia Sibir'* 24 (30 January 1925): 4.

proposal originally focused on biological material and explicitly harkened back to the imperial period when foreign scientists had to enter a competitive seller's market to outbid one another for Siberian scientific collections.[23] With input from Siberia's state trade firm, Sibtorg, the project quickly expanded to include ethnography and then archaeology.[24] Some regional Sibtorg representatives were particularly well qualified to judge the profitability of trade in Siberian science, since they had personal experience with it during their years in czarist exile: one former exile working on fur exports in a Tomsk Sibtorg branch in 1929 recalled how he and other exiles made a lot of money selling their herbaria, insects, and osteological collections to foreigners before the Revolutions of 1917.[25]

Scientific crude's historical significance lies in its co-option of scientific and cultural artifacts into industrial-style state investment and planning, by assuming the possibility of collecting uniform sets of nearly identical objects on a five-year plan coordinated with USSR-wide socialist construction.[26] The scheme was grounded in contemporary standards of international science and reputable markets in scientific goods insofar as only trained museum workers and other experts were entrusted with collecting and labeling things.[27] In a study of recent valuation processes related to compensation for ecological damage, Marion Fourcade concluded, "Modern social institutions spend considerable time and effort measuring what seems unmeasurable and valuing what seems beyond valuation in the service of enhancing their own capacities for calculation, crafting new opportunities for profit, or expanding their jurisdictional authority."[28] The case of scientific crude in Soviet Siberia presents a particularly stark example of a characteristic common to modern parastatal agencies: the institutional creators of scientific crude had opportunistic motives for inventing this category of value insofar as it expanded their funding, their potential power over other scientific agencies, and their access to desired things. Based on its initiative in creating this export category, the Novosibirsk Society proposed that its oversight over scientific crude be extended from Siberia to the Far East, Kazakhstan, and even socialist Mongolia.[29] In this way, their proposal created an export category that only they (i.e., the trained naturalists affiliated with the Society for the Study of Siberia) were competent to assemble and assess, not unlike Lee Vinsel's psychologists vis-à-vis safe driving, described elsewhere in this volume.[30]

The society did not neglect the importance of advertising. Translated and internationally circulated Soviet research in the sciences represented by scientific crude would speak to the sophistication and significance of the "raw materials" for sale. Short of funds for publication since their origins in the 1880s, Siberian museum orga-

[23] Society for the Study of Siberia to Sibkraigostorg, 9 March 1929, GANO, f. 41, op. 1, d. 259, l. 250.

[24] Sibtorg to the Society for the Study of Siberia, 26 May 1929, GANO, f. 41, op. 1, d. 259, l. 270; Stat'ia "Zabytyi uchastok," in "Informatsionnyi biulleten' eksportnykh organizatsii" #7(101) ot 15.XI.1929, cited in Sibtorg to Novoeksport, 16 November 1929, GANO, f. 41, op. 1, d. 259, l. 234.

[25] Sibtorg Tomsk office to Sibtorg, 29 May 1929, GANO, f. 41, op. 1, d. 259, l. 273.

[26] Protokol tret'ego zasedaniia eksportnoi komissii Obshchestva izucheniia Sibiri, 13 May 1930, GANO, f. R-217, op. 1, d. 123, l. 125.

[27] Perepiska (cit. n. 19), l. 57ob.

[28] Marion Fourcade, "Cents and Sensibility: Economic Valuation and the Nature of 'Nature,'" *Amer. J. Sociol.* 116 (2011): 1721–77, on 1723.

[29] Perepiska (cit. n. 19), l. 250ob.

[30] Lee Vinsel, "'Safe Driving Depends on the Man at the Wheel': Psychologists and the Subject of Auto Safety, 1920–55," in this volume.

nizations hoped that this proposal for sales promotion would convince Moscow that it was in the state's economic interest to fund such publications to maximize export profits.[31] Publication would, of course, also benefit the scientist-authors, who would retain copyright of authorship [*avtorskoe pravo*] on research drawn from any collections they assembled for export.[32]

This outline of the scientific/museum crude concept offers a direct window onto the multiplicity of value registers occupied by these (yet only imagined) collections.[33] Scientific personnel saw the potential for a return in the form of professional capital through publications, thus building their reputations. This and all other potential dividends rested on convincing the project's investor (the Soviet state) that these collections had a hard currency value to the foreign researchers and institutions at whom they were marketed. This monetary value would only come from buyers' belief that the collections were responsibly documented and could thus generate new natural or social knowledge, or that the collections would effectively display meaningful knowledge to a foreign paying public (or to a nonpaying but counted public, in the case that the museum in question received provisional public funds dependent on visitation statistics). Contemporary Soviet people, saturated with the chronopolitics of early Stalinism, may have understood the value of the ethnographic component of museum crude in yet a different way: as the First Five-Year Plan aimed to wipe out "remnants of the past," these collections might constitute a final archive of the "old ways of life," as well as providing a starting point of measurement for the degree of successful Soviet modernization. Finally, some Soviet people might—and by 1934 certainly did—see especially the archaeological collections as national patrimony.

A shift toward stricter state control over the economy over the course of the Five-Year Plan was accompanied by an ideological turn in Soviet life toward more patriotic (and xenophobic) values.[34] Epitomized by the barriers that ultimately kept museum crude from leaving the USSR and entering international markets in this case, the concept of "national patrimony" allowed Stalinists to redefine a wider range of objects as "priceless" than the "unique objects" that even scientific crude's proponents would not sell. In invoking national patrimony, Stalin-era state and scientific actors began to control flows of museum collections in a way that was broadly in conformity with contemporary international norms—far more so than Soviet sales of artworks had been.[35] For ordinary people who understood scientific crude as national heritage and believed it should not leave the country, submitting these objects to international markets would have been a kind of profanation. On the other hand, others may have seen marketization of these objects—regardless of their cultural meaning—as a kind of patriotic sacrifice in the cause of obtaining currency to build the socialist future.[36] The ambiguities

[31] Perepiska (cit. n. 19), l. 250ob.

[32] Protokol tret'ego zasedaniia, GANO, f. R-217, op. 1, d. 123, l. 125.

[33] For a discussion of multiple values registers, see Claes-Fredrik Helgesson and Fabian Muniesa, "For What It's Worth: An Introduction to Valuation Studies," *Valuation Stud.* 1 (2013): 1–10, on 7.

[34] Nicholas Timasheff, *The Great Retreat: The Growth and Decline of Communism in Russia* (New York, 1946).

[35] On archaeology, national patrimony, and debates about keeping these objects in the fatherland before the Revolution, see Ekaterina Pravilova, *A Public Empire: Property and the Quest for the Common Good in Imperial Russia* (Princeton, N.J., 2014), 129–78.

[36] This understanding of how flexibly marketization can be related to the profane and the sacred is indebted to the discussion on putting a monetary value on a loved one's death in Zelizer, "Human Values" (cit. n. 8), 594.

of this extreme case of scientific commodification only underscore the variety of registers of value that have long defined museum and scientific objects as they move into and out of markets.

The Society for the Study of Siberia needed to keep these diverse potential value registers in play to convince collectors, funders, and potential customers to participate in scientific crude. The society's activists had to convince the state trade bureau (Gostorg) and New Export (a state trading firm founded in Moscow in early 1929) of two things. First, they made the case that this was the kind of creative and profitable innovation in export thinking that was being asked of them to "liquidate export-illiteracy." Second, they had to offer reassurance that the society had the organizational power to supervise state investment capital without prompting responses of "local patriotism"—collectors aware that objects bound for export would disappear from Siberia forever with the destruction of the old life already under way might need to be prevented from hoarding these objects instead of turning them over for sale. To this end, the society emphasized to its commissioned collectors that scientific crude was only a continuation of a process already familiar from the imperial period but unnaturally interrupted by war and revolution.[37] European explorers, merchants, and middlemen had been coming to Siberia as scientific buyers and entrepreneurs since the eighteenth century. This foreign trade, the society reminded collectors, had halted artificially because of the geopolitical upheaval of 1914–22: its stoppage was not due to any reversal in proper values surrounding science and national heritage.

Citing the initiative of Sibtorg's "Miscellaneous Export" department in coming to Gostorg with the society's satisfactory proposals for "scientific crude," Moscow promised money and scarce (deficit) goods for archaeological digs and to exchange with national minorities for objects that would have ethnographic value to foreigners. Although the expeditions—including expensive archaeological collecting trips on the Angara River and in the Minusinsk and Altai regions—were to receive rubles from the Siberian and central offices of New Export, both Novosibirsk and the Siberian localities preferred to use deficit goods, rather than cash, to barter with local populations as a result of the state's capital shortage. In addition, keeping currency out of the domestic exchange relationship in acquiring scientific crude was intended to address anxieties surrounding the archaeological expeditions' potential to stimulate rural populations to dig for material profit, that is, without proper scientific documentation.[38] In other words, scientific crude's extractors faced the same potential competition from rural archaeological entrepreneurs as Russian archaeologists had for the past half-century.

To complete the organizational outline, final decisions and actions were transferred to the centers: the State Academy of the History of Material Culture in Leningrad was responsible for approving the archaeological collections for export, having verified

[37] E. N. Orlova, "Etnograficheskie kollektsii, kak novyi vid eksport dlia Sibkraia (plan organizatsii zagotovok etn. kollektsii na 1929–1930 gg.)," State Archive of Irkutsk Region (hereafter cited as "GAIO"), f. R-565, op. 2, d. 44, ll. 19–26; Protokol vtorogo zasedaniia, GANO, f. R-217, op. 1, d. 123, ll. 128, 138; Doklad Nikolaia Auerbakha, GANO, f. R-217, op. 1, d. 123, ll. 146, 154; Stenogramma soveshchaniia predstavitelei uchrezhdenii po voprosu ob organizatsii eksporta estestvenno-nauchnykh kollektsii, 1930, GANO, f. R-217, op. 1, d. 192, l. 3.

[38] This came up both in the May meeting in Novosibirsk and in the "instructions" put together in the East Siberian Geographic Society in Irkutsk on how to conduct these archaeological expeditions in a way that would discourage local populations from digging on their own. GAIO, f. R-565, op. 2, d. 44, ll. 37–41; 52ob.

that they contained no unica; Moscow State University and other central scientific institutions would mount collections for display in foreign museums; and the New Export firm in Moscow would price and sell them.[39] New Export reminded Novosibirsk that it—New Export—was the "exclusive manager of the export of scientific crude" and opened negotiations with the state bureaucracies in charge of trade with the West. The agency for trade with the United States (Amtorg) dealt with interested buyers like the Brooklyn Museum, where director Henry Foch wanted to create an exhibit room of Soviet nationalities and was eager to occupy an advantageous position in this market once it reopened as the Soviet Union stabilized.[40] A German firm responded for budgeting purposes that customs duties on imported "museum crude" into Germany would be thirty marks per kilogram, thus almost completely erasing the distinction of individual objects and underscoring the inherent difficulty of valuation of scientific commodities.[41]

Capitalization of museum crude rested on the assumption that Siberia's scientific goods would receive a major boost in price as a result of scarcity resulting from Northern Asia's inaccessibility to foreign researchers during and since World War I, the Revolution, and the Russian Civil War.[42] New Export was so sanguine about the prospective profits of these exports to Americans and Europeans that the collecting institutions (Siberian local museums) were promised a return of 10 percent of state profits on these collections to go toward purchasing museum and laboratory equipment and publications from abroad.[43] In this way, scientific sales to capitalist countries would facilitate modernization of Soviet scientific infrastructure just as they would help the state purchase tractors and industrial equipment.

COLLECTING SCIENTIFIC CRUDE

To maximize exchange value in ethnography, the Society for the Study of Siberia reminded its ethnographic collectors of Westerners' perceived tastes:

> Aside from art objects, these things should all show traces of daily use. That is, they should not be completely new, although very dirty and tattered things (especially clothes) should be avoided. We have to obtain primitive things, characteristic of the given people: for example, a Russian jacket or trousers worn by an Ostyak [current name: Khanty] are not going to create any interest abroad.[44]

Soviet collectors assumed—correctly—that Westerners were considerably more preoccupied with a constructed quality of authenticity than Soviet audiences might have been. Indeed, imperial ethnographic collecting for Russian audiences since the turn of the twentieth century had not emphasized authenticity-as-cultural-purity but instead explicitly included signs of material-cultural change or assimilation as marks of the Russian civilizing mission's success in the Empire—a mission conceived along uni-

[39] Ibid., ll. 52–4; P. Smidovich and A. Tsivin, "O sodeistvii eksportu: vsem kraevedcheskim organizatsiiam i uchrezhdeniiam," *Sovetskoe kraevedenie* 7–8 (1930): 52–3, on 53.
[40] Novoeksport to the Society for the Study of Siberia, 5 April 1930, GANO, f. R-217, op. 1, d. 206, l. 173; Amtorg to Novoeksport, 24 February 1930, ibid., l. 174.
[41] Raznoeksport to Sibtorg, GANO, f. 41, op. 1, d. 259, l. 257.
[42] I. M. Zalesskii, "Ob eksporte zoologicheskogo materiala," n.d., GANO, f. R-217, op. 1, d. 217, l. 1.
[43] Smidovich and Tsivin, "O sodeistvii eksportu" (cit. n. 39), 53.
[44] GAIO, f. R-565, op. 2, d. 44, l. 30ob.

versalist and chronopolitical rather than racial lines.[45] If Western preoccupation with authenticity may be explained as a response to rampant commodification, threatening objects' ability to maintain singularity, in the less market-saturated Russian Empire and USSR objects moved much more fluidly from one status to another, generating less anxiety about authenticity. In the scenario created by scientific crude's invention, singularization and commodification were thus occurring simultaneously and in mutual dependence. At the very same time that cultural revolutionaries and state museum officials were pressuring local museums to collect everyday objects showing Soviet modernization and contemporary material culture, collection for export was set to capitalize on discarded and suddenly rare objects of the past just as this past underwent total destruction except in museum exhibits.

In 1930, teams from Siberian museums collecting for export prepared to set out for the Altai Mountains, along the Angara River, into the Tomsk hinterlands and the Minusinsk-Khakass steppe, and to the northern parts of the Irkutsk and Krasnoyarsk regions. The ethnographic team in Tomsk and the Angara archaeologists had to cancel their activities entirely because the funds did not come through until they had missed seasonally limited river transport departures for the current expedition year. Even after the New Export firm did transfer credits, teams came up against local problems: local financial institutions could not or would not pay out on the bills from Moscow.[46] The Altai ethnographic collectors found that the indigenous clothes and shamanist accoutrements that European and American museums craved were not to be found, as communist activists had just confiscated them in the process of dekulakization. Having run up against this unexpected rival collector—Party and state representatives of collectivization and Cultural Revolution—the export collectors had to settle for purchasing directly from an existing museum, counter to the very idea of "crude."[47] In Minusinsk, kulaks' confiscated property went on sale at auction precisely when the museum's director and archaeologist were away on export expeditions. The Minusinsk museum's entomologist quickly drafted an official letter declaring the museum's right to this "ethnographic commodity" over the rights of "some other trade establishment."[48] As these cases demonstrate, there was fierce competition—usually seen as a distinguishing feature of market coordination of an economy[49]—in demand for these goods, even though the competition was delineated by politics rather than economics and took place on the near side of these objects' commodification rather than under market conditions.

[45] This viewpoint was part of the plan and collecting policy for Russia's first national ethnographic museum established at the turn of the twentieth century, the Ethnographic Section of the Russian Museum of Alexander III, now called the Russian Ethnographic Museum. Dmitrii Klements, "Otdel'noe mnenie starshego etnografa Imperatorskoi Akademii Nauk D. Klementsa, chlena podkomissii dlia vyrabotki proekta razdeleniia Etnograficheskogo otdela Russkogo muzeia Imperatora Aleksandra III," Sankt Peterburgskii filial Instituta Vostokovedov Rossiiskoi Akademii Nauk (SPbfIV RAN) f. 28, op. 1, d. 197, l. 23.

[46] Yelena Orlova (director of the museum of the Society for the Study of Siberia), Otchët po eksportnym etnograficheskim ekspeditsiiam provedënnym v Sibirskom krae letom 1930 g., GANO, f. R-217, op. 1, d. 202, l. 7ob. A series of telegrams from Novosibirsk to Minusinsk during the summer of 1930 offers a more frantic picture of the translation of planning into funding. Minusinskii regional'nyi kraevedcheskii muzei (hereafter cited as MRKM) d. 242, ll. 168; 179.

[47] Protokol zasedaniia komissii Obshchestva izucheniia Sibiri po organizatsii sbora nauchnykh kollektsii dlia eksporta, 2 January 1931, GANO, f. R-217, op. 1, d. 206, l. 86.

[48] Aleksandr Kharchevnikov to Fedosii Kravchenko, MRKM, d. 242, l. 145.

[49] Aspers and Beckert, "Value in Markets" (cit. n. 16), 5.

Even for successful export expeditions like those to the Khakass steppe and Mountain Shoria, plans were not as synchronized with Moscow as the head of the Society for the Study of Siberia, Nikolai Auerbakh, had presented them when he personally traveled to different cities to handpick and instruct archaeologists.[50] Without response to their claims to additional space in Novosibirsk, the society could not centralize the standardization of collections, thus complicating their valuation and advertising abroad, and also giving New Export a justification for stopping funds on the basis of Novosibirsk's "disorganization."[51] Then, within a year of approving the five-year plan for Siberian museums to collect scientific/museum crude for export, the following things happened: the Society for the Study of Siberia was shut down in May 1931 as part of the early Stalinist assault on independent regional organizations; one of the society's New Export contacts in Leningrad was arrested on political charges; and the society's head, Nikolai Auerbakh, died of a heart attack on a train from Novosibirsk to Krasnoyarsk to recruit the latter's museum workers. Had Nikolai Auerbakh not died in November 1930, he would have begun writing to foreign institutions and individuals to begin the process of turning the collections into currency (a colleague's letter listing archaeologists in Berlin, Paris, Helsinki, Stockholm, and Harbin was in the post to Auerbakh at the time of his death).[52] However, given the demise of the Society for the Study of Siberia, the New Export expeditions would have been short-lived in any case, and their collections would have had an uncertain future.

A group in Irkutsk met in February 1931 and planned to bypass the difficulties with New Export's funding of archaeology and ethnography, instead limiting themselves to collecting nephrite, reindeer pelts and antlers, and other mineralogical and zoological items for export.[53] Irkutsk archaeology and ethnography professor Bernard Petri participated in this meeting. In 1934, this same Professor Petri became the scapegoat for a way of thinking of objects and their value that had fallen into disfavor since 1930: the very conception of objects of science as properly exchangeable for money was declared un-Soviet.[54] Although export of Siberian minerals and biomaterial continued, by the mid-1930s the entire scientific/museum crude endeavor was being recast as emblematic of a particularly Siberian, provincial, and pernicious attitude toward Soviet cultural and natural heritage.[55]

DENOUNCING SCIENTIFIC CRUDE

The chronology of the fine arts trade—peaking in 1930 and declining sharply after 1932—exactly matches that of the scientific exports. In the case of art, the stanched demand created by a flooded market and economic depression abroad were central

[50] Nikolai Auerbakh to East Siberian Section of the Russian Geographic Society (Irkutsk), 1 July 1930, GAIO, f. R-565, op. 2, d. 44, l. 8.

[51] Yelena Orlova to Sibtorg, GANO, f. R-217, op. 1, d. 206, l. 116.

[52] Nikolai Makarov, Aleksandr Vdovin, and Yekaterina Detlova, "About the History of Krasnoyarsk Archaeologists' International Relations," *J. Siberian Federal Univ. Hum. Soc. Sci.* 3 (2009): 336–48, on 343.

[53] "Protokol rasshirennogo soveshchaniia po novomu eksportu pri Vostsibkraitorgotdele," 14 February 1931, GAIO, f. R-565, op. 2, d. 44, ll. 1–1ob.

[54] G. Ershon, "Bezobrazie v muzee," *Vostochno-sibirskaia pravda* 32 (8 February 1937): 4. For Petri's biography, see A. A. Sirina, "Zabytye stranitsy sibirskoi etnografii: B. E. Petri," in *Repressirovannye etnografy*, ed. D. D. Tumarkin (Moscow, 2002), 57–80.

[55] On the continued channeling of unusual exports into the New Export project, see "Novye vidy eksporta," *Ezhegodnik vneshnei torgovli* (1931), 633.

to this trade's decline.[56] In the scientific crude export case, on the other hand, the sequence of slowdown, failure, denial, renunciation, de facto confiscation of the collections, and later punishment of those involved was primarily motivated by changes within Soviet politics, as socialism was redefined from an internationalist-economic ideology to Russo-Soviet patriotism. These shifts—prompted by a complex and interconnected series of Soviet domestic policy failures and fear-inducing international events—had a direct impact on the range of Soviet approaches to a moral economy of international scientific exchange.

In 1934, it suddenly came to the attention of the administrative and scholarly apparatus for archaeology that the trading house Mezhkniga had purchased a large collection of prehistoric archaeological artifacts from Irkutsk professor Bernard Petri. The Academy of the History of Material Culture—scientific crude's oversight authority for archaeology—was immediately galvanized into action, notifying the firm that their purchase of these objects and any future sales abroad were illegal.[57]

The Academy of the History of Material Culture was represented in this case by Georgii Sosnovsky, an expert in Siberian archaeology who also happened to be a former protégé of Bernard Petri in Petrograd before World War I and his graduate student at Irkutsk University after the Civil War. Sosnovsky's fieldwork, moreover, had been conducted in Krasnoyarsk during the mid-1920s, side by side with Nikolai Auerbakh, the same head of the Society for the Study of Siberia who spearheaded the coordination of the scientific crude project with Siberian museums in 1929–30. In his explanation to the director of the academy of how Petri's collections could have come to be sold to Mezhkniga, Sosnovsky asserted that he had raised the alarm on this export-oriented approach to science in 1930, on the grounds that "a commercial approach to this material is not permissible."[58] Sosnovsky's objections cannot have been too strident, however, given that he gave his explicit expert approval to continued excavation work by Siberian archaeologists whose thorough annual reports openly acknowledged that the collections were earmarked for export.[59] In signing off on these reports, Sosnovsky was following his agency's policy: Nikolai Marr, head of the Academy in 1930, had approved excavation-for-export on the condition that no unica left the USSR, a condition that Siberian collectors had also insisted on.[60] Even in 1934, as Sosnovsky declared that the entire project of collecting for export had been wrong-headed, the Museum of Archaeology and Ethnography—right across the Neva from Sosnovsky's office in Leningrad—was completing the mounting of some collections already sold to the Museum of Peoples in Ankara to join similar sales already exported to Turkey.[61]

Over the course of the First Five-Year Plan, archaeology and ethnography had briefly been branded object-fetishizing bourgeois sciences by cultural revolutionaries before

[56] On the periodization of art exports, see Anne Odom and Wendy Salmond, "From Preservation to the Export of Russia's Cultural Patrimony," in Odom and Salmond, *Treasures into Tractors* (cit. n. 1), 3–31, on 20.

[57] Institute of the History of Material Culture of the Russian Academy of Sciences (hereafter cited as "IIMK"), f. 2, 1934, l. 22.

[58] Georgii Sosnovsky to directorship of the State Academy for the History of Material Culture, 8 October 1934, IIMK, f. 2, 1934, no. 225, ll. 99–101ob.

[59] Multiple examples in IIMK, f. 2, 1931, no. 799.

[60] Nikolai Marr to the Society for the Study of Siberia, 22 March 1930, GANO, f. R-217, op. 1, d. 206, l. 185.

[61] Institute of Archaeology and Ethnography to Feliks Kon in Narkompros, 15 September 1934, State Archive of the Russian Federation (hereafter cited as "GARF") A-2306, op. 70, d. 1054, l. 30.

a moderated, state-supported line rehabilitated them as thoroughly Marxist approaches to studying society's past and present through its material culture. Sosnovsky's sudden, impassioned defense of national archaeological patrimony in 1934 occurred in a moment when Soviet scientists were keen to demonstrate their right thinking in relation to the current Party line on these sciences.[62] During the intervening years since the height of scientific crude, several of the participants in the New Export expeditions experienced a first arrest on political charges. Bernard Petri, however, was arrested only in May 1937 and later shot as part of the roundup of "German specialists" in Irkutsk (Petri was a Russian imperial citizen, a Baltic German born in Switzerland to an exiled Populist revolutionary). The accusations against him, though not unveiled until 1937, related directly to his participation in collecting museum crude for export in the early 1930s. Accused of spying, Petri was asserted to have passed information about the USSR to the German government via Mr. Grosskopf, the German consul in Novosibirsk. Petri admitted under interrogation—almost certainly involving torture—that this had indeed been the nature of his relationship with Grosskopf. Grosskopf had, in fact, acted as an intermediary between Petri and Germany, but the former had conveyed not state secrets, but some of Petri's ethnographic collections for profitable sale.[63] In this way, the networks sketched out by Soviet scientists' participation in international scientific markets functioned much like other networks connecting Soviet citizens to one another or to foreigners under Stalin: they could be read as contagion vectors for anti-Soviet activity.

In 1934, the majority of the museum crude collections had been stopped in Leningrad and remained in the Museum of Archaeology and Ethnography (formerly the Kunstkamera of Peter the Great).[64] In effect, an initiative from a regional center (Novosibirsk), intended to signal provincial cooperation and enthusiastic participation in the centralized drive to think creatively about resources, ended as simply another way of using local labor and expertise to remove objects of value from the regions to the center. Since most of the things remained in the Soviet Union, the director of the regional museum in Novosibirsk wrote to the People's Commissariat of Enlightenment (Narkompros) requesting the return of some of the collections, citing the government policy obliging major museums in the capitals to distribute museum collections to the localities in cases in which the major museum had duplicates.[65] Given that the ethnographic collections had been assembled in almost identical sets, Novosibirsk could point to this clause in order to stake a claim to these things that its employees had helped to collect for export, but which were absent in their own museum. Though Novosibirsk's

[62] "O vreditel'stve v oblasti arkheologii i o likvidatsii ego posledstvii," *Sovetskaia etnografiia* 3 (1937): 5–10.

[63] Sirina, "Zabytye stranitsy sibirskoi etnografii" (cit. n. 54), 76.

[64] This inference is based on the fact that the only references to the products of this collecting drive occur in the context of later ethnographers' work with these collections in Leningrad museums: V. P. D'iakonova and N. I. Kliueva, "Etnograficheskie kollektsii 30-x godov po altaitsam i khakasam v sobranii MAE," *Sbornik Muzeia antropologii i etnografii* 35 (1980): 73–6; N. I. Kliueva and E. A. Mikhailova, "Katalog s"emnykh ukrashenii narodov Sibiri (po kollektsii MAE)," *Sbornik Muzeia antropologii i etnografii* 42 (1987): 206–7; M. A Demin and T. K Shcheglova, eds., *Etnografiia Altaia i sopredel'nykh territorii* (Barnaul, 1998), 180.

[65] Director of Novosibirsk regional museum to Feliks Kon in Narkompros, 15 August 1934, GARF A-2306, op. 70, d. 1054, l. 27. See the decree issued by Soviet president Mikhail Kalinin and A. Kiselev, "Postanovlenie Prezidiuma VTsIKa 'o sostoianii i zadachakh muzeinogo stroitel'stva RSFSR'," in *Sbornik postanovlenii po muzeinomu stroitel'stvu RSFSR 1931–1934* (Moscow, 1934), 19–20.

claim was forwarded and supported by a patron of Siberian museums—the head of the Museum Section of the People's Commissariat of Enlightenment was a veteran of Siberian museums from his exile days—the Museum of Archaeology and Ethnography offered only to cobble together some displays on early man and the peoples of the USSR to send to Novosibirsk's museum, rather than simply returning the Siberian collections intended for export.[66]

CONCLUSIONS

Lukas Rieppel has shown not only how the business of ore extraction overlapped technically with extraction of dinosaur fossils, but also how an entire set of understandings about how to maximize profits for one's find when savvy buyers were far away and risk averse carried over from prospecting for minerals to prospecting for dinosaurs.[67] In Rieppel's treatment, the case of dinosaurs illustrates rising tensions over profits between specimen dealers and research scientists once commodification of museum specimens had reached a high level in the late nineteenth century.[68] As with all goods circulating as gifts or commodities, objects of scholarly interest invoke certain "moral economies of exchange."[69] Wherever there is mutual interpenetration of markets and science, uncertainties have existed around questions such as who decides when a thing enters or exits commodity status, and how a world of noncommodity things is to be culturally demarcated. Scholarship suggests that markets themselves perform this boundary work by defining in practice what is and is not a commodity, thus implying the existence of a noncommodity status beyond that boundary.[70] In other words, a thing's potential for people pointedly to exempt it from market exchange depends on the robust presence of markets to suggest the option of a noncommodity alternative.

The seeming lack of initial anxiety surrounding commodification in the creation of scientific crude can arguably be explained by two things: the thinness of market penetration in Russian and especially Soviet culture, and the distinction between international markets (essential to Soviet industrialization) and domestic markets (corrupting, anti-Soviet influences on socialist citizens).[71] With its peculiar real and trumpeted antagonism to cultures of capitalism, the Soviet Union eagerly participated in and exploited long-established international norms surrounding the market circulation of scientific objects. What, then, was distinctively Soviet about scientific crude?

In many regards, the nineteenth-century American declaration that "bird skins are capital" indicates the existence in substance of similar attitudes toward science-as-resource in nineteenth-century Western specimen dealing. To the extent that the Soviet project aimed to bind these collections to a commodity fate the moment they were

[66] Director of Novosibirsk regional museum to Kon, 15 August 1934 (cit. n. 65), l. 30.

[67] Rieppel, "Prospecting for Dinosaurs" (cit. n. 16).

[68] On this tension, see also Steve Ruskin, "The Business of Natural History: Charles Aiken, Colorado Ornithology, and the Role of the Professional Collector," *Hist. Stud. Nat. Sci.* 45 (2015): 357–96, esp. 381.

[69] Anne Secord, "Corresponding Interests: Artisans and Gentlemen in Nineteenth-Century Natural History," *Brit. J. Hist. Sci.* 27 (1994): 391–8.

[70] See, e.g., the essays in Appadurai, *The Social Life of Things* (cit. n. 7); Pierre Bourdieu, *The Field of Cultural Production* (New York, 1993), pt. 1.

[71] The Soviet Union in the 1920s did still have markets in many goods, though this was a politically and culturally fraught issue. Scientific crude was imagined at precisely the moment when new state policy was annihilating this market culture. See Alan M. Ball, *Russia's Last Capitalists: The Nepmen, 1921–1929* (Berkeley and Los Angeles, 1990).

harvested, scientific crude was a logical outgrowth of nineteenth-century businesses dealing in natural history and ethnographic specimens. I contend, however, that scientific crude's distinction stems from, first, the Soviet state's capitalization of these prospecting expeditions, thus treating scientific goods as a nationalized resource, and, second, the application of industrial-style planning to scientific crude's extraction, processing, and sale.[72] The even more tellingly Soviet (more so, Stalinist) aspect of the story is the fact that these collections were approved and trumpeted only for most of them to be stopped at the border by inconsistent state policy: blocked from full, internationally viable commodification, their monetary value remained unrealized because of the political (not market) barriers encountered once they reached Moscow and Leningrad. Instead of becoming part of a profitable international trade run by the Soviet state, scientific crude's accumulation in Moscow and Leningrad became a kind of de facto requisitioning of value from the periphery to the center, not unlike state treatment of grain and other resources. With the Depression and associated shrinking of international capitalist markets on which the Soviet Union was dependent for currency, self-prospecting for export became less profitable than self-extraction in the form of starving peasants and workers, confiscating citizens' property, and establishing total central control over national (i.e., all) resources.

The historical actors in this article saw themselves as building socialism in one country based on Marxist principles. They were well informed about capitalist processes of commodification, valuation, and exchange but were eager to use capitalist regimes of value creation in international trade for the good of the Soviet state, even while rejecting these practices in Soviet domestic life. Indeed, Stalinist scientific capitalism retained the opposite goals of the West's: while the Metropolitan art museum's founders had cultivated wealthy collectors to "convert pork into porcelain, grain and produce into priceless pottery, the rude ores of commerce into sculpted marble, and railroad shares and mining stocks . . . into the glorified canvas of the world's masters," First Five-Year Plan activists sought to convert museum specimens into food and steel.[73] State investment in scientific crude channeled resources into entrepreneurial scientific institutions claiming to have a workable scheme to turn a profit on an unwanted cultural past, in order further to capitalize the creation of the socialist future of material abundance.

The scientific crude campaign involved acquiring the seemingly worthless (the shaman's old shirt, the cow skull, the potsherd) and turning it into a profitable commodity through the alchemy of "scientific value." Consistent with principles of profits from big data, scientific crude represented the reprocessing and repackaging of things that were plentiful but individually of no use value. Through ideological activism, the economic authority of the Soviet state and its allocation of resources, and the historical inevitability of capitalism's decline, Stalin aimed to turn the Soviet Union into a giant factory, destroying unwanted cultural and economic legacies of the past by forging them into a productive post-present. Siberian scientific personnel volunteered to identify and collect this metaphorical factory's "scraps," using their modest resale value to fund further state capitalization of socialist modernity. The capitalism of scientific crude was actually a microcosm of Soviet socialism's most extreme expression.

[72] For further discussion of capitalization of scientific research, see Courtney Fullilove, "Microbiology and the Imperatives of Capital in International Agro-Biodiversity Preservation," in this volume.

[73] Roy Rosenzweig and Elizabeth Blackmar, *The Park and the People: A History of Central Park* (Ithaca, N.Y., 1992), 358.

ENTANGLED CIRCUITS

Making the Chinese Copycat:
Trademarks and Recipes in Early Twentieth-Century Global Science and Capitalism

*by Eugenia Lean**

ABSTRACT

This essay examines early twentieth-century international disputes over alleged Chinese copying of the trademarks and brand recipes of Burroughs Wellcome and Company's Hazeline Snow vanishing cream. By doing so, it explores the complex back-and-forth that occurred between metropole manufacturers and actors in the colonial periphery in negotiating the parameters of a newly emerging global trademark regime. The essay does not present Chinese adapters of brand trademarks and recipes as simply unethical counterfeiters or passive victims of imperial aggression but treats them as full participants in a global debate over questions of ownership of commercial marks and manufacturing and chemical knowledge. Furthermore, because of Chinese adaptation of marks and circulation of brand recipes as "common knowledge," Burroughs Wellcome and Company mobilized the trademark law of the newly emerging industrial property regime to halt the travel of adapted marks and recipes. The company's deployment of trademark law thus serves as an example of how a capitalist corporation sought to ensure its advantage in competitive pharmaceutical markets by obstructing the purportedly "free markets" of capitalism and to stymie any open circulation of chemical and manufacturing knowledge. Such findings allow us to refine the recent emphasis on "circulation" often used in the historical analysis of modern science and capitalism.

AUDACIOUS FRAUD EXPOSED

> *Chinese Imitation of "'Hazeline' Snow"*
> After prolonged investigations, all members of a gang of Chinese (six in number) have been convicted on a charge of forging the Trade Mark "'Hazeline' Snow." A most elaborate fraud, in which pots, labels and advertising matter were slavishly copied, has thus been exposed. Fines amounting to 600 dollars and terms of imprisonment ranging from four to six months have

* Department of East Asian Languages and Cultures, Columbia University, New York, NY 10027; eyl2006@columbia.edu.

Thanks to William Deringer, Lukas Rieppel, Paize Keulemans, and two anonymous reviewers for reading drafts of this essay. I would also like to thank participants at the "Science and Capitalism: Entangled Histories" conference, sponsored by the Weatherhead East Asian Institute, the Center for Science and Society, the Society of Fellows, and the Heyman Center at Columbia University, and the audiences at related talks that I have given at Tel Aviv University, MIT, and National Taiwan University.

> been imposed. Fraudulent packages have been confiscated and plates, etc., destroyed.
> —Burroughs Wellcome and Company announcement, ca. 1923[1]

> Recently, some have conducted research on Hazeline Snow to determine what it contains. To manufacture it, one only needs to mix the following ingredients, hazeline, stearic acid, glycerin, sodium bicarbonate and soda water. This is the [manufacturing] method. Because the procedure is slightly complicated, . . . it has not been well understood.
> —From "Method for Manufacturing Cosmetics," 1915[2]

In 1892, Burroughs Wellcome and Company (hereafter BW&C)[3] introduced Hazeline Snow, the first commercial stearate cream.[4] The company boasted that witch hazel extract, a key ingredient in the product, could treat a range of ailments. According to one 1912 BW&C advertisement, the witch hazel extract could be "prescribed in cases of haemorrhage from the nose, lungs, womb, rectum, &c. Is a valuable agent in the treatment of bruises, sprains, inflammation, peritonitis, piles, fistula, anal fissure, ulcers, varicose veins, eczematous surfaces, tonsillitis pharyngitis, nasal and post-nasal catarrh, stomatitis, leucorrhoea, nasal polypi, &c."[5] The vanishing cream was considered a toiletry item that was also medicinal. It was sold in Britain and abroad. Given its immediate and far-reaching success, both the cream's trademark and its manufacturing recipe were widely copied and circulated.[6] To stem the widespread imitation, BW&C pursued alleged copiers around the world through a globally emerging trademark regime. Yet as we see here, the Chinese copier in particular caught BW&C's attention. By reviewing trademark disputes between BW&C and alleged Chinese copycats, this essay sees these disputes not merely as a matter of law, but instead as instances of global entanglements over issues of ownership and copying of both brand names and manufacturing formulas. In turn, these debates over exactly what constituted ownership and industrial property gained pointed geopolitical significance. In counterfeiting brand marks and unapologetically circulating brand recipes and their manufacturing knowledge as "common knowledge," persistent Chinese copiers earned the wrath of companies like BW&C in part because their actions threatened to expose how BW&C and similar companies sought to impede the circulation of chemical and manufacturing knowledge in order to promote asymmetric accumulation of capital in their favor. By defiantly casting BW&C brand recipes as common knowledge, Chinese adapters fundamentally challenged the globalizing pretensions of imperial science by exposing how brand rec-

[1] This announcement can be located among a collection of Burroughs Wellcome and Co. announcements featuring manufacturers around the world that the company had caught in what it perceived to be fraudulent use of its marks. This one is item 16. See Wellcome Library, Special Collection, WF/L/06/097.

[2] Shen Ruiqing, "Method for Manufacturing Cosmetics," *Funü zazhi* 1 (May 1915): 18–25, on 25.

[3] Burroughs Wellcome & Co. was also known as the Wellcome Foundation Limited. Henry Wellcome and Silas Burroughs founded BW&C in 1880 as a pharmaceutical company. BW&C started to expand overseas at the turn of the century and by 1912 had a branch in Shanghai.

[4] See Cosmetics and Skin's " 'Hazeline' Snow" page, http://www.cosmeticsandskin.com/aba/hazeline-snow.php (accessed 21 February 2016).

[5] Ibid.

[6] BW&C obtained registration for the " 'Hazeline' Snow" trademark on toilet goods in the British Patent Office on 7 October 1903. The imagery, consisting of a snow-capped mountain, a stream, a tree, and rocks and featured on Hazeline Snow packaging, was trademarked on 24 October 1904. See *Patent and Trade Mark Review* 17 (December 1918): 25.

ipes (and their chemical and manufacturing know-how) were treated as objects of property to be jealously guarded and owned only by proper corporate agents, and not as vehicles that share and circulate scientific knowledge openly.

Adaptation of BW&C trademarks and packaging and the circulation of its brand recipes occurred worldwide. To stem this tide of global copying, BW&C and other British and multinational pharmaceutical companies worked in conjunction with their governments to erect a global regime of industrial property rights that turned on both trademark law and convention. Trademark infringement laws were meant to curb not only any fraudulent use of trademarks but also the unauthorized circulation of recipes. Because recipes for patent medicines and cosmetics tended to be straightforward prescriptions of how to assemble ingredients rather than the creation of new chemical compounds and materials, they were not subjected to patent protection. Pharmaceutical companies like BW&C thus protected their manufacturing recipes by branding them and targeting their unapproved circulation as violations of brand infringement.

If copying of Hazeline Snow occurred in all corners of the world, the Chinese counterfeiter quickly came to assume the dubious distinction of being a particularly egregious perpetrator. By the 1910s, Hazeline Snow advertisements were appearing with regularity in Chinese journals and newspapers. Almost as quickly, Chinese copycats started attracting BW&C's attention. As the first epigraph above suggests, by the 1920s, BW&C was painting Chinese merchants as "audacious" in their counterfeiting of Hazeline Snow's trademark and assorted packaging. The question of how and why the "Chinese copycat" came to occupy such a prominent position in the global imaginarium of the twentieth century, is a central concern of this essay. BW&C's targeting of Chinese merchants as consummate perpetrators of fraud stemmed in part from the vibrant and unruly copying that was indeed taking place in China. But, as this essay argues, the attempt to discipline domestic offenders infringing in the British pharmaceutical company's own backyard was just as significant a reason. By contrasting legitimate British merchants with audacious Chinese committers of fraud, companies like BW&C sought to identify an effective foil against which they could define the parameters of proper commercial behavior and pursue alleged counterfeiters both in their own country and abroad. The venal Chinese copycat was effectively marshaled to defend a modern intellectual property (IP) regime that would only further the interests of British firms.[7] By creating the "Chinese counterfeiter," BW&C was implicitly identifying itself (and like-minded players) as the legitimate purveyor of sanctioned products and knowledge.

If imperialist interests informed how BW&C and other Western actors promoted modern trademark infringement and targeted alleged Chinese copiers, the Chinese were hardly passive victims. Rather, they engaged with the emerging new property rights regime in complex ways. To start, "the Chinese" were a heterogeneous group, pursuing a variety of interests through issues of duplication, infringement, and ownership. If they

[7] By identifying the rise of the Chinese copycat in an era of modern capitalism and science, my essay provides a parallel case example to Lee Vinsel's contribution to this volume, which sheds light on the constitution of the "accident-prone driver." In both cases, we see how new identities are created in part as a foil against which to secure certain capitalist interests. If the Chinese copycat functioned to justify a trademark regime that would advance imperialist industrial interests like BW&C, in Vinsel's case, the accident-prone driver was mobilized to exonerate auto manufacturers from accountability in instances of automative injury and death. See Vinsel, "'Safe Driving Depends on the Man at the Wheel': Psychologists and the Subject of Auto Safety, 1920–55," in this volume.

did not always perceive "copying" as an act of infringement, they cannot simply be understood as exceptionally unethical perpetrators, as much of the conventional West-based legal narrative that was emerging at the time (and that persists into the contemporary period) would have it. Nor were they mere victims of an imperialist conspiracy, as Chinese nationalist interpretations continue to claim.[8] At times, they acted in concert against British merchants. At times, they were a fractured, diverse group, pursuing their own individual interests in regard to trademark use and legislation. For certain actors, international tension over the proper use of trademarks could quickly became a matter of imperialism and profoundly violate Chinese sovereignty. For others, trademark regulation was to be pursued domestically. Chinese manufacturers who found their brands and recipes subjected to infringement by copycat firms drew from the evolving trademark regime to present themselves as "modern" industrialists as opposed to tradition-mired merchants. And for still others, these matters were simply issues to be worked around, in practical pursuit of profit in their day-to-day operations.

By focusing on how Chinese compilers rendered the recipes for Hazeline Snow and other brand objects as common knowledge, this essay furthermore sheds light on how Chinese "copiers" defied Western corporate interests that sought to universalize certain norms of capitalism and science. As the second epigraph above shows, Hazeline Snow's trademark and packaging were not the only part of the commodity that was being shared in China. The cream's ingredients and related manufacturing tips circulated in the early twentieth-century Chinese press and were indicative of a more general trend whereby manufacturing formulas and the basic chemistry behind the production of both domestic and foreign brand-name patent medicines and toiletry items were printed with little hesitation as "common knowledge" [*changshi*]. Common knowledge was not "pure" scientific knowledge being pursued in China's newly established academic universities. Nor was it the "Mr. Science" being promoted by highbrow intellectuals in the New Culture Movement (1915–9). Part of the slogan of "Mr. Science and Mr. Democracy," "Mr. Science" was an abstract and lofty notion of universal science that, along with democracy, was identified as one of the pillars to building Chinese modernity. Instead, common knowledge was unapologetically commodified. It was know-how on manufacturing and chemistry and included brand recipes, all of which were disseminated widely in the commercial press. Hardly basic or ordinary, the "common" aspect of common knowledge and its value derived from it being shared without regard to brand ownership. As such, it helped engender the savvy and informed consumer in an era when global capitalism introduced untrustworthy markets into China, filled with faulty, suspect, and even possibly counterfeit goods. It was also crucial for nativist industrialists and budding manufacturers invested in building Chinese industry to compete in a highly competitive global market of cosmetics and pharmaceutics. If such strategic adaptation of manufacturing knowledge and chemical knowledge often defied ideas of trademark infringement being promoted by BW&C, for Chinese compilers and manufacturers, common knowledge was invaluable.

[8] PRC nationalist historiography often views the history of trademarks in China as one of many examples of China's victimization vis-à-vis Western imperialism. See Zuo Xuchu, *Zhongguo jindai shangbiao jianshi* [The brief history of modern trademarks in China] (Shanghai, 2003), esp. 51–6, who has characterized Western intervention in compiling China's modern Trademark Law in 1923 as an example of Western imperialist encroachment into China.

Like Hallam Stevens's contribution to this volume, this essay historicizes the process by which the modern intellectual property rights (IPR) regime emerged.[9] In doing so, it challenges the legal conceits of "proper" ownership that have been normalized and institutionalized by that very regime. Legal narratives emerging globally at the turn of the twentieth century were crucial in promoting the idea that trademarks were crucial to help the public avoid confusion in the market, creating and enforcing clarity, and thus were there to protect the interests of the consumers. They have also generated the strong impression that the Chinese copycat was among the most audacious counterfeiters. However, in this study, the politics of empire behind these legal conceits are exposed. The modern trademark regime—and its attendant counterfoil, the Chinese copycat—was a strategic resource used to create and open foreign markets such as China's by eliminating competition for large corporations such as BW&C. Moreover, trademark law (and IP more generally) was used to cast certain kinds of epistemic and commercial exchange as illegitimate. It functioned to obstruct the free circulation of key knowledge to hobble emerging manufacturers and create favorable conditions for well-established capitalist corporations.

Finally, by shedding light on this dynamic between companies like BW&C and Chinese copiers, this essay shows that despite uneven relations of global power, actions by Chinese actors impinged, even if unintentionally and obliquely, on the evolving shape of the new regime of trademark infringement and, by extension, the parameters of global capitalism. One might argue that modern IP emerged as it did in ways crucially dependent upon the action of periphery actors like Chinese adapters. Frequent and vibrant, adaptation of both the mark of a brand and its manufacturing know-how in China (and other peripheral locales) spurred BW&C and related companies to develop and normalize trademark laws. These strategies, and the threat of becoming like the venal Chinese copycat, were in turn used to domesticate local copycats as well. From this perspective, we can better understand why British companies such as BW&C were so invested in propagating the idea of the Chinese copycat. Much of it had to do with compensating for its own considerable lack of control over the situation, both domestically and abroad.

A MARK OF EMPIRE: THE MODERN TRADEMARK AND INDUSTRIAL PROPERTY

With advances in modern chemistry and industrial manufacturing, the nineteenth-century world had become saturated to an unprecedented degree with mass-produced near-uniform goods, patent medicines and drugs among the most popular. By the early twentieth century, the rise of a modern drug and pharmaceutical industry that was increasingly predicated on laboratories and clinical science helped render trademarks and patents into what it deemed to be ethically and legally legitimate means to protect both the consumer and the company in ways familiar to us today.[10] Seeking to sell their prod-

[9] Hallam Stevens, "Starting up Biology in China: Performances of Life at BGI," in this volume.

[10] Joseph M. Gabriel, *Medical Monopoly: Intellectual Property Rights and the Origins of the Modern Pharmaceutical Industry* (Chicago, 2014), 196. In the first decade of the twentieth century, European manufacturers, like the German manufacturer Bayer, led the way in patenting and trademarking drug products. In the United States, manufacturers only started to utilize this new framework by World War I, at which point they began to defend their scientific innovations by pursuing product patents and taking out trademarks to protect their innovation at the level of the product itself. Prior to the first decade of the twentieth century, U.S. manufacturers were ambivalent about the role of trademarks and patents, and many held that both perpetuated unethical forms of monopoly. See also Adrian Johns, *Pi-*

ucts at an unprecedented economy of scale and scope, manufacturers quickly came to regard the modern trademark as an indispensable means to carve out distinction for their product among a sea of indistinguishable or nearly indistinguishable products. It was often difficult, if not impossible, for even the most discerning consumer to determine the quality of the commodity or the danger of the remedy based on the mere appearance of the item. A trademark guaranteeing the reputation of a brand was to aid consumers to make such evaluations. And as the authority of trademarks spread, so too did the copying of trademarks. "Counterfeiting" successful marks and packaging designs, along with adapting brand-name manufacturing recipes, quickly became highly lucrative endeavors that copycats around the world undertook to sell their wares.

The new ways of understanding trademarks were entangled with the politics of empire from the start. By the nineteenth century, regulation and policing of fraudulent claims of ownership were aimed both at domestic (Western metropole-based) copycats and at copycats in far-flung corners of the colonial world. On 20 March 1883, the Paris Convention for the Protection of Industrial Property was signed. As one of the first intellectual property treaties, the convention sought to codify trademark enforcement and criminalize brand infringement globally. As the trademark was presented as a mark of modern capitalism and civilized business practices, large-scale manufacturers of mass-produced goods increasingly worked along with their respective governments to deploy mechanisms of international law, diplomatic pressure, and direct force to influence the colonial world to revise indigenous legal systems to abide by and adopt the legal norms and practices that benefited the modern industrial property rights regime. Pharmaceutical companies such as BW&C and Unilever were particularly aggressive in lobbying their respective governments to utilize legal institutions, political means, and diplomatic channels to police potential fraud around the world.[11] For them, trademarks were not simply to clarify market confusion but could also serve a more insidious yet crucial function. By demarcating what constituted legitimate goods, they could also be used to obstruct the movement of purportedly illegitimate goods, including those whose "illegitimacy" stemmed from their national origin.

Implementation of this global regime of industrial property rights was not easily achieved. China, in particular, became an ideal site where the emerging system could be vigorously tested and subjected to interrogation and challenge. The early Republic was a period when the central state was exceptionally weak in the face of internecine warfare among militarists and ongoing imperialist pressures. Extraterritoriality in China's treaty ports meant that the political and commercial presence of multiple Western powers and Japan in China served to fragment Chinese political authority and control, as well as that of any single imperialist force. Such fragmentation characterized the commercial sphere as well. Multiple views of brand ownership and product authenticity commingled in China's markets. Competitors from around the world saw their practices and views on proper commercial protocol come into direct conflict with each

racy: The Intellectual Property Wars from Gutenberg to Gates (Chicago, 2009), chap. 5. Johns argues that the first pharmaceutical patent emerged in eighteenth-century Britain, when the medical market grew more anonymous and it was increasingly difficult for consumers to assay the chemical composition of potentially dangerous remedies. The widespread practice of relying on patents, however, did not take root at that point.

[11] Eugenia Lean, "Marking and Copying from Qing to China: Late Imperial Culture in Trademark Infringement Cases of the Early Twentieth-Century" (paper presented at "From Qing to China: Rethinking the Interplay of Tradition and Modernity, 1860–1949," Tel Aviv University, 20–22 May 2012).

other. The presence of multiple national players competing in China's commercial market also meant that a vibrant and unruly market of goods and things emerged where copying was rampant. Copiers were not just Chinese merchants but consisted of a range of nationalities, all of whom were seeking to carve out advantage in the competitive environment of China's highly cosmopolitan markets. Whether national and provincial officials, alleged copycat companies (Chinese and otherwise), or Chinese consumers, participants on the ground were driven by an array of motivations with respect to trademark infringement. Chinese treaty-port markets were thus in many respects ideal forums for international debates on trademark law, property rights, counterfeiting, and notions of legal ownership of brand and reputation.

In fact, it was precisely this context that made early Republican China an ideal place in which to bring rigorous trademark law into being, rather than the metropole. Timothy Mitchell has argued that modern statecraft emerged in areas of the world that experienced imperialism. Egypt, he claims, was where Great Britain "performed" the modern economy.[12] The economy as a knowable and quantified entity had to be ontologically conjured and brought into being through technical means (such as national income measures), and because the colonial state did not have enough power to regulate economic activity to the extent it did in the metropole, this conjuring was initially accomplished in the colonial periphery in Egypt. One might see certain parallels in the history of modern IP. It was in China during the unruly "warlord era" where Great Britain honed its modern trademark law, which it billed as a mechanism for bringing order to an unruly and confusing (and thus dangerous) marketplace. In order for legally protected trademarks to be brought into being at home, they first had to be rigorously enforced abroad.

THE EMERGENCE OF THE CHINESE COPYCAT: BW&C TRADEMARK CASES

In examining a set of 1910s and early 1920s BW&C announcements regarding the company's prosecution of trademark infringement by its in-house patent office, this section is less concerned with whether Chinese merchants had in fact engaged in fraud. Instead, it uses the alleged copying and the announcements to illuminate what was a global dialogue over what constituted an authentic mark and the nature of its ownership. It considers how in an era of waning yet persistent imperialism in China, Chinese copycats made their presence felt, spurring firms like BW&C to invest in controlling trademark infringement around the world. The texts reveal how savvy tactics of copying and borrowing by Chinese merchants, along with prevailing Orientalist assumptions in Europe of Chinese corruption and shady business dealings, allowed pharmaceutical companies like BW&C to render the Chinese "counterfeiter" as a significant foil in their efforts to eradicate domestic British copying. As the uncivilized counterfeiting Other to modern, Western law-abiding merchants, the Chinese copycat became a compelling counterexample. It also came to stand at the center of the making of modern IP.

BW&C actively promoted and enforced what it deemed to be proper trademark use and practice in a variety of ways (fig. 1). It sought to register its popular marks with newly established administrative institutions throughout the world and acted under

[12] See Timothy Mitchell, *The Rule of Experts: Egypt, Techno-Politics, Modernity* (Berkeley and Los Angeles, 2002).

Figure 1. *Photograph of an "authentic" BW&C "'Hazeline' Snow" jar that was sold on the Chinese market around 1923–4. From Wellcome Library London, Archives, Image no. L0050258.*

the assumption that these registrations were enforceable through law.[13] The company also poured resources into policing trademark usage by opening up in-house offices to deal with trademark infringement and patent abuse cases. Run by the Wellcome Foundation secretariat, A. E. Warden, the office monitored markets for brand counterfeiting from London to Bombay and from Ningbo to Johannesburg. It aggressively pursued companies and individuals it perceived as infringing on its trademark, relying on reports of counterfeited products or sending their own representatives to scour stores and markets for look-alike products.[14] These tactics were fairly effective. From 1917

[13] The earliest registration of a trademark with the word "Hazeline" for the chemical and toilet article classes of goods (class 3) was by BW&C in the United Kingdom on 10 May 1880 (no. 22,388) and was renewed in 1894. Ensuing registration of BW&C trademarks with the word "Hazeline" in the United Kingdom took place in the years following. For a review of this history, see a set of legal documents prepared by A. E. Warden, who was in charge of BW&C's in-house department dedicated to trademark infringement, especially the document titled "Hazeline," 14 March 1907, Wellcome Library, Special Collection, WF/L/06/158.

[14] Monitoring tactics were as prosaic as sending undercover agents into stores suspected of selling items with forged trademarks. BW& C archives include examples of reports filed by "undercover agents" visiting unsuspecting stores asking for counterfeit products. See, e.g., the memo titled "HAZOL," 22 March 1907, Wellcome Library, Special Collection, WF/L/06/158.

to 1937, BW&C secured over fifteen public apologies for trademark infringement of their brands in Great Britain alone. In the India market they prosecuted at least eleven forgers, and in the Asia market (China, Japan, and Formosa), they acted against at least six.[15] Most counterfeiters were charged with fraudulently copying either BW&C's famous trademark "Tabloid" or the brand name and packaging of "'Hazeline' Snow."

Once trademark offenders were uncovered, BW&C's patent office would prosecute the offending firms and publicly announce its success. Such announcements would often include an apology from the copycat company. In the case of the mark or design of the packaging being copied, a picture of the offending version was often positioned next to the original to remind readers of the announcement of the genuine article. These announcements were not necessarily intended to deter copycats in far-flung markets. Written in English, the apologies targeted English-reading audiences and appeared in British trade journals.[16] Possible audiences for these announcements included fellow British chemical and pharmaceutical companies, small and large, and their publication was meant to reinforce and disseminate domestically (or within the British Empire) the new ideals of trademark and brand ownership.

In order to drive home that these new practices of trademark ownership were emblematic of a modern nation such as Great Britain, having examples of copycat merchants from colonial and semicolonial outposts defying such practices was highly effective. Chinese copycat firms constituted a compelling foil against which these "modern" practices could be articulated. By the nineteenth century, European actors had come to identify Qing China as an empire lacking civilized law, which allowed them to legitimate their extraterritorial claims in China.[17] The actual situation was much more complicated. In terms of commercial law, although the Qing did not provide much in the way of explicit codified law, it did protect marks through customary practice.[18] And as guild mechanisms and customary protection of marks worked only for domestic producers, foreign traders and merchants entering into China's markets in the nineteenth

[15] For these announcements, see Wellcome Library, Special Collection, WF/L/06/097.

[16] One announcement, "Before the Mixed Court Shanghai: Burroughs Wellcome & Co. *Plaintiffs* versus Zing Tsa-nan trading as The Nanyang Medical Co. *Defendants*," included the image of a newspaper clipping from the 5 April 1919 issue of "The Chemist and Druggist." This clipping announced BW&C's success in securing a "perpetual injunction" restraining the defendants, the Nanyang Medical Co., from selling "Nanyang Snow," which it claimed was a "toilet preparation simulating their well-known 'Hazeline Snow.'" See "Before the Mixed Court," Wellcome Library, Special Collection, WF/L/06/097. By featuring an announcement within an announcement, BW&C's announcement illustrates quite literally how the news of the injunction traveled in different forms of English-language media and was meant to be disseminated widely.

[17] The British government had long rendered commercial and legal concerns in China into diplomatic and political issues, using them as pretexts for imperialistic pursuits. It is well documented, for example, how the British presence in China during the nineteenth century was in large part motivated and informed by Britain's desire to avoid and subvert Chinese legal prohibition of opium trading. See Li Chen, *Chinese Law in Imperial Eyes: Sovereignty, Justice, and Transcultural Politics* (New York, 2016).

[18] The Qing code exhibited some concern with preventing in broad terms unfair economic behavior, such as monopolization and unfair trading. See Robert Heuser, "The Chinese Trademark Law of 1904: A Preliminary Study in Extraterritoriality, Competition and Late Ch'ing Law Reform," *Oriens Extremus* 22 (1975): 183–210, on 187. But, overall, legal protection for forms of "intellectual property" in imperial China was primarily aimed at bolstering the state's interests in preserving imperial power and maintaining political order and social stability, rather than protecting private interests and ownership. See William P. Alford, *To Steal a Book Is an Elegant Offense: Intellectual Property Law in Chinese Civilization* (Stanford, Calif., 1995), esp. 24–5. In terms of protecting private marks, however, formal codification of state protection was relatively absent.

century found their products quickly and deftly counterfeited. To protect their trade names, foreign merchants and their governments began pressing for reform in China's commercial law.[19]

Britain led the charge. As its merchants and traders muscled their way into the Chinese market in the late nineteenth and early twentieth centuries, it increasingly placed pressure on successive Chinese governments to draft regulations to police counterfeit trademarks. With the Qing state on its last legs and intellectuals and reformers questioning all aspects of the regime's past in terms of "what went wrong," Chinese reformers came to believe that the very concept of law in China had to be changed to appear to adhere more closely to Western legal standards (many of which were themselves evolving). These reformers worked with successive early Republican states in the face of imperialist pressure to codify trademark infringement and establish a sense of legal equivalence with the West, which at the time was regarded as superior.[20] China's first complete trademark law was drafted in 1923 and served as the basis of the Nationalist government's law until 1930, when the Nationalists offered their own revision of the trademark law.

It was in this context that BW&C's identification of Chinese copycats took place and gained momentum. For an example of how the company targeted Chinese merchants, consider its 1922 announcement regarding the Chinese Fo Meng Factory of Ningbo. The announcement started with the heading "Warning" and then featured a translation of the Chinese factory's apology:

Warning
ANOTHER "'HAZELINE' SNOW" IMITATION SUPPRESSED
20,000 Chinese labels, 5,000 Foreign labels, 1,000 cartons—DESTROYED!
TRANSLATION OF APOLOGY
I, R. B. King, representing the Fo Meng Factory of Ningpo, do hereby promise and guarantee that the said firm will not produce any label, carton or circular in connection with their product which in any way resembles or imitates the Designs or Trademarks used or registered by BURROUGHS WELLCOME & CO. for their well-known products.
Dated this 11th day of February, 1922
(Signed) R. B. WILLIAM KING.
We prosecute offenders rigorously in the interests of the trade, the public and ourselves[21]

With the apology, BW&C not only secured a public declaration that the company would never imitate its designs but also reminded the English-reading public that it

[19] Alford, *To Steal a Book* (cit. n. 18), 34–5.

[20] The first registered marks were with the Imperial Maritime Customs, but since customs was unable to enforce these registrations, the British Foreign Office pressed the matter during negotiations over the Boxer Uprising protocols. A series of commercial treaties seeking to deal with protection of foreign marks followed at the turn of the twentieth century. By 1904, the Qing government promulgated a set of provisional regulations in accordance with British demands, titled "Experimental Regulations for the Registration of Trademarks." These regulations were to be the foundation of the trademark code slotted for future codification, which British and other imperialist powers pushed for into the Republican period. See Heuser, "Chinese Trademark Law" (cit. n. 18).

[21] See the announcement titled "Audacious Fraud Exposed," n.d., Wellcome Library, Special Collection, WF/L/06/097; emphasis in the original. This announcement is undated, but it is filed between an announcement regarding a Parisian copycat issued on 7 February 1922 and one featuring a German counterfeiter issued on 7 March 1924. The one featuring "Chinese gangs" probably appeared sometime between these two.

was the sole owner of this trademark, and that it had disciplining mechanisms with which it could deter potential copycat companies. With the statement in boldface type, BW&C sought to portray itself as not merely acting out of self-interest, but in fact representing the interests of fair trade, and the consuming public.

BW&C also pursued more illicit entities, and again, Chinese actors served as powerful examples. The first epigraph at the beginning of this essay comes from a BW&C announcement that specifically cites BW&C's prosecution of "audacious fraud" by a "gang of Chinese" engaged in "[a] most elaborate fraud, in which pots, labels and advertising matter were slavishly copied."[22] It identified the vague entity of "gangs" to remind readers of BW&C's capacity to investigate and convict such illicit conspirators. Just as important, of course, was its promotion of BW&C as the rightful owner of the lawful and "authentic" trademark (fig. 2). The announcement sought to authenticate the registered trademark "'Hazeline' Snow" as a BW&C product, and not that of dubious actors, arguing that not only the owners of the "'Hazeline' Snow" trademark (i.e., BW&C, which was invested in the prosecution of those who fraudulently used the mark), but also traders and customers who purchased the "real" product wanted this clarity.

From these announcements one can detect the on-the-ground copying that had prompted BW&C to begin policing infringement. That such in-house policing was even necessary speaks to what were fairly nascent disciplinary mechanisms of brand enforcement both in Great Britain and beyond. Chinese counterfeiters were hardly alone. Among the copycats identified in this archive of BW&C announcements, a number were Western. Local British ones were the most numerous, a few were German and French, and one was Canadian. Among those from outside Europe, alleged offenders from India were most numerous, and except for one from South Africa, the rest were East Asian, with Chinese cases being most numerous.[23] Yet if copying was spread around the world, China stood out for its "egregious" infringement. The nature of China's treaty ports, where multiple yet fragmented imperialist powers coexisted alongside weak, if intact Chinese sovereignty, might very well have allowed for substantial Chinese copying of Hazeline Snow. And in subsequent years, Chinese copycat soap and cosmetic companies were to continue to be a thorn in the sides of British pharmaceutical firms like BW&C.[24] It was for these reasons that Chinese counterfeiters provided a colorful example of unlawful, fraudulent copiers and came to be deemed illicit and deleterious. As such, Chinese copiers could be placed in sharp contrast to the lawful commercial merchant that recognized the authenticity and sole corporate ownership of certain marks, an ideal that BW&C was invested in establishing in both far-flung markets and the heart of the British metropole.

FRAUDULENT APOLOGIES AND CHINESE AGENDAS

At first glance, these BW&C announcements might appear to indicate considerable BW&C sway over Chinese actors. But, when read carefully, they shed light on how trademark infringement cases became opportunities for Chinese actors to pursue their

[22] Ibid.

[23] It is not surprising that the largest number of offenders came from India and South Africa. These were direct colonies and easy markets for BW&C to police.

[24] Lean, "Marking and Copying" (cit. n. 11) discusses some of these cases from the 1930s.

The registered Trade Mark "'Hazeline' Snow" is a brand which denotes specifically a Burroughs Wellcome & Co. product. The substitution of goods of other manufacture is an infringement and unlawful. Offenders are prosecuted rigorously in the interests of trader, customers and the owners of the Trade Mark.

"'HAZELINE' SNOW"

(Trade Mark)

Original and supreme non-greasy toilet preparation. A product whose exceptional purity and delicate fragrance win permanent customers everywhere.

THE AUTHENTIC PACKINGS

Vanishes without leaving a trace of greasiness

Figure 2. *The visual of the Hazeline Snow packing comes from a 1910s Burroughs Wellcome brochure and is not the same as the one in the original announcement. See http://www.cosmeticsandskin.com/aba/hazeline-snow.php (cit. n. 4).*

own agendas. Take, for example, one 1918 case: a Shandong official turned being forced to issue a proclamation against Chinese merchants selling counterfeit products into an opportunity to assert an anti-Japanese agenda. In a fierce fight between British and Japanese governments over mastery of the Chinese import trade in the early twentieth century, the British came to consider Sino-Japanese counterfeiters to be a real threat between 1913 and 1923 and sought to target both Japanese manufacturers and cooperative Chinese merchants for trademark infringement.[25] This context informed BW&C's

[25] For more on Anglo-Japanese negotiations over counterfeiting in China and the emergence of Chinese trademark law, see Eiichi Motono, "Anglo-Japanese Trademark Conflict in China and the Birth of the Chinese Trademark Law (1923), 1906–26," *East Asian Hist.* 37 (2011): 9–26. Heuser also discusses the long history of Western complaints about Japanese imitations of foreign goods (i.e., Western goods) in China, and about the failure of the 1904 trademark regulations to address non-Chinese infringement of Western marks in China; see Heuser, "Chinese Trademark Law" (cit. n. 18), 190, 199. He argues that these regulations favored the Japanese, who were in fact advisers to the Chinese drafting the regulations, and that this fact was particularly bothersome to Western merchants, who felt disadvantaged in trade as a result (201–3).

bare-knuckle commercial diplomacy in its prosecution of Hasoda, Matsui, Imayeda & Co. for selling fraudulent Hazeline Snow in China. In April 1918, BW&C succeeded in prosecuting the Japanese firm through the Japanese Patent Tribunal and then proceeded to print the following public announcement:

> A Japanese imitation of "'Hazeline' Snow" was offered for sale in China. In preparation for foisting this despicable fraud upon the Chinese and in an endeavor to legalise it, the infringers effected registration of the three trade marks here shown. . . . The fictitious name "Nippon Toilet & Co.," it will be observed, is slavishly followed by the addresses which indicate the Foreign and Colonial Houses of B. W. & Co.[26]

Even as BW&C claimed that the Japanese imitation product was harming the Chinese public, this announcement proceeded to target not only the Japanese counterfeiters, but also Chinese merchants who were selling the allegedly fraudulent merchandise. It worked "with the assistance of one of H. B. M. Consuls" to take local action and convinced the Chinese Special Commissioner for Foreign Affairs in Shandong to crack down on Chinese trade of the Japanese counterfeit.

Since China's defeat in the Opium Wars in the mid-nineteenth century, the unequal treaty system meant that imperialist powers could muscle their way into China's markets with legal, political, and commercial privilege. By the early twentieth century, national humiliation in the face of economic imperialism had reached a fever pitch, and patriotic consumption had become an imperative, with boycott movements against imperialist powers gaining momentum. Patriotic Chinese consumers, aware of how Western merchants had long put pressure on Chinese officials to yield to their wishes, no doubt found this announcement infuriating. Seeking to crack down, the British ensnared Chinese merchants as well as the original Japanese counterfeiters and forced a Chinese official to condemn the Chinese merchant shops. Such actions only reinforced the sentiment that within the fragmented environment of China's treaty ports, where non-Chinese copycat companies were at work and multiple imperialist powers were present, profound humiliation and exploitation of the Chinese could occur.

In a follow-up announcement titled "A Japanese Imitation," a proclamation by the Special Commissioner for Foreign Affairs in Shandong was published, in which Commissioner Tang offered a condemnation of both Japanese counterfeit manufacturers and Chinese merchants. At first blush, this passage might be cited as a humiliating example of a Chinese official submissively accepting blame for the putative wrongdoing of the Chinese merchants. Yet when read more carefully, we can see how the proclamation reveals an agency on the part of the Chinese Special Commissioner that was complex and strategic. The commissioner's statement reads as follows:

> WHEREAS in accordance with instructions issued by Board of Commerce and by the Governor several proclamations have been issued prohibiting Chinese from . . . the sale of goods with fraudulent trade marks, . . .
>
> AND WHEREAS I have received a dispatch from the British Consul at Tsinan stating that the British firm of Burroughs Wellcome & Co., of Shanghai, accuse four Tsinan shops . . . of selling imitations of their Company's "'Hazeline' Snow,"

[26] See the announcement titled "Before the Japanese Patent Tribunal: Burroughs Wellcome & Co. versus Hosoda, Matsui, Imayeda & Co.," n.d., Wellcome Library, Special Collection, WF/L/06/097. This document is undated, but the patent trial, the announcement notes, took place on 30 April 1918, and the announcement was most likely printed shortly thereafter.

> AND WHEREAS I . . . carefully examined the proprietors of these four shops, all of whom bore witness that they had bought the imitation products in other ports and had no intention of imitation foreign trade marks with a view to deceive their customers; . . . and solemnly warned them against continuing in their evil ways;
>
> NOW recently . . . the Chinese tradesman not noticing whether the article was genuine or imitation has also been deceived and ordered a stock of such goods for sale, thereby not only injuring the genuine article to a great degree, but also bringing discredit on Chinese trade . . .
>
> . . . A Special Proclamation: (Signed) T'ANG K'O SAN,
> Concurrently appointed . . . as Special Commissioner for Foreign Affairs in Shantung.[27]

In a period when China was struggling against imperialist incursions, Chinese officials like Tang, who aligned themselves, at least on the surface, with the trademark law, served to endorse the ideological argument that trademark enforcement was an effective mechanism for reducing confusion in the marketplace. By doing so, they appeared to play into the hands of the British by attempting to leverage the latter's rigorous enforcement of IP law in their own conflicts with Japanese importers. Yet at the same time, this announcement became an opportunity for the Chinese commissioner to pursue a rather complicated agenda. By recounting how he personally went to the four shops to investigate the situation and warned the shop owners "against continuing in their evil ways," Commissioner Tang was seeking to demonstrate his authority and efficacy as an administrator who was able to discourage the sales of imitation pieces in Chinese shops, even if he was effectively sanctioning the principle that infringement had to be policed. Where a more obviously subversive tone emerges is midway through where the passage shifts toward defending the Chinese merchants by claiming that they were actually victims who had been duped by the original counterfeiters, the Japanese. With the phrase, "the Chinese tradesman not noticing whether the article was genuine or imitation has also been deceived . . . thereby not only injuring the genuine article to a great degree, but also bringing discredit on Chinese trade," the Commissioner portrays the Chinese tradesmen as being guilty at most for their naïveté. He establishes the injury and deception done to them. The Chinese merchants were victims, and Japanese manufacturers were the true perpetrators.

The fact that it was a Japanese merchant company that was the original counterfeiter, and that the location of the counterfeiting was Shandong, made the case particularly fraught and Commissioner Tang's assertion compelling. Three years earlier, in 1915, with the outbreak of World War I and the weakening of Germany's hold in North China, Japan had issued the 21 Demands to the Chinese government. A subset of these demands allowed Japan to expand its sphere of influence into the railways and major cities of Shandong. The public was outraged when Yuan Shikai's government acquiesced to Japanese demands. In June 1919, Shandong was completely ceded to the Japanese by the Treaty of Versailles. News of this development sparked modern China's most iconic

[27] See the announcement titled "A Japanese Imitation of One of Burroughs Wellcome & Co.'s Packings," n.d., Wellcome Library, Special Collection, WF/L/06/097. The special commissioner was appointed by the Ministry of Foreign Affairs. The longer quotation describes the array of institutional mechanisms that were already on the ground policing trademark violation in China. The Board of Commerce and the governor of Shandong were portrayed here as having dispensed instructions prohibiting sales of products with offending trademarks. BW&C had also deployed diplomatic dispatches from the British Consul in the Shandong city of Jinan to put pressure on the commissioner directly. Finally, the announcement mentioned the Shanghai Custom House, which was where trademarks were registered.

modern political movement, the May Fourth Movement. By this time, Japan's "influence" in Shandong was keenly felt, and this case of a Chinese official being forced to kowtow to Western demands for violations committed by the Japanese would have been particularly volatile. Given the specific colonial situation in Shandong, the text cleverly used the occasion to evoke popular sentiment that the Chinese once again were victims of Japanese counterfeiting and related excesses.

Even as he faced inhospitable relations of power vis-à-vis British imperialists, Tang was not merely acceding to BW&C's wishes. His voice here was far more nuanced. He sought to bolster his official status by displaying his ability to handle the case effectively as well as to use this proclamation to promote an image of an efficient Chinese officialdom in bringing order to the affair. He deftly directed attention away from the Chinese shop owners to the Japanese, whom he sought to portray as the real offenders. Finally, one wonders whether the apology was not simply a pro forma matter that once taken care of had little influence on actual commercial practice. The commissioner's apology could well have been nominal and meant to appease BW&C. It could have forestalled any real concessions or allowed the Chinese commissioner to avoid engaging in concrete efforts to crack down on alleged offenders. Indeed, even while seeking to exhibit BW&C's ability to discipline imitators worldwide, much in these announcements points to a rather turbulent situation. These announcements likely served more as an indication of how BW&C felt it was necessary to respond to something not quite yet under its control, rather than a sign of any success in containment.

BW&C not only fought counterfeiting via English-language announcements. The company also sought to target Chinese audiences directly by publishing in Chinese media sources. Take, for example, the full-page ad for Hazeline Snow that appeared in the January 1915 issue of the *Ladies' Journal* (Funü zazhi). Splashed across the top of the ad was "'HAZELINE' 夏士蓮雪花 SNOW" in large block letters. Smaller block letters for "TRADEMARK" and *shangbiao* 商標, the Chinese characters for trademark, bookended the large "'HAZELINE' 夏士蓮雪花 SNOW" to indicate that it was a genuine trademark. The bottom part of the ad included the image of the trademarked container used to package Hazeline Snow and detailed copy in Chinese, which touted the virtues of the cream for the skin as well as its medicinal functions. The copy also warned readers and potential consumers about fakes, letting them know how to distinguish the genuine cream and trademark from fraudulent versions. It informed the consumer what the original should feel like when used and also how to recognize the "lofty decoration" that was most elegant, whereas both the feel and the packaging of counterfeit products could not be more divergent. Finally, the bottom left corner of the ad included a voucher that the reader could cut out and mail in for a lengthier booklet that would explain in detail how to distinguish the genuine product from fakes. The ad is remarkable for the lengths to which it went to explain how to identify fraudulent brands. Notably this ad appeared in a journal that published articles (which I discuss below) that provided specific information for non-BW&C manufacturers wanting to make their own versions of Hazeline Snow, the very fakes against which the ad was warning.[28]

We need to keep in mind that internally, China was not simply a site uniquely teeming with unethical counterfeiters and hapless officials. Different views and practices of copying marks existed domestically, as Chinese themselves grappled with changing

[28] See the discussion of Shen, "Method for Manufacturing Cosmetics" (cit. n. 2) in the next section.

ideas about what ownership of a brand might mean. With the chaos that characterized the early Republic, the central Chinese state was not particularly interested in or able to (should it even want to) enforce trademark legislation that was only gradually being drafted into code. We saw with Commissioner Tang the conditions under which a provincial level official might have at least superficially cracked down on infringement. At the local level, nonstate entities often stepped into the vacuum created by a distracted state to promote trademark enforcement, often to their own advantage. Take, for example, pharmaceutical industrialist Chen Diexian (1879–1940), a powerful Shanghai editor and leading cosmetic industrialist. Chen's most famous product was Butterfly Brand tooth powder, unique in its ability to double as face powder. As the Butterfly Brand took off, copycat brands quickly emerged. Chen sought to crack down through a variety of methods, including notably using his access to the media as a bully pulpit. He published notices that identified alleged counterfeits of his "Butterfly" trademark in key journals, such as the *Jiangsu Industrial Monthly*.[29] In these pieces, Chen underscored the role of provincial level state agencies in institutionalizing trademark enforcement, described new institutional apparatus for policing trademark ownership, and outlined in elaborate detail how these agencies pursued alleged counterfeiters. Much like the BW&C announcements discussed above, these notices sought to shame domestic companies into abiding by these new practices of enforcing corporate ownership over trademarks. They also functioned to spur local officials to act and pursue copycats. As domestic Chinese manufacturers saw their trademarks, recipes, and manufacturing processes disseminated, the emerging trademark regime became a useful means for them to stave off potential adapters.

CIRCULATING RECIPES: "COMMON KNOWLEDGE" OR TRADEMARK INFRINGEMENT?

Just as marks spread worldwide, so too did recipes, trade secrets, and formulas for manufacturing branded commodities. Emerging manufacturers around the world (and not just in China) sought access to manufacturing processes of successful brand-name items, and those who were manufacturing successfully increasingly became wary of the spread of their formulas. The recipes and ingredients of Hazeline Snow circulated globally in pharmaceutical gazettes at the turn of the twentieth century, and by the 1910s were being translated and compiled into Chinese and printed in China's periodical press. BW&C thus employed trademark laws to accuse alleged copiers of infringing upon their brand (given the simplicity of products like Hazeline Snow, their recipes and formulas were not protected by patent laws). They sought to stem what they saw as fraudulent circulation of their recipes, both on the periphery and in the metropole, and curb the possibility of what some today might call industrial espionage.

Circulation of manufacturing know-how was not always a matter of legal or ethical violation in early twentieth-century China. Even as some Chinese manufacturers were starting to support the crackdown on trademark violation led by foreign companies

[29] See, e.g., Zhang Yi'ou, ed., "Shanghai jiating she Wudipai camian yanfen jiamao yinggai gai gan weibian" [Shanghai Household Industry Peerless Brand Face and Toothpowder trademark cannot be copied], *Jiangsu shiyeyuekan* [Jiangsu industrial monthly] 29 (1921): 44–5. For a closer analysis of this particular notice, see Eugenia Lean, "The Butterfly Mark: Chen Diexian, His Brand, and Cultural Entrepreneurism in Republican China," in *The Business of Culture: Cultural Entrepreneurs in China and Southeast Asia, 1900–65*, ed. Christopher Rea and Nicolai Volland (Vancouver, 2014), 62–91, esp. 80–2.

like BW&C, just as many heralded such shared information as "common knowledge" necessary for strengthening the nation (including some of the very same individuals arguing for trademark laws, including the aforementioned Chen Diexian). The circulation of common knowledge was not only commercially profitable but also deemed virtuous in a period when a movement to buy and manufacture Chinese goods, known as the National Products Movement, was starting to gain momentum.[30] It functioned to establish a new kind of consumer who read newspaper columns for common knowledge about manufacturing and basic chemistry. By consuming such knowledge, the reader was in a better position to navigate an increasingly fraught capitalist market of pharmaceutical goods, replete with fraudulent and inauthentic goods. The circulation of manufacturing knowledge of brand items clearly defied any legal hegemony of the global trademark law. Whereas IP law notably sought to reduce confusion in the marketplace by arresting circulation (of goods), the emphasis on common knowledge in China did precisely the opposite: it sought to reduce confusion by promoting the circulation of knowledge as well as to undercut capitalist interests of imperialist companies like BW&C.

As noted earlier, by the late nineteenth century, as modern patent law started to emerge, it soon became clear that simpler manufacturing processes, including those behind daily commodities like vanishing cream, patent medicine, and cosmetics, were often not subjected to patent protection. By the 1880s, for example, U.S. courts made a distinction between mere recipes or simple formulas, where ingredients retained their discrete characters despite assemblage with other ingredients, and formulas for a specific product that included the manufacturing of new ingredients in the process of making the new product. Patents were granted only in the latter cases.[31] When this occurred, pharmaceutical companies like BW&C relied on trademarking to stake a legal claim over their non-patentable manufacturing processes and halt their widespread circulation. Recipes were considered to be part of the brand and trademark and thus subjected to trademark laws.

Such tactics could prove quite effective. Take, for example, the St. Louis–based journal *National Druggist*, which published an entry on Hazeline Snow and its formula in its August 1907 issue.[32] The entry noted explicitly that it had been translated from the French journal *Journal de Phamacie d'Anvers*, speaking to the global transmission of such information. In the December issue, however, the *National Druggist* published "Hazeline Snow a Trade-Mark, and Name Not Public Property," in which it cited the August issue's inclusion of the extract from the *Journal de Phamacie d'Anvers*. Essentially a retraction and apology, the announcement states:

> Our attention has been called to fact that the title Hazeline Snow is the trade mark Burroughs Wellcome & Co and hence being their exclusive property can not be employed by any other person or firm. We take pleasure in publishing this notice and we will ask our readers to take care not to use this title to designate any preparation of their own or anybody else's besides that of Burroughs Wellcome & Co in order that they may avoid the imputation of trade mark infringement and a possible suit for injunction and damages.[33]

[30] For more on the National Products Movement, see Karl Gerth, *China Made: Consumer Culture and the Creation of the Nation* (Cambridge, Mass., 2003).
[31] Gabriel, *Medical Monopoly* (cit. n. 10), 121–2.
[32] "Hazeline Snow, a Toilet Cream," *National Druggist* 37 (1907): 272.
[33] "Hazeline Snow a Trade-Mark, and Name Not Public Property," *National Druggist* 37 (1907): 411.

BW&C had learned of the earlier publication of its recipe, had intervened with the threat of legal action, and the *National Druggist* was forced to publish this notice.[34]

By the 1910s, pieces offering manufacturing tips for Hazeline Snow were appearing in Chinese journals. Featured in a 1915 issue of the *Ladies' Journal*, the article, "The Method for Manufacturing Cosmetics," argued that women needed to know how to make cosmetics and stated that such knowledge was crucial for the making of the modern household.[35] To inform the targeted genteel female reader, the piece provided concrete information on how to make a long list of cosmetics. In the entry on vanishing cream, it noted that face cream was a commodity readily available for purchase and identified three brands that dominated the market. Two were Chinese brands, the Three Star Brand (Sanxing) and Shuangmeimo. The third was Hazeline Snow (Xiashilian).[36] As seen with the second epigraph at the beginning of this essay, the entry divulged that the ingredients for Hazeline Snow included hazeline, stearic acid, glycerin, sodium bicarbonate, and soda water.

To better understand the manner in which the sharing of manufacturing know-how of brand commodities might have been appreciated in early twentieth-century China, it is worthwhile to examine how such knowledge was often deemed "common knowledge." Common knowledge was published in China's burgeoning print media and presented as tasteful knowledge for learned readers to help them navigate emerging capitalist markets for material goods in China. This category of knowledge often included manufacturing information and recipes for trademarked commodities and brands, both domestic and foreign. As such, it was often portrayed as both commodified knowledge and crucial information necessary for building Chinese industry. Take, for example, "Common Knowledge for the Household" ("Jiating changshi"), a column featured in the highly regarded literary supplement, "Free Talk" ("Ziyoutan") in Shanghai's largest newspaper, the *Shenbao*, that was compiled by Chen Diexian, the manufacturer of Butterfly Brand Toothpowder who also happened to be a powerful Shanghai editor and, ironically, an advocate for the domestic policing of the infringement of trademarks and packaging, especially of his own. Running from 1917 to 1927, the column featured a wealth of information, including tips on healing different conditions, keeping foods fresh, and maintaining hygiene, as well as science and manufacturing. These tips were digestible yet informative nuggets of knowledge meant to improve everyday life and to be consumed in a leisurely fashion by learned generalist readers of the literary supplement, who might have included merchants, intellectuals, and urbane citizen-readers. Featuring tips on how to manufacture soap, dyes, camphor, matches, bleach, and soy sauce, along with entries on engraving copper plates, welding together aluminum, and producing gadgets such as the chemical foam-based fire extinguisher, the column became known specifically for its coverage of manufacturing and industrial knowledge, along with chemistry and physics. By doing so, the column helped engender a new understanding of the learned Chinese consumer as someone fully capable of evaluating and grasping production and manufacturing technologies.

[34] Earlier accounts of how to manufacture Hazeline Snow did not seem to invite similar scrutiny. The 4 December 1897 issue of *Chemist and Druggist: The Newsweekly for Pharmacy* (51: 884) included a brief description of how to improve Hazeline Snow, and the information was shared without any sense that it might constitute a legal or ethical violation.

[35] Shen, "Method for Manufacturing Cosmetics" (cit. n. 2).

[36] Ibid., 24–5.

"Common knowledge" also appeared in journals that were intended not only for general consumption but also to serve as the basis of emerging expertise. Take, for example, the new-style industry periodical, *Huaxue gongyi* [The chemical industry journal]. Published from 1922 to 1923 by the Shanghai School of Chemical Industries, this publication regularly presented brand-name recipes and reports on factory and production conditions of famous and emerging companies.[37] There appeared to be little regard for any sense of exclusive ownership over the manufacturing knowledge or specific formulas and recipes, including for brand and trademarked items, and such knowledge was unapologetically presented as crucial for readers and students to emerge as experts in the chemical industry. The journal featured articles on large-scale manufacturing processes, surveys and reports on factories and companies, as well as pieces on chemical research, some of which included specific information about the production processes of famous brands and companies. Typical titles were "The method to make foreign candles," "Common knowledge in testing soap," and "The general state of Guangxi's manufacturing of cinnamon oil." The article "Common Knowledge Chemists Should Possess" detailed what common knowledge entailed.[38] Serving the same purpose was an article describing the basic knowledge in cosmetics production that a student needed to acquire, including how to manufacture soap, vanishing cream, tooth powder, scented powders, hair tonic, and perfumes.[39] Some pieces included recipes for famous brands of cosmetic items, both foreign and domestic. Others featured manufacturing processes of well-known factories. One report was an investigation of the ingredients of Xiangmao soap, a popular British brand.[40] A review of a visit to the Great Five Continents soap factory described in detail the kind of machinery being used, including the filter press, the automatic soap dryer, milling machines, and the compressor machine.[41]

In the October 1922 issue, the manufacturer of Butterfly Brand Toothpowder, Chen Diexian, found his company's manufacturing processes featured in a detailed survey.[42] Readers learned the exact steps in mixing Butterfly tooth powder and manufacturing magnesium carbonate, a key ingredient in the tooth powder and other powder-based cosmetics. The piece describes two methods to manufacture magnesium carbonate, providing the exact reaction sequences as chemical formulas in its explication. The first process involved adding sulfuric acid to magnesite powder, which would turn the solution to magnesium carbonate, and then, upon adding water and mixing in sodium carbonate, a sediment that was magnesium carbonate would be produced. The second method described involved the extraction of sodium chloride from seawater to leave magnesium chloride, which the reader is then instructed to mix with sodium carbonate in order to produce magnesium carbonate. These processes were not manufacturing secrets exclusive to the young company, but rather common knowledge that was to

[37] The school was an example of vocational education focused on industry and commerce to train and educate lower middle-class youths; such schools were starting to appear in large cities such as Shanghai. For more on the school's goals, curricula, and student requirements, see the advertisement for the school in the inner cover of the October 1922 issue (vol. 1, no. 2) of *Huaxue gongyi*.

[38] Yu Ziming, "Huaxuejia yingyou zhi changshi" [Common knowledge chemists should possess], *Huaxue gongyi* 1 (May 1923): 57–9.

[39] Lü Heng, "Yong huazhuangpin yingju zhi changshi" [Common knowledge one needs to use cosmetics], *Huaxue gongyi* 1 (October 1922): 18–20.

[40] Fang Chaoheng, "Xiangmao pai feizao tian shiliao zhi diaocha" [Investigation of Xiangmao Soap's ingredients], *Huaxue gongyi* 1 (October 1922): 44–6.

[41] See *Huaxue gongyi* 1 (May 1922): 49–52.

[42] See *Huaxue gongyi* 1 (October 1922): 41–4.

sow the grounds for the growth of native industry in the face of economic imperialism. Common knowledge was thus hardly a violation, but a precious commodity that was invaluable both for savvy consumers who were skillful in discerning among brands and for adaptive and innovative producers able to compete in the burgeoning pharmaceutical industry.

Precedents for innovative nativist adaptation of foreign technologies and industrial and scientific knowledge for strengthening China can be found in the late nineteenth century. During the Self-Strengthening Movement (ca. 1861–95), the concept of adapting or duplicating [*fangzhi*] in the process of weapons production played a key role in the manufacturing pursuits in arsenals erected to strengthen Qing China in the face of unrelenting imperialist encroachment. At the Jiangnan Arsenal, a key arsenal that the Qing state had established as part of the movement, the central task was technological learning and training with the ultimate goal of manufacturing modern armaments. If there was considerable failure at the arsenals, with bombs notoriously failing to explode, they nonetheless were the site of considerable technological transfer, adaptation, and innovation.[43] Creativity in technological production at the arsenal was heavily dependent upon duplication and emulation [*fangzhi*] of Western knowledge and technology. Research at the arsenal, sample making (of parts of a weapon), and testing for efficacy turned on the practice of *gaizao*, or "remaking," and it was with this process of remaking that creativity and innovation occurred.[44]

By the early twentieth century, with the decline of state-supported Self-Strengthening arsenals and the rise of private light industry in China's treaty ports, the adaptation of foreign technologies to manufacture commodities and daily use objects remained pressing. The "science" behind the manufacturing of commodities such as soap and cosmetics was predicated on basic chemical knowledge, and not particularly challenging. Such knowledge could be transferred quite easily, and little capital was required in the manufacturing of such items. For Chinese merchants and industrialists seeking to build nativist industry, the sharing and strategic adaptation of industrial manufacturing know-how as common knowledge in China's burgeoning mass media was not just unproblematic, but in fact patriotic and urgent. That this manufacturing information was associated with successful brands—both foreign (e.g., Hazeline Snow) and Chinese (e.g., Butterfly Brand)—made their dissemination all the more valuable.

As these recipes started to circulate in Chinese publications, they would have had the effect of exacerbating anxieties about Chinese copying that players such as BW&C might have already started to feel with the widespread emulation of their marks and packaging. It was no accident that this was around the time that foreign entities started to pursue trademark infringement in China so aggressively. In addition to the aforementioned cases in which BW&C pursued alleged Hazeline Snow counterfeiters, American lawyer Norman Allman published *Handbook on the Protection of Trade-Marks, Patents, Copyrights, and Trade-names in China* in 1924.[45] By the 1930s, his Shanghai-based litigation firm, Allman & Co., represented British companies in cases of trade-

[43] Meng Yue, "Hybrid Science versus Modernity: The Practice of the Jiangnan Arsenal, 1864–1897," *EASTM* 16 (1999): 13–52. See also Benjamin Elman, *On Their Own Terms: Science in China, 1550–1900* (Cambridge, Mass., 2004), 355–95.

[44] Meng, "Hybrid Science" (cit. n. 43).

[45] Norman Allman, *Handbook on the Protection of Trade-Marks, Patents, Copyrights, and Trade-names in China* (Shanghai, 1924).

mark infringement in the courts of China. This included representing BW&C in its attempt to prosecute the Chinese firm, Liya, for counterfeiting the iconic design of Hazeline Snow face cream to sell Liya's Himalaya soap. In another case in 1934, Allman & Co. was to represent Lever Brothers (China) Ltd. against a certain Li Cheng factory, for allegedly infringing upon Unilever's famous brand marks (the English name "Lux" and the Chinese name for Lux soap, *Lishi*) along with the packaging and design of Lux soap cartons, to sell its face cream. Although both cases involved the British foreign office seeking diplomatic means to put pressure on the Nationalist regime to discipline the offenders, they did not do so to particularly good effect. Neither saw the light of day in a Chinese courtroom.[46] Yet despite the failure to prosecute the alleged Chinese copycat firms, both were indicative of the continued vigilance and aggression with which BW&C and other pharmaceutical companies based in the West employed modern trademark law to pursue Chinese adaptation of marks and recipes.

CONCLUSION

Much writing on the history of IPR has tended either to focus solely on the West or to start with the assumption that the modern regime of intellectual property originated wholly in the West and was then exported to the rest of the world.[47] China scholars, too, have tended to see the emergence of the formal legal treatment of trademarks as property in China as uniquely modern and originating from abroad. William Alford, for example, argues that intellectual property rights had not developed indigenously in imperial China because of its particular political culture, which focused not on ownership and private interest but on political order and the preservation of imperial power.[48] In contrast, this piece has adopted a global approach and assiduously avoids viewing the modern trademark as "Western" and seeing China playing a "catch-up" role, lagging behind the West in its efforts to erect a modern property rights regime. It takes seriously how Chinese actors were active agents in a worldwide debate over the ownership of marks and recipes and the making of a modern property rights regime. Chinese manufacturers, merchants, officials, and editors of industrial journals and magazines were hardly passive and insisted on playing by their own rules. They actively challenged, blithely ignored, and appropriated savvy ideas of ownership that BW&C and other pharmaceutical companies were seeking to promote. They ingeniously adapted a host of strategies and counterstrategies to adjust to newly emerging logics of trademark infringement in international disputes. As a result, they had a hand in shaping what was hardly a fixed aspect of capitalism in the early twentieth century, defining (if, at times, through acts of resistance or defiance) the parameters of a regime of trademark infringement that was becoming a key legal mechanism of control in modern global commerce, industry, and science.

[46] For a discussion of these cases, see Lean, "Marking and Copying" (cit. n. 11).

[47] There is a large corpus of historical scholarship on IPR that focuses on the West. For excellent examples, see the work by Mario Biagoli and his collaborators, including Biagoli and Peter Galison, eds., *Scientific Authorship: Credit and Intellectual Property in Science* (New York, 2003); and Biagoli, Martha Woodmansee, and Peter Jaszi, eds., *Making and Unmaking Intellectual Property* (Chicago, 2011). For an example of a study that is global in scope but identifies the rise of modern IPR in early modern Europe and traces its exportation to the rest of the world into the modern period, see Johns, *Piracy* (cit. n. 10).

[48] Alford, *To Steal a Book* (cit. n. 18).

This essay converges with several contributions in this volume that place transnational entanglements at the forefront of their analysis of modern science and capitalism. It is similar, for example, to Sarah Milov's essay in its concern with showing how uneven power relations in the twentieth century that seemingly favored Western capitalist corporations often resulted in unintended consequences.[49] Milov illustrates how even as American tobacco companies helped constitute the "Japanese smoker," data based on wives of Japanese smokers, served as the basis of a transnational grassroots regulatory movement against American tobacco. In a similar vein, this essay sheds light on how even as BW&C sought to constitute the "Chinese copycat," Chinese copying served to impose and expose limits to BW&C's attempts to erect an effective global trademark infringement regime. The powerful corporation was not fully able to advance purportedly "universal" views of corporate ownership of trademarks and recipes as widely as it would have liked.

With this in mind, this essay argues for a more refined understanding of the motif of circulation that takes into account the historical contingencies behind the transmission and travel of knowledge and things, and their obstruction. Conventional accounts of modern science often take as a given that Western science was built on a foundation of openness, sharing, and transparency. Proponents of capitalism similarly make the ideological claim that a core trait of capitalism is free and open markets. Here, however, we see how "common knowledge," unapologetically commercial and derivative in its inclusion of domestic and foreign brand recipes and manufacturing tips, helped expose the hypocrisy behind purportedly ethical claims of both the universal nature of modern science and the ideal of the free market promoted by advocates of modern capitalism. Chinese compilers, editors, and readers found common knowledge, manufacturing and chemical information shared by all alike and hence owned by no one in particular, to be invaluable. The shared knowledge was to arm the new Chinese consumer with simple yet crucial manufacturing and chemical knowledge necessary to navigate a burgeoning and treacherous domestic market of pharmaceuticals filled with questionable products. It was also intended for emerging Chinese manufacturers who were seeking to compete in domestic and global pharmaceutical markets. Common knowledge and Chinese copying of marks defied the ideals of ownership being pursued by advocates of the new IP regime. Deeply troubled, Western corporations aggressively deployed trademark law to discourage the spread of trade names and brand recipes. As noted in the introduction to this volume, recent trends in the history of science that have focused on the global movement of knowledge in the making of modern science have generated an overly rosy picture of the seamless circulation of knowledge in the making of science, which serves to reinforce the neoliberal claims of "universal" scientific knowledge and free and open capitalist markets. In contrast, this essay provides us with a powerful instance of how modern IP functioned to arrest circulation flows, exploring how this occurred and why.

Finally, this essay has shed light on the manner in which China's reputation as a brazen copier might have emerged. While hardly unique in its copying, Chinese adaptation of Hazeline Snow's marks and the circulation of its recipe were savvy and often tremendously successful. This success invited attention and vitriol, and BW&C condemned Chinese copiers as ethically dubious and particularly venal in behavior. Such

[49] Sarah Milov, "Smoke Ring: From American Tobacco to Japanese Data," in this volume.

condemnation helped ensure that the Chinese copycat would emerge as an iconic counterfeiter. This reputation persists to this day, as Chinese manufacturers are regularly accused of manufacturing knockoff Gucci handbags or fraudulent versions of the iPhone. Hallam Stevens's article in this volume troubles the easy distinction between imitation and innovation in such an argument by showing how copying and innovation go hand in hand in contemporary Shenzhen.[50] Similarly, this essay shows that the assumption that the two—imitation and innovation—are mutually exclusive has a long history and was created by European manufacturers over time to protect their capital by slowing down the free circulation of things and ideas into China. Chinese adaptation of brand-name items continues to generate global anxiety today, which, in turn, continues to engender claims that China can only copy and not innovate. This twenty-first-century reality only makes the historical investigation into the making of the Chinese copycat in the early twentieth century, and its crucial entanglement with the rise of global commerce, science, and modern imperial power, all the more urgent.

[50] Hallam Stevens, "Starting up Biology in China: Performances of Life at BGI," in this volume.

Microbiology and the Imperatives of Capital in International Agro-Biodiversity Preservation

*by Courtney Fullilove**

ABSTRACT

This essay considers the political economy of transnational scientific research based on global collection of biota for laboratory manipulation, focusing on a program to develop pest-resistant wheat using fungal endophytes common in a range of wild but closely related grasses. This effort extends long-standing efforts to commoditize living substances of increasing scope and complexity, and it is supported by efforts to collect and preserve biological diversity. The essay explores how imperatives of capital shape biodiversity as a policy category and determine which forms of life are saved, materially altering our records of life on earth. These newly legible and malleable organisms become more perfect commodities, suitable for standardization and transmutation into finance capital. Yet endophytes are also of interest in part because of their resistance to such control, throwing into sharp relief the reductive imperatives of commoditization while also provoking new ways of justifying capital accumulation and flow. This essay questions the extent to which histories of capitalism and science as conjoined projects rooted in the biological species concept can explain contemporary practices of biodiversity preservation and the microbiological research they support. Microbiological research provides new renderings of life on earth that may challenge or reconfigure metaphors and practices common to capitalism and science.

Rumination is a virtue, and ruminants are the unsung heroes of science. Slow and steady, committed to their single purpose, animals are nevertheless always finding things as they graze. Take the Lewis Chessmen, a set of diminutive and apparently bewildered twelfth-century warriors now housed in the British Museum: according to some reports, a cow roaming the shore of the Western Isles of Britain conducted their excavation. The central drama of this essay originates in the accidental poisoning of cattle and sheep. American cattle grazing on certain fescue became ill with fever, rapid breathing, and excessive salivation. Meanwhile, New Zealand sheep grazing on certain ryegrass were

* Department of History, Wesleyan University, Middletown, CT 06459; cfullilove@wesleyan.edu.
In addition to the *Osiris* volume workshop at Columbia University in June 2016, I am grateful to the members of the Department III colloquium at the Max Planck Institute for the History of Science in Berlin, Germany, and especially to commentary and questions from David Sepkoski, Dagmar Schäfer, Dan Bouk, and BuYun Chen. This essay also benefits from participation in Matthias Rillig's "Fungal Biology and Ecology" master course at Freie Universität in Berlin.

afflicted with the staggers, developing tremors, incoordination, and collapse. Farmers had long lamented the damage to their flocks caused by the disease and associated susceptibility to accident and death. So when researchers in the United States and New Zealand independently traced toxicity to the *Epichloë* fungal endophyte in the late 1970s and early 1980s, they summarily set about liberating the plant of its affliction.[1]

Removing the endophyte from the plant did improve the health of the animals—but not that of the plants. Without the fungus, the grasses were devoured by pests and battered by the elements.[2] Clearly, the plant derived benefit from the microorganisms colonizing its tissues. Researchers took note. Over the next two decades, scientists in the United States and New Zealand worked to develop *Epichloë* endophytes that conferred protective traits on plants without harming animals.[3] In the United States, Max Q was released for use with tall fescue in 2000. In New Zealand AR 1 was released for use with perennial ryegrass in 2001. These pasture grasses distinguished themselves by containing endophytic fungi "beneficial" to the plant hosts and their animal consumers. Thus a bevy of commercially bred pasture grasses, and a new research field, were born.

This narrative, suitable for enthusiasts of interspecies agency and serendipity as tropes in the history of science, provides one story of origins for research into fungal endophytes in pasture grasses in the 1980s. This essay provides another kind of exploration, for seeds and plants are not stable objects but contested artifacts, classified according to variable logics of science, heritage, and property. Their utility and value hinges on the conditions of their preservation, circulation, and reproduction: laboratory and field, public and private, commercial and communitarian. In bred pasture grasses, endophytic fungi confer pest resistance and other potential benefits on their hosts. Breeders exploit these benefits to raise fat cattle and sheep. Contemporary researchers aim to replicate these traits in cultivated cereals for human consumption. Since the 1990s, research has expanded into the beneficial properties of endophytes in a range of plant species consumed by animals and humans, introducing new questions about the potentials and exploitation of microorganisms for human prosperity.

Literally meaning "in" (*endo*) "plant" (*phyte*), an endophyte can be a virus, another plant, a fungus, or bacteria. Although research targets microbes that may aid agriculture, endophytes can range from mutualistic to pathogenic. Fungi constitute major threats to crops but also provide beneficial properties for plant growth and health, important sources of food products, and model organisms for investigating biological questions.[4] In spite of a now vast body of research on endophytes and their manipulation, however,

[1] C. W. Bacon, J. K. Porter, J. D. Robbins, and E. S. Luttrell, "*Epichloë typhina* from Toxic Tall Fescue Grasses," *Appl. Environ. Microbiol.* 34 (1977): 576–81, esp. 576–8; L. R. Fletcher and I. C. Harvey, "An Association of a Lolium Endophyte with Ryegrass Staggers," *N. Zeal. Vet. J.* 29 (1981): 185–6.

[2] J. H. Bouton, R. N. Gates, D. P. Belesky, and M. Owsley, "Yield and Persistence of Tall Fescue in the Southeastern Coastal Plain after Removal of Its Endophyte," *Agron. J.* 85 (1993): 52–5; C. P. West et al., "Endophyte Effects on Growth and Persistence of Tall Fescue along a Water-Supply Gradient," *Agron. J.* 85 (1993): 264–70; A. J. Popay, "Argentine Stem Weevil Response to Variable Endophyte Infection in Grasslands Greenstone Ryegrass," *Proc. N. Zeal. Plant Protection Conf.* 50 (1997): 69–72.

[3] Linda J. Johnson, Anouck C. M. de Bonth, Lyn R. Briggs, John R. Caradus, Sarah C. Finch, Damien J. Fleetwood, Lester R. Fletcher et al., "The Exploitation of Epichloae Endophytes for Agricultural Benefit," *Fungal Diversity* 60 (2013): 171–88.

[4] On crop pathogens, see Matthew C. Fisher, Daniel A. Henk, Cheryl J. Briggs, John S. Brownstein, Lawrence C. Madoff, Sarah L. McCraw, and Sarah J. Gurr, "Emerging Fungal Threats to Animal, Plant and Ecosystem Health," *Nature* 484 (2012): 186–94.

these microorganisms are not fully understood, and the way they interact with different plants is highly variable.[5]

In the following, I present an account of a development program based in New Zealand that aims to create wheat resistant to pests and abiotic stresses through the introduction of fungal endophytes. Cereal endophyte researchers seek wild relatives of wheat harboring specific fungal endophytes beneficial in bred pasture grasses. Research thus relies on the global collection of grasses classified as wild relatives of cereal crops. Subsequently screened for endophytes in the laboratory, these grasses provide material for researchers. Lodged in nationally and internationally managed gene banks at the sponsoring institutions and the countries of origin, the collected seeds remain accessible to researchers worldwide in compliance with the Convention on Biodiversity (1992) and the International Treaty on Plant Genetic Resources (1995). These seeds represent part of an ambitious and often fraught program to preserve world biodiversity against the encroachments of modern agricultural methods, development, conflict, and climate change.[6]

This essay analyzes the political economy of transnational scientific research based on global collection of biota for laboratory manipulation. In historical terms, Linnaean taxonomy devised in the context of European maritime exploration undergirded the definition and production of commodities based on global seeds and plants, and provided the organizing principle for botanic gardens and the gene banks that succeeded them. Reduction of global biodiversity to the commodity form continues to drive research. Yet although microbiota represent the largest proportion of biodiversity on the planet, they are not easily characterized by conventional species concepts derived from Linnaean taxonomy. In contrast to the sexual reproduction of macroorganisms Linnaeus used to order plant species, microorganisms reproduce asexually, transfer genetic material from other microbial communities, and form symbiotic associations with one another.[7] Inasmuch as fungi and other microorganisms challenge norms of classification and conservation predicated on the species as a reproductively isolated population, international seed banks ordered by plant species make an unlikely home for their research.[8] In the last thirty years, many research foci have shifted from botanic to fungal

[5] For a summary of current research into the properties of endophytic fungi, see Charles W. Bacon and James F. White, "Functions, Mechanisms and Regulation of Endophytic and Epiphytic Microbial Communities of Plants," *Symbiosis* 68 (2016): 87–98.

[6] Research for this piece entailed participation in seven collecting expeditions targeting pasture grasses and wild relatives of cereal crops in the Caucasus and Central Asia between 2010 and 2016, as well as laboratory site visits and interviews at AgResearch headquarters in Palmerston North, New Zealand, in January 2016. Collection took place in July and August in Armenia and Georgia (2010), Tajikistan (2011), northwest Russia (2013), Kabardino-Balkaria, Adygea, and Karachai-Cherkessia (2014), and Kazakhstan (2015). Participating international institutions were AgResearch (Palmerston North, New Zealand), the Vavilov Research Institute of Plant Industry (VIR, St. Petersburg, Russia), and the International Center for Agricultural Research in the Dry Areas (ICARDA, Aleppo, Syria, and Rabat, Morocco).

[7] Annette Davison, Christine Yeates, Michael Gillings, and Jan de Brabandere, "Microorganisms, Australia and the Convention on Biological Diversity," *Biodiversity & Conserv.* 8 (1999): 1399–1415.

[8] On the institutional history of research in molecular biology, see Michael R. Dietrich, "Paradox and Persuasion: Negotiating the Place of Molecular Evolution within Evolutionary Biology," *J. Hist. Biol.* 31 (1998): 85–111; Bruno J. Strasser and Soraya de Chadarevian, "The Comparative and the Exemplary: Revisiting the Early History of Molecular Biology," *Hist. Sci.* 49 (2011): 317–36; Carl R. Woese, "How We Do, Don't and Should Look at Bacteria and Bacteriology," in *The Prokaryotes*, vol. 1, *Symbiotic Associations, Biotechnology, Applied Microbiology*, ed. Martin Dworkin et al. (New York, 2006), 3–23.

and microbiological taxonomy, and from species to genome as a locus of diversity.[9] The capacity of gene banks ordered by plant species to encompass microbiological research suggests the ways novel classifications of life and rubrics of diversity can sustain capitalized research and development while remaining governed by conventions and techniques of natural science some four centuries old.[10]

Cereal endophyte research extends long-standing efforts to commoditize living substances of increasing scope and complexity. That is, although research into fungal endophytes could suggest new concepts of value and interdependency, the industrial imperative of research creates path dependency that forecloses such investigations. For all the marvels of the fungal symbiont, the cereal endophyte program primarily aims to build a better wheat plant, and researchers pursue only those fungi that are responsible for conferring pest resistance and abiotic stress tolerance in related grasses. Ironically, collected material supports development efforts that may further attenuate the diversity on which they rely. Imperatives of capital thus shape biodiversity as a policy category and determine which forms of life are saved, materially altering our records of life on earth. In turn, these newly legible and malleable organisms become more perfect commodities, suitable for standardization and transmutation into finance capital.

This bias of selection is a variant of what some historians of science have dubbed "agnotology," or the cultural production of ignorance.[11] Shaped by the dictates of political economy, researchers pursue topics and methods that reproduce logics of capital accumulation: in the process, they shed other forms and meanings of life. In science and technology studies (STS), these political rationalities have been analyzed as material assemblages of social relations and objects, with the nonhuman exerting as much control as intentional human subjects over the terms of life.[12] In the past decade, STS scholars have paid special attention to multispecies interactions, and to the ways knowledge produced from them is situated in particular social and economic contexts.[13] Others have called for renewed or sustained attention to the social and ideological features of networks and institutions that produce or support technoscience.[14] Hannah Landecker's formulation of the "biology of history" bridges these approaches to political economy

[9] Two summations of the debate over the locus of biodiversity preservation in species, genome, or ecosystem are provided by Timothy Farnham, "A Confluence of Values: Historical Roots of Concern for Biological Diversity," in *The Routledge Handbook of Philosophy of Biodiversity*, ed. Justin Garson, Anya Plutynski, and Sahotra Sarkar (New York, 2017), 11–25; and by James Maclaurin and Kim Sterelny, *What Is Biodiversity?* (Chicago, 2013).

[10] On the persistence of institutional networks and practices in the history of science, see, e.g., Bruno J. Strasser, "Laboratories, Museums, and the Comparative Perspective: Alan A. Boyden's Quest for Objectivity in Serological Taxonomy, 1924–1962," *Hist. Stud. Nat. Sci.* 40 (2010): 149–82.

[11] E.g., Robert N. Proctor and Londa Schiebinger, *Agnotology: The Making and Unmaking of Ignorance* (Stanford, Calif., 2008).

[12] Bruno Latour, "One More Turn after the Social Turn: Easing Science Studies into the Non-modern World," in *The Social Dimensions of Science*, ed. Ernan McMullin (Notre Dame, Ind., 1992), 272–92; Latour, *We Have Never Been Modern* (Cambridge, Mass., 1991).

[13] Donna Haraway, *When Species Meet* (Minneapolis, 2008); Haraway, "Situated Knowledges: The Science Question in Feminism and the Privilege of Partial Perspective," *Feminist Stud.* 14 (1998): 575–99; Heather Paxson and Stefan Helmreich, "The Perils and Promises of Microbial Abundance: Novel Natures and Model Ecosystems, from Artisanal Cheese to Alien Seas," *Soc. Stud. Sci.* 44 (2014): 165–93; Helmreich, with contributions from Sophia Roosth and Michele Friedner, *Sounding the Limits of Life: Essays in the Anthropology of Biology and Beyond* (Princeton, N.J., 2015).

[14] See, e.g., Sheila Jasanoff, ed., *States of Knowledge: The Co-production of Science and Social Order* (London, 2004); Jasanoff and Sang-Hyun Kim, eds., *Dreamscapes of Modernity: Sociotechnical Imaginaries and the Fabrication of Power* (Chicago, 2015).

and epistemology by focusing on how past practices of science have shaped the material worlds in which today's researchers operate. "At the same time that we now know more," Landecker observes, "we come to inhabit the material future produced by what we thought we knew." Science itself drives biological change.[15]

Yet if the object of microbiology has primarily been to "harness" the "productive or reproductive capacities" of cells "to human intention," as Landecker has argued, endophytes are of interest in part because of their resistance to such control.[16] *Epichloë* endophytes require the plant host to survive, and vice versa. They refuse removal from the bodies of their hosts, with which their life cycles are coterminous. Unlike other forms of cell culture, endophytes refuse to "live differently" in space and time.[17] The fungus's lack of interest in reducibility or its own market value provides an opportunity to examine the drive toward commoditization that fuels international research and development.[18] The fungal symbiont's way of life throws into sharp relief the reductive imperatives of commoditization. It may also provoke new ways of justifying capital accumulation and flow.

To what extent do histories of capitalism and science as conjoined projects to organize resources help us understand contemporary practices of biodiversity preservation and microbiological research? Theorizations of capitalism as a world system derive power from biological imaginaries, and vice versa. Increasingly, metaphors of industrialization and standardization seem ill fitted to microbiological research, with its orientation toward interdependency and flux.[19] The essay proceeds along historical and ethnographic lines, moving from a history of nature collection to a study of contemporary laboratory process in microbiology. First, it characterizes the history of natural history collection and species preservation as adjuncts to capitalism as a world system, exploring the transition from Linnaean hierarchical taxonomy to more novel forms of organization in the twentieth-century life sciences. Second, it considers the cereal endophyte development program as a vehicle to understand the material, political economic, and conceptual implications of contemporary biotechnology and agriculture, including transnational flows of research capital and investigation of organisms that supersede established classificatory schemes for global plant genetic resources. In many respects, applied microbiological research represents a new direction in the conception,

[15] Hannah Landecker, "Antibiotic Resistance and the Biology of History," *Body & Soc.* 22 (2016): 19–52, on 37. Landecker characterizes the biology of history as "a recursive structure in which knowledge is produced in and through matter that has itself been altered by previous modes of thought" (37).

[16] Hannah Landecker, *Culturing Life: How Cells Became Technologies*. (Cambridge, Mass., 2010), 1.

[17] Hannah Landecker, "Living Differently in Time: Plasticity, Temporality and Cellular Biotechnologies," *Culture Machine* 7 (2005), http://www.culturemachine.net/index.php/cm/article/view/26 (accessed 19 July 2017).

[18] I use the term "interest" self-consciously. In studies of capitalism as a human social system, interest typically refers to the self-oriented motives and behaviors of individuals or groups to amass resources. In science and technology studies, interest is more frequently regarded as multisited and operating through structures of problematization, communication, translation, and institutionalization of knowledge and materials. Two influential explorations of these processes are Michel Callon, *Some Elements of a Sociology of Translation: Domestication of the Scallops and the Fishermen of St. Brieuc Bay* (n.p., 2000); Susan Leigh Star and James R. Griesemer, "Institutional Ecology, 'Translations' and Boundary Objects: Amateurs and Professionals in Berkeley's Museum of Vertebrate Zoology, 1907–39," *Soc. Stud. Sci.* 19 (1989): 387–420.

[19] On industrial modernity as the paradigm for genetics, see Phillip Thurtle, *The Emergence of Genetic Rationality: Space, Time, and Information in American Biological Science, 1870–1920* (Seattle, 2007).

organization, and research of biological diversity on earth.[20] Microbiological research provides new renderings of life on earth that may challenge or reconfigure metaphors and practices common to capitalism and science. Does natural diversity defy the reductive logics of capital, or are the latter so malleable that they subsume all forms of natural diversity?

HISTORY OF AGRO-BIODIVERSITY PRESERVATION

Histories of capitalism and natural science have emphasized their mutual development. European nation-states supported the accumulation of foreign territories, and naturalists supported the classification of mineral and biological resources with an eye to their profit-making potential.[21] Concepts of nature's diversity emerged in tandem with capitalism as a world system, with Linnaean taxonomy of genus and species fitted to the commodification of flora for the enrichment of nation-states.[22] From their inception, seed banks were instruments of capital accumulation. Twentieth-century gene banks elaborated the model of European botanic gardens forged through empire. The interest of plant breeders in exploiting global plant genetic resources for twentieth-century agricultural modernization contributed to their ex situ preservation in centralized seed banks.[23]

This long history of biodiversity preservation suggests strong continuities between early modern European maritime exploration and twenty-first-century agricultural research. Even as both are collecting projects of encyclopedic ambition and global scale, they have each remained defined by service to capitalist interests. In the coterminous development of market and nation-state, the commodity is a stable and essential unit pursued for utility and profit, ordering the natural world to support the governance of populations.[24] In diverse histories of capitalism ranging from Immanuel Wallerstein's world systems theory to Michel Foucault's biopower, the commodity form remains the basis of capitalist development, historically driven by competition between European nation-states.[25]

[20] Applied microbiological research builds on the development of microbiology as a discipline in the preceding centuries. See J. Sapp, "The Structure of Microbial Evolutionary Theory," *Stud. Hist. Phil. Biol. Biomed. Sci.* 38 (2007): 780–95; Sapp, *Where the Truth Lies: Franz Moewus and the Origins of Molecular Biology* (Cambridge, 1990); Judy Johns Schloegel, *From Anomaly to Unification: Tracy Sonneborn and the Species Problem in Protozoa, 1954–1957* (Dordrecht, 1999).

[21] On European botanic gardens and tropical agriculture, see Richard Harry Drayton, *Nature's Government: Science, Imperial Britain, and the "Improvement" of the World* (New Haven, Conn., 2000); E. C. Spary, *Utopia's Garden: French Natural History from Old Regime to Revolution* (Chicago, 2000); Richard Grove, *Green Imperialism: Colonial Expansion, Tropical Island Edens, and the Origins of Environmentalism, 1600–1860* (Cambridge, 1995).

[22] On the species as commodity form, see Lisbet Koerner, *Linnaeus: Nature and Nation* (Cambridge, Mass., 2001); Federico Marcon, *The Knowledge of Nature and the Nature of Knowledge in Early Modern Japan* (Chicago, 2015), 296–7; Staffan Mueller-Wille, "Philosophy of Biology beyond Evolution," *Biol. Theory* 2 (2007): 111–2.

[23] Tiago Saraiva, "Breeding Europe: Crop Diversity, Gene Banks, and Commoners," in *Cosmopolitan Commons: Sharing Resources and Risks across Borders*, ed. N. Disco and E. Kranakis (Cambridge, Mass., 2013), 185–212; Marianna Fenzi and Christophe Bonneuil, "From 'Genetic Resources' to 'Ecosystems Services': A Century of Science and Global Policies for Crop Diversity Conservation," *Cult. Agricult. Food Environ.* 38 (2016): 72–83.

[24] On the mutual development of nation-state and market, see also Karl Polanyi, *The Great Transformation* (Boston, 1957).

[25] Immanuel Maurice Wallerstein, *The Modern World-System* (New York, 1974); Michel Foucault, *The Order of Things: An Archaeology of the Human Sciences* (New York, 1971).

Structural renderings of capitalism nevertheless have certain limits. Historians confronting the *longue durée* of economic development have emphasized capitalism's variability and adaptability. Giovanni Arrighi, following Fernand Braudel, made his subject the historical evolution of capitalism, identifying systemic cycles of accumulation and hegemonic transition between nation-states.[26] According to Arrighi's model, when material expansion reaches a natural limit, interstate competition for mobile finance capital presents the greatest possibility of expansion, and states strive to assert control over global resources through managerial authority and intellectual property rights rather than simple material accumulation. Although Arrighi backdates the origins of capitalism to the mercantile activity of thirteenth- and fourteenth-century Italian city-states, European maritime exploration, territorial acquisition, and surveys of mineral and biological resources in successive centuries constituted a primary phase of material accumulation. Even as Arrighi emphasized geo-economic processes as essential movers of wealth, however, he identified the pairing of finance capital with interstate competition as the primary feature of capitalism. Territory and money capital are in a constant push-pull cycle, mediated by geography, colonialism and its aftermath, and rising inequality in centers of production.[27]

On the face of it, botanical collection is ill-suited to a model such as Arrighi's, which prioritizes financialization as a motor of capitalist development. Yet contemporary biotechnology operates through the fictions of finance, dependent on money capital and redefinition of plant matter as raw material for research and development. While centralized seed banks are the successor institutions to European botanic gardens, they function as reservoirs for research and development possible only through infusions of finance capital, facilitating the technological manipulation of biological material and its subjection to intellectual property rights and commercialization.[28] Scientific knowledge functions as a form of capital, providing intellectual, technological, and legal instruments for profit-making applications.

While crops have moved with human hosts since their genesis, movement and interdependence escalated dramatically after 1500, as European nation-states invaded, appropriated, and integrated new geographies into maritime trade and plantation agriculture. American agricultural expansion in the eighteenth and nineteenth centuries was one manifestation of a *longue durée* of plant transfers resulting from human migration, escalated by European maritime activity from the sixteenth century. The purposeful and incidental transfer of plants from Eurasia to America from the sixteenth to eighteenth centuries supported European settler colonies in the Americas and dramatically

[26] Giovanni Arrighi, *The Long Twentieth Century: Money, Power, and Origins of Our Times* (New York, 1994); Arrighi, "The Winding Paths of Capital," *New Left Rev.* 56 (2009): 61–94; Fernand Braudel, *Capitalism and Material Life, 1400–1800* (New York, 1973); Braudel, *The Mediterranean and the Mediterranean World in the Age of Philip II* (New York, 1972).

[27] Arrighi, *The Long Twentieth Century*; Arrighi, "The Winding Paths of Capital" (both cit. n. 26). A succinct summary of Arrighi's theory of systemic cycles of accumulation and hegemonic transition is William I. Robinson, "Giovanni Arrighi: Systemic Cycles of Accumulation, Hegemonic Transitions, and the Rise of China," *New Polit. Econ.* 16 (2011): 267–80.

[28] On the commodity logic of preserving crop wild relatives, see Miguel A. Altieri, M. Kat Anderson, and Laura C. Merrick, "Peasant Agriculture and the Conservation of Crop and Wild Plant Resources," *Conserv. Biol.* 1 (1987): 49–58; Maywa Montenegro de Wit, "Stealing into the Wild: Conservation Science, Plant Breeding and the Makings of New Seed Enclosures," *J. Peasant Stud.* 44 (2017): 169–212.

altered ecologies on both sides of the Atlantic.[29] Meanwhile the importation of tropical biota to metropolitan Europe fueled colonial expansion and provided an international infrastructure of nature collection and preservation consisting of ship holds, vented cases, ledgers, naturalists' notebooks, and herbaria.[30]

Systematic collections of dried plants, generally assembled in a file, box, or cabinet, herbaria are one of the most durable forms of documentation and preservation. They originated in travelers' accounts and devotional books, repurposed and expanded for natural history study in sixteenth-century Europe. Like the herbals of earlier centuries, herbaria draw on reclaimed Greek and Latin texts on medicinal plants, as well as contemporary field specimens, but they differ in observing, cataloging, and describing nature for its own sake. Gradually herbaria evolved from memory aids to tools of study and centers of documentation, supplementing an infrastructure of botanic gardens oriented toward transplantation and cultivation.[31] These organized and expanded sites preserved the labor of collecting linked to European maritime exploration, colony, and empire. The knowledge systems they devised provided the basis for the expansion of commodity culture through transplantation of biota and mass cultivation for international markets.

Modes of resource control originated in plantation agriculture were extended in capitalist forms of labor. Along with rapid industrialization and private property rights in invention, institutions of public agricultural research were handmaidens of capitalist development.[32] Over the course of the nineteenth century, public research boosted private enterprise through federally consolidated research and development. In the United States, the U.S. Department of Agriculture, land grant colleges, and experiment stations pursued improved seeds, mechanization, and chemical applications on the farm beginning in the 1870s and 1880s. In the economies of scale they supported, farmers filled grain elevators and railroad cars, yoking east to west and producing an agricultural surplus connecting the United States to international markets.

[29] On plant movements and human migration, Alfred W. Crosby, *The Columbian Exchange: Biological and Cultural Consequences of 1492* (Westport, Conn., 2003); Crosby, *Ecological Imperialism: The Biological Expansion of Europe, 900–1900* (Cambridge, 1986). Judith Carney has revised Crosby's rendering of the Americas as neo-Europes by focusing on the purposeful and incidental transplantation of African crops and agricultural knowledge. See Carney and Richard Rosomoff, *In the Shadow of Slavery: Africa's Botanic Legacy in the New World* (Berkeley and Los Angeles, 2011).

[30] There is a large literature on the history and epistemic practices of the life sciences, including studies of early modern European natural philosophy, classification, and the experimental sciences. Staffan Müller-Wille and Sara Scharf, "Indexing Nature: Carl Linnaeus (1707–1778) and His Fact-Gathering Strategies," in Working Papers on the Nature of Evidence: How Well Do "Facts" Travel?, ed. Jon Adams (Department of Economic History, London School of Economics and Political Science, 2009), http://eprints.lse.ac.uk/47386/ (accessed 25 July 2018). On practices and institutions of botanical collection, see, e.g., Harold John Cook, *Matters of Exchange: Commerce, Medicine, and Science in the Dutch Golden Age* (New Haven, Conn., 2007); Paula Findlen, *Possessing Nature: Museums, Collecting, and Scientific Culture in Early Modern Italy* (Berkeley and Los Angeles, 1994); Londa L. Schiebinger, *Plants and Empire: Colonial Bioprospecting in the Atlantic World* (Cambridge, Mass., 2004); Jim Endersby, *Imperial Nature: Joseph Hooker and the Practices of Victorian Science* (Chicago, 2008).

[31] Brian W. Ogilvie, *The Science of Describing: Natural History in Renaissance Europe* (Chicago, 2006), 139–264.

[32] William Cronon, *Nature's Metropolis: Chicago and the Great West* (New York, 1991); Margaret W. Rossiter, *The Emergence of Agricultural Science: Justus Liebig and the Americans, 1840–1880* (New Haven, Conn., 1975); Deborah Kay Fitzgerald, *Every Farm a Factory: The Industrial Ideal in American Agriculture* (New Haven, Conn., 2003); Jack Ralph Kloppenburg, *First the Seed: The Political Economy of Plant Biotechnology, 1492–2000* (Madison, Wis., 2004); Alan L. Olmstead and Paul Webb Rhode, *Creating Abundance: Biological Innovation and American Agricultural Development* (New York, 2008).

New breeding methods consolidated monocultural production and hastened the erosion of agro-biodiversity, incidentally inspiring projects to preserve it. Modern agriculture consists of efforts to select and improve plants according to novel rules and systems of organization, including controlled cross-pollination, hybridization, mutation, marker-assisted selection, and genetic modification. Among the most celebrated of these efforts was the development of high-yielding semidwarf hybrids exported to Asia in the 1960s: the "Green Revolution" alternately credited with averting famine on the Indian subcontinent and ushering in an era of unsustainable agricultural practice.[33]

Aiming to build on the alleged successes of this so-called Green Revolution, the Food and Agriculture Organization of the United Nations (FAO) supported programs of agricultural modernization and the free exchange of germplasm between countries for the use of breeders. The Consultative Group on International Agricultural Research (CGIAR) is an international public organization funded by the UN Food and Agriculture Organization, the Rockefeller Foundation, and the World Bank, among others, that oversees fifteen international agricultural research organizations. These include the Center for Maize and Wheat Improvement (CIMMYT) in Mexico, the International Rice Research Institute (IRRI) in the Philippines, and the International Center for Agricultural Research in the Dry Areas (ICARDA).[34] Since the 1970s, international agricultural research organizations have prioritized preservation initiatives to offset losses of biodiversity, while breeders survey genetically diverse material to produce new seeds tolerant of heat, drought, and salinity.

Biodiversity preservation initiatives continued to support plant-breeding projects fitted to the production of high-yielding cereal crops for large-scale production, ultimately restricting the scope of their commitment. When international agricultural research centers turned their attention to biodiversity loss, it was to argue that public and private breeders should have access to global plant genetic resources: moving seed stocks out of the field and into banks for circulation to countries with the capital to pursue research.[35] While the 1992 Convention on Biodiversity (CBD)[36] incorporated varied forms of traditional and indigenous knowledge, its framing in terms of stakeholders nevertheless elevated the interests of those with the most capital: states, biotechnology companies, and large NGOs. The formulation of Trade Related Intellectual Property Rights (TRIPS) at the Uruguay Round of the Global Agreement on Tariffs and Trade (GATT) in 1998 made the supremacy of industry apparent.[37]

Late twentieth-century international agreements in the spheres of nature preservation and economic development support Arrighi's contention that we are in a period of hegemonic transition. The nature of the current hegemonic crisis is less clear, as are the features of the cycle of financial accumulation. Arrighi concurred with Braudel and

[33] E.g., Nick Cullather, *The Hungry World: America's Cold War Battle against Poverty in Asia* (Cambridge, Mass., 2010); John H. Perkins, *Geopolitics and the Green Revolution: Wheat, Genes, and the Cold War* (Oxford, 1997).

[34] Warren C. Baum, Michael L. Lejeune, and World Bank, *Partners against Hunger: The Consultative Group on International Agricultural Research* (Washington, D.C., 1986).

[35] Robin Pistorius and International Plant Genetic Resources Institute, *Scientists, Plants, and Politics: A History of the Plant Genetic Resources Movement* (Rome, 1997).

[36] Aichi Nagoya Protocol on Access and Benefit-Sharing (ABS), adopted 29 October 2010 at the tenth meeting of the Conference of the Parties (COP 10) to the Convention on Biological Diversity (CBD).

[37] On RAFI, see P. R. Mooney, *Seeds of the Earth: A Private or Public Resource?* (San Francisco, 1983); Cori Hayden, "From Market to Market: Bioprospecting's Idioms of Inclusion," *Amer. Ethnol.* 30 (2003): 359–71; Hayden, "Taking as Giving: Bioscience, Exchange, and the Politics of Benefit-Sharing," *Soc. Stud. Sci.* 37 (2007): 729–58.

Joseph Schumpeter's belief that capitalism was quick to adapt to new forms of wealth accumulation.[38] Further, Arrighi credits Max Weber with the idea that interstate competition for mobile capital creates the conditions for financial expansion that distinguish modern history.[39] But in the case of twenty-first-century research and development in agriculture and microbiology, the forms of finance are less predictable and traceable to the coffers of individual states. Hegemonic crisis thus facilitates new configurations of capital ancillary to the nation-state as an institution.

Theorists of laissez-faire have embraced the ecosystem as a model of capitalist dynamism, using the metaphor to justify and naturalize the movement of capital.[40] The invocation of biological metaphors is not novel to the twentieth century. The study of human and natural economies emerged in tandem in Western Europe, and so it is little surprise that economists have turned to the natural sciences for metaphors of human behavior.[41] Eighteenth-century models of a well-ordered economy made frequent reference to natural hierarchies. Darwin's theory of evolution did not so much upend these ordering impulses as provide a historical reinterpretation of them according to the premise of common descent. Subsequent students of human economy looked to these lessons for inspiration. While some neoclassical economists turned to physics for its embrace of universal laws of motion, others veered toward biology for its approach to the world as an interconnected system.[42] Alfred Marshall drew inspiration from Darwin's theory of evolution, and perhaps more significantly Herbert Spencer's application of it to human social life. Friedrich Hayek toyed with the application of evolutionary theory to law, politics, and markets.[43] Since the 1970s, economists have looked to ecology and population biology for models of competitive behavior.[44] As with most attempts to apply theories to new objects, however, evolution takes on hazy and metaphorical significance in these models, reverting to the most general and simplified renderings of natural selection emphasizing competition between individuals.

Yet if nature provides evidence of competition between organisms for scarce resources, it just as readily provides examples of interdependence and flux. Characterized by flexible configurations of finance capital, late capitalist forms of research and development mimic the protean forms of diversity and complexity represented by mi-

[38] Thomas E. Reifer, "Histories of the Present: Giovanni Arrighi and the Long Duree of Geohistorical Capitalism," *J. World-Systems Res.* 15 (2009): 249–56, on 253.

[39] G. Arrighi, "Financial Expansions in World Historical Perspective: A Reply to Robert Pollin," *New Left Rev.* 224 (1997): 154–9, on 156.

[40] On imagery of capital flows as an alternative to industrial organization as descriptive of capitalism, see Philip Mirowski, *The Reconstruction of Economic Theory* (Boston, 1986); Yuval Yonay, *The Struggle over the Soul of Economics: Institutionalist and Neoclassical Economists in America between the Wars* (Princeton, N.J., 2001); Margaret Schabas, *Natural Origins of Economics* (London, 2011).

[41] This conjoined history is emphasized by Margaret Schabas, "The Greyhound and the Mastiff: Darwinian Themes in Mill and Marshall," in *Natural Images in Economic Thought*, ed. Philip Mirowski (Cambridge, 1994), 322–35.

[42] Camille Limoges and Claude Menard provide a summary of work to date analyzing metaphors from the physical and natural sciences in economic thought, with Philip Mirowski emphasizing the former, and Neil B. Niman the other, analyzing Alfred Marshall's embrace of population biology and evolution for theories of organization. Limoges and Menard, "Organization and the Division of Labor: Biological Metaphors at Work in Alfred Marshall's *Principles of Economics*," in Mirowski, *Natural Images* (cit. n. 41), 336–59.

[43] Geoffrey M. Hodgson, "Hayek, Evolution, and Spontaneous Order," in Mirowski, *Natural Images* (cit. n. 41), 408–50.

[44] Sharon E. Kingsland, "Economics and Evolution: Alfred James Lotka and the Economy of Nature," in Mirowski, *Natural Images* (cit. n. 41), 231–48.

croorganisms. Metaphors can be turned to many different purposes, and the remainder of the essay explores their operation in one instance of international agricultural research focused on fungal symbionts in wild grasses.

POLITICAL ECONOMY OF TWENTY-FIRST-CENTURY INTERNATIONAL AGRICULTURAL RESEARCH

As Hannah Landecker has noted, it is never wise to wait for the historians. Ethnography can serve as a proxy for histories of late twentieth-century and twenty-first-century sciences that have yet to be written, illuminating core practices and assumptions at work in the production of new knowledge.[45] The cereal endophyte development program disrupts conventional historiographies oriented toward divisions between nation-states, metropole/periphery, and North/South, revealing more hybrid forms and flows of research capital in international agricultural research. In the following sections I aim to analyze the political economy and technological practices of contemporary international agricultural research, considering how modes of organization fitted to microbiological research provide new metaphors for human economy, and vice versa. First, I discuss the economic organization of the research in question, and specifically, the imperatives of AgResearch as a public-private institution charged with agricultural improvement and biodiversity preservation. Next, I follow a seed from point of collection in the Pamir Mountains of Tajikistan to its progress through AgResearch laboratories in Palmerston North, New Zealand.

The cereal endophyte program is representative of international agricultural research and biodiversity preservation projects in several respects. First, the public-private nature of the collaboration is typical for late twentieth- and early twenty-first-century research and development, much of which is conducted under the aegis of contemporary agro-biodiversity preservation initiatives. Multiple institutions direct the seed-collecting expeditions providing raw material for the cereal endophyte development program. These organizations include the public-private institute of AgResearch (New Zealand), the national gene bank of the Vavilov Research Institute of Plant Industry (VIR, St. Petersburg, Russia), and the international public organization of ICARDA, Aleppo, Syria, and Rabat, Morocco). Funders for each of these institutions consist of international public organizations, philanthropic and private foundations, and for-profit corporations. In spite of drawing money from the same streams, breeding and biodiversity preservation initiatives coexist uneasily. Projects shift between environmentalist and commercial registers, with seeds collected, catalogued, and stored for posterity and immediate research. While the collecting expeditions lodge material at nationally and internationally managed gene banks charged with the preservation of biodiversity, a sample is screened for the presence of Epichloë endophytes at AgResearch in Palmerston North.

In form, AgResearch is a profit-making science institute, but it draws substantial public funding and hosts the national gene bank of New Zealand (the Margot Forde Germplasm Centre).[46] Nevertheless, its overall movement has been toward increased

[45] Landecker, "Living Differently in Time" (cit. n. 17).

[46] General information about the cereal endophyte breeding program is drawn from interviews and observation with breeders and gene bank managers at AgResearch, Palmerston North, New Zealand, in January 2016, including Stuart Card, Anouck de Bonth, Warren Williams, David Hume, Wayne Simpson, Alan Stewart, Phil Rolston, Valerio Hoyos-Villegas, and Suliana Teasdale.

industry and commercial investment. As the balance of AgResearch's funding has shifted in the last twenty-five years, arguments for public funding within the organization have given way to a discussion of whether majority funding should be commercial. In 1992, the same year the Convention on Biodiversity attempted to recognize local communities and indigenous peoples as stakeholders in the preservation of biodiversity, agricultural research in New Zealand was reconfigured in ways that portended its increasing involvement with the private sector. AgResearch was named a Crown Research Institute, and its charter dictated that it return a small profit to its owner, the New Zealand Government, to demonstrate financially viability. Institutional priorities shifted accordingly, as research relied increasingly on industry-good and corporate funding. The cereal endophyte program is exceptional for the amount of non-government money it receives, with substantial funding coming from Grain Research and Development Corporation (GRDC) in Canberra, Australia, as well as Grasslandz Technology, a subsidiary of AgResearch charged with investing in the development of plant-based technologies for licensing to commercial companies. Royalties are invested in further research.

Second, the continuation of research depends on foreign germplasm collection and exchange facilitated by networks of international gene banks. This phenomenon is intensified in New Zealand, where a large proportion of the agricultural and forestry industries is based on transplanted biota from Europe, North America and China. Until the mid-twentieth century, New Zealand's agricultural history was coterminous with its status as a colony of Great Britain, producing for international markets.[47] Moreover, while collecting missions proceed on geographic theories of biological diversity that mute geopolitics, political concerns remain salient for the cereal endophyte team that I have accompanied. Its collection centered in post-Soviet republics characterized by enormous geographic scope and ecological diversity, imperial legacies of cooperation with St. Petersburg, and recent histories of conflict and rural flight. Collectors amass material and raw data in underdeveloped regions that support highly capitalized research in other locales. These conditions raise complex questions about how preservation projects engage with underdeveloped regions that are identified as centers of biodiversity.

Finally, even as international agricultural research is characterized by flexible and geographically far-flung networks of research capital, it remains rooted in European imperial science. In New Zealand, European interests shaped the development of a pasture economy oriented toward meat, wool, and dairy. In fact, it is New Zealand's status as a center of pasture grass breeding that explains its current research into common fungal endophytes in food crops. There, the growth of the pasturing industry, combined with a native deficit of plant diversity suited to consumption by ruminants, stimulated research and development. British immigrants introduced perennial ryegrass for pastures beginning in the nineteenth century. By the late nineteenth century, farmer-breeders selected, sowed, and traded predominantly locally adapted ecotypes of ryegrass derived from European ones. Beginning in the 1930s, government plant breeders led breeding efforts. Since productive temperate grasses were not native to New Zealand, breeding

[47] On New Zealand's industry of animal breeding as a British colony, see Rebecca Woods, *The Herds Shot Round the World: Native Breeds and the British Empire* (Chapel Hill, N.C., 2017); Woods, "From Colonial Animal to Imperial Edible: Building an Empire of Sheep in New Zealand, ca. 1880–1900," *Comp. Stud. South Asia, Africa, Middle East* 35 (2015): 117–36.

relied on persistent plant introduction efforts, including Mediterranean and Northern European material suited to winter growth.[48]

The current wave of research into cereals and other crops for human consumption builds on the success of endophyte breeding in pasture grasses within the past thirty years, especially in New Zealand. And here we return to our poisoned sheep. When researchers in the late 1970s traced the ryegrass staggers in sheep to the *Epichloë* fungal endophyte, the most parsimonious approach was to remove the endophyte. But this approach rendered the grass susceptible to pests, especially the Argentine stem weevil. The first attempts to breed pasture grass resistant to the stem weevil, meanwhile, proved toxic to ruminants. While "Endosafe" eliminated the staggers, grazing animals continued to suffer from heatstress.[49] Endosafe was recalled, and breeders went back to the drawing board.

By the mid-1980s, breeders in New Zealand successfully identified which endophyte strains were toxic and nontoxic to livestock, allowing them to select strains that were resistant to pests without harming ruminants. The resulting endophyte strain, AR 1, lacked the alkaloids toxic to animals but retained the alkaloid peramine, which deters the Argentine stem weevil. For farmers, this solution was superior to prior attempts to dilute animal consumption by sowing clover and other pasture species amid the ryegrass or otherwise physically manage animal intake of toxic endophytes. Newly developed strains, including AR 37, promise resistance to other exotic and native insects such as the black beetle and porina.[50]

Cereal endophyte researchers aim to build on the successes in pasture grasses by extending endophyte research into a range of other crops for animal and human consumption. Extending research to human food plants, however, raises a range of new issues regarding safety and suitability for human consumption. Deliberate and accidental selection of food plants extend backward some 20,000 years, long before the advent of settled agriculture. Needless to say, there is no clear documentary record of these events. Perhaps many fungal endophytes were selected out of the food supply from an early stage of domestication because of their toxicity to human beings. Only within the past few years have researchers managed to form a synthetic association of fungal endophyte strains with cultivated cereal grasses. One may question the wisdom of putting fungi in, or back in; and yet, perhaps, as for cattle and sheep, breeders can find the ones they want and leave the others behind.

Moreover, if endophytes confer environmental hardiness, perhaps understanding these microorganisms can help humans adapt to future climate change. Certain tall fescue and perennial rye grasses have coevolved with symbiotic fungal endophytes that resist environmental stressors, especially pests. In New Zealand, Australia, and the United States, where forage grasses provide the feed for grazing ruminants, selection of these

[48] Alan B. Stewart (PGG Wrighton Seeds, Plant Breeding, Christchurch, New Zealand), "Genetic Origins of Perennial Ryegrass (*Lolium perenne*) for New Zealand Pastures," in *Breeding for Success: Diversity in Action*, ed. C. F. Mercer, Proceedings of the 13th Australasian Plant Breeding Conference (Christchurch, New Zealand, 2006), 11–20; Stewart, "Progress in Domesticating New Zealand Native Grasses," unpublished paper, in the possession of the author; Riddet Institute, *Floreat Scientia: Celebrating New Zealand's Agrifood Innovation* (Auckland, 2011).

[49] Johnson et al., "The Exploitation of Epichloae Endophytes" (cit. n. 3), 178.

[50] C. A. Young, D. E. Hume, and R. L. McCulley, "Forages and Pastures Symposium: Fungal Endophytes of Tall Fescue and Perennial Ryegrass: Pasture Friend or Foe?," *J. Anim. Sci.* 91 (2013): 2379–94.

endophytes has benefited pastoral agriculture enormously.[51] There is also some evidence that toxic endophyte infection promotes water circulation during plant growth, improves recovery after drought, and increases resistance to elevated temperatures: qualities prized in a warming world.[52]

Regardless, the success of endophyte inoculation into pasture grasses does not spell immediate success for comparable advances in cereal crops. The asexual *Epichloë* (formerly *Neotyphodium*) endophytes are found widely distributed in populations of the grasses *Lolium perenne* and *Festuca arundinacea*. In contrast, fungal endophytes have not been found in cultivated cereals (e.g., wheat, barley, rye), or in close relatives of the wheat plant. It is as yet unclear whether it is possible to produce a consumable cereal hosting pest-resistant endophytes.

Meanwhile, VIR, ICARDA, AgResearch, and assorted post-Soviet gene banks support collecting expeditions across Eurasia, including the post-Soviet republic of Tajikistan. In the section that follows, I track the transit of a seed from a field in Tajikistan to the gene bank and laboratory in Palmerston North during a 2011 expedition to the Pamir Mountains comprised of collectors from AgResearch, ICARDA, VIR, and the Tajik national gene bank. My focus is on the mismatch between protocols of collection and preservation for wild flora and the body of microbiological research they support, and the significance of this rift for the relation between science and capital.

MICROBIOLOGY AND PROBLEMS OF CLASSIFICATION: FIELD, GENE BANK, AND LABORATORY

The Pamir Mountains of Tajikistan and Afghanistan, identified as a center of biological diversity by the pioneer Russian plant geneticist Nikolai Vavilov in the 1920s, remain a focus of plant genetic resource collection, and a site of collection. Tajikistan, bordered by Afghanistan, Uzbekistan, Kyrgyzstan, and China, is the poorest of the fifteen post-Soviet republics, with the lowest GDP, a lack of employment opportunities, and industrial and agricultural production crippled by the civil war of the 1990s. Tajikistan remains dependent on international aid agencies for basic subsistence, and the most active of these agencies is the Aga Khan Foundation, directed by the Aga Khan IV, an international business magnate who is also the imam of the Ismaili Shiites, the majority population in the Pamirs. Agriculture in the area was decimated by conflict. Much of the area has only recently been cleared of land mines, its fields dotted with warning signs and Russian tanks collapsed over embankments. One of the largest sectors of Tajikistan's economy may be the black market traffic in opium from Afghanistan to

[51] G. E. Aiken and J. R. Strickland, "Managing the Tall Fescue-Fungal Endophyte Symbiosis for Optimum Forage-Animal Production," *J. Anim. Sci.* 91 (2013): 2369–78; H. S. Easton, "Ryegrass Endophyte: A New Zealand Grassland Success Story," *Proceedings of the Conference—New Zealand Grassland Association* 63 (2001): 37–46; J. P. J. Eerens, Kris Miller, J. G. H. White, H. S. Easton, Richard J. Lucas, "Ryegrass Endophyte and Sheep Reproduction," *56th New Zealand Grassland Association Conference*, 1994, https://researcharchive.lincoln.ac.nz/handle/10182/4591 (accessed 19 July 2017).

[52] D. E. Hume and J. C. Sewell, "Agronomic Advantages Conferred by Endophyte Infection of Perennial Ryegrass (*Lolium perenne* L.) and Tall Fescue (*Festuca arundinacea* Schreb.) in Australia," *Crop Pasture Sci.* 65 (2014): 747–57; Johnson et al., "The Exploitation of Epichloae Endophytes" (cit. n. 3); David E. Hume, Geraldine D. Ryan, Anaïs Gibert, Marjo Helander, Aghafakhr Mirlohi, and Mohammad R. Sabzalian, "Epichloë Fungal Endophytes for Grassland Ecosystems," *Sustainable Agricult. Rev.* 19 (2016): 233–305.

neighboring countries: unmarked trucks careen along the roads near the Panj River bordering Afghanistan.

Four days and fourteen sites into the expedition, the collecting team paused for a break in Vitchkut village, in the Gorno-Badakhshan region of Ishkashim, adjacent to some hot springs. While the bulk of the team wandered off to the baths, the team's most dedicated botanist, Josephine Piggin, continued to collect. Among her collection was an alpine species of *Hordeum brevisubulatum* subsp. *turkestanicum*, a wild relative of barley. The same species was found five sites later in a mountain pasture near the village of Yazba. Screened a year hence at AgResearch in Palmerston North, the plants from Yazba were found to harbor the fungal endophyte *Epichloë*, which is responsible for pest resistance in many pasture grasses.[53] These collections complemented previous collections of *Hordeum*, *Elymus*, and other genera found to harbor *Epichloë*.

Twenty-first-century researchers employ an array of technological practices with long histories. Instruments used to preserve and record plant matter, including herbaria, Linnaean binomials, breed names, and genetic sequences create different grammars for seeds through varied modes of parsing, labeling, and representation. If these are grammars that order natural resources for exploitation and improvement, they convey assumptions about natural and human creativity/agency, and the proper political economy for managing material and intellectual resources. On one hand, the persistence of early modern European collecting practices in contemporary biodiversity preservation projects is an indication of continuity in natural sciences forged to exploit the natural world. The Linnaean binomial is one realization of this universalizing system, maintained into the twenty-first century as the lingua franca of plant scientists, and indeed their only common language. In collecting expeditions staffed by Tajik, Kazakh, Armenian, Georgian, Russian, New Zealand, Greek, and Syrian scientists, we generally bellowed across the field in Latin, condensing a welter of information into a single identifier: *Aegilops crassa! Aegilops tauschii! Hordeum brevisubulatum!* Bagged, threshed, cleaned, and lodged in a gene bank, a plant's local identity, condensed to its "passport data," becomes subsidiary to the binomial and accession number in the database.

On the other hand, although collecting teams proceed on the basis of plant genus, species, and subspecies, for cereal endophyte researchers, the true objects of investigation are microorganisms within the plant. This mismatch is not unique to cereals endophyte research, or to the twenty-first century. In spite of the entrenchment of Linnaean taxonomy in the practice of collecting and preserving biological diversity, the targets of conservation have long been a matter of dispute.[54] Even within the purview of population biology, genes and ecosystems vie with species as the locus of conservation. These have received much less attention than the genomes of multicellular organisms

[53] Expedition tag TJK11, Site 14, Accession 1, *Hordeum brevisubulatum*, ssp. *turkestanicum*, collected 16 August 2011 by Josephine Piggin, Vitchkut, near hot springs, Gorno-Badakhshan, Ishkashim, Tajikistan; Site 19, Accession 1, *Hordeum brevisubulatum*, ssp. *turkestanicum*, collected 17 August 2011, Yazba mountain pasture, Gorno-Badakhshan, Ishkashim, Tajikistan.

[54] On foundational concepts of biodiversity preservation, see, e.g., Timothy J. Farnham, *Saving Nature's Legacy: Origins of the Idea of Biological Diversity* (New Haven, Conn., 2007); Farnham, "A Confluence of Values"; David Sepkoski, "Extinction and Biodiversity: A Historical Perspective," in Garson, Plutynski, and Sarkar, *The Routledge Handbook*, 26–40; Maclaurin and Sterelny, *What Is Biodiversity?* (all cit. n. 9). Many argue that a species approach is too narrow, provoked by a popular concern with species extinction and the endangerment of certain flora and fauna. Nevertheless, species remains a proxy for other forms of diversity in contemporary policy.

such as our own, and they suggest that the "tree of life" is something other than what we have imagined.[55] Ecosystem approaches provide more contextual and holistic approaches to the conceptualization of natural diversity. Detractors observe the difficulty of classifying or preserving individual ecosystems. Many advocate the gene as a more suitable unit of conservation, while others argue that a myopic focus on the genome has attenuated broader and more contextual conceptions of life on earth. Since the 2003 sequencing of the human genome, criticisms of "genocentricism" have abounded, resisting simplistic accounts of evolutionary theory and assertions that genetics can explain complex psychological traits.[56] Even so, the assumption of a unique relationship between organism and genome has proved tenacious, with the genome acting as a kind of "species 'barcode,'" to use John Dupré's term. But as Dupré also observes, such an approach, oriented toward multicellular organisms, does not take into account fungi hosting multiple distinct genomes and capable of moving within and among their hosts.

In practice, species remains a surrogate for other forms of diversity, well-suited to ex situ administration and subsequent subjugation to molecular biological and genetic analysis. In the field, moreover, a knowledge of plant morphology trumps expertise in genetic sequencing or molecular biology, and the tools of the trade remain a magnifying glass, a jeweler's loupe, and a good knife. For all the enthusiasm surrounding the genome as a locus of conservation, one cannot collect without identifying a plant phenotypically, drawing on centuries of work by taxonomists. That is, collectors still have to know which grasses they are looking for, and how a *Hordeum marinum* compares to a *murinum* to a *brevisubulatum*. In expedition vehicles, field guides in multiple languages lie piled on dashboards and wedged between seats and doors. Routinely, collectors cluster around the seats with magnifying glasses, examining whole plants for correct identification. As one collector groups bags by species-subspecies and assigns collecting numbers, another logs each sample's data in a spreadsheet or database, along with passport data for the site consisting of GPS coordinates, regional and local names, aspect, soil pH and salinity, and collecting area.

Somehow envelopes are always scarce, with cargo reduced to a minimum of camping gear and collecting equipment in the expedition vehicles. It follows that collectors routinely argue over the conservation of envelopes, with one party asserting that all samples of the same plant from a single site be mixed, and others demanding that they remain separate to distinguish phenotypic variation and minimize errors in sampling due to misidentification of species. This argument has recurred in every collecting expedition I have attended, and it has never been successfully resolved, at least not until the team members consumed generous amounts of vodka at campsites. Then, on occasion, the team runs out of packets. Once the collecting team spent two hours in a provincial Russian city searching for replacements. Only the wasted hours not spent collecting, and the late start due to overconsumption of vodka, saved the expedition from running out of envelopes entirely.

The collecting team traveled for a total of six weeks in the Pamir Mountains and river valleys of the Gorno-Badakhshan region of Tajikistan. Afterward, some 900 brown paper bags of seeds from forty-five sites were laid out and filed on the floor of a field shed at the Tajik national gene bank outside the capital city of Dushanbe. Over the course of

[55] John Dupré, *Processes of Life: Essays in the Philosophy of Biology* (New York, 2012), 116–7.
[56] Barry Barnes and John Dupré, *Genomes and What to Make of Them* (Chicago, 2008).

four twelve-hour days, the collecting team and three technicians and fieldworkers from the Tajik national gene bank unbagged the seeds and then threshed, cleaned, and divided them among the participating institutions. AgResearch and ICARDA selected seeds in their mandate and target areas, while the Tajik national gene bank retained a duplicate of every collected accession. Finally, 364 samples identified by species and subspecies were logged with "passport data" mandated in an addendum to the Standard Material Transfer Agreement (SMTA) of the 2001 International Treaty on Plant Genetic Resources for Food and Agriculture, the mandatory model for parties who provide and receive material under the Multilateral System. Passport data consist of a persistent, unique numerical identifier, collecting institute codes, country of origin, date of collection, and taxonomic and site data.[57]

For the most part, the existing infrastructure of international gene banks supports the extension of biodiversity preservation programs to explore fungal diversity. This capacity is an indication of the robust and malleable characteristics of international networks of research capital drawing on dispersed governmental, nongovernmental, and corporate coffers, mingling profit-making and utilitarian claims to preserve natural diversity. Nevertheless, the organization of gene banks and international plant genetic resource policies introduces additional obstacles to cereal endophyte research. Gene banks treat accessions with fungicide to maintain plant health and maximize the production of viable seed. Depending on the fungicide and dosage used, a potential and inadvertent side effect is the loss of the endophyte. While the Margot Forde Germplasm Centre uses endophyte friendly fungicides, other institutions may not. Thus researchers cannot rely fully on previously collected material from other gene banks for their investigations.

Moreover, prevailing classificatory systems in internationally managed gene banks fail to represent the complexity of contemporary scientific practice. Currently no national or international plant genetic resource policy is organized to account for microorganisms. Even as microorganisms constitute the greatest share of biodiversity on the planet, the Convention on Biodiversity was drafted with reference to plants and animals. In the entire text of the Convention, the words "microorganism" and "microbial" each appear only once.[58] Although researchers may identify the gene as the locus of diversity and focus increasing attention on microorganisms and their relations, they retain Linnaean taxonomy to organize their vaults and databases. Moreover, collection depends on the taxonomic identification of the genus, species, and subspecies of a plant. In practice, genotype and phenotype remain tethered.

National laws regarding invasive species and biosafety also complicate collection. Although fungi are the true objects of the cereal endophyte funded collecting expeditions, researchers contend with a policy framework in which "good" diversity of flora is often distinguished from "bad" diversity of pathogens and pests. Even sanctioned projects may run into trouble at national borders, where bureaucrats enforcing regulations against invasive plants, pests, and diseases may confiscate and quarantine collected material. On one collaborative expedition in 2010, Syrian customs blocked ICARDA's

[57] "Standard Material Transfer Agreement: International Treaty for Plant Genetic Resources for Food and Agriculture (ITPGRFA)," www.wipo.int/tk/en/databases/contracts/texts/smta.html (accessed 19 September 2017); FAO/Bioversity Multi-Crop Passport Descriptors V.2.1, December 2015, https://www.bioversityinternational.org/fileadmin/user_upload/online_library/publications/pdfs/FAOBIOVERSITY_MULTI-CROP_PASSPORT_DESCRIPTORS_V.2.1_2015_2020.pdf.

[58] Davison et al., "Microorganisms" (cit. n. 7), 1407.

portion of the collections from Armenia and Georgia. In some cases, specimens are destroyed. Gene bank personnel cringe over a recent instance of Australian customs incinerating unique lichen specimens on loan from France and New Zealand according to biosecurity protocols.[59] In yet another recent case, the New Zealand Ministry for Primary Industries is the target of a class action suit by kiwifruit growers, who argue that an import of pollen with vine-killing disease would have been prevented had the protocols of the national Biosecurity Act been properly enforced.[60] The court ruled in favor of the growers.[61] In this climate, and as journalists politicize agricultural science in debates over genetically modified organisms, researchers leery of misunderstanding may also fail to articulate different kinds of diversity imperative to their investigations.

In some expeditions, seeds are bagged and shipped via DHL or international courier. This creates a certain amount of anxiety on the part of collectors fearing mishandling or destruction of their specimens. The situation may be no different should the collector carry the seeds in his or her checked luggage, but at least he or she can attend the crisis in person. Such interventionism can be to the benefit of the collectors, as when the New Zealand gene bank manager phoned a senior adviser at the Ministry of Primary Industries (MPI) late at night to secure special exemptions for importing seeds for the cereal endophyte project. Although the seeds had not yet been cleaned according to standard biosecurity protocols, the manager was able to arrange their transfer directly into quarantine for final processing under MPI supervision. Anecdotal evidence suggests that not all collectors are so scrupulous. On the collections I have attended, casual talk about circumventing the onerous paperwork and clearance rules has been met with ominous disapproval, with the trigger word being "smuggle." These utterances threaten to cross the porous boundary between biopiracy and multilateral sharing of seeds for international research.

It is easy to see how these parcels could resemble trash or booty. Envelopes used in field collection are repurposed in the ultimate division of samples between sponsoring parties, relabeled in ink with their accession number, gene bank destination, and species-subspecies, and reduced to a minimum size through folding and stapling. Thereafter they are jammed into cloth sacks, again aiming to reduce bulk as much as possible, and then stuffed into collapsible bags or suitcases along with dirty laundry and other artifacts of travel. Safely conveyed from Dushanbe to Palmerston North in the luggage of the New Zealand gene bank manager via transfer through Istanbul, two days in Dubai, and onward flights to Sydney and Auckland, the seeds from Vitchkut and Yazba, along with the rest of the those from Tajikistan, traveled with official permission past the enforcers of New Zealand's biosecurity protocols, and finally to AgResearch facilities in Palmerston North. There, still in their collecting envelopes, now triple folded, stapled, and labeled in permanent marker, the seeds begin another life (fig. 1).

Obligate endosymbionts cannot exist apart from the plant host, but many can be isolated and cultured in the laboratory with standard microbiological techniques. To convey the time- and labor-intensive quality of these techniques, I describe them in some

[59] "Australian Customs Destroys Unique Lichen Specimens in Quarantine Mix-Up," *Sydney Morning Herald*, 8 May 2017, http://www.smh.com.au/environment/australian-customs-destroys-unique-lichen-specimens-in-quarantine-mixup-20170508-gw0fui.html.

[60] "Class Action Lawsuit over Kiwifruit Psa Outbreak," *Newshub*, 8 July 2017, http://www.newshub.co.nz/home/money/2017/08/class-action-lawsuit-over-kiwifruit-psa-outbreak.html.

[61] "Kiwifruit Growers Claim Victory in High Court Psa Case | Stuff.Co.Nz," https://www.stuff.co.nz/business/farming/105113232/kiwifruit-growers-win-partial-victory-in-high-court (accessed 24 July 2018).

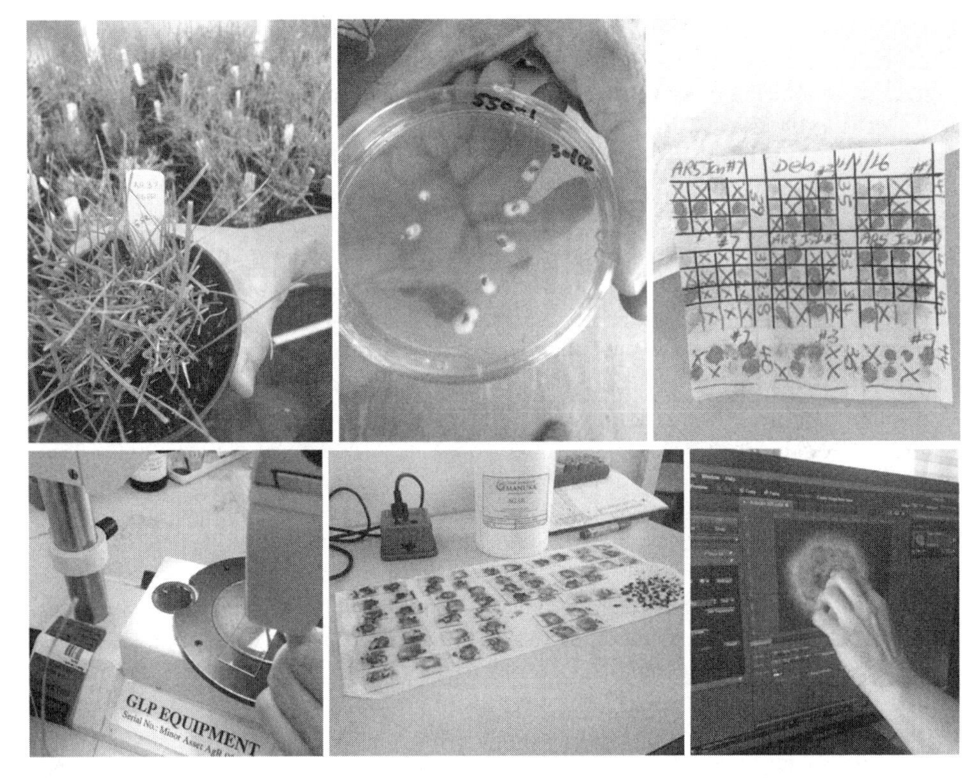

Figure 1. *Immunoblot: red chromogen-bound tiller imprints from endophyte-infected plants and unbound imprints from endophyte-free plants (top, left to right); aniline blue stain of infected seedlings using microscopy (bottom, left to right); magnification of fungal endophyte culture, AgResearch, Palmerston North, New Zealand. Photographs by Courtney Fullilove.*

detail below.[62] The seeds collected in the field reach the gene bank in their paper packets, notated with accession number and species-subspecies. Gene bank managers import or key in data from spreadsheets collated in the field, including GPS coordinates, site data, and individual notes on the plant encompassing maturity, habitat, and unusual phenotypic characteristics. While the majority of collected specimens are logged in the Margot Forde Germplasm Centre database, repackaged, and transferred to cold storage vaults, a minority are selected for the working material of the cereal endophyte researchers. These are grown out as seedlings in containment glasshouses, where they are inspected by the MPI for pests and diseases and subjected to further examination.

Researchers use multiple methods for detecting endophytes: epidermal leaf peel, seed squash, and immunoblot detection using rabbit generated antibodies. In a leaf peel, sheath tissue from a live plant is removed and viewed at 16 × magnification, adaxial epidermis up, cut transversely, and mounted on a slide prepared with a drop of

[62] My account of the following techniques is drawn from observation of technicians and partially documented in W. R. Simpson, J. Schmid, J. Singh, M. J. Faville, and R. D. Johnson, "A Morphological Change in the Fungal Symbiont *Neotyphodium lolii* Induces Dwarfing in Its Host Plant *Lolium perenne*," *Fungal Biol.* 116 (2012): 234–40.

aniline blue stain. Heated over a flame, then cooled and reexamined, infected plants display fungal hyphae running along the leaf axis. In a seed squash, grains are covered in sodium hydroxide solution overnight before being decanted, rinsed, covered with an aniline blue solution, and boiled on a hot plate. After cooling, the upper and lower bracts of the floret are discarded and the true seed mounted on a microscope slide with aniline blue and "squashed" gently. In infected plants, stained blue hyphae are visible when examined at 100 × and 400 × magnification under a compound light microscope.

In immunodetection, grass tillers are cut from live plants at soil level, with the freshly cut end placed onto a nitrocellulose membrane. The circular blots left from the imprint form a grid on the nitrocellulose sheet, with positive and negative controls included. Researchers coat blotted sheets with a milk protein-blocking solution, shake them for two hours using miniorbital shakers, decant, rinse twice with more blocking solution, and then add 25 microliters of a solution containing rabbit antibodies to endophytes. Antibodies are the active ingredients in this particular cocktail. After fifteen minutes of shaking, incubation overnight at 4°C, and another round of decanting and rinsing in blocking solution to remove excess rabbit antibody, a secondary goat antibody solution is added and subjected to similar processing. Ultimately the test blot membrane is immersed in a chromogen solution. Shaken another fifteen minutes, the positive control blot, and any others manifesting infection, turn red. Of all the tests, this is the most rapid and efficient. Results are available immediately. But the immunoblot test also has the highest ratio of false positives, meaning positive results are typically rescreened using another method.

The foregoing are techniques for detecting fungal endophytes in seedlings. Cereal endophyte research also entails techniques for the inoculation of seedlings with isolated fungal endophytes. To isolate the fungi, technicians remove tillers from endophyte-infected grass to a length of around 5 centimeters. They then disinfect the plant tissue in an ethanol rinse, a bleach soak, and two rinses in sterile water. Technicians then section the tillers transversely, with sheath rings plated in a base of antibiotic potato dextrose agar and incubated in the dark for three to five weeks. Seedlings to be inoculated with the fungal endophyte are subjected to a similar regimen, with seeds sterilized in sulfuric acid, water, and bleach before being dried, arranged on Petri plates, and germinated in the dark for five to seven days. Technicians inoculate the seedling with the aid of a microscope and a scalpel, making a narrow slit in the base of the seedling plant and placing cultured mycelium from the infected plant into the incision. After another week in the incubator and another week after that under white fluorescent bulbs, seedlings are planted and grown for six weeks in a glasshouse.

Genetic sequencing is also a usable technique for identifying the presence of fungal endophytes in host grasses. Tillers cut at soil level and transferred to DNA extraction vials are processed using Q-BIOgene FastDNA kits, amplified by PCR (polymerase chain reaction) to identify two markers (B10 and B11) polymorphic across strains of *Epichloë* endophyte.

Several things are noteworthy about these techniques. First, they are extremely time- and labor-intensive, requiring growth of seedlings, manual inoculation, microscopy, and multiple stages of processing. Second, they require an enormous amount of commercially produced laboratory equipment, including microscopes, slides, and chemical solutions for sterilization, staining, and protein detection. Notably, many of the Q-BIOgene FastDNA kits are branded and produced by American biotech companies for international markets. The potato dextrose agar is supplied by Difco Becton, Dickinson and Co. Third, detection methods rely on infrastructures of experimentation that depend on the continued exploitation of flora and fauna: here not simply host grasses, but small animals whose antibodies aid in the detection of fungi. For example, AgResearch uses an antibody for detecting endophytes developed using rabbits in Massey University's Small Animal Production unit.

Finally, the primary methods of detection are not new, but rather built on generations of scientific research. Growing out experimental plots is a practice standard to eighteenth-century botanical gardens, and arguably to millennia of agricultural practice in less formal capacities. In the cereal endophyte program, wild relatives are grown out in containment facilities. Microscopy too is centuries old, and the models used in cereal endophyte research are old workhorses rather than state-of-the-art. The "Western blot," which distinguishes itself by the use of animal antibodies for protein detection, has been in use since the late 1970s. W. Neal Burnette devised this protein immunoblot while he was a postdoc at the Fred Hutchinson Cancer Center in Seattle, Washington. The name, and the method, was a play on the "Southern blot" method, devised by Ed Southern several years earlier in 1975.[63] Immunoblots continue to be used in testing for a wide range of diseases, including hepatitis, Lyme disease, and HIV-AIDS. Such established methods of detection are still preferred to the identification of simple sequence repeats (SSRs), in spite of the attention to genetics in the practice of molecular biology.

How will networks of international research forged in relation to plant species change as the diversity of microorganisms becomes a target of investigation? According to a recent estimation by Hawksworth et al., there are "at least 1.5 million species" of fungi, and "probably more than 3 million." Other estimates indicate up to 5.1 million, only a handful of which are described.[64] Fungal taxonomy has been upended by genomic sequencing analysis, putting phylogenetic hypotheses in flux. Fungi belong to two kingdoms in the Eukarya domain, but the movement of the kingdoms makes domain a more reliable criterion. Within the Eumycota kingdom in particular, there have been many recent changes, including the rise of a new phylum in 2001 and the dissolution of the "Zygomycota" based on genome scale comparisons (of 192 conserved proteins).[65] Entirely new subphyla and classes are being identified using sequencing methods, as with active under-snow fungi and soil fungi constituting a new class within the Ascomycota. New research also suggests that Archaeorhizomycetes may be present in soil in forests and grasslands from the tundra to the tropics.[66]

[63] For an account of the Western blot's origins by its creator, see W. N. Burnette, "'Western blotting': Electrophoretic Transfer of Proteins from Sodium Dodecyl Sulfate-Polyacrylamide Gels to Unmodified Nitrocellulose and Radiographic Detection with Antibody and Radioiodinated Protein," *Anal. Biochem.* 112 (1981): 195–203.

[64] D. L. Hawksworth, "Global Species Numbers of Fungi: Are Tropical Studies and Molecular Approaches Contributing to a More Robust Estimate?," *Biodiversity Conserv.* 21 (2012): 2425–33.

[65] Joseph W. Spatafora, Ying Chang, Gerald L. Benny, Katy Lazarus, Matthew E. Smith, Mary L. Berbee, Gregory Bonito, et al., "A Phylum-Level Phylogenetic Classification of Zygomycete Fungi Based on Genome-Scale Data," *Mycologia* 108 (2016): 1028–46. A summary is provided by J. E. Stajich et al., "The Fungi," *Curr. Biol.* 19 (2009): R840–45.

[66] Christopher W. Schadt, Andrew P. Martin, David A. Lipson, and Steven K. Schmidt, "Seasonal Dynamics of Previously Unknown Fungal Lineages in Tundra Soils," *Science* 301 (2003): 1359–61; Anna Rosling, Filipa Cox, Karelyn Cruz-Martinez, Katarina Ihrmark, Gwen-Aëlle Grelet, Björn D. Lindahl, Audrius Menkis, and Timothy Y. James, "Archaeorhizomycetes: Unearthing an Ancient Class of Ubiquitous Soil Fungi," *Science* 333 (2011): 876–9.

In spite of these possibilities, a much narrower conception of fungal diversity supports international agricultural research. The production of a select few commodities, including grain and meat, continue to drive research agendas. It is striking that the problem of morbidity in commercially raised cattle and sheep is what inspired researchers to disrupt the existing plant-host endophyte relationship in pasture grasses. Were it not for industry interest in growing fat ruminants, the relation between plant and fungus would have been considered stable and symbiotic, defending the plant against consumption by both insects and mammals. Endophytes have been detected in a range of wild relatives of modern cereals in the Triticeae tribe, but funders prioritizing wheat production are eager for more immediate progress directly related to the wheat plant. Wheat, along with maize and rice, represent the bulk of the global diet.

Unlike seed grain, which is the ultimate target product of cereal endophyte research, fungal endophytes themselves are not a perfect commodity. They resist reduction to experimental norms, requiring researchers to conduct time-consuming and laborious investigations toward their potential domestication. Grain is small, dry, and uniform, making it easily transported, exchanged, and preserved over long periods of time. Although there are different types and qualities of wheat, grading systems reduce these to standard values. There is a one-to-one correspondence between the input and product. These measures in turn facilitate the development of futures markets in grain, and the circulation of finance capital divorced from the material it represents. In short, staple grains earn an "A" grade for commensurability, the standard by which commodities are judged to be more or less perfect.[67]

Endophytes make less compliant subjects, exercising their own agency in the life cycle of the plant. The complexity of microbiological interactions resists imperatives of standardization fundamental to the commodity production of cereal crops, making them challenging objects of development. While the traits of common fungal endophytes are heritable, they nevertheless depend on the life cycle of the plant, living and dying with their hosts. Although they can select harmful and beneficial endophytes by isolating and culturing material from an infected plant, researchers do not understand all of the mechanisms conferring agronomic and ecological effects. Endophyte strains determine the type of secondary metabolite produced, while the host plant determines their quantity. But the behavior of endophytes appears to vary based on genotype and environmental conditions. Researchers are still working to understand these microorganisms and their overall symbiotic relationships with plants. On the whole, the cereal endophyte research process remains laborious and time-consuming, with the organic and symbiotic qualities of the plant-fungus relationship defying reduction to commodity form. Even as agriculture has reshaped global nutrient flows and metabolic capacities according to human selection, we cannot take for granted that contemporary biology can be reduced to the logics and processes of industrialization.[68] For the

[67] On grading systems and futures markets, see William Cronon, *Nature's Metropolis: Chicago and the Great West* (New York, 1991); Jonathan Ira Levy, "Contemplating Delivery: Futures Trading and the Problem of Commodity Exchange in the United States, 1875–1905," *Amer. Hist. Rev.* 111 (2006): 307–35. On commensurability, see Wendy Nelson Espeland and Mitchell L. Stevens, "A Sociology of Quantification," *Eur. J. Soc.* 49 (2008): 401–36.

[68] On the "biology of industrialization . . . that hugely magnifies some metabolic capacities and feeds them onward, reshaping genomic time and space," see Landecker, "Antibiotic Resistance" (cit. n. 15). On genetics as a technology of modernization, see Thurtle, *Emergence of Genetic Rationality* (cit. n. 19); Fenzi and Bonneuil, "From 'Genetic Resources' to 'Ecosystems Services'" (cit. n. 23).

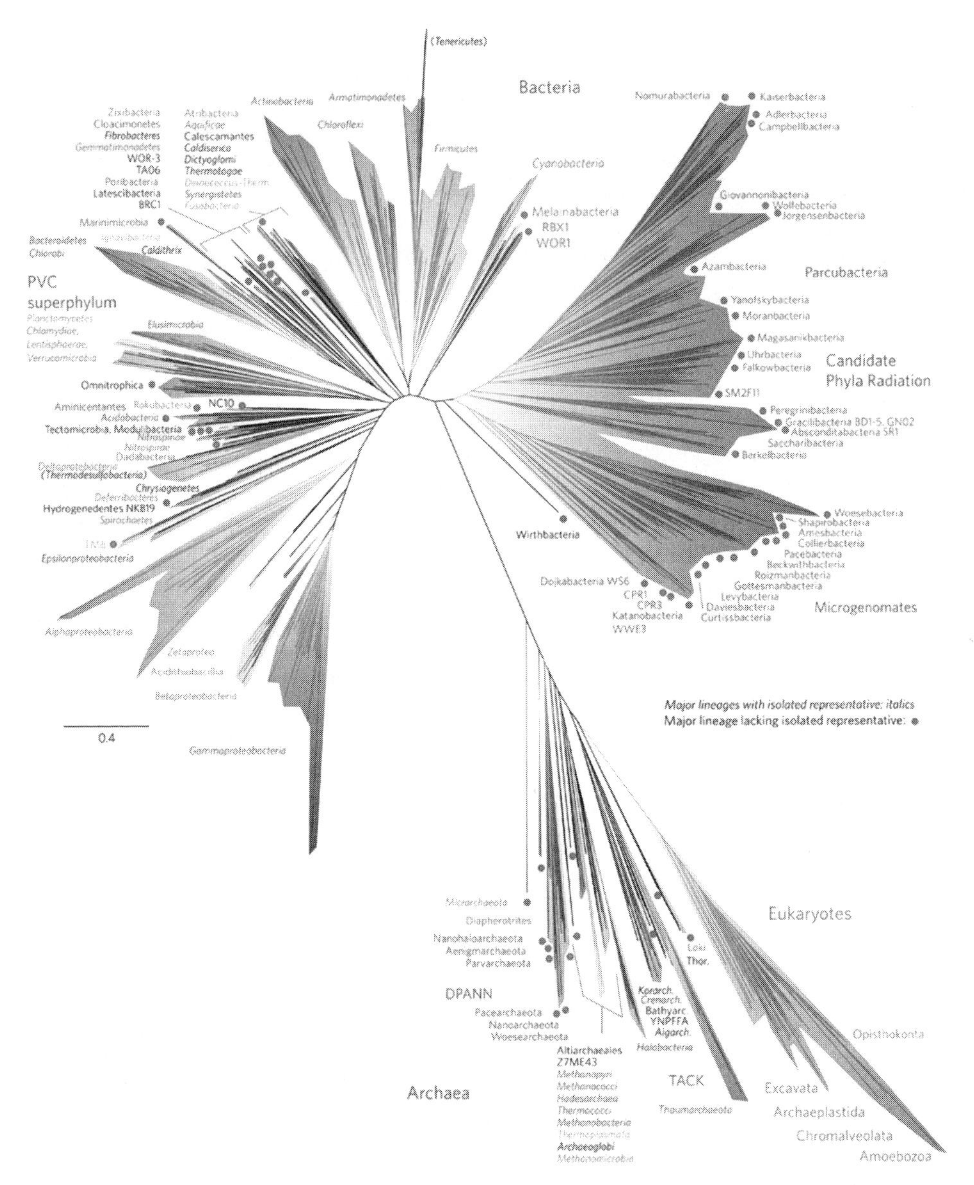

Figure 2. *"A current view of the tree of life, encompassing the total diversity represented by sequenced genomes," from Laura A. Hug, Brett J. Baker, Karthik Anantharaman, Christopher T. Brown, Alexander J. Probst, Cindy J. Castelle, Cristina N. Butterfield et al., "A New View of the Tree of Life,"* Nature Microbiol. *1, article no. 16048, 11 April 2016, https://www.nature.com/articles/nmicrobiol201648#f1.*

moment, endophyte is not legible as a mechanism for harnessing the diversity of life (fig. 2).

THE CAPITAL OF MICROBIOLOGICAL DIVERSITY

If we inhabit a system in which the imperatives of capital delimit our imagination, and indeed the very forms of life on earth, it is worthwhile to understand how. Competition between nation-states abetted by imperial natural science gives us one framework for understanding capitalist development, based on commodification and the practices of science that supported it. In this view, the path dependency of current research is an outcome of an economic system that has for five centuries, as Jason Moore has recently argued, relied on "Cheap Nature" strategies that extract and exploit labor power, energy, food, and raw materials.[69] Extractive agriculture entailed the broad geographic movement of plants, animals, and microbes and the intensification of production with resort to chemical fertilizers, precipitating what John Bellamy Foster has dubbed a "metabolic rift" between humans and their environments.[70] This intensification of plant and animal energy precipitated a series of agricultural revolutions reconfiguring the global countryside in service of metropolitan growth, setting the stage for industrialization reliant on fossil fuel extraction. Intensified energy regimes unleashed continuous habitat destruction and biodiversity erosion that portend species extinction, including our own.

But we must also consider the ways that knowledge systems have abetted the exploitation of nature. These consist not merely of early modern European classificatory schemes and projects so thoroughly investigated by historians of science, but also a more protean array of microbiological techniques and investigations. When the activist Vandana Shiva opposed European and American patents on turmeric and neem, she decried not simply the economic impact of these practices on unindustrialized countries, but also the fundamental "theft of nature's creativity" they require. With other critics, she rejects not merely an inequitable distribution of resources, but a political economy of Western science that construes nature as a raw material for economic development.[71] The charge of neo-imperialism is nevertheless an inadequate characterization of twenty-first-century science on both political economic and epistemological grounds. International networks of research capital may be products of empire, yet they are not identical to their antecedents, and understanding their operations requires approaches that eschew the nation-state and its archives as primary movers. Old categories of center and periphery obscure new networks of exchange intertwined with imperial legacies of removal and settlement and pursuing novel resources for profit.

Examining the values of biodiversity as an operational category in international agricultural research provides another view into this history and its outcomes. A critical history of biodiversity preservation should acknowledge its complex and contradictory array of logics and imperatives for environmental sanctity, poverty reduction, and growth. It should examine the technological, legal, and administrative apparatuses governing global plant genetic resources and identify the possibilities and limits of concepts in biodiversity, food security, and rural development as rubrics for political action and allocation of rights and resources on a global scale.

As they are currently framed, biodiversity preservation projects in international agricultural research can only tweak inputs to adapt to extremity: fabricating seed grain resistant to the biotic and abiotic stressors associated with climate change. When new aspects of biodiversity have become visible, it has been largely through observations,

[69] Jason W. Moore, *Capitalism in the Web of Life: Ecology and the Accumulation of Capital* (London, 2015).

[70] John Bellamy Foster, *Marx's Ecology: Materialism and Nature* (New York, 2000).

[71] Vandana Shiva, *Biopiracy: The Plunder of Nature and Knowledge* (Boston, 1997); Carolyn Merchant, *The Death of Nature: Women, Ecology, and the Scientific Revolution* (New York, 1980).

investigations, and techniques of visualization and manipulation that prioritize human interests. In the context of agro-biodiversity, capitalized breeding interests fund collection that in turn preserves material for the use of researchers. The history of such investigations is thus a conceptual and a material one, constitutive of both our ideas about biological diversity and the very fabric of life on earth. Its analysis requires attention to political economy, scientific practice, and the imaginaries of life common to both.

Viewed through the periscope of microbiological diversity and buttressed by new methods in genome sequencing, eighteenth- and nineteenth-century taxonomies and geopolitical orderings seem to misrepresent the complexity, fluidity, and interdependence of life on earth. Biological metaphors of diversity and flow may be invoked to justify global reconfigurations of finance capital, masking continued reliance on the commodity form as the basis of profit accumulation. These reconfigurations, and the imaginaries that inform them, portend structures of governance beyond the nation-state as drivers of world capitalist development and call for new attention to the organization of mobile capital in the twenty-first century.

But there are other uses for biodiversity, including its ability to demonstrate alternative ways of organizing life to support mutuality and subsistence. Differently deployed, studies of biodiversity have the potential to dissolve whole taxonomies, as they have done for fungi, and to indict the commodity form as a primitive reduction of nature to human. To the extent that we need histories of economic thought attentive to its animating fictions, we also need analyses of the sciences that demonstrate how habits and assumptions harden into ideologies and technologies. To probe the possibilities and limits of science and capital as world-making instruments, we should consider not simply what the farmer knew about the ruminants, about the grass, about the fungus, about the soil, but also what the soil knew about the fungus, about the grass, about the ruminants, about the farmer.

Smoke Ring:
From American Tobacco to Japanese Data

*by Sarah Milov**

ABSTRACT

This article explores the political, legal, and social history of the 1981 "Japanese Smokers' Wives Study." This large-scale cohort study, led by Takeshi Hirayama, chief epidemiologist at the National Cancer Institute in Tokyo, found that the nonsmoking wives of smokers were themselves at greater risk for developing lung cancer. The study was successfully used by American anti-tobacco activists to regulate indoor smoking during the 1980s. Taking a transnational approach, the article explores the complex, multilayered relationship between American tobacco production, Japanese epidemiology, and American tobacco restriction. It argues that smoking and anti-smoking in the United States and Japan have produced each other through flows of tobacco and epidemiological data. In the postwar era, Japanese cigarettes were increasingly filled with American-grown tobacco as part of aid packages, or sold to the Japanese Tobacco Monopoly. The subject population of the study was made, in part, by American tobacco producers. Meanwhile Americans who lived under an expanding umbrella of tobacco ordinances in the 1980s were made by Japanese tobacco consumers. The circulation of tobacco from American farms to Japanese bodies, from leaf to epidemiological data, is a window into the coproduction of science and capitalism in the second half of the twentieth century.

In the early 1980s, city councils and state legislatures across the United States began considering legislation that would restrict indoor smoking. From tiny Helena, Montana, to medium-sized heartland communities like Overland Park, Kansas, to large, diverse metropolitan areas like Los Angeles, elected officials responded to organized, anti-tobacco activism by passing laws that limited the places where people could smoke.[1] At public hearings, one name rolled perhaps a bit uneasily off the tongues of supporters of smoking restrictions: Hirayama. In 1981, the Japanese epidemiologist Takeshi Hirayama published a groundbreaking study on the carcinogenic effects of smoking on nonsmokers.[2] From a fourteen-year longitudinal study of the nonsmoking

* Corcoran Department of History, Nau Hall–South Lawn, University of Virginia, Charlottesville, VA 22904; smilov@virgina.edu.

I wish to thank William Deringer, Eugenia Lean, and Lukas Rieppel for their stewardship of this volume. Thanks also to Claire Edington, Dan Navon, and participants in the Science Studies Colloquium at the University of California–San Diego for helping to improve an early draft of this essay.

[1] "Restrictions on Smoking Spreading across US," *New York Times*, 1 March 1984.

[2] Takeshi Hirayama, "Non-smoking Wives of Heavy Smokers Have a Higher Risk of Lung Cancer: A Study from Japan," *Brit. Med. J.* 282 (1981): 183–5; Allen Brandt, *The Cigarette Century* (New York, 2007), 284–5; Richard Kluger, *Ashes to Ashes: America's Hundred-Year Cigarette War, the Public Health, and the Unabashed Triumph of Philip Morris* (New York, 1996), 499–500.

wives of smoking husbands, Hirayama concluded that the wives of smokers were significantly more likely to develop lung cancer themselves. This research vindicated and substantiated nonsmokers' long-standing complaints that cigarette smoke was not just annoying to those around them, it was also deadly.

In 1986, for example, New York City Mayor Ed Koch proposed a bill to restrict smoking. In hearings that included testimony from the then surgeon general C. Everett Koop and four former U.S. surgeons general, the Japanese Smokers' Wives Study was cited in expert testimony, forming part of the basis for the committee's conclusion that "involuntary, passive smoking . . . is also hazardous to health."[3] The following year, the New York City Council approved the city's first comprehensive anti-smoking bill, restricting smoking in most enclosed public spaces—businesses, stores, restaurants, schools, sports venues, and taxis.[4] By 1986, forty states and the District of Columbia had passed legislation restricting smoking in public places.[5] More than eighty cities and counties had also passed smoking ordinances, further cordoning off spaces where smoke wafted in the air among nonsmokers.

Hirayama's research was also cited in federal reports. In 1985, Alaska Congressman Ted Stevens, who had introduced legislation to establish separate smoking lobbies in federal buildings, requested that the Office of Technology Assessment (OTA) make a study of the research and laws surrounding passive smoking—as secondhand smoke was sometimes called. The OTA's report cited Hirayama's as "the first major study linking passive smoking to lung cancer in nonsmokers" before noting that the half dozen other studies of similar design that followed shared relatively consistent findings of secondhand smoke as a risk factor for lung cancer.[6] Stevens's bill died in committee, but Hirayama's findings on the effects of smoking on nonsmokers played a star role in the 1986 Surgeon General's Report on Secondhand Smoke, which concluded, among other findings, that "involuntary smoking is a cause of disease, including lung cancer, in nonsmokers."[7] The 1986 report, like the landmark 1964 report announcing a definite link between smoking and lung cancer in men, was intended to prompt political action: "The scientific case against involuntary smoking as a health risk is more than sufficient to justify appropriate remedial action, and the goal of any remedial action must be to protect the nonsmoker from environmental tobacco smoke."[8] Hirayama was one of just a handful of non-American experts to be thanked in the acknowledgments of the 1986 report, and the only non-Western researcher so cited.

Japanese epidemiology formed the evidentiary basis for American laws that restricted smoking in public places—pushing smokers out of the spatial commons, and the practice of smoking toward the realm of social unacceptability.[9] In this way, American laws,

[3] "Report of the Mayor's Committee on Smoking and Health," 1 July 1986, R. J. Reynolds Records, https://www.industrydocumentslibrary.ucsf.edu/tobacco/docs/hqkk0087 (accessed 12 July 2018).

[4] "New York Council Enacts Tough Law against Smoking," *New York Times*, 24 December 1987.

[5] U.S. Department of Health and Human Services, *Health Consequences of Involuntary Smoking: A Report of the Surgeon General* (Washington, D.C., 1986), xi.

[6] Office of Technology Assessment, *Passive Smoking in the Workplace: Selected Issues* (Washington, D.C., 1986), 17–9.

[7] U.S. Department of Health and Human Services, *Health Consequences of Involuntary Smoking: A Report of the Surgeon General* (Washington, D.C., 1986), 27, 67–76.

[8] Ibid., x.

[9] As Kluger notes, the Hirayama study specifically—and the attempts to quantify harm posed by secondhand smoke generally—was subject to a great deal of uncertainty. Uncertainty about how to measure exposure to cigarette pollution did not deter anti-smoking activists from citing the study

attitudes, and practices surrounding smoking—not to mention the physical constitution of American bodies—are the product of the transnational currents in scientific research. In the United States, state and local laws govern indoor smoking. The Japanese Smokers' Wives Study highlights the way that even local governance—the democratic hurly burly of city council meetings, citizen protest, and earnest assertion of rights—has been constituted by flows of information derived from far-flung places, deterritorialized and universalizable through shared disciplinary practice and peer review. The more than 240,000 bodies that formed the subject population in Hirayama's study were Japanese citizens. But they, too, were biologically constituted by foreign matter. Beginning in the 1960s, Japanese cigarettes were made with increasing amounts of American tobacco in supplement to domestically produced leaf. Japanese smokers and the nonsmokers who inhaled the sidestream smoke of a cigarette were tied to the United States through flows of tobacco and capital.

The subject population of Takeshi Hirayama's study was made, in part, by American tobacco producers; Americans that lived under an expanding umbrella of tobacco ordinances in the 1980s were made, in part, by Japanese tobacco consumers—both those who consumed by choice and those who did not. This article explores the complex, multilayered relationship between American tobacco production, Japanese epidemiology, and American tobacco restriction between the 1960s and the 1980s. I argue that smoking and anti-smoking in the United States and Japan have produced each other through flows of tobacco and epidemiological data. The tobacco that the United States exported to Japan was later imported to the United States in the form of data—having been processed in the bodies of a cohort of Japanese smokers and their wives. In this way, the tobacco that the United States exported to Japan later became important evidence for arguments to restrict tobacco consumption in the United States. The Smokers' Wives Study is a cautionary tale against viewing consumption as "the end of the road for goods and services," in the words of cultural critic Arjun Appadurai.[10] Diseases—particularly diseases of industrialization—can illuminate the afterlives of consumption, the knock-on effects of a postwar political economy of American abundance. Routed through Japan, tobacco had a boomerang effect in the United States, reshaping the social, cultural, and economic meaning of smoking and of smokers.

The circuit of tobacco, knowledge, and power is not just a mechanistic reflex action of a commodity entering a marketplace and changing global patterns of production and consumption. Rather, American tobacco in Japan set into motion a set of unintended consequences and unpredictable processes that refashioned the relationship between American citizens and their government. The data about nonsmoker mortality from exposure to ambient cigarette smoke added urgency and empirical heft to the claims of anti-tobacco activists. Public smoking bans proliferated in the United States during the 1980s. Yet despite being one of the most important nodes of tobacco knowledge in the 1970s and 1980s, Japanese anti-tobacco activists were not as successful as their American counterparts in securing regulation on behalf of nonsmokers. It is only

---

in support of restrictive legislation; Kluger, *Ashes to Ashes* (cit. n. 2), 499–503. As Robert Proctor has pointed out, some of the uncertainty surrounding Hirayama's study was, in fact, generated by the tobacco industry; Proctor, *Golden Holocaust: The Origins of the Cigarette Catastrophe and the Case for Abolition* (Berkeley and Los Angeles, 2011), 438.

[10] Arjun Appadurai, *Modernity at Large: Cultural Dimensions of Globalization* (Minneapolis, 1996), 66.

by following the unexpectedly intertwined biographies of American tobacco and the Japanese Smokers' Wives Study that national policy differences can be meaningfully explained without resorting to culturalist explanations. The nation-state was the most important site of contentious politics for American and Japanese anti-tobacco activists.[11] A transnational perspective is essential for understanding how nationally particular political institutions and ideas mattered. This article demonstrates the permeability and mutual constitution of American and Japanese anti-tobacco politics, along with their differential outcomes.

Historians of global capitalism have been the most forceful in reminding us that commodity histories illuminate connections between people and places that would otherwise be siloed within nationalist histories.[12] "As a tool for tracking the idiosyncratic operations of capitalist social relations as they bind together distant locales," writes historian Paul Kramer, commodities serve "as a kind of traceable dye through complex circulatory systems."[13] In its complex route through eastern North Carolina, Washington, D.C., Tokyo, and Los Angeles, tobacco serves as diagnostic tracer of power and knowledge: highly territorialized capital in the form of government subsidies and monopolies; highly territorialized knowledge in the form of a national population study; deterritorialized, freely circulating epidemiological data; and, finally, hyperlocal conflicts over the substance and exercise of rights.

Tobacco is an especially compelling commodity to follow in the postwar era. The manufactured cigarette was first an emblem of American power and global modernity, and also a symbol of its underside: imperialism and disease. But the circulation of tobacco from the United States to Japan was not a one-way trade foisted upon an unwilling nation. The Japanese state was a monopolist and an eager manufacturer and merchant of tobacco, which became an important generator of state revenues. Neither is this a story of "dependence"—whether Japanese dependence on the United States or the dependence of Japanese state budgets on a national cigarette habit. By following the metabolism of American tobacco once it became a Japanese cigarette, a habit of a Japanese smoker, and the secondhand smoke inhaled by a Japanese housewife, this article joins the political economy of tobacco production and the sociolegal history of its consumption.

Historians of science have resoundingly shown that the cigarette industry in the twentieth century was pioneering in its use of doubt to subvert regulation.[14] No political-industrial cabal has produced doubt "more effectively than the tobacco mongers, the masters of fomenting ignorance to combat knowledge."[15] Tobacco companies were pi-

[11] Sidney Tarrow, *The New Transnational Activism* (Cambridge, 2005).

[12] Sven Beckert, *Empire of Cotton: A Global History* (New York, 2014); April Merleaux, *Sugar and Civilization: American Empire and the Cultural Politics of Sweetness* (Chapel Hill, N.C., 2015); John Soluri, *Banana Cultures: Agriculture, Consumption, and Environmental Change in Honduras and the United States* (Austin, Tex., 2006); Heidi Tinsman, *Buying into the Regime: Grapes and Consumption in Cold War Chile and the United States* (Durham, N.C., 2014); Richard P. Tucker, *Insatiable Appetite: The United States and the Ecological Degradation of the Tropical World* (Lanham, Md., 2007).

[13] Paul Kramer, "Embedding Capital: Political-Economic History, the United States and the World," *J. Gild. Age Prog. Era* 15 (2016): 341.

[14] Erik M. Conway and Naomi Oreskes, *Merchants of Doubt: How a Handful of Scientists Obscured the Truth on Issues from Tobacco Smoke to Global Warming* (New York, 2010); Proctor, *Golden Holocaust* (cit. n. 9); Robert N. Proctor and Londa Schiebinger, eds., *Agnotology: The Making and Unmaking of Ignorance* (Stanford, Calif., 2008).

[15] Robert N. Proctor, "Agnotology: A Missing Term to Describe the Cultural Production of Ignorance (and Its Study)," in Proctor and Schiebinger, *Agnotology* (cit. n. 14), 1–36, on 11.

oneering in their exploitation of reasonable doubt, lavish in their spending on advertising intended to "reassure consumers that the hazard had not yet been proven."[16] And for other polluting and poisoning industries—from energy corporations to agribusiness—the tobacco industry has been a model of diversionary tactics and subversion of science. Tobacco corporations paid scientists to maintain "controversy" over the cigarette's harms—and then pointed to the fabricated controversy as evidence that further evidence was needed before cigarettes could be regulated.[17] This essay does not take issue with these important claims about the politics of knowledge production. However, it does suggest that to understand the construction of knowledge about tobacco—including nakedly mercenary denials of the hazards of cigarettes—it is necessary to explore the broader political context of tobacco's political economy and knowledge economy.

Expanding our gaze beyond the perfidy of corporations, attorneys, and PR flacks, we begin to see a political economy in which both the production of tobacco and the epidemiological research that ultimately curtailed its consumption were matters of state. These states had independent goals and capacities of their own: they were not just neutral terrain through which tobacco oligopolies operated.[18] Businesses have shaped the knowledge that informs personal choice and public policy. But this should not obscure an equally important dynamic: governments, particularly agriculture and finance bureaucracies, have been willing and decisive coconspirators in making tobacco. Nor should a focus on doubt science distract from the political context of "legitimate" and peer-reviewed science. Japanese tobacco epidemiology was, in part, the product of decades of geopolitical interaction with the United States. The Smokers' Wives Study was so influential in the United States because, on the one hand, it resonated with the claims of a preexisting nonsmokers' rights movement and, on the other hand, because American doctors and public health officials were familiar with conclusions the Japanese longitudinal study had already yielded. By following tobacco from promotion to regulation, this article expands our understanding of the institutions responsible for embedding the cigarette in modern life, while specifying the routes by which epidemiology was disseminated and made politically powerful in the United States. Protobacco policies were not made by corporate capture of the state; anti-tobacco policies were not achieved through the natural victory of good science, of truth over doubt-mongering darkness. This essay brings the state back into the history of tobacco, implicitly edging the major tobacco corporations out of the center of the commodity's history.

Tobacco policy in the United States was also an ecological, biological policy. U.S. historian Paul Kramer has recently called for historians of capitalism to see capitalism "ecologically" by bringing "the externalities back in."[19] The study of the smoke ring

[16] Ibid.

[17] Ibid., 11–4; Brandt, *Cigarette Century* (cit. n. 2), esp. chap. 6.

[18] For literature on the role of the state in shaping American political development, see Brian Balogh, *A Government Out of Sight: The Mystery of National Authority in Nineteenth-Century America* (Cambridge, 2009); Balogh, *The Associational State: American Governance in the Twentieth Century* (Philadelphia, 2015); Daniel Carpenter, *Reputation and Power: Organizational Image and Pharmaceutical Regulation at the FDA* (Princeton, N.J., 2010); Carpenter, *The Forging of Bureaucratic Autonomy: Reputations, Networks, and Policy Innovation in Executive Agencies, 1862–1928* (Princeton, N.J., 2001); Meg Jacobs, William J. Novak, and Julian E. Zelizer, *The Democratic Experiment: New Directions in American Political History* (Princeton, N.J., 2003); William J. Novak, "The Myth of the 'Weak' American State," *Amer. Hist. Rev.* 113 (2008): 752–72; Novak, *The People's Welfare: Law and Regulation in Nineteenth-Century America* (Chapel Hill, N.C., 1996); Peter B. Evans, Dietrich Rueschemeyer, and Theda Skocpol, eds., *Bringing the State Back In* (Cambridge, 1985).

[19] Kramer, "Embedding Capital" (cit. n. 13), 334.

connecting the United States and Japan goes perhaps a step further than Kramer's call for transnational, ecological histories of capitalism. One externality of tobacco—measured in terms of disease and debility to nonsmokers—reshaped tobacco regulations and the social meaning of smoking in the United States. On Japanese bodies—the disease suffered by the nonsmoking wives of Japanese smokers—one of the unanticipated costs of tobacco consumption was made visible. Routed through Japan in the 1960s, tobacco had a boomerang effect in the United States in the 1980s, as harms to nonsmokers were cited as a primary reason for the regulation of indoor smoking. The circulation of tobacco from American farms to Japanese bodies, from leaf to epidemiological data, is a window into the coproduction of science and capitalism in the second half of the twentieth century. When tobacco is viewed solely through a nation-shaped keyhole, or through the dark lens of Big Tobacco, the profound interconnection between U.S. and Japanese political economy and public health are all but obscured. By casting a wide gaze, this article emphasizes two seemingly contradictory features of postwar capitalism: embeddedness and contingency.

In both the United States and Japan, tobacco was embedded in nationalist and nation-building structures. But the trade policies of nation-states were hardly determinative of the market and meaning of tobacco. This article follows the route of tobacco across space and time, as tobacco was metabolized through institutions and through human beings. It begins in the southeastern United States, where tobacco producers looked to the U.S. government to sell their tobacco leaf. Tobacco farmers succeeded in the postwar era by using aid and trade programs as a means of surplus disposal and to shape consumer preferences. As Japan was a primary recipient of such tobacco aid, Japanese smoking rates, particularly among men, were among the highest in the world.[20] The second part of this essay turns to the efforts of Tokyo-based epidemiologist Takeshi Hirayama to assess the population-level impacts of the nation's cigarette consumption. Hirayama's large-cohort study of nearly 250,000 Japanese citizens, from which the Smokers' Wives Study was drawn, was itself a product of transnational developments in public health methods, as well as factors specific to Japan's atomic and industrial history. Epidemiologists and activists operating within transnational and highly local networks sought to make visible the human toll of tobacco. The final section of the essay turns to the reception of the Smokers' Wives Study in the United States. Ironically, Hirayama's work proved far more influential in changing American than Japanese law in the 1980s. The achievements of anti-tobacco activists in the United States underscore the importance of national institutional differences in shaping regulatory policy.

CULTIVATING A TASTE FOR AMERICAN TOBACCO

For most of the twentieth century, the production and sale of cigarette tobacco in the United States was rigidly defined by law. The New Deal–era Agricultural Adjustment Act established a program of supply restriction and price supports for farmers.[21] Despite the sea change in social and legal perspectives on tobacco that occurred in the last four decades of the twentieth century, the program remained intact through the end of

[20] Judith Mackay and Michael Eriksen, *Tobacco Atlas* (Geneva, 2002), 24.

[21] Anthony Badger, *Prosperity Road: The New Deal, Tobacco and North Carolina* (Chapel Hill, N.C., 1980); Pete Daniel, *Breaking the Land: The Transformation of the Cotton, Tobacco, and Rice Cultures since 1880* (Urbana, Ill., 1985).

the twentieth century: farmers in the southeastern United States were guaranteed minimum prices in exchange for restricting supply.

In 1945, American tobacco farmers feared abundance: they had seen the havoc that too much leaf could wreak. In illustration of what economist Mancur Olson would later call a collective action problem, tobacco farmers facing low prices after World War I had few options but to produce to the hilt and hope that by sheer volume of sale they would make enough money to survive.[22] The result of these individually rational choices was chronic overproduction and chronically low prices. After World War II, farmers were in a dramatically different position: they were organized by the federal government, their economic security underwritten by the welfarist commitments of New Deal agricultural policy. Overproduction could ruin this newfound prosperity. If farmers produced too much tobacco, it would accumulate at taxpayer expense, reducing the economic and political viability of the federal tobacco program. Farmers feared that the high prices and brisk sales volume of 1945 were not likely to last. "History repeats itself," one agricultural official warned leaders. "It looks at the present time as if conditions faced by growers following World War I are upon us again."[23]

The state of North Carolina, which was the leading producer of flue-cured (cigarette) tobacco, was especially concerned with maintaining the viability of the federal tobacco program. In 1946, the General Assembly passed a law that allowed all farmers to tax themselves ten cents per tobacco growing acre to fund an organization called Tobacco Associates.[24] Tobacco Associates' goal was to cultivate global demand for the taste of American flue-cured tobacco. It would not sell leaf directly but would grease the channels of international trade by maintaining contact with "private and public agencies in the United States and in foreign countries in which American flue-cured tobacco is being used or might be used."[25] With the help of state and federal agencies, numerically constricted, geographically concentrated American tobacco producers financed global tobacco promotion.

Tobacco Associates promoted the general virtues of American flue-cured tobacco—"mildness" and "inhalability," two traits that, in fact, make flue-cured tobacco the most deadly form of the leaf.[26] American cigarette manufacturers were the only segment of the tobacco economy left out of the export equation. Farmers and businessmen were not selling manufactured American cigarettes abroad; they sought to sell raw leaf to the foreign tobacco monopolies that produced nationally specific tobacco brands. Whereas many may have seen mounting stocks of tobacco as evidence of tobacco overproduction, Tobacco Associates reframed the issue as one of global tobacco "un-

[22] Mancur Olson, *Logic of Collective Action: Public Goods and Theory of Groups* (Cambridge, Mass., 1971).

[23] "The Price Outlook for Flue-Cured Tobacco," *Progressive Farmer*, June 1947, 82.

[24] "Memorandum in Connection with Meeting of Producers," 17 February 1947, Folder 1, Box 1, North Carolina Farm Bureau Records, North Carolina State University. Tobacco Associates incorporated in South Carolina, Virginia, Georgia, and Florida the following year, where referenda were also held. See, e.g., "Farmers Cast Votes in Leaf Referendum Today," *Dispatch* (Lexington, N.C.), 23 July 1949.

[25] Tobacco Associates, Articles of Incorporation, 1947, North Carolina Department of the Secretary of State, https://www.sosnc.gov/online_services/search/by_title/_Business_Registration (accessed 12 July 2018). The Articles of Incorporation are available to download by entering "Tobacco Associates" in the "Search For" field at this URL.

[26] John Flannagan (Va.), "Tobacco and the European Recovery Program in General," U.S. Congress, House, *Congressional Record*, 80th Congress, 2nd Session, 31 March 1948, 3881.

derconsumption." The organization used reconstruction and aid laws like the Marshall Plan and Public Law 480, better known as the Food for Peace program, to expand the global palate for American leaf.[27] Of the $13 billion in Marshall Plan assistance that was provided to Europe, a staggering one billion dollars went to the procurement of tobacco, a figure that represented a third of total food aid.[28] The Marshall Plan was a temporary measure: it had an expiration date of 1952, after which U.S. taxpayers would no longer so directly underwrite Europe's tobacco consumption—and U.S. tobacco production.

Tobacco Associates wanted to make sure that Europeans would continue to be supplied with American tobacco long after direct aid dollars ran out. As one Tobacco Associates official told a Rotary dinner in eastern North Carolina, the organization's most pressing need was to "devise some means of supplying countries with some flue-cured tobacco in order to preserve the taste and desire for American-style cigarettes until these nations become self-supporting."[29] Taste, tobacco growers realized, was political, and they were in a special position to alter the politics of taste. The official went on in a defense of growers' collective production rights—their cartel. "Manufacturers of automobiles cannot, under existing law, come together and decide the quantity of automobiles to be built in any period. Nor can the auto group prevent a new plant from engaging in the manufacture of [cars]." But tobacco growers, by contrast, "are permitted to control the production of tobacco and can deny the right to engage in production to others." Tobacco Associates realized that it had a good thing going: it was a self-financed cartel facing a market of unlimited potential. By leveraging their proximity to the U.S. government, they could cultivate a global taste for an American-style cigarette produced under the auspices of national brands and monopolies.

The passage of the Food for Peace program established a permanent role for Tobacco Associates in the global commodity trade. Public Law 480 authorized export sales of American surplus commodities in exchange for foreign currencies.[30] This program was especially suited to the task of shifting consumer preferences, encouraging private organizations like Tobacco Associates to use local currencies to engage in "market development activities."[31] Working alongside the Departments of Agriculture and Commerce and the U.S. Information Agency in trade fairs and advertising campaigns, Tobacco Associates assisted in the manufacture and advertising of new brands produced under national monopoly labels but made with U.S. flue-cured tobacco.[32]

The organization achieved its most pronounced success in the 1950s and 1960s in its collaboration with the Japan Tobacco Monopoly (JTM). In 1956, the first year of

[27] "Remarks of J. B. Hutson before Board of Governors, Tobacco Association of the United States," 30 January 1948, Folder 1c, Box 1, Hutson Papers, Eastern Carolina University, Greenville, N.C. (hereafter cited as "ECU").

[28] Proctor, *Golden Holocaust* (cit. n. 9), 46.

[29] "Untitled Speech at Washington, N.C.," n.d. (1947), Folder "Speech Folder," Box 1, Papers of J. Con Lanier, ECU.

[30] Robert R. Sullivan, "The Politics of Altruism: An Introduction to the Food-for-Peace Partnership between the United States Government and Voluntary Relief Agencies," *West. Polit. Quart.* 23 (1970): 762–8; Mitchel B. Wallstein, *Food for War–Food for Peace: United States Food Aid in a Global Context* (Cambridge, Mass., 1980).

[31] The Public Law 480 Market Development Program, 11 July 1958, Folder 1g, Box 1, Hutson Papers, ECU.

[32] Ibid.; Tobacco Associates, "Annual Report," 25 February 1969, Tobacco Institute, http://legacy.library.ucsf.edu/tid/kor59b00/pdf (accessed 12 July 2018); Tobacco Associates, "Annual Report," 7 March 1961, Tobacco Institute, http://legacy.library.ucsf.edu/tid/qor59b00 (accessed 12 July 2018).

Japan's participation in Public Law 480, the nation imported 5.6 million pounds of American flue-cured tobacco. By 1968, that figure had skyrocketed to nearly 42 million pounds—an increase of more than 700 percent.[33] By the late 1970s, after Japan had ceased participation in the program, it still imported more American flue-cured tobacco than any other country. The intervening years saw numerous trips by Tobacco Associates officials to Japan—as well as longer excursions taken in the other direction. The JTM even established a permanent office in Raleigh.

Beginning in 1956, the organization used part of the yen it received from the sale of flue-cured tobacco under Food for Peace to contribute to local magazine, radio, television, and newspaper advertising campaigns for brands that contained American flue-cured leaf.[34] Tobacco Associates focused, in particular, on the "Peace" brand of cigarettes, helping to finance an advertising boot camp for two officials from the JTM. After two weeks learning the latest principles of advertising from Madison Avenue firms, JTM officials spent time with the sales departments of American cigarette manufacturers. Back in Tokyo, these two officials "initiated a new sales training program for Japan Tobacco Monopoly personnel and revamped their entire sales techniques leaning heavily on the knowledge they gathered" while in the United States.[35] Over the next ten years, Tobacco Associates' work in Japan "evolved from a subordinate position in [the Monopoly's] advertising activities to one of major importance in its annual planning operations."[36] This was reflected in the fact that JTM's marking program was overseen by J. Walter Thompson, the famed advertising agency that handled accounts for Liggett & Myers, Lorillard, and the American Tobacco Company. By working to develop not only specific Japanese brands, but also American-style advertising practices within Japan, Tobacco Associates' efforts reached well beyond the surplus-disposal mandates of Public Law 480. The goal was always to increase the use of flue-cured tobacco in Japanese brands, which, according to the trade journal *Tobacco Reporter*, "contained about two-thirds American flue-cured" by 1969.[37] A decade later, Japanese citizens were reportedly spending more money on cigarettes than on rice or meat.[38]

MODERNITY'S DATA: EPIDEMIOLOGY IN THE POSTWAR ERA

The American cultivation of the Japanese tobacco market had unforeseen consequences in the United States in subsequent years. For as tobacco flowed overseas from the southeastern United States, physicians, epidemiologists, and statisticians around the world developed new techniques for investigating the relationship between tobacco and disease. After World War II, epidemiologists focused increasingly on chronic rather than infectious diseases.[39] Agents of chronic disease were increasingly associated with

[33] Tobacco Associates, "Annual Report," 25 February 1969 (cit. n. 32).
[34] The Public Law 480 Market Development Program (cit. n. 31).
[35] "Problems Selling Tobacco in the Export Markets," n.d., Folder 1g, Box 1, ECU.
[36] Tobacco Associates, "Annual Report," 7 March 1967, Tobacco Institute, http://legacy.library.ucsf.edu/tid/mor59b00/pdf (accessed 12 July 2018).
[37] "USDA Trade Mission Reports on Overseas Markets," *Tobacco Reporter* 96 (1969): 70.
[38] "Japan Smokers' Puff Away despite Some Warnings," *Nashua Telegraph*, 15 May 1981; Tobacco Institute Newsletter, 28 July 1981, American Tobacco Records, https://www.industrydocuments library.ucsf.edu/tobacco/docs/rymm0002 (accessed 12 July 2018).
[39] For more on the epidemiological transition, see James C. Riley, *Rising Life Expectancy: A Global History* (Cambridge, 2001); George Weisz, *Chronic Disease in the Twentieth Century: A History* (Bal-

environmental and lifestyle factors, rather than bacteria, parasites, and viruses. Methods for understanding the multiple and sometimes interacting agents of chronic disease were developed and refined. A cohort study selects a population to be studied and then follows them with an eye toward particular diseases or health outcomes. The study tracks exposure over time to risk factors such as smoking, alcohol consumption, or occupational risks in order to determine a relationship to disease outcomes such as cancer or cardiovascular disease.

The cohort study developed as an exemplary method for understanding chronic disease because it had the singular virtue of following a large number of participants over time, allowing researchers to understand health events as they unfold in relation to the particular risk factors identified. The Framingham Heart Study was perhaps the most famous and long-lived cohort study, yielding thousands of research papers based on its data, and continuing to this day with a third generation of study participants.[40] The study began in 1947 as an attempt to better understand the causes of cardiovascular disease, which was responsible for 44 percent of deaths in the United States. The first 5,200 disease-free volunteers enrolled in the study—residents of the largely white, middle-class Boston suburb—underwent their first physical examinations in 1948. The study's designers planned a five- to ten-year follow-up to investigate "factors suspected of causing predisposition to coronary heart disease."[41] By definition the Framingham Study could not yield immediate fruit in terms of establishing a link between heart disease and risk factors such as obesity, stimulant use, or high cholesterol. But its design was immediately influential—even if the resources required for such large-scale studies posed budgetary strains.[42]

If the large-cohort study was a paradigmatic postwar method, lung cancer was a paradigmatic disease outcome, and smoking the paradigmatic exposure. In 1951 Dr. Richard Doll and Sir Austin Bradford Hill, epidemiologists in the Statistical Research Unit at the London School of Hygiene, mailed a simple questionnaire to registered physicians living in the United Kingdom. Physicians were asked to give their name, address, age, and sex, and to classify themselves as smokers, former smokers, or never-smokers. Of the roughly 60,000 physicians listed on the Medical Register in 1951, approximately 40,000 responded to the inquiry. With the cooperation of the U.K. Registrars-General, Doll and Hill tracked the death certificates of the men and women whose occupations were recorded as "medical practitioner." After three years of data collection, Doll and

timore, 2014); Mark Parascandola, "The Epidemiologic Transition and Changing Concepts of Causation and Causal Inference," *Rev. Hist. Sci.* 2 (2011): 243–62; Abdel Omran, "The Epidemiologic Transition: A Theory of Epidemiology of Population Change," *Milbank Memorial Fund Quart.* 49 (1971): 509–38.

[40] Robert Aronowitz, "The Framingham Heart Study and the Emergence of the Risk Factor Approach to Coronary Heart Disease, 1947–1970," *Rev. Hist. Sci.* 64 (2012): 263–95; Sejal Patel, "Methods and Management: NIH Administrators, Federal Oversight, and the Framingham Heart Study," *Bull. Hist. Med.* 86 (2012): 94–121; Thomas R. Dawber, *The Framingham Study: The Epidemiology of Atherosclerotic Disease* (Cambridge, Mass., 1980); Jonathan M. Samet and Álvaro Muñoz, "Evolution of the Cohort Study," *Epidemiologic Rev.* 20 (1998): 1–14; Richard Doll, "Cohort Studies: History of the Method. I. Prospective Cohort Studies," *Soz. Praventivmed.* 46 (2001): 75–86; Doll, "Cohort Studies: History of the Method. II. Retrospective Cohort Studies," *Soz. Praventivmed.* 46 (2001): 152–60; Mervyn Susser, "Epidemiology in the United States after World War II: The Evolution of Technique," *Epidemiologic Rev.* 7 (1985): 147–77.

[41] Quoted in Gerald M. Oppenheimer, "Becoming the Framingham Study, 1947–1950," *Amer. J. Public Health* 95 (2005): 602–10, on 606.

[42] Patel, "Methods and Management" (cit. n. 40).

Hill published their preliminary findings in the *British Medical Journal*. They found a statistically significant relationship between lung cancer and smoking—"and steadily rising mortality from deaths due to cancer of the lung as the amount of tobacco smoked increases."[43] So new was this research design that Doll and Hill invoked no less an authority than the *Oxford English Dictionary* in explaining their "prospective" approach. In 1956, Hill and Doll published a follow-up study, which confirmed the prediction that smoking was a predisposing risk for lung cancer, as well as a number of other diseases.[44] The prospective methodology developed for the British Doctors Study was explicitly intended to avoid some of the limitations of earlier case-control studies—including their own—which were seen as less than authoritative in their demonstration of an association between cigarette smoking and lung cancer because of the potential for biases on the part of both participants and researchers. Examining the effects of risk exposures over a long period of time, and requiring not only the cooperation of volunteers, but also very frequently that of governmental agencies, the cohort study emerged as an important technique for the study of chronic disease—the study of the effects of modernity upon aged bodies.

Across the Atlantic, statisticians E. Cuyler Hammond and Daniel Horn conducted a similarly designed prospective study—and did so with the intentions of eliminating the recall and interviewer biases associated with the case-control methods. Operating through the American Cancer Society, Hammond and Horn followed the health outcomes of nearly 200,000 fifty- to sixty-year-old men over four years. Their study, also published in 1954, echoed the British Doctors Survey in finding that lung cancer deaths were stunningly prevalent among smokers—twenty-four times more common than among nonsmokers.[45] Smokers also constituted a large percentage of those who had died from circulatory and heart disease. The Hammond and Horn study thus underscored the causal relationship between smoking and lung cancer, while also suggesting a wider range of diseases associated with cigarette consumption.[46]

Dr. Takeshi Hirayama followed developments in Anglo-American epidemiology avidly. Indeed, he was training in the United States for part of the 1950s, earning a master of public health from Johns Hopkins in 1952 and researching the association between smoking and lung cancer at Sloan Kettering in 1959. Hirayama was steeped in disciplinary debates about methods, research design, and the nature of causality, as well as the political implications of tobacco research.[47] And he knew Cuyler Hammond and Richard Doll well: both men gave "precious guidance," in Hirayama's words, to

[43] Richard Doll and A. Bradford Hill, "The Mortality of Doctors in Relation to Their Smoking Habits," *Brit. Med. J.* 1 (26 June 1954): 1455.

[44] Richard Doll and A. Bradford Hill, "Lung Cancer and Other Causes of Death in Relation to Smoking," *Brit. Med. J.* 2 (10 November 1956): 1071–81.

[45] Brandt, *Cigarette Century* (cit. n. 2), 145.

[46] Ibid.; E. Cuyler Hammond and Daniel Horn, "The Relationship between Human Smoking Habits and Death Rates: A Follow-Up Study of 187,766 Men," *J. Amer. Med. Assoc.* 155 (1954): 1316–28. For current practitioners of public health, these early cohort studies are considered foundational to knowledge about the harms associated with smoking. Michael J. Thun and S. J. Henley, "The Great Studies of Smoking and Disease in the Twentieth Century," in *Tobacco: Science, Policy, and Public Health*, ed. Peter Boyle, Nigel Gray, Jack Henningfield, John Seffrin, and Witold Zatonski (2004; repr., Oxford, 2010).

[47] For a discussion of clinical versus epidemiological conceptions of disease causality as they shaped biomedical knowledge of tobacco, see Brandt, *Cigarette Century* (cit. n. 2), 131–58.

"planning and conducting the study."[48] To an even greater extent than in the United States or England, tobacco consumption was a ubiquitous part of daily life in Japan, where 85 percent of men smoked—significantly higher than the 50 percent peak smoking prevalence for American men.[49]

The design of Hirayama's large-cohort study was shaped not only by the expanding methodological influence of cohort studies generally but also by the publication of the 1962 Report of the Royal College of Physicians and the 1964 Surgeon General's Report. In concluding that smoking was a cause of lung cancer, both documents leaned hard upon the handful of cohort studies that had been published in the wake of the British Doctors Survey. The 1964 Surgeon General's Report highlighted the "consistency of findings" among the seven prospective studies published since 1951. These illustrated "the nature and potent magnitude of the smoking-health problem." Indeed, the 1964 report suggested that in the absence of findings from cohort studies, causal statements linking smoking and cancer would have been more tenuous, as the case-control studies had been attacked on the basis that they harbored biases. Doll and Hill undertook their prospective doctors survey precisely because case-control studies seemed "unlikely to advance our knowledge materially or to throw any new light upon the nature of the association" between smoking and cancer.[50] Lauded as necessary to confirm the results of older approaches, prospective studies were praised in the pages of the 1964 report: "Certain shortcomings to the retrospective survey approach, some real and some exaggerated, led several courageous investigators to undertake the necessarily protracted, expensive, and difficult prospective approach."[51] The 1964 Surgeon General's report was, from the perspective of evidentiary revelation, an anticlimax. The combination of case-control and prospective cohort studies had built a scientifically solid case against cigarettes by the mid-1950s.[52] But as a political document, the 1964 report was a watershed. It stated in no uncertain terms that the health consequences of smoking were serious enough to warrant "remedial action" by government. The report's warm support of the cohort method highlighted the power of such studies to mobilize public attention, resources, and power.

"LIFESTYLE AND MORTALITY": THE JAPANESE CENSUS-BASED COHORT STUDY

Perhaps more than any other country, Japan bears the environmental and bodily scars of modernity. In the "toxic archipelago," to use environmental historian Brett Walker's evocative phrase, the epidemiology of chronic disease was grounded in three pillars of Japanese modernity: nuclear catastrophe, industrialization, and consumerism.[53] Hirayama's "lifestyle" study was the first large-scale inquiry into the latter. The Japanese Smokers' Wives Study existed within a longer national tradition of both indus-

[48] Takeshi Hirayama, *Life-Style and Mortality: A Large-Scale Census-Based Cohort Study in Japan* (Basel, 1990), acknowledgments, 1–5.

[49] Donald R. Shopland, David M. Burns, Lawrence Garfinkel, and Jonathan M. Samet, *Changes in Cigarette-Related Disease Risks and Their Implication for Prevention and Control* (Bethesda, 1997), 13–5.

[50] Doll and Hill, "The Mortality of Doctors" (cit. n. 43), 1451.

[51] *Smoking and Health: Report of the Advisory Committee to the Surgeon General of the Public Health Service*, PHS Publication no. 1103 (Washington, D.C., 1964), 162.

[52] Robert N. Proctor, "The History of the Discovery of the Cigarette–Lung Cancer Link: Evidentiary Traditions, Corporate Denial, Global Toll," *Tobacco Control* 21 (2012): 87–91.

[53] Brett Walker, *Toxic Archipelago: A History of Industrial Disease in Japan* (Seattle, 2010).

trial disease and the bureaucratic surveillance of bodies. Japanese bodies were indices of exposure to industrial and atomic processes. In Japan, postwar public health focused on environmental and pollution-related diseases.[54] This was in keeping with a broader shift in epidemiology, but the atomic and industrial history of Japan meant that Japanese epidemiologists confronted explicitly political and economic phenomena in their research populations. This was especially true in the case of the Atomic Bomb Casualty Commission (ABCC), an American agency established in 1948 to study the long-term effects of radiation on the survivors of the atomic bombs dropped on Hiroshima and Nagasaski.[55] The design of the ABCC was guided less by a cool assessment of contemporary developments in statistics and epidemiology than it was by the terrifying exigencies of atomic catastrophe, and the political calculi of American and Japanese medical professionals.[56] The commission outlived the formal U.S. military occupation of Japan, and its investigations into the effects of atomic radiation continue to this day under the auspices of the Radiation Effects Research Foundation, jointly administered by the United States and Japan. Unlike the studies designed to investigate the lifestyle choices of one large cohort, the Atomic Bomb Casualty Commission compared the health outcomes of atomic bomb survivors to the health outcomes in an unexposed control city. With a particularly harsh commitment to experimental purity—born of a desire not to admit American wrongdoing in dropping the bombs—the ABCC officially adopted a no-treatment policy for bomb survivors.[57] The irradiated Japanese bodies as well as the corpus of knowledge those bodies produced (and continue to produce) reflected a deeply asymmetric relationship between the United States and Japan—or, more precisely, between U.S. military personnel and the Japanese civilian population.

As the ABCC performed diagnostic check-ins around Hiroshima and Nagasaki, another large cohort of exposed subjects was becoming increasingly visible. Residents living in and around Kumamoto Prefecture were poisoned by methylmercury, an industrial by-product of fertilizer and plastics manufacture. The Chisso Corporation dumped tons of the chemical in Minamata Bay, where it bioaccumulated in the aquatic food chain, poisoning residents who consumed local fish and shellfish.[58] Minamata disease, as methylmercury poisoning came to be known, was painful, characterized by neurological symptoms such as confusion, seizures, loss of muscle and motor control, and paralysis. And because mercury affects fetal brain development, many women exposed to mercury underwent abortions—by choice and by coercion. Hirayama's large-cohort study was thus undertaken in the context of other inquiries into how man-made disasters unfolded upon the body.

[54] Itsuzo Shigematsu, "Epidemiology in Japan and Future Problems," *J. Epidemiol.* 6 (1996): S3–S7.

[55] Susan Lindee, "Atonement: Understanding the No-Treatment Policy of the Atomic Bomb Casualty Commission," *Bull. Hist. Med.* 68 (1994): 454–90; Lindee, *Suffering Made Real: American Science and the Survivors at Hiroshima* (Chicago, 1994).

[56] As Lindee has argued, "the interaction between scientists and research subjects . . . was dictated primarily by cultural or social location rather than by the needs of research"; Lindee, "Atonement" (cit. n. 55), 458.

[57] In practice, however, both Japanese and American physicians "promoted diagnosis as treatment, provided occasional chemotherapy, and overlooked the actions of individual physicians who chose to ignore the official policy"; ibid., 473.

[58] For more on Minamata disease, see Walker, *Toxic Archipelago* (cit. n. 53); Timothy S. George, *Minamata: Pollution and the Struggle for Democracy in Postwar Japan* (Cambridge, Mass., 2001).

Hirayama's study was perhaps even more ambitious than the American Cancer Society, Framingham, or British Doctors Survey. Those projects were undertaken to explore specific risk factors hypothesized as associated with disease outcomes: cardiovascular disease and lung cancer. Hirayama explicitly intended his study to "confirm associations" that had been established by case-control studies, but the risk factors and disease outcomes he surveyed were more expansive.[59] He sought to capture the relationship between "lifestyle"—of which cigarette smoking was just one component—and an array of diseases that would only reveal themselves over time. The Japanese large-cohort study included other lifestyle variables such as "alcohol drinking, diet, occupation, socioeconomic status, marital status, and reproductive history." The inclusion of these variables allowed Hirayama to draw conclusions about a wider range of individual and social relationships that contributed to disease. It is why he was able to detect a relationship between husbands' smoking status and the incidence of lung cancer in their wives. This study design thus allowed Hirayama to draw environmental and relational conclusions, going beyond the individualized understanding of lung cancer risk that had largely shaped the understanding of the disease and of its regulation in the form of warning labels and educational campaigns.[60]

The study was remarkable not just for its scope, but also for its size. Between 1 October and 31 December 1964, 265,118 adults over age forty were interviewed and enrolled. These figures represented 95 percent of the census population in the six prefectures selected as the study areas, and 0.2 percent of the entire population of Japan. An army of 1,500 nurses and midwives employed at regional public health clinics were dispatched to the homes of residents living in the survey area. There, they administered a Likert scale questionnaire that—aside from collecting demographic data on gender, age, previous medical history, number of children, and marital status—asked about subjects' frequency of consumption of cigarettes, meat, fish, milk, yellow-green vegetables, pickles, soybean paste soup, green tea, and alcohol. As in the American and British cohort studies, these lifestyle variables—whether the consumption of cigarettes, green tea, or leafy vegetables—could be considered the equivalent of a medical or therapeutic intervention in a clinical trial. Only instead of imposing an intervention upon a randomly selected group, Hirayama observed the effects of these lifestyle "interventions" forward in time: they were captured as "cause of death" on subjects' death certificates. In fact, the earliest results yielded from the large-cohort study were the markedly high mortality rates among Japanese smokers. In 1970, Hirayama published preliminary results from the first few years of the study. His results demonstrated a causal relationship between smoking and lung cancer, and a dose-response relationship between numbers of cigarettes smoked and likelihood of developing cancer. These results garnered significant attention in the United States: his study was extensively discussed in the 1972 Surgeon General's Report on Smoking and Health, and the *New*

[59] Hirayama, *Life-Style and Mortality* (cit. n. 48), 3. For more on the idea of "risk factors" and behavior in modern public health, see Allan M. Brandt, "Just Say 'No': Risk, Behavior, and Disease in Twentieth-Century America," in *Scientific Authority and Twentieth-Century America*, ed. Ronald G. Walters (Baltimore, 1997), 82–98.

[60] Carsten Timmermann, *Lung Cancer: A History of the Recalcitrant Disease* (Basingstoke, 2013). For more on the regulatory regime implied by the "risk-to-self" and the "risk-to-others" framing of tobacco, see Robert Rabin and Stephen Sugarman, eds., *Smoking Policy: Law, Politics, and Culture* (New York, 1993), 6–15.

York Times even took note of Hirayama's conclusions.[61] His was the first investigation of the association between cigarette smoking and disease in a non-Caucasian population. Hirayama's results confirmed the associations observed in studies conducted in the United States, Great Britain, and Canada.[62] The very differences between Western and Japanese bodies, histories, cultures, and everyday lifestyles helped to establish the universality of the conclusion that cigarettes caused lung cancer. Lung cancer's association with smoking grew stronger as cancer crossed the color line, dissolving the importance of the line itself.[63]

JOINING RISK AND NONSMOKERS' RIGHTS

Though less dramatic than the burns of the bomb or the convulsions of an acute poisoning, Hirayama's findings received widespread attention in the United States. In an article published in the *British Medical Journal*, Hirayama reported that the susceptibility of nonsmoking wives to lung cancer was linked to the smoking habits of their husbands. There was a twofold increase in the mortality rate from lung cancer for nonsmoking women who were captive to their husbands' cigarette smoke. The wives of the heaviest smokers had an even greater risk of developing lung cancer, while those married to partners who smoked less had less of a risk. This dose-response relationship was particularly significant for suggesting a causal relationship between secondhand smoke and cancer, which garnered instant attention in the American media.[64] The *New York Times*, *Los Angeles Times*, and *Chicago Tribune* all reported on the study within a day of its publication; the *New York Times* even ran a summary of the study on its front page. From there, reports were syndicated in newspapers all over the United States—from Daytona to Cape Girardeau. Headlines emphasizing the "high risk" of "mates' smoking" simplified the technical details of Hirayama's study, suggesting that Hirayama's conclusions resonated with Americans' informal experience of secondhand tobacco smoke.[65]

The leaf that American tobacco farmers sold to Japan would now come back to the United States in the form of processed data. The regulatory regime that allowed farmers to cultivate foreign tobacco consumption as a means of safeguarding their monopoly production rights gave rise to calls for a different vision of regulation. This new vision demanded that the long-standing privileges bestowed on politically favored economic groups be weighed against the quantifiable harms inflicted on others—namely, the figure of the nonsmoker.

Across the United States, anti-smoking activists seized on the Hirayama study as vindication of a decade of complaint and struggle against the social default of public smoking. During the 1970s, the nonsmokers' rights movement had generated a vocabulary for talking about the effect of tobacco smoke on nonsmokers, but this vo-

[61] "Smoking Peril Seen in Japanese Study," *New York Times*, 28 March 1972.
[62] U.S. Department of Health, Education, and Welfare, *A Report of the Surgeon General: 1972* (Washington, D.C., 1972), 74–5.
[63] Keith Wailoo, *How Cancer Crossed the Color Line* (New York, 2011).
[64] Takeshi Hirayama, "Non-smoking Wives" (cit. n. 2).
[65] "Study Says Cancer Risk Is High for Smokers' Wives," *Chicago Tribune*, 17 January 1981; "Wives' Lung Cancer Linked to Mates' Smoking," *Los Angeles Times*, 17 January 1981; "How to Maintain a Healthy Marriage: Please Don't Smoke," *New York Times*, 18 January 1981.

cabulary had not yet yielded substantial restrictions on smoking. Two distinct but allied approaches gave birth to the nonsmoker in the 1970s: Group Against Smokers' Pollution (GASP), which represented a chapter-based, environmentally conscious, local-ordinance-focused approach to the enactment of public smoking restrictions, and Action on Smoking and Health (ASH), which billed itself as the "legal arm of the nonsmoker's rights movement" and was almost entirely the operation of a brash young public law professor named John Banzhaf. As the acronym-driven names of these groups suggest, the nonsmokers' rights movement possessed an irreverent sensibility common to other public interest movements of the 1970s. Using humor and experiential arguments about the discomfort of sharing space with smokers, GASP advocated for the voluntary adoption of smoking restrictions at restaurants, grocery stores, and doctors' offices.

Prior to the Hirayama study, many of GASP's most ambitious initiatives failed. In 1978 and 1980, California activists proposed ballot initiatives that would have curbed smoking in a variety of public places. To defeat these initiatives, tobacco interests massively outspent nonsmoking groups, pouring millions into their defeat.[66] The industry tapped into the same antigovernment rhetoric that had resulted in the passage of Proposition 13, the "tax revolt" that slashed state property taxes: "We don't need any more big business laws—vote no on 5." Stoking the electorate's fears that Proposition 5 would be costly to implement, the tobacco industry defeated the initiative, with 54 percent of Californians voting against it.[67] In 1979, the Miami Chapter of GASP launched a ballot initiative in support of a citywide anti-smoking ordinance that would have restricted smoking in offices and public places and would have required businesses to designate separate nonsmoking areas. Despite being outspent by the tobacco industry at a ratio of ninety to one, the ordinance was defeated by just 820 votes—a margin of 0.4 percent.[68]

Just as the nonsmokers' rights movement launched these campaigns, the tobacco industry began to take serious stock of groups like GASP and ASH. Since 1975, the Tobacco Institute had paid the Roper Organization, a public opinion polling firm, to conduct surveys on consumers' attitudes toward the tobacco industry. Roper's 1978 poll called out a warning to the industry: the public smoking issue was an albatross. "What the smoker does to himself may be his business, but what the smoker does to the non-smoker is quite a different matter," the report began. Describing the rise of the nonsmokers' rights movement as "the most dangerous development to the viability of the tobacco industry that has yet occurred," the report recommended "developing and widely publicizing clear-cut, credible, medical evidence that passive smoking is not harmful to the non-smokers' health."[69] Hirayama's study was seen as a threat to the tobacco industry, since it showed that secondhand smoking was in fact quantifiably harmful and deadly, not just a nuisance to scolding killjoys.

[66] Stanton A. Glantz and Edith Balbach, *Tobacco War!: Inside the California Battles* (Berkeley and Los Angeles, 2000), 10–20.

[67] Ibid., 14–7; Stanton Glantz, John Slade, Lisa Bero, Peter Hanauer, and Deborah Barnes, eds., *The Cigarette Papers* (Berkeley and Los Angeles, 1996), 417–31.

[68] Michael S. Givel and Stanton Glantz, "Tobacco Control and Direct Democracy in Dade County, Florida: Future Implications for Health Advocates," *J. Public Health Policy* 21 (2000): 276–8.

[69] 1978 Roper Organization Report, Carton 4, Folder 31, Records of Americans for Non-Smokers Rights, Kalmanovitz Library, University of California, San Francisco (UCSF).

The tobacco industry did not take the Hirayama study lying down. In what appears now as a classic example of Big Tobacco's cultivation of doubt through the purchase of mercenary scientists, the industry responded with a public relations campaign against Hirayama intended to discredit the scientist as well as the study design.[70] In its hallmark fashion, the Tobacco Institute quietly hired scientists to dispute the technical basis of the study's conclusions and then endlessly quoted their hired guns in press releases as "evidence" of controversy. Nathan Mantel, a renowned biostatistician and professor at George Washington University, was hired by the Tobacco Institute to dispute the emerging science of environmental tobacco smoke.[71]

Letters poured in to the *British Medical Journal*, some at the request of the Tobacco Institute. Many letters were positive, praising the study as vital to public health and noting the research avenues that it opened up. Others, however, took Hirayama to task for a variety of sins: not considering the possibility of assortative mating, ignoring Japanese women's carcinogenic exposure to kerosene stoves, and making simple mathematical errors. Hirayama's thorough, data-based response to these concerns made no difference to the Tobacco Institute, which continued its campaign against the study through press releases and statements by industry spokesmen on television and radio. The *British Medical Journal*, for its part, took the unusual step of publishing correspondence not sent to the journal—including a press release from the Tobacco Institute—in order to let Hirayama respond to his critics. The scientist noted that his study's statistical validity had been "confirmed by prominent statisticians in many institutes, including the US National Cancer Institute and MIT."[72] Even as he stood by his data, Hirayama did concede that his results should be viewed as a starting point for further biological study of secondhand smoke.

Such studies mounted quickly. In 1986, Surgeon General C. Everett Koop issued the first report on the health consequences of involuntary smoking—a document whose very existence demonstrated the political salience of the nonsmokers' rights issue.[73] At a press conference announcing its release, Koop highlighted the major conclusion of the report, which drew on the research of more than sixty scientists in the United States and abroad (including Hirayama): involuntary smoking was a cause of lung cancer in nonsmokers, a death toll that Koop estimated to be 2,400 per year. Koop had already announced in 1984 that he sought "a smoke-free society by the year 2000. . . . We have the scientific basis for such a goal . . . and we have more than enough public understanding and sympathy for such a goal," he told an annual meeting of the

[70] Mi-Kyung Hong and Lisa Bero, "How the Tobacco Industry Responded to an Influential Study of the Health Effects of Secondhand Smoke," *Brit. Med. J.* 325 (2002): 1413–6; Glantz et al., *Cigarette Papers* (cit. n. 67), 413–6; Elisa Ong and Stanton Glantz, "Hirayama's Work Has Stood the Test of Time," *Bull. World Health Org.* 78 (2000): 938–9; Proctor, *Golden Holocaust* (cit. n. 9), 438; "State TAN Directors," 7 July 1981, Tobacco Institute, Roswell Park Cancer Institute (RPCI), https://industrydocuments.library.ucsf.edu/tobacco/docs/jnpx0031 (accessed 12 July 2018).

[71] Privately, industry researchers acknowledged what Mantel was paid to dispute. As an executive at Brown and Williamson acknowledged in 1981, research associates in Germany had concluded that "Hirayama was correct, that the Tobacco Institute knew it, and that the Tobacco Institute published . . . statement[s] about Hirayama knowing [that his] work was correct"; Glantz et al., *Cigarette Papers* (cit. n. 67), 415.

[72] "Correspondence," *Brit. Med. J.* 283 (1981): 916–7.

[73] *Health Consequences of Involuntary Exposure to Tobacco Smoke: A Report of the Surgeon General* (Washington, D.C., 1986).

American Lung Association. "That should be clear enough by the rising level of accepted non-smoker militancy."[74]

The Hirayama study was one of the best pieces of ammunition the nonsmokers' rights movement possessed, for it suggested that smoking was not just annoying to nonsmokers but deadly. And, by extension, it legitimated the social standing of the anti-tobacco movement: nonsmokers were not finger-wagging Carrie Nations, scolding, single-minded, and wielding "no smoking" signs like hatchets. Mounting evidence of smoking's concrete harms to nonsmokers made GASP look less moralizing and hysterical—the gendered epithet thrown at it by the tobacco companies—and more coldly, even antiseptically rational. As GASP pursued its strategy of local activism, the Hirayama results made their way into testimony at city council meetings and in reports produced by state departments of health. As a distressed Tobacco Institute official reported in 1982, "The Hirayama study continues to be relied upon by advocates of restrictive legislation on public smoking."[75] These advocates achieved a domino effect of successes after 1981. Beginning with the tiny town of Ukiah, 150 miles north of San Francisco, the California GASP achieved a notable string of victories passing local ordinances that required nonsmoking areas in most public places. In the next two years, twenty-one cities or counties passed similar clean indoor air ordinances.[76] In 1984, the California anti-smoking movement achieved its greatest victory yet with the passage of the Los Angeles workplace smoking ordinance requiring the establishment of smoke-free areas in all office buildings that employed more than four individuals. Hirayama's name and study were mentioned within the first two minutes of testimony at hearings on behalf of the ordinance.[77]

THE RIGHT TO HATE (BUT NOT TO REGULATE) TOBACCO SMOKE

Nonsmokers in Japan found themselves with fewer institutional levers with which to cordon off smokers. Though Japan became an important node on the transnational nonsmokers' rights network, its own anti-smoking movement was weak. In 1978, Ken-en Ken was founded. Roughly translated as "the right to hate tobacco smoke," Ken-en Ken, like ASH, called itself the legal arm of the nonsmokers' rights movement. Unlike ASH, Ken-en Ken had to battle a foe even more formidable than the American tobacco companies, for when it took on the Japan Tobacco Monopoly, it took on the Japanese government. As a state agency, the JTM occupied a central role in creating health policy. In the United States, anti-tobacco activists exploited the contradictory mandates of the American bureaucracy—the state subsidized tobacco even as it discouraged smoking. In Japan, there was frustratingly little hypocrisy to exploit. The message from the government was to smoke.

More than two decades after a tobacco-friendly Congress mandated that all cigarette packs carry a warning label, the Japanese government had yet to issue any com-

[74] C. Everett Koop, "Julia M. Jones Memorial Lecture Presented to the Annual Meeting of the American Lung Association," Miami Beach, Fla., 20 May 1984, Koop Papers, National Library of Medicine (NLM), Bethesda, Md.

[75] T. I. Hirayama, 20 July 1982, Tobacco Institute, https://www.industrydocumentslibrary.ucsf.edu/tobacco/docs/tqkg0062 (accessed 12 July 2018).

[76] Glantz and Balbach, *Tobacco War!* (cit. n. 66), 22.

[77] Hearing on an Anti-Smoking Ordinance for Los Angeles County, 8 December 1983, R. J. Reynolds, https://www.industrydocumentslibrary.ucsf.edu/tobacco/docs/gyfb0101 (accessed 12 July 2018).

parable warning. Indeed, the message on the side of Japanese cigarettes was darkly delicate: "For the sake of health, let's be careful not to smoke too much." Well after Hirayama's study began changing U.S. tobacco regulations on the grounds that smoke harmed nonsmokers, no Japanese government agency had gone so far as to deem smoking hazardous even to the smoker. Hirayama had his own explanation for his country's laissez-faire attitude toward smoking: "There are many among the influential leaders who have had their nerve centers for making judgments disturbed by nicotine. Politicians puffing away on cigarettes are like specimens of fools."[78]

In late 1987, Japan hosted the World Conference on Smoking and Health—an international gathering of hundreds of anti-tobacco activists from the fields of law, medicine, public health, government, social work, and advocacy. Just a few months earlier, the Tokyo District Court had dealt a blow to the Ken-en Ken movement. Even as Tokyo subways went smoke-free on a limited and experimental basis, a judge threw out Ken-en Ken's suit demanding smoke-free cars on the Japan National Railway. As delegates from around the world descended on Tokyo for the conference—at least a few of them anxious about finding smoke-free accommodations in the country—the Japanese anti-tobacco movement looked toward American activists for inspiration and tactics. Japanese anti-smoking groups solicited materials such as posters, booklets, pamphlets, and reports from American movement activists.[79] It was in this context of transnational anti-tobacco activism that the *New York Times* declared that the "US Is Exporting Nonsmoking to Japan."[80]

The headline erred in two ways. First, nonsmoking activism in the United States was immeasurably infused with the public health activism of Takeshi Hirayama—a kind of Japanese export to the United States. And second, for Japanese anti-tobacco activists, the United States was to blame for the global tobacco scourge, even as it also nurtured an effective anti-tobacco movement. In 1985 the Reagan administration threatened Japan with retaliatory sanctions if it did not "open" its markets to American-manufactured cigarettes—those produced by R. J. Reynolds, Philip Morris, and Brown & Williamson. Prior to 1985, the 28 percent tariff levied on imported cigarettes kept the foreign market share at less than 2 percent of all cigarettes consumed in Japan. Between 1985 and 1987, however, imports of American cigarettes rose more than threefold.[81] But perhaps more noticeable from the perspective of smokers and nonsmokers in Japan was the surge in television advertising. American cigarettes announced their arrival on the TV airwaves, which were soon choked with cigarette ads as the Japan Tobacco Monopoly ramped up its own television advertising to remain competitive. Cigarette advertisements surged from fortieth place in advertising volume on Japanese television to second place in less than ten years.[82]

The American anti-smoking movement had yielded results, as measured in terms of declining smoking rates among American teens and adults. Facing a shrinking domestic market, tobacco companies were unabashed in their desire to seek profits abroad,

[78] "ASH Newsletter," Luther L. Terry Papers, Box 6, Folder 13, NLM, Bethesda, Md.

[79] H. Nagami to Donna Shimp, August 1987, Donna Shimp Papers, UCSF.

[80] "US Is Exporting Nonsmoking to Japan," *New York Times*, 10 May 1987.

[81] Frank J. Chaloupka and Adit Laixuthai, "US Trade Policy and Cigarette Smoking in Asia" (Working Paper 5543, National Bureau of Economic Research, 1996).

[82] Stan Sesser, "Opium War Redux," *New Yorker*, 13 September 1993; A. Lambert, J. D. Sargent, S. A. Glantz, and P. M. Ling, "How Philip Morris Unlocked the Japanese Cigarette Market: Lessons for Global Tobacco Control," *Tobacco Control* 13 (2004): 379–87.

which they sought to accomplish through hardball trade negotiations. The Japanese nonsmokers' rights movement feared that they were living in the worst of all worlds: one in which competition for smokers—which meant more advertising and more promotion—would be even worse for public health than the world in which the state served as the single merchant of death.

In both the United States and Japan, the postwar tobacco regime was essentially nationalist in its administration: price supports and supply restrictions for American farmers produced leaf that the Japanese cigarette monopoly sold to Japanese citizens. This regime was superseded by a freer flow of cigarettes made by the major American tobacco firms. These cigarettes were shipped to more places and were made with much less American-grown leaf. As a domestic corollary to the "opening" of Asian markets, the United States had eased import controls on foreign tobaccos under Reagan, resulting in less and less American-grown leaf going into an American-style cigarette. Articulated in terms of trade deals, globalization appeared to threaten national boundaries. Transnational anti-tobacco activism was also born of globalizing forces: the flow of commodities abroad, international networks of knowledge, a circuit of international public health conferences. In the United States, anti-tobacco activists had never been wedded to the nation-state as the site for tobacco regulation. They continued—and continue—to tighten the noose around public smoking at the state and local levels, as state and local governments are the only legal venues in which smoking restrictions are promulgated. To this day, there is no federal anti-smoking law. In his address at the 1987 World Conference in Tokyo, Bunguku Watanabe, the lawyer who founded the Ken-en Ken movement, expressed a more cosmopolitan if less concrete vision for his movement. Critiquing the long-standing dependence of Japanese smokers on American tobacco, Watanabe declared that "it is hardly permissible for any country to export hazardous products to other countries, while attempting to protect the life and health of its own citizens from the harm of those same products." Referencing Surgeon General Koop's pledge, Watanabe continued, "I believe we should attain the common goal of creating a smokefree society, not in one country but in all the countries of the world."[83]

Localism, which resulted in a patchwork of differing laws and protections for Americans, was a rocky, inefficient, and incomplete route to a smoke-free society. But it was a route. Japan, by contrast, was still deemed a "smoker's paradise" and "the land that time forgot" in the 1990s.[84] A kind of public health developmentalism undergirded the perception that, eventually, industrialized, capitalist nations would converge upon American approaches to tobacco control. According to this train of thought, Japan was only in the Rostovian takeoff stage of snuffing-out. But the idea of a U.S.-led modernization of the Japanese anti-smoking movement missed the deep bodily connections between the two nation-states. And it also missed the institutional political differences that allowed American nonsmokers to succeed in asserting rights but stymied Japanese anti-tobacco activism. The United States did not so much "export anti-smoking to Japan," as the *New York Times* put it in 1987, as Japan exported anti-tobacco knowledge to the United States. Tobacco and anti-tobacco in Japan and the

[83] Buguku Watanabe at World Conference on Smoking or Health, 1987, Carton 6, Folder 10, Shimp Papers, UCSF.

[84] Roddy Reed, *Globalizing Tobacco Control: Anti-Smoking Campaigns in California, France and Japan* (Bloomington, Ind., 2005), 201.

United States were conjoined, born of commodity flows shaped by the United States' overwhelming global economic power and Japanese institutional structures, such as the strength of the Ministry of Finance and the importance of tobacco revenues in the national budget. Japan was a source, but not a site, of anti-tobacco activism.

Although the policies of the United States and Japanese nation-states were determinative in routing tobacco from the southeastern United States to (largely male) Japanese consumers, nations were not the only venues in which tobacco's meaning was forged. The design of Hirayama's large-cohort study derived from transatlantic innovations in epidemiological methods and the capacity of the Japanese public health bureaucracy, shaped by that nation's nuclear and industrial relationship with the United States and the world. "Just as bodies can become industrialized," writes Brett Walker, "so too can they be nationalized." For Walker, the carcinogenic chemicals in the lungs of Japanese smokers and their wives should be considered "physical inscriptions of the nation's policies on the body."[85] But national policies like the promotion of cigarette consumption as a symbol of culture and modernity—not to mention an important source of revenue for the state—existed within a global economy in which the United States was a hegemon. The bodies of Japanese smokers and their wives were transnationalized—chemically deterritorialized through their inhalation of U.S. tobacco in a Japanese-made, U.S.-style cigarette. So too would be the bodies of American nonsmokers who used the Hirayama study to articulate and defend a right to smoke-free air. The American activists who seized upon the study as vindication of their longstanding complaints against indoor smoking (and smokers) succeeded in regulating tobacco not at the federal level, but through a patchwork of local and state regulations. The divided, decentralized nature of American legal institutions—which can frequently impede public health regulation—was crucial for the cause of nonsmokers' rights in the United States.

American tobacco circulated through channels of trade, inside bodies, and within marriages, producing biomedical knowledge that, in the hands of activists, chiseled away at the practice and acceptability of smoking. Paradoxically, the American tobacco regime of the twentieth century brought both the expansion of smoking abroad, and the regulation of cigarette consumption in the United States. American tobacco producers had a hand in making the postwar world a place where they could prosper by selling their tobacco, but they could not control the economic, legal, and scientific forces unleashed by their efforts.

[85] Walker, *Toxic Archipelago* (cit. n. 53), 11.

Notes on Contributors

Harold J. Cook is John F. Nickoll Professor of History at Brown University and was previously Director of the Wellcome Trust Centre for the History of Medicine at UCL. Recipient of the Pfizer Prize of the History of Science Society and the Welch Medal of the American Association for the History of Medicine, he is author of five books and coeditor of six others, together with well over sixty articles. His primary research interests are in the emergence of the new medicines and sciences of early modern Europe; global knowledge exchanges; the coproduction of science and commerce; and processes of translation.

William Deringer is Leo Marx Career Development Assistant Professor of Science, Technology, and Society at the Massachusetts Institute of Technology. His research excavates the history of economic and political knowledge practices. His first book, *Calculated Values: Finance, Politics, and the Quantitative Age* (Cambridge, Mass., 2018), reconstructs how numerical calculation became an authoritative mode of public reasoning in Anglophone political culture. His new project, *Discounting: A History of the Economic Future (in One Calculation)*, traces "present value" calculations from their early modern beginnings to contemporary debates about climate change.

Julia Fein is a historian of Modern Russia, specializing in the intersecting histories of science and empire. She spent two years as a Mellon Postdoctoral Associate at Rutgers University, where she was affiliated with the "Networks of Exchange" seminar at the Rutgers Center for Historical Analysis. She has also taught modern European history at Macalester College.

Courtney Fullilove is Associate Professor of History, Environmental Studies, and Science in Society at Wesleyan University. She is the author of *The Profit of the Earth: The Global Seeds of American Agriculture* (Chicago, 2017) and of articles on claims to natural and cultural heritage. She is currently writing a history of international biodiversity preservation.

Arunabh Ghosh is Assistant Professor in the History Department at Harvard University. A historian of modern China, his interests include social and economic history, history of science and statecraft, and transnational history. His book, *Making It Count: Statistics and Statecraft in the Early PRC, 1949–1959*, is forthcoming from Princeton University Press. Articles have appeared in the *Journal of Asian Studies*, *BJHS-Themes*, and the *PRC History Review*. He is currently working on two new projects: a history of Chinese dam building in the twentieth century and a history of China-India scientific networks, ca. 1900–1960.

Martin Giraudeau is Assistant Professor in the Department of Sociology at Sciences Po, in Paris, and a researcher in the Centre de Sociologie des Organisations at the Centre National de la Recherche Scientifique. He has published articles on the history of business plans, the invention of entrepreneurship, and the roles of accounting in organizations and society. He recently coedited, with Frédéric Graber, a volume on the modern history of projects: *Les Projets: Une histoire politique (16e–21e siècles)* (Paris, 2018).

Eugenia Lean is Associate Professor of Chinese History in the Department of East Asian Languages and Cultures at Columbia University. She is the author of *Public Passions: The Trial of Shi Jianqiao and the Rise of Popular Sympathy in Republican China* (Berkeley and Los Angeles, 2007). Her forthcoming book, *Manufacturing China's Vernacular Industrialism: Nativist Tinkerer and Toothpowder Magnate, Chen Diexian (1879–1940)*, employs the figure of Chen Diexian, a professional editor, science enthusiast, and pharmaceutical industrialist, to examine the practices of nativist tinkering, innovative adaptation, and knowledge work in the building of early twentieth-century Chinese industry. Her latest research examines China's involvement in shaping modern global regimes of intellectual property from the early twentieth to twenty-first centuries.

Victoria Lee is Assistant Professor in the Department of History at Ohio University, where she teaches and writes about modern science and technology. She is currently writing a book about Japanese society's engagement with microbes in science, industry, and environmental management. She has published in *Historical Studies in the Natural Sciences*, *New Perspectives on the History of Life Sciences and Agriculture* (ed. Denise Phillips and Sharon Kingsland), and *Studies in History and Philosophy of Biological and Biomedical Sciences*. She was a postdoctoral fellow at the Max Planck Institute for the History of Science.

Paul Lucier is a historian of science, business, and the environment. He is the author of *Scientists and Swindlers: Consulting on Coal and Oil in America, 1820–1890* (Baltimore, 2008) and is currently finishing a book on the geology of gold, silver, and copper mining in the American West. The goal of his research is to understand whether and how science-driven capitalism can be ethical, innovative, intellectually creative, and environmentally responsible.

Sarah Milov is Assistant Professor in the Corcoran Department of History at the University of Virginia. Her first book, *Smoke and Ashes: From Corporatism to Neoliberalism in Tobacco's Twentieth Century*, is forthcoming. Her recent research focuses on whistle-blowing, particularly the relationship among gender, credibility, and bureaucracy.

Emily Pawley is Assistant Professor of History at Dickinson College, where she teaches environmental history, the history of capitalism, and the history of science. Her research focuses on cultivated landscapes as sources of knowledge; recent publications examine analytic tables and the invention of nutritional value, cattle portraiture and markets in blood, and varietal description and counterfeit fruit. Her book project, the *Balance-Sheet of Nature: Agriculture and Speculative Science in the Antebellum North*, examines the kinds of speculative and futuristic knowledge that emerged to make sense of the rapidly commercializing landscape of post–Erie Canal New York and is under advance contract at the University of Chicago Press. She is also interested in the transatlantic history of moon farming.

Lukas Rieppel is David and Michelle Ebersman Assistant Professor of History at Brown University, where he teaches courses on the history of science and the history of capitalism. His forthcoming book, *Assembling the Dinosaur*, uses the history of paleontology as a means to examine how the ideals, norms, and practices of modern capitalism shaped the way scientific knowledge was made, certified, and distributed during North America's Long Gilded Age. In addition, Rieppel has written several essays about the material culture of the earth sciences, the history of museums, the valuation of fossils, and the authentication of specimens.

David Singerman is Assistant Professor of History and American Studies at the University of Virginia. His work has appeared in *Radical History Review*, the *Journal of the Gilded Age and Progressive Era*, and the *Journal of British Studies*. He has also written on the sugar industry for the *New York Times* and on U.S.-Canada trade wars for *The Atlantic*.

Hallam Stevens is Associate Professor of History and Biology and the Head of History at the School of Humanities at Nanyang Technological University (Singapore). He is the author of *Life Out of Sequence: A Data-Driven History of Bioinformatics* (Chicago, 2013), *Biotechnology and Society: An Introduction* (Chicago, 2016), and the coeditor of *Postgenomics: Perspectives on Biology after the Genome* (Durham, N.C., 2015).

Lee Vinsel is Assistant Professor of Science, Technology, and Society at Virginia Tech, where he teaches and writes about government and technology. He is a cofounder of The Maintainers, a global, interdisciplinary research network that studies maintenance, repair, and mundane work with things. His forthcoming book with Johns Hopkins University Press examines the history of automobile regulation in the United States from 1893 to today's dreams of a driverless future.

Index

L

M

N

O

SUGGESTIONS FOR CONTRIBUTORS TO OSIRIS

OSIRIS is devoted to thematic issues, conceived and compiled by guest editors who submit volume proposals for review by the OSIRIS Editorial Board in advance of the annual meeting of the History of Science Society in November. For information on proposal submission, please write to the Editors at pmccray@history.ucsb.edu and ss536@cornell.edu.

1. Manuscripts should be submitted electronically in Rich Text Format using Times New Roman font, 12 point, and double-spaced throughout, including quotations and notes. Notes should be in the form of footnotes, also in 12 point and double-spaced. The manuscript style should follow *The Chicago Manual of Style*, 16th ed.

2. Bibliographic information should be given in the footnotes (not parenthetically in the text), numbered using Arabic numerals. The footnote number should appear as superscript. "Pp." and "p." are not used for page references.

a. References to books should include the author's full name; complete title of book in *italics*; place of publication; date of publication, including the original date when a reprint is being cited; and, if required, number of the particular page cited (if a direct quote is used, the word "on" should precede the page number). *Example*:

[1] Mary Lindemann, *Medicine and Society in Early Modern Europe* (Cambridge, 1999), 119.

b. References to articles in periodicals or edited volumes should include the author's name; title of article in quotes; title of periodical or volume in *italics*; volume number in Arabic numerals; year in parentheses; page numbers of article; and, if required, number of the particular page cited. Journal titles are abbreviated according to the journal abbreviations listed in *Isis Current Bibliography. Example*:

[2] Lynn K. Nyhart, "Civic and Economic Zoology in Nineteenth-Century Germany: The 'Living Communities' of Karl Möbius," *Isis* 89 (1999): 605–30, on 611.

c. All citations are given in full in the first reference. For succeeding citations, use an abbreviated version of the title with the author's last name. *Example*:

[3] Nyhart, "Civic and Economic Zoology" (cit. n. 2), 612.

3. Special characters and mathematical and scientific symbols should be entered electronically.

4. A small number of illustrations, including graphs and tables, may be used in each volume. Hard copies should accompany electronic images. Images must meet the specifications of The University of Chicago Press "Artwork General Guidelines" available from the Editor.

5. Manuscripts are submitted to OSIRIS with the understanding that upon publication copyright will be transferred to the History of Science Society. That understanding precludes consideration of material that has been previously published or submitted or accepted for publication elsewhere, in whole or in part. OSIRIS is a journal of first publication.

OSIRIS (ISSN 0369-7827) is published once a year.

Single copies are $35.00.

Address subscriptions, single issue orders, claims for missing issues, and advertising inquiries to *Osiris*, The University of Chicago Press, Journals Division, 1427 E. 60th Street, Chicago, IL 60637-2902.

Postmaster: Send address changes to *Osiris*, The University of Chicago Press Subscription Fulfillment, 1427 E. 60th Street, Chicago, IL 60637-2902.

OSIRIS is indexed in major scientific and historical indexing services, including *Biological Abstracts, Current Contexts, Historical Abstracts*, and *America: History and Life*.

Paperback edition, ISBN 978-0-226-53877-8

A RESEARCH JOURNAL DEVOTED TO THE HISTORY OF SCIENCE AND ITS CULTURAL INFLUENCES

A PUBLICATION OF THE HISTORY OF SCIENCE SOCIETY

W. Patrick McCray
Co-Editor, Osiris
Department of History
University of California, Santa Barbara
Santa Barbara, CA 93106-9410 USA
pmccray@history.ucsb.edu

Suman Seth
Co-Editor, Osiris
Department of Science & Technology Studies
321 Morrill Hall
Cornell University
Ithaca, NY 14853 USA
ss536@cornell.edu